INTERACTIONS BETWEEN PHYSICS AND DYNAMICS OF SOLAR SYSTEM BODIES

INTERACTIONS BETWEEN PHYSICS AND DYNAMICS OF SOLAR SYSTEM BODIES

Proceedings of the International Astronomical Symposium held in Pléneuf-Val-André (France) from June 21 to June 28, 1992

Edited by

E. BOIS
Observatoire de la Côte d'Azur, Grasse, France

P. OBERTI
Observatoire de la Côte d'Azur, Nice, France

and

J. HENRARD
Département de Mathématique, FNDP, Namur, Belgium

Reprinted from Celestial Mechanics and Dynamical Astronomy
Volume 57, Nos. 1-2, 1993

SPRINGER-SCIENCE+BUSINESS MEDIA, B.V.

A C.I.P. Catalogue record for this book is available from the Library of Congress.

ISBN 978-94-010-4840-8 ISBN 978-94-011-1902-3 (eBook)
DOI 10.1007/978-94-011-1902-3

Printed on acid-free paper

TABLE OF CONTENTS

Preface ix

Asteroids

PETIT J-M and FARINELLA P. / Modelling the outcomes of high-velocity impacts between small solar system bodies 1

BENDJOYA P. / A comparison between $2D$ and $3D$ wavelet analysis method for asteroid family dynamical study 29

CELLINO A. and ZAPPALÀ V. / Asteroid "clans": super-families or multiple events? 37

PAOLICCHI P., VERLICCHI A. and CELLINO A. / Catastrophic fragmentation and formation of families: preliminary results from a new numerical model 49

NAKAMURA T. / Structure of the asteroid belt for small members 57

MILANI A. / The Trojan asteroid belt: Proper elements, stability, chaos and families 59

HADJIDEMETRIOU J.D. and VOYATZIS G. / Long term evolution of asteroids near a resonance 95

ROBUTEL Ph. / The stability of the planetary three-body problem: influence of the secular resonances 97

MOONS M. and MORBIDELLI A. / The main mean motion commensurabilities in the planar circular and elliptic problem 99

Comets

BENEST D. and GONCZI R. / Stochasticity of two comets in resonance with Jupiter 109

NAKAMURA T. and YOSHIKAWA M. / Orbital evolution of giant comet-like objects 113

FROESCHLÉ C. and PETIT J-M. / On a temporary confinement of chaotic orbits of four dimensional symplectic mapping: a test on the validity of the synthetic approach 123

CHAMBERS J.E. / A simple mapping for comets in resonance 131

Artificial Satellites

EXERTIER P. / Geopotential from space techniques 137
DELHAISE F. and MORBIDELLI A. / Luni-solar effects of geosynchronous orbits at the critical inclination 155
METRIS G., EXERTIER P., BOUDON Y. and BARLIER F./ Long-periodic variations of the motion of a satellite due to non-resonant tesseral harmonics of a gravity potential 175
FERRÁNDIZ J.M., SANSATURIO M.E. and CABALLERO R. / On the roto-translatory motion of a satellite of an oblate primary 189
FLORÌA: Canonical elements and Keplerian-like solutions for intermediary orbits of satellites of an oblate planet 203
VOKROUHLIKÝ D., FARINELLA P. and LUCCHESI D. / Albedo perturbation models: general formalism and applications to LAGEOS 225
STEICHEN D. / Study of a Moon's artificial satellite dynamics valid for all eccentricities and inclinations 245

Geodynamics

LEGROS H. and LEFFTZ M. / On the fluid and viscoelastic deformations of the planets 247
FERRÁNDIZ J.M. and GETINO J. / On the tidal variation of the geopotential 279

Attitude Dynamics

LASKAR J. and JOUTEL F. / Orbital, rotational and climate interactions 293
BOIS E. and JOURNET A. / Lunar and terrestrial tidal effects on the Moon's rotational motion 295
ECKHARDT D. / Passing through resonance: the excitation and dissipation of the Lunar free libration in longitude 307
CELLETTI A. / Stability of the synchronous spin-orbit resonance by construction of librational trapping tori 325
WINTER O.C. and SESSIN W. / Nebular drag and capture into spin-orbit resonance 329

Natural Satellites

STEVES B.A., VALSECCHI G.B., PEROZZI E. and ROY A.E. / Significant high number commensurabilities in the main Lunar problem II: The occurence of Saros-like near periodicities 341
KINOSHITA H. / Motion of the orbital plane of a satellite due to a secular change of the obliquity of its mother planet 359
LOHINGER E. and FROESCHLÉ C. / Fourier analysis of local Lyapunov characteristic exponents for satellite-type motions 369

Dust Dynamics

SICARDY B. , BEAUGÉ C., FERRAZ-MELLO S., LAZZARO D. and ROQUES F. / Capture of grains into resonances through Poynting-Robertson drag 373

SPAHN F., PETIT J-M. and BENDJOYA Ph. / The gravitational influence of satellite Pan on the radial distribution of ring-particles in the region of the Encke-division in Saturn's A ring 391

Abstracts

CAMPO BAGATIN A., FARINELLA P. and PAOLICCHI P. / Collisional evolution of the asteroid size distribution: a numerical simulation 403

BARKIN Yu.V. / Resonant and secular phenomena in the libration of the Moon 404

BONNEFOND P. and EXERTIER P. / Precise orbit determination with a short-arc technique 405

BOIS E. and BREITER S. / Spherical coordinates perturbations for small eccentricity orbits 406

BRUNINI A. / Dynamical friction induces perturbations on Oort cloud comets 406

KLAČKA J. and KAUFMANNOVÁ J. / Poynting-Robertson effects 407

LEFFTZ M. and LEGROS H. /Influence of internal load on the non-hydrostatic form of the Moon 408

SOLOVAYA N.A. and PITTICH E.M. / Evolution of asteroidal orbits with high inclinations 409

Participants in the International Astronomical Symposium held in Pléneuf-Val-André (France) from June 21 to June 28, 1992.

PREFACE

Fans of "Asterix the Gallic" know well that the only fear of people in Brittany is that the sky falls upon their head. So it must have been a shock for them (the fans of Asterix) to learn that a horde of Physicists and Dynamicists (some of them being actually Roman — ils sont fous ces Romains!) invaded the bay of Saint-Brieuc and spend a full week conjuring all the nastiness that the sky has in reserve, revelling in the horrors hidden beyond the blue dome; they talked with delight about "asteroids", "comets" and "meteor streams"; they grinned at the idea of "artificial satellites", these pots and pans of space always ready to fall upon you; some of them said strange things about the Moon, the planets, and evoked the "rings" of Saturn or of some other of their gods. One evening, a Roman from Pisa went as far as cornering some inhabitants in the large hut they used for their witchcraft and filled them with terror by describing the fate of the poor dinosaurs victims of a particularly nasty asteroid (or was it a comet?).

You will be surprized to learn that Bretons did not exact a spectacular revenge for these offenses. On the contrary. They made a point to paint the sky solid blue and to pin a jolly yellow sun onto it; they unfurled an immense tapestry of sandy and pristine beaches right in front of the "Grand-Hôtel du Val-André" where the barbarians were gathering. The sea was as blue as the sky (unfortunately someone forgot to heat the water!); The food was splendid; the wine, from the nearby Val-de-Loire, sparkling like . . . crémant.

This reaction of the Bretons made us think that maybe what was said in the large hut "du Guémadeuc" would be of interest to some other people and could be collected into this volume. We thank the participants who made available the LaTex file of their contribution, permitting us to present a "camera ready" volume of high quality (as for the printing *and* the content). We thank the chairmen of the sessions (D. Eckhardt, P. Farinella, C. Froeschlé, A.W. Harris, J. Henrard, H. Kinoshita, A. Milani, H. Rickman and J. van der Ha) who took care of having the papers reviewed by competent referees, resulting in many improvements in the form and content of the papers.

So many people are to be thanked! The Mayor of Pléneuf-Val-André, Monsieur Guédo, and its council; Jean-Marie Guigot and his friends; the local organizing committee: Françoise Beranger, Yvette et Loïc Bois, Anne-Christine Bois, Louisanne Guigot, François Lebarque, Anne-Marie Masson, and Marie-Pierre Pradeau; Jean-François Rollin in charge of the hall "du Guémadeuc"; the "Maison Départementale du Tourisme des Côtes d'Armor"; the "Office du Tourisme de Pléneuf-Val-André"; the "Inter-Marché du Val-André"; the "Planétarium du Trégor"; the "Vignerons de la Noëlle" and Monsieur Jean-Louis Dejean who made us discover the "Crémant de Loire"; the "Confrérie des Chevaliers de la Coquille St-Jacques"; the "Rotary-Club de Draguignan"; the ARY 173 Association.

For financial support we should thank: the "Municipalité de Pléneuf-Val-André",

Celestial Mechanics and Dynamical Astronomy **57**: ix-x, 1993.

the "Conseil Général des Côtes d'Armor", the "Municipalité de Saint-Brieuc", the "E.D.F. G.D.F. Services Côtes d'Armor", the "Agence Spatiale Européenne", the "Département CERGA de l'Observatoire de la Côte d'Azur", the "Centre National de la Recherche Scientifique", the "Centre National d'Études Spatiales".

Last but not least we thank our colleagues who opened the Symposium to the general public; André Brahic spoke (with his hands and his enthusiasm) about "les dernières nouvelles de l'exploration du Système Solaire", Andrea Milani and his herd of dinosaurs told the public of Pléneuf-Val-André what happens "Quand le ciel nous tombe sur la tête"; Philippe Baille trilled the schoolchildren with "L'Astronomie, ses Mystères et ses Défis" and Alain Le Gué made available his exhibit "Voyage Lumière dans l'Univers".

E. Bois P. Oberti J. Henrard

MODELLING THE OUTCOMES OF HIGH–VELOCITY IMPACTS BETWEEN SMALL SOLAR SYSTEM BODIES

JEAN–MARC PETIT

OCA – Observatoire de Nice, B.P. 229, 06304 Nice Cedex 04, France

and

PAOLO FARINELLA

Dipartimento di Matematica, Università di Pisa, Via Buonarroti 2, 56127 Pisa, Italy

Abstract. We present a self–consistent numerical algorithm aimed at predicting the outcomes of high–velocity impacts between asteroids (or other small bodies of the solar system), based on a set of model input parameters which can be estimated from the available experimental evidence, and including the possible gravitational reaccumulation of ejected fragments whose velocity is less than a suitably defined escape velocity. All the fragment mass distributions are modelled by truncated power laws, and a possible correlation between fragment ejection velocity and mass is taken into account in different ways, including a probabilistic one. We analyze in particular the effectiveness of the gravitational reaccumulation process in terms of different choices of the collisional parameters and the assumed relationship between fragment speed and mass. Both the transition size beyond which solid targets are likely to reaccumulate a large fraction of the fragment mass and the collision energy needed to disperse most of the fragments are sensitive functions of the assumed fragment velocity versus mass relationship. We also give some examples of how our algorithm can be applied to study the origin and collisional history of small solar system bodies, including the asteroid 951 Gaspra (recently imaged by the Galileo probe) and the asteroid families.

Key words: Asteroids – physics of the solar system – high-velocity impacts – numerical simulation

1. Introduction

As a consequence of their non–negligible eccentricities and mutual inclinations, the orbits of most asteroids can occasionally intersect each other, opening the way to the occurrence of high–velocity impacts. A typical main–belt asteroid undergoes collisions with other (normally, much smaller) asteroids at an average velocity of 5.8 km/s and at a rate of the order of 3×10^{-18} $(R/km)^2 N_{pr}$ yr^{-1}, where R and N_{pr} are the assumed target radius and number of projectiles (Farinella and Davis, 1992).

As a consequence, mutual impacts — whose outcomes range from small–scale cratering events to catastrophic target break–up — have played a crucial role in the evolution of the asteroid belt (the pioneering work in this field dates back to the 50s, e.g. Öpik, 1951, and Piotrowski, 1953; for a recent a review, we refer to Davis *et al.*, 1989). In the 80s it has been also realized that disruptive impacts may have been important in the origin and evolution of other types of small solar system bodies, like some icy satellites of the outer planets (Farinella *et al.*, 1983), the planetary rings (Harris, 1984), and the interplanetary dust complex (Sykes *et al.*, 1989).

In order to model the outcomes of collisional events occurring in the solar system, hypervelocity impact experiments have been carried out in the laboratory at

Celestial Mechanics and Dynamical Astronomy **57**: 1–28, 1993.

a much smaller scale (for reviews, see Davis *et al.*, 1986; Fujiwara *et al.*, 1989), and from the observed results some simple empirical relationships have been derived to describe the collisional outcomes in terms of several basic *input parameters*, such as the projectile and target masses, the impact velocity, some coefficients describing the impact response of the target's material, and sometimes the incidence angle (when it is relevant to distinguish between "central" and "grazing" impacts). For the sake of simplicity, both the target and the projectile are normally considered as nearly homogeneous and spherical bodies (see Chapman *et al.*, 1989, for a discussion of possible complications).

It is important to keep in mind that at least one physically fundamental difference exists between small laboratory targets and real celestial bodies, which are tens or hundreds of km across: in the latter case, self–gravity is not negligible, and this has two important consequences. First, self–compression can strengthen the interior of the bodies making more difficult to fracture and shatter them, and second — if fragmentation occurs — mutual gravity can cause the partial reaccumulation of the ejected fragments into a so–called *pile of rubble*.

From the theoretical point of view, our understanding of the complex physical processes which take place after an energetic impact is still quite limited. Research work on this subject is currently in progress, based either on semiempirical working hypotheses about the fracture criteria and the geometry of fragment velocity fields (Paolicchi *et al.*, 1989, 1992), or on the integration of suitable hydrodynamical equations in two or three dimensions (Ryan *et al.*, 1990; Ryan and Melosh, 1992; Nolan *et al.*, 1992; Asphaug *et al.*, 1992).

However, the corresponding numerical experiments are still so complicated and demanding (in terms of CPU time and memory) that they cannot be used to explore the full range of the relevant parameters, in order to interpolate and extrapolate to actual solar system conditions, nor to simulate the long–term collisional evolution of systems like the asteroid belt or the planetary rings.

Therefore, the results of both the laboratory and the numerical experiments need to be incorporated into suitable scaling algorithms, requiring a number of input parameters (which depend on the "initial conditions" of the impact and the detailed physical processes following it) and capable of predicting in a fast and effective manner the outcome of any possible impact occurring in the system under scrutiny.

In this paper (Secs. 2 to 5) we present such a numerical algorithm, which is physically simple (but still fully self–consistent) and at the same time fairly realistic, as it includes models for the effects of self–gravity and fragment reaccumulation. This algorithm is a follow–up and a refinement of those used in the past by Greenberg *et al.* (1978), Davis *et al.* (1979, 1985, 1989), and Farinella *et al.* (1992), as it includes some additional physical effects.

In Sec. 6 we shall discuss how much the predictions of our algorithm are sensitive to different choices of some collisional parameters, and how the effectiveness of the gravitational reaccumulation process depends upon the assumed relationship between the ejection velocity and the mass of the fragments. We shall also give some examples of how this algorithm can be applied to model the origin of asteroid families and the collisional evolution of different types of small solar system bodies.

2. The Fragmentation Threshold

In this and the next three sections we are going to describe our algorithm to predict the outcome of a collision between two (spherical and homogeneous) bodies of masses M_i, densities ρ_i, diameters $D_i = (6M_i/\pi\rho_i)^{1/3}$, impact strengths S_i (the index i is always equal to either 1 or 2), when the relative velocity is V_{rel} and the relative kinetic energy is $E_{rel} = M_1 M_2 V_{rel}^2/2(M_1 + M_2)$. The impact strength S is defined as the mimimum specific energy delivered to each body to cause catastrophic fragmentation of it. We are implicitly assuming that an abrupt transition always occurs between the cratering and shattering regimes. This is supported by the experimental results, which show that the transition from localized target damage to widespread fragmentation is associated with a modest increase of the impact energy (see e.g. Fujiwara *et al.*, 1977, Fig. 10). For targets having a significant self–gravity, recent hydrocode modelling work by Nolan *et al.* (1992) and Greenberg *et al.* (1992) suggests that the transition may be less distinct than indicated by laboratory experiments, in the sense that a very large crater can form and remain on the surface even though a significant fraction of the target material is shattered and reaccumulated. While we shall keep our "catastrophic transition" assumption, we shall treat the formation of giant craters in a way that does not conflict with the above results (see Sec. 4).

We further assume that in the shattering case the available energy is partitioned in equal parts to the two colliding bodies (Hartmann, 1988), so for each of them fragmentation occurs if $E_{rel} > 2SM/\rho x_{cr}$ (here x_{cr} is a numerical coefficient which will be justified later). The value of S is obtained from that observed in laboratory impact fragmentation experiments (S_0) through the scaling formula:

$$S_i = \left(S_{0,i} + \frac{\pi\gamma}{15} G\rho_i^2 D_i^2\right) \left(\frac{D_i}{0.2\ m}\right)^{-1/4}, \tag{1}$$

where the term proportional to D^2 is due to self-compressional strengthening of the interior of each body (see Davis *et al.*, 1985), while the $D^{-1/4}$ factor is consistent with strain–rate scaling with size (Housen and Holsapple, 1990). The numerical coefficient γ is possibly material–dependent and has to be determined by experiments (see e.g. Housen *et al.*, 1991). For solid materials like basalt or concrete, abundant experimental evidence indicates that $S_0 \approx 3 \times 10^6\ J/m^3$.

Given S_i and the other impact parameters, our goal is to compute the mass distribution of the bodies generated by the impact. Since (i) this is is possible only for a finite number of target–projectile pairs (M_1, M_2), and (ii) it is normally useless to keep track of all the masses of the individual fragments, in our numerical computations we normally grouped the fragments into a limited number (ranging between $\approx$ 20 and 100) of discrete, logarithmic mass bins, having their limits (and central values) scaled by a constant amount (typically, a factor $g = 2$). In other words, we considered a geometric sequence of N_{bin} "central" mass values $m_{bin,k} = g^{k-1} m_{bin,1}$ $(k = 1, ..., N_{bin})$, with every bin spanning the range $(m_{bin,k}/\sqrt{g}, m_{bin,k}\sqrt{g})$. In order to compute the amount of mass reaccumulated by self–gravity after an impact, we also considered the fate of the low–mass "tail" of the distribution, ranging between the assumed lower cutoff of the fragment mass (M_{dust}, see later) and the lower limit of the first bin $(m_{bin,1}/\sqrt{g})$.

3. Shattering Events

When one body is shattered, the mass of the largest remaining fragment is given by $M_{max,i} = M_i f_{l,i}$, where

$$f_{l,i} = \frac{1}{2}\left(\frac{S_i M_i}{\rho_i E_{rel}/2}\right)^{1.24}, \tag{2}$$

in agreement with the experimental relationship derived by Fujiwara *et al.* (1977), which predicts that when the critical energy threshold for fragmentation is just reached, a "core" half the mass of the original body survives, while this fraction decreases rapidly for increasing impact energies. Actually, the previous formula is valid for head–on impacts; if we take into account that oblique impacts are on the average less effective in shattering the involved bodies, $f_{l,i}$ can be averaged over the incidence angles leading to

$$\overline{f_{l,i}} = 3 f_{l,i}^{2/3} - 2 f_{l,i} \tag{3}$$

(Davis *et al.*, 1985). Hereinafter by $f_{l,i}$ we will always mean the averaged value $\overline{f_{l,i}}$ given by Eq.(3). In order to keep the definition of threshold collision as that generating a largest fragment with half the mass of the target, the condition for fragmentation given in Sec. 2 includes a numerical factor $x_{cr} = 4^{-1/1.24} \cong 0.327$, derived by noting that $\overline{f_{l,i}} = 1/2$ when $f_{l,i} = 1/8$. This implies that the effective strength of the targets versus impacts from random directions is about 3 times higher than the value found in the laboratory for head–on impacts.

Next, we need to model the mass distribution of the fragments. Following many previous investigations (e.g., Hartmann, 1969; Kresàk, 1977; Greenberg *et al.*, 1978; Zappalà *et al.*, 1984; Fujiwara *et al.*, 1989), we shall assume here a simple, albeit somewhat artificial, mathematical model: a *Pareto* power–law distribution truncated at a minimum cut–off mass M_{dust} and consistent with the fact that $M_{max,i} = f_{l,i} M_i$ is the mass of the largest fragment. More precisely, we define $N_i(\geq m)$ to be the number of fragments of body i having mass larger than m and

$M_i(\leq m)$ and $M_i(> m)$ the total mass of fragments of body i with an individual mass respectively smaller and larger than m; we assume that $N_i(\geq m)$ has a discontinuity at $m = M_{max,i}$ and that we know that there is just one fragment of mass $M_{max,i}$. If $\Theta(x)$ is the Heaviside step function (= 0 for $x < 0$ and = 1 for $x \geq 0$) and $\delta(x)$ is the Dirac delta function ($\delta(x) = d\Theta(x)/dx$), then we have

$$\begin{aligned} N_i(\geq m) &= B_i m^{-b_i}\Theta(M_{max,i} - m)\Theta(m - M_{dust}) \\ &\quad + B_i(M_{dust})^{-b_i}[1 - \Theta(m - M_{dust})] \end{aligned} \tag{4}$$

where b_i is the characteristic exponent of the cumulative size distribution, and $B_i = (M_{max,i})^{b_i}$, so that $N_i(\geq M_{max,i}) = 1$. The differential (or incremental) fragment size distribution is then given by:

$$\begin{aligned} n_i(m)dm &= -dN_i(\geq m) \qquad (5) \\ &= \left[b_i B_i m^{-b_i-1}\Theta(M_{max,i} - m) + \delta(m - M_{max,i})\right]\Theta(m - M_{dust})dm \end{aligned}$$

while for the total mass accounted by fragments of mass smaller than any given value m we obtain (provided $b_i \neq 1$):

$$\begin{aligned} M_i(\leq m) &= \int_0^m m\, n_i(m)\, dm \\ &= \frac{b_i B_i}{1 - b_i}\left[m^{1-b_i} - (M_{dust})^{1-b_i}\right][1 - \Theta(m - M_{max,i})] \\ &+ \left[\frac{b_i B_i}{1 - b_i}\left[(M_{max,i})^{1-b_i} - (M_{dust})^{1-b_i}\right] + M_{max,i}\right]\Theta(m - M_{max,i}) \end{aligned} \tag{6}$$

Of course, the cumulative mass accounted by fragments larger than m will be $M_i(> m) = M_i - M_i(\leq m)$. Since mass conservation implies $M_i(\leq M_{max,i}) = M_i$, from Eq. (6) we get the condition

$$M_i(1 - f_{l,i}) = \frac{b_i B_i}{1 - b_i}\left[(M_{max,i})^{1-b_i} - (M_{dust})^{1-b_i}\right], \tag{7}$$

which allows us to rewrite Eq.(6) as

$$\begin{aligned} M_i(\leq m) &= \frac{b_i (M_{max,i})^{b_i}}{1 - b_i}\left[m^{1-b_i} - (M_{dust})^{1-b_i}\right][1 - \Theta(m - M_{max,i})] \\ &\quad + M_i \Theta(m - M_{max,i}) \,. \end{aligned} \tag{8}$$

Eq.(7) is important because from it, once M_i, M_{dust} (which are "input parameters") and $f_{l,i}$ (or $M_{max,i}$, to be derived from Eqs.(2,3)) are known, the exponent b can be derived in a unique way. For this purpose, Eq.(7) can be rewritten as

$$b = \left[\frac{M}{M_{dust}}(1 - f_l)\right] \Big/ \left[\frac{M}{M_{dust}} - \left(\frac{M_{max}}{M_{dust}}\right)^b\right] \tag{9}$$

and solved numerically by an iterative method. As pointed out with simplified formulae (ignoring e.g. the lower cutoff M_{dust}) by Greenberg *et al.* (1978) and Farinella *et al.* (1982), the assumption that the fragment mass distribution is given by a single–exponent power law allows us to uniquely specify its "free parameters" B_i and b_i. However, we have to stress here that this assumption is certainly an oversimplification of reality. Indeed, many experiments show that a much better fit is obtained by using different exponents in different mass ranges (Fujiwara *et al.*, 1989); the same property is shared by the real asteroid population (Cellino *et al.*, 1991). On the other hand, the physical meaning of the transition masses (or sizes) for which the best–fitting exponents change is not currently well understood; the corresponding values probably depend both on the material properties and on the impact parameters. As a consequence, we did not feel justified at this stage to introduce in our algorithm such a complicating feature as a two– or three–exponents power law to represent the fragment mass distribution — implying several more free parameters to be given in input, whose choice would be almost arbitrary.

It is easy to see that the critical value $b_i = 1$ separates the case in which most of the mass is in the largest bodies ($b_i < 1$) from the opposite one. Actually, note that when $M_{dust} \Rightarrow 0$, $b_i \Rightarrow 1 - f_{l,i}$ (rather than the commonly used relationship $b = (1+f_l)^{-1}$, see e.g. Greenberg *et al.*, 1978). For $M_{dust} > 0$, the "supercatastrophic" cases when $f_l \ll 1$ can yield $b > 1$.

How many fragments are going to escape from the gravitational well of the two colliding bodies? A detailed answer would require the knowledge of the original position of the fragments inside their parent bodies, and of their motion immediately after the ejection. Although this problem could be tackled with models such as that of Paolicchi *et al.* (1989), its treatment is well beyond the scopes of the present work. Instead, we are going to derive an "effective escape velocity" V_{esc} defined by analogy with the usual definition of surface escape velocity of a celestial body, and to assume that only the fragments ejected with speeds exceeding V_{esc} will escape "to infinity" (i.e., will achieve independent heliocentric orbits). We use the energy balance equation

$$\frac{1}{2}(M_1 - M_{max,1} + M_2 - M_{max,2})V_{esc}^2 + W_{tot} = W_{fr,1} + W_{fr,2} + W_h \ , \tag{10}$$

where we have assumed that after the impact both the colliding bodies are shattered (if this is not the case for body i, $W_{fr,i}$ will have to be changed into $W_{crat,i}$, to be derived in Sec. 4) and that the largest fragment from either of them has negligible kinetic energy (this is consistent with experimental results by Fujiwara and Tsukamoto, 1980, Nakamura and Fujiwara, 1991, and others). The various W terms in Eq.(10) stand for the following potential energies: W_{tot} is the total potential energy of the two bodies just before fragmentation, that is

$$W_{tot} = -\frac{3GM_1^{5/3}}{5Q_1} - \frac{3GM_2^{5/3}}{5Q_2} - \frac{GM_1M_2}{Q_1M_1^{1/3} + Q_2M_2^{1/3}}, \tag{11}$$

where $Q_i = (4\pi\rho_i/3)^{-1/3}$; $W_{fr,i}$ is the sum of the self-gravitation potential energy of each fragment of body i, derived form the mass distribution (5):

$$W_{fr,i} = \frac{3G}{5Q_i}\int_0^\infty m^{5/3}\, n_i(m)\, dm = -\frac{3G}{5Q_i}\frac{5M_{max,i}^{5/3} - 3b_iM_{max,i}^{b_i}M_{dust}^{5/3-b_i}}{5-3b_i} \tag{12}$$

and W_h is an estimate of the potential energy when the escaped fragments are separated by a distance of the order of the Hill's radius of the total colliding mass in the gravitational field of the central mass (for the asteroids, the Sun) of mass M_0 and orbital distance R_0:

$$W_h = -\frac{3G(M_1 + M_2)^{5/3}}{5}\frac{(3M_0)^{1/3}}{R_0}. \tag{13}$$

4. Cratering Events

Assume now that for body i, $E_{rel} < 2S_iM_i/\rho_ix_{cr}$, so that it is not shattered by the collision. In this case a crater is formed on it. If the case of a small–scale collision, resulting into a crater of mass smaller than 1% of the target mass, we assume that the mass M_{crat} excavated from the crater (and converted into ejected fragments) is just proportional to the impact energy, through a *crater excavation coefficient* α to be given in input and depending on the material properties (for "soft" and "hard" materials, laboratory data yield $\alpha \approx 4 \times 10^{-4}\ s^2/m^2$ and $\alpha \approx 10^{-5}\ s^2/m^2$, respectively; see Stöffler *et al.*, 1975, and Dobrovolskis and Burns, 1984, p. 467). For larger craters, we still assume a linear dependence of the excavated mass on E_{rel}, but with coefficients such that the largest possible crater — created when E_{rel} is just less than the fragmentation threshold $2S_iM_i/\rho_ix_{cr}$ — has a mass of 1/10 that of the target. Note that this is much more than the largest crater predicted by conventional scaling theories, which has a diameter of the order of the target radius and a volume ≈ 3% of the target volume; rather, our assumption is in agreement with the recent results of numerical hydrocode simulations of asteroid impacts, suggesting that such giant craters (or concavities) can be formed without complete break–up of the target, but damaging the material structure enough for generating a substantial amount of regolith (Greenberg *et al.*, 1992).

Imposing continuity for $M_{crat} = M_i/100$, we obtain the relationships

$$\begin{aligned} M_{crat} &= \alpha E_{rel} && \text{for} \quad E_{rel} \leq \tfrac{M_i}{100\alpha} \\ M_{crat} &= \frac{9\rho_i x_{cr}\alpha}{200S_i\alpha - \rho_i x_{cr}} E_{rel} + \frac{M_i}{10}\frac{\rho_i x_{cr} - 20S_i\alpha}{\rho_i x_{cr} - 200S_i\alpha} && \text{for} \quad \tfrac{M_i}{100\alpha} < E_{rel} \end{aligned} \tag{14}$$

In the cratering case, the largest surviving body has always a mass $M_i - M_{crat}$; we shall again assume a single–exponent power law mass distribution of the ejected fragments, but with an exponent b_i having the constant value 0.8, instead of being related to the impact parameters (as in the shattering case). As a consequence, the largest fragment from the crater has a mass $f_{l,i}M_{crat}$, with $f_{l,i}$ given by the implicit equation

$$f_{l,i} = 1 - b_i + b_i f_{l,i}^{b_i}\left(\frac{M_{dust}}{M_i}\right)^{1-b_i}, \tag{15}$$

which can be easily solved numerically. Also, in the cratering case the post–impact potential energy is given by

$$\begin{aligned} W_{crat,1} &= -\frac{3G}{5Q_i}\int_{m=0}^{m=\infty} m^{5/3}\, n_i(m)\, dm - \frac{3G(M_i - M_{crat})^{5/3}}{5Q_i} \\ &= -\frac{3G}{5Q_i}\frac{5M_{max,i}^{5/3} - 3b_i M_{max,i}^{b_i} M_{dust}^{5/3-b_i}}{5 - 3b_i} - \frac{3G(M_i - M_{crat})^{5/3}}{5Q_i} \end{aligned} \tag{16}$$

which has to be substituted in Eq.(10) for V_{esc} instead of $W_{fr,i}$.

5. Ejection Velocity vs. Mass Relationship and Reaccumulation

Most models of asteroid collisional evolution (e.g., Davis *et al.*, 1985, 1989) neglect any possible correlation between the ejection velocity and the mass of the fragments (both from shattering and for cratering events), and just assume that the fraction of the total fragment mass ejected with speeds exceeding some given value V is $(V/V_{min})^{-k}$, V_{min} being a fixed lower cutoff to the ejection velocity. Such a relationship was observed for crater ejecta (with $k \approx 9/4$) by Gault *et al.* (1963), and later confirmed by other experiments. On the other hand, recent experimental evidence (Davis and Ryan, 1990; Nakamura and Fujiwara, 1991) indicates that some correlation exists, with the largest fragments normally going somewhat slower; this is also predicted by the semiempirical models of Paolicchi *et al.* (1989, 1992). It is easy to show (see e.g. Nakamura and Fujiwara, 1991) that in this case the relationship given above for the mass fraction ejected at speeds $> V$ still holds provided the fragment speeds are proportional to the power $-r$ of their mass, with $r = (1 - b)/k$.

Here we shall assume that there is a definite correlation between the mass and the velocity of a fragment, which is modelled either in a deterministic way (e.g., through a definite formula relating the two quantities), or stochastically (see later). We define $E_i = E_{rel}/2$ as the energy received by body i at impact, and $E_{fr,i} = f_{ke} E_i$ as the total kinetic energy of ejecta from body i, where f_{ke} is an input parameter specifying which fraction of the impact energy is partitioned into kinetic energy of the fragments (for a detailed discussion about this *anelasticity parameter*, see Davis *et al.*, 1989). In order to avoid any possible divergence of the kinetic energy of the fragments, we set a high velocity cutoff V_{max} to the distribution (a reasonable physical motivation for this is that the fragment velocity is unlikely to exceed the sound velocity in the material, of the order of some km/s). To give the explicit relationship between the velocity v and the mass m of a fragment, we must distinguish two cases according to the sign of r_i. If $r_i > 0$:

$$\begin{array}{lll} v = C_i m^{-r_i} & \text{for} & \bar{M}_i \le m \le M_{max,i} \\ v = V_{max} & \text{for} & m < \bar{M}_i \,. \end{array} \tag{17}$$

where $\bar{M}_i = (V_{max}/C_i)^{-1/r_i}$. On the other hand, if $r_i < 0$ (this case is treated for consistency, see later):

$$\begin{array}{lll} v = C_i m^{-r_i} & \text{for} & M_{dust} \le m \le \bar{M}_i \\ v = V_{max} & \text{for} & \bar{M}_i < m \,. \end{array} \tag{17}$$

The constant coefficient C_i is obtained from an energy conservation equation. Consistently with the definition of V_{esc}, in the shattering case the largest fragment is assumed to have a negligible kinetic energy in the reference frame of the center of mass. On the other hand, this is not true in the cratering case. Therefore, we will insert the corresponding term in the conservation equation multiplied by a factor λ_i, with $\lambda_i = 0$ for shattering and $\lambda_i = 1$ for cratering. We also define $\bar{V}_i = C_i (M_{max,i})^{-r_i}$. We now have two possibilities: either for all fragments $C_i m^{-r_i}$ is smaller than V_{max}, in which case we just solve the equation:

$$\begin{aligned} E_{fr,i} &= \lim_{\epsilon \to 0} \int_{M_{dust}}^{M_{max,i}-\epsilon} \frac{v^2}{2} m n_i(m) dm + \lambda_i \frac{\bar{V}_i^2}{2} M_{max,i} \\ &= \lambda_i C_i^2 \frac{M_{max,i}^{1-2r_i}}{2} + \frac{C_i^2}{2} \frac{b_i M_{max,i}^{b_i}}{1 - b_i - 2r_i} \left[M_{max,i}^{1-b_i-2r_i} - M_{dust}^{1-b_i-2r_i} \right] \end{aligned} \tag{18}$$

and obtain

$$C_i = \sqrt{\frac{2E_{fr,i}}{\lambda_i M_{max,i}^{1-2r_i} + \frac{b_i M_{max,i}^{b_i}}{1-b_i-2r_i} \left[M_{max,i}^{1-b_i-2r_i} - M_{dust}^{1-b_i-2r_i} \right]}}, \tag{19}$$

or for some fragments (i.e. those having mass M_{dust} if $r_i > 0$ or $M_{max,i}$ if $r_i < 0$) $C_i m^{-r_i} > V_{max}$, in which case the correct equations are different depending on the sign of r_i: for $r_i > 0$ we have

$$E_{fr,i} = \lim_{\epsilon\to 0} \int_{\bar{M}_i}^{M_{max,i}-\epsilon} \frac{v^2}{2} m n_i(m) dm + \frac{V_{max}^2}{2} M(\le \bar{M}_i) + \lambda_i \frac{\bar{V}_i^2}{2} M_{max,i} \text{ ,(20)}$$

$$M(\le \bar{M}_i) = \frac{b_i}{1-b_i} M_{max,i}^{b_i} \left[\bar{M}_i^{1-b_i} - M_{dust}^{1-b_i}\right] \quad (20)$$

and

$$\begin{aligned} E_{fr,i} &= \lambda_i C_i^2 \frac{M_{max,i}^{1-2r_i}}{2} + \frac{V_{max}^2}{2} \frac{b_i}{1-b_i} M_{max,i}^{b_i} \left[\bar{M}_i^{1-b_i} - M_{dust}^{1-b_i}\right] \\ &+ \frac{C_i^2}{2} \frac{b_i M_{max,i}^{b_i}}{1-b_i-2r_i} \left[M_{max,i}^{1-b_i-2r_i} - \bar{M}_i^{1-b_i-2r_i}\right] \\ &= +C_i^{k_i} b_i \frac{M_{max,i}^{b_i} V_{max}^{2-k_i}}{2} \left[\frac{1}{1-b_i} - \frac{1}{1-b_i-2r_i}\right] \quad (20) \\ &- \frac{V_{max}^2}{2} \frac{b_i}{1-b_i} M_{max,i}^{b_i} M_{dust}^{1-b_i} + C_i^2 \frac{M_{max,i}^{1-2r_i}}{2} \left[\lambda_i + \frac{b_i}{1-b_i-2r_i}\right] ; \end{aligned}$$

on the other hand, for $r_i < 0$ we have:

$$E_{fr,i} = \int_{M_{dust}}^{\bar{M}_i} \frac{v^2}{2} m n_i(m) dm + \frac{V_{max}^2}{2} M(> \bar{M}_i) , \quad (21)$$

$$M(> \bar{M}_i) = \frac{b_i}{1-b_i} M_{max,i}^{b_i} \left[M_{max,i}^{1-b_i} - \bar{M}_i^{1-b_i}\right] + \lambda_i M_{max,i} , \quad (21)$$

$$\begin{aligned} E_{fr,i} &= \frac{V_{max}^2}{2} \frac{b_i M_{max,i}^{b_i}}{1-b_i} \left[M_{max,i}^{1-b_i} - \bar{M}_i^{1-b_i}\right] + \lambda_i \frac{V_{max}^2}{2} M_{max,i} \\ &+ \frac{C_i^2}{2} \frac{b_i M_{max,i}^{b_i}}{1-b_i-2r_i} \left[\bar{M}_i^{1-b_i-2r_i} - M_{dust}^{1-b_i-2r_i}\right] \\ &= C_i^{k_i} b_i \frac{M_{max,i}^{b_i} V_{max}^{2-k_i}}{2} \left[\frac{1}{1-b_i-2r_i} - \frac{1}{1-b_i}\right] \quad (21) \\ &+ \frac{V_{max}^2}{2} M_{max,i} \left[\lambda_i + \frac{b_i}{1-b_i}\right] - C_i^2 \frac{M_{max,i}^{b_i}}{2} \frac{b_i}{1-b_i-2r_i} M_{dust}^{1-b_i-2r_i} \end{aligned}$$

Thus, C_i is the solution of the equation

$$a C_i^{k_i} + b - C_i^2 = 0 , \quad (22)$$

where

$$a = M_{max,i}^{2r_i+b_i-1} V_{max}^{2-k_i} \left[\frac{2b_i r_i}{[(1-2r_i-b_i)\lambda_i + b_i]\,(1-b_i)} \right] \tag{22}$$

$$b = 2M_{max,i}^{2r_i-1} \left[\frac{1-b_i-2r_i}{(1-2r_i-b_i)\lambda_i + b_i} \right] \left[E_{fr,i} + \frac{V_{max}^2}{2} \frac{b_i}{1-b_i} M_{max,i}^{b_i} M_{dust}^{1-b_i} \right]$$

for $r_i > 0$, and:

$$a = M_{dust}^{2r_i+b_i-1} V_{max}^{2-k_i} \left[\frac{2r_i}{1-b_i} \right] \tag{22}$$

$$b = 2M_{dust}^{2r_i+b_i-1} M_{max,i}^{-b_i} \left[\frac{b_i+2r_i-1}{b_i} \right] \left[E_{fr,i} - \frac{V_{max}^2}{2} M_{max,i} \left(\lambda_i + \frac{b_i}{1-b_i} \right) \right]$$

for $r_i < 0$. Our program first tries the explicit formula (19); then, if the V_{max} cutoff turns out to be relevant, it solves numerically Eqs. (22). Also, we keep the freedom of giving as an input parameter either r_i or k_i, and then deriving the other one through the simple relationship

$$r_i = \frac{1-b_i}{k_i} \,. \tag{23}$$

This implies that in the "supercatastrophic" case $b_i > 1$, when most of the fragment mass is concentrated at small sizes, we obtain the somewhat counterintuitive result that either r_i or k_i has a negative value.

All the previous formulae have been derived for a *deterministic* relationship between v and m, as given by Eqs. (17). A more realistic treatment may assume that the same relationship has instead a probabilistic nature, i.e., it holds for the r.m.s. value $V_0 = \sqrt{\langle v^2 \rangle}$ of a Maxwellian distribution function for v:

$$P(v; V_0) = \sqrt{\frac{2}{\pi}} \frac{3\sqrt{3}v^2}{V_0^3} \exp\left(-\frac{3v^2}{2V_0^2} \right) . \tag{24}$$

Note that the mean kinetic energy for bodies of a given mass m is still given by $mV_0^2/2$. Therefore, neglecting statistical fluctuations, we can keep the same energy conservation equations obtained before to compute the constant C_i in this case (provided V_{max} is now interpreted as the higher cutoff to V_0). Also in this *probabilistic* case, we have the choice of fixing in input either r_i or k_i.

The self–gravitational reaccumulation process is then treated as follows. The impact is assumed to always generate a remnant body formed by the two largest remnants from the colliding bodies (either the largest shattered fragment(s) or the cratered target(s)), plus the non–escaping fraction of the smaller fragments. Let us define $M_{esc,i} = (C_i/V_{esc})^{1/r_i}$, namely the mass corresponding to the escape velocity. In the deterministic case, if $r_i > 0$, the reaccumulated (non–escaping) fragments are those with mass in the range between $\max(M_{dust}, M_{esc,i})$ and $M_{max,i}$; if $r_i < 0$, the reaccumulation range is between M_{dust} and $\min(M_{esc,i}, M_{max,i})$. The corresponding total mass is easily found by integrating between these limits $m \cdot n_i(m)dm$, with $n_i(m)dm$ given by Eq.(5).

In the probabilistic case, the whole range between M_{dust} and $M_{max,i}$ is considered. For the mass bins containing ≤ 100 bodies, the non–escaping fragments are drawn at random by using the velocity distribution (24) — more precisely, given V_0 (derived from m through equations similar to (17)), a number is drawn at random according to the distribution (24), and then compared to V_{esc}. For the mass bins containing more than 100 fragments, we just compute the fraction of non–escaping fragments f from

$$f = \int_0^{V_{esc}} \sqrt{\frac{2}{\pi}} \frac{3\sqrt{3}v^2}{V_0^3} \exp -\frac{3v^2}{2V_0^2} dv \, ; \tag{25}$$

calling $\mathrm{erf}(v)$ the error function

$$\mathrm{erf}(v) = \frac{2}{\sqrt{\pi}} \int_0^v \exp -x^2 dx \, , \tag{26}$$

we get

$$f = -\frac{2}{\sqrt{\pi}} \sqrt{\frac{3}{2}} \frac{V_{esc}}{V_0} \exp -\frac{3V_{esc}^2}{2V_0^2} + \mathrm{erf}\left(\sqrt{\frac{3}{2}} \frac{V_{esc}}{V_0} \right) \, . \tag{27}$$

As for the "tail" of the mass distribution (i.e., the fragments having masses between M_{dust} and the lower limit of the smallest mass bin, $m_{bin,1}/\sqrt{g}$), we keep the option of dealing with it "deterministically" or in agreement with the probabilistic procedure. In the former case, the reaccumulated tail mass is the integral of $m \cdot n_i(m)dm$ between the values $\max(M_{dust}, M_{esc,i})$ and $\min(M_{max,i}, m_{bin,1}/\sqrt{g})$ (provided $r_i > 0$), or between M_{dust} and $\min(M_{esc,i}, M_{max,i}, m_{bin,1}/\sqrt{g})$ (if $r_i < 0$).

On the other hand, the "probabilistic" tail treatment is done by taking into account the velocity distribution (24). This is made by evaluating the double integral

$$I = bM_{max}^b \int_{M_{dust}}^{m_{bin,1}/\sqrt{g}} \int_0^{V_{esc}} P(v; V_0(m)) m^{-b} dv \, dm \, . \tag{28}$$

If $V_0 = V_{max}$, the integral over the mass can be done analytically and that over the velocity reduces to (25); otherwise a numerical calculation is carried out.

6. Some Results and Applications

In this Section we shall illustrate some results of our model by means of appropriate plots. We are not going here to explore systematically the multi–dimensional space of our input model parameters in order to study how the collision outcomes depend on them, but only to touch on some specific issues, which we believe are important from the physical point of view. Therefore, we have assumed a set of *nominal* input values as follows: density $\rho = 2500\ kg/m^3$; impact velocity $V_{rel} = 5.81\ km/s$; lower cutoff of the fragment mass $M_{dust} = 10^{-5}\ kg$; higher cutoff of the fragment ejection velocity $V_{max} = 5\ km/s$; pre–scaling impact strength $S_0 = 3 \times 10^6\ J/m^3$ (S was then derived by means of Eq. (1), with $\gamma = 1$); anelasticity parameter $f_{ke} = 0.1$; crater excavation coefficient $\alpha = 10^{-4}\ s^2/m^2$. For our numerical computations we used 50 logarithmic mass bins spaced by a factor 2 (i.e., a factor 1.26 in size), with central values of the corresponding radii ranging from $1.211 \times 10^{-2}\ km$ to $10^3\ km$.

Figs. 1 and 2 give an idea of the mass distribution predicted by the model for shattered bodies of radii ranging from 1 km to $10^3\ km$ (roughly, the size range where most observable asteroids and natural satellites are found). We have plotted in the same frame the mass ratio f_l between the largest fragment and the target (or the total mass excavated in the crater) and the exponent b of the cumulative fragment mass distribution (see Eqs. (2), (3) and (9)) versus the logarithm of the radius R of the parent body, by taking a fixed value (either 10^{-4} or 10^{-3}) of the projectile–to–target mass ratio M_2/M_1. As shown by Fig. 1, both b and f_l approach the limit value $1/2$ when M_2/M_1 and R are such that the threshold between shattering and cratering impacts is reached (this occurs going towards larger target radii, due to the increase of S with size (see Eq.(1)). b is of course closer to 1 (and f_l to zero) for more disruptive collisions, owing to a lower strength or a larger projectile–to–target mass ratio. For craters, on the other hand, both b and f_l are assumed to have constant values (see Sec. 4). Notice that for the nominal values of S_0 and V_{rel}, a collision involving a projectile of $1/10$ the size of the target normally causes extensive break–up, and generates a swarm of fairly small fragments. For an asteroid 200 km across, such as the parent bodies of several asteroid families, the average collision rate quoted in Sec. 1 implies that such collisions have a probability of order unity of occurring over the solar system lifetime (taking into account that several thousands of bodies of diameter $\approx 20\ km$ exist in the main asteroid belt, see Cellino *et al.*, 1991).

The curves shown in Figs. 1 and 2 are not significantly changed when the lower mass cutoff of the fragments M_{dust} is modified within reasonable bounds. Actually, as discussed in Sec. 3, this parameter becomes relevant only for *supercatastrophic* impacts, namely when b is close to 1. This is illustrated in Fig. 3, where the relationship between f_l and b is plotted for three values of M_{dust} (differing from each other by orders of magnitude) in the supercatastrophic case $M_2/M_1 = 1$. Note that for b smaller than unity the approximation $f_l \cong (1 - b)$, which corresponds to

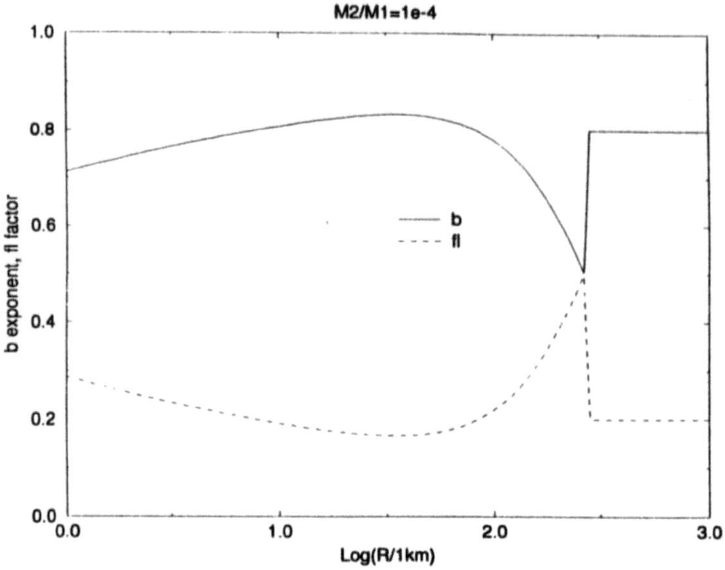

Fig. 1. b and f_l versus the logarithm of the radius of the parent body for the nominal values of the impact parameters and a projectile–to–target mass ratio of 10^{-4}.

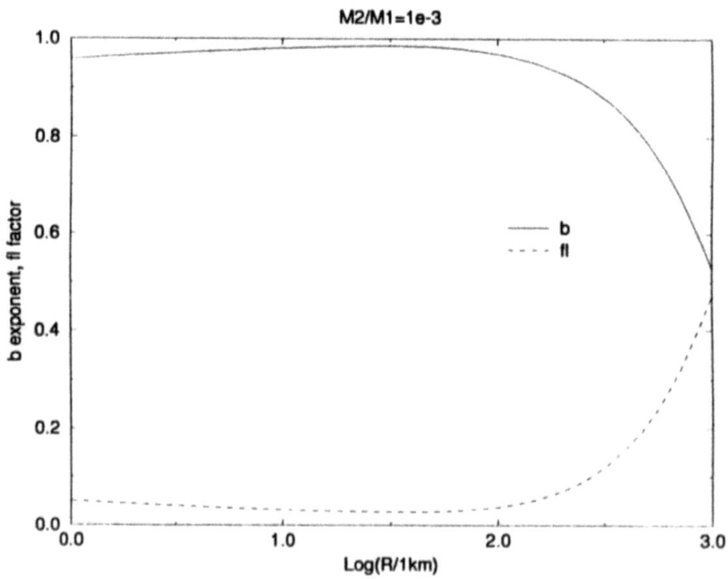

Fig. 2. The same as Fig. 1, but for $M_2/M_1 = 10^{-3}$.

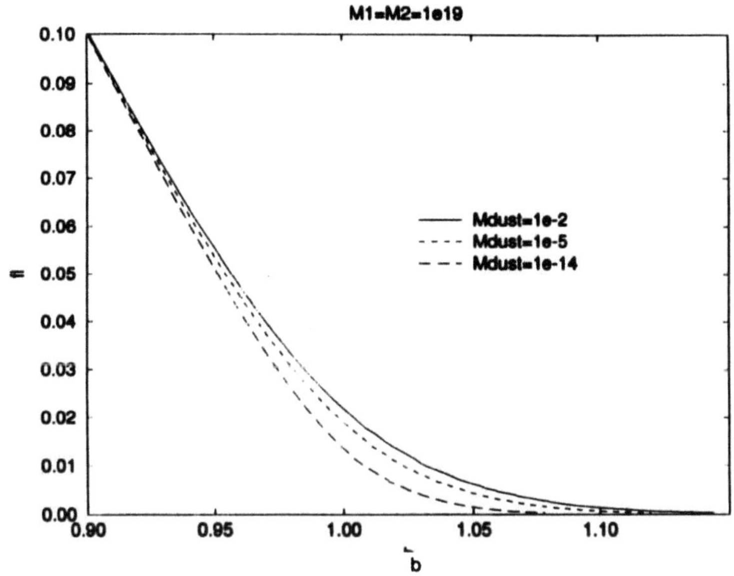

Fig. 3. f_l versus b for $M_1 = M_2 = 10^{19}\ kg$ and different values of M_{dust}.

the limit $M_{dust} \Rightarrow 0$, is fairly accurate.

Fig. 4 shows, for three different values of the projectile–to–target mass ratio, how the averaged escape velocity V_{esc} defined by Eq.(10) (normalized to the surface escape velocity of M_1) varies as a function of b. As shown by Farinella *et al.* (1988), when $M_2 \ll M_1$ and b is close to 1, the limiting value of the normalized escape velocity is about 0.77. The curves shown in Fig. 4 are somewhat different from that appearing in Fig. 1 of the above–mentioned paper, because (i) we now take into account the mass of the projectile (so the curves are shifted upward for higher values of M_2/M_1), and (ii) we have made the assumption that the largest fragment, whose mass is higher for smaller values of b, has always a negligible kinetic energy. Fig. 4 shows that taking a constant value ≈ 0.8 for the normalized escape velocity after shattering impacts is normally a good approximation. As shown by Fig. 5, using $M_{dust} > 0$ has little influence on the escape velocity.

One of the most interesting aspects of the algorithm presented in this paper is that it allows us to model in several alternative ways the gravitational reaccumulation process, through the use of different fragment velocity vs. mass relationships, as discussed in Sec. 5. Qualitatively speaking, the reaccumulation process is due to the fact that, for targets having escape velocities of tens to hundreds of m/s, the amount of energy sufficient for fragmenting them is normally not enough to accelerate all the fragments to speeds exceeding V_{esc}. Thus the slower fragments fall back and reaccumulate onto the largest one, giving rise to a so–called *pile of rubble.* The relevance of this process for asteroid evolution has been discussed in detail more than a decade ago by Davis *et al.* (1979) and Farinella *et al.* (1982). But of course the quantitative aspects of it depend in a critical manner on the conversion

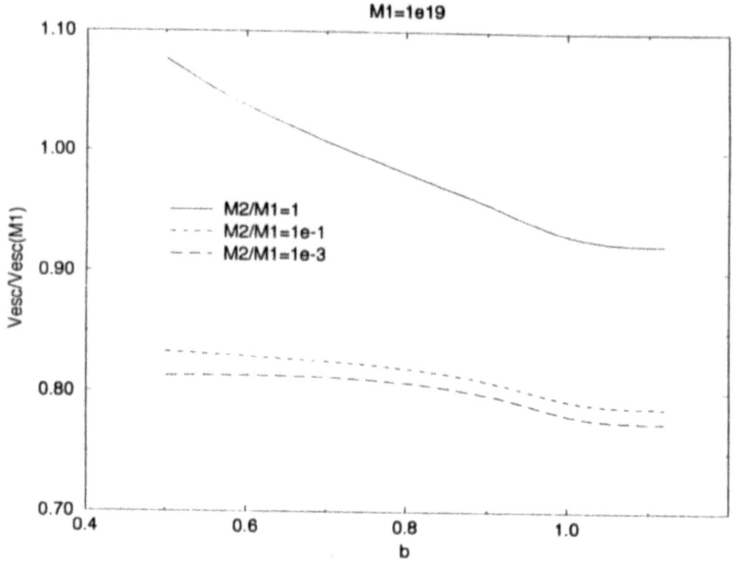

Fig. 4. The average fragment escape velocity, normalized to the surface escape velocity of M_1, versus b for three different values of the projectile–to–target mass ratio.

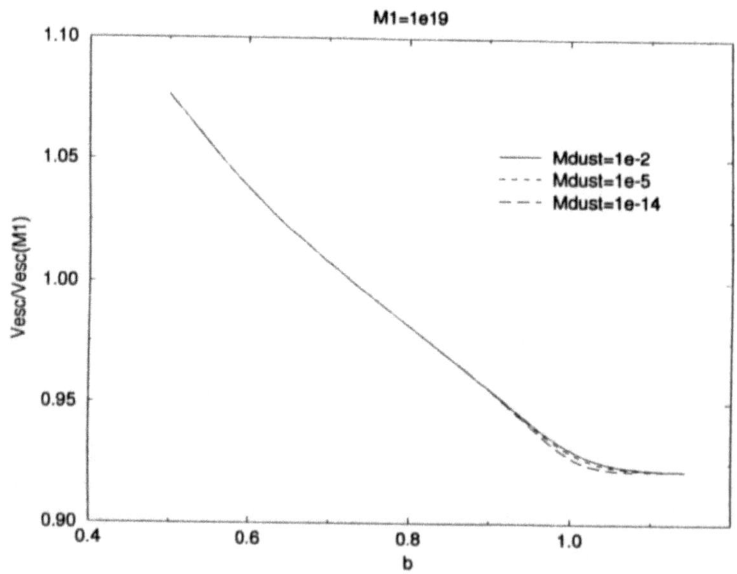

Fig. 5. The same as Fig. 4 for $M_2/M_1 = 1$, but for three different values of M_{dust}.

efficiency of the projectile's kinetic energy into kinetic energy of fragments (i.e., the *anelasticity coefficient* f_{ke}), as well as on the way the available kinetic energy is partitioned among fragments of different masses. We shall focus here on the latter issue.

Fig. 6 shows, for a target 100 km in radius and $M_2/M_1 = 10^{-3}$, the fraction of reaccumulated mass versus the logarithm of the impact energy, normalized to the critical energy for fragmentation. Clearly, with our nominal parameter choice, there is a gap of more than two orders of magnitude between the fragmentation threshold and the energy needed to disperse to infinity a significant fraction of the target mass. As a consequence, most asteroids of radius $\approx$ 100 km have been probably shattered but not *disrupted* (i.e., have reaccumulated) as a consequence of the most energetic collisions they have undergone over the history of the solar system. Fig. 6 also shows that the reaccumulation process is confined to a narrow energy range if we assume $k = 9/4$, i.e. we require that for every velocity V (exceeding a lower cutoff) the total mass ejected with higher speeds is $\propto V^{-9/4}$. The reason for this is that since $r = (1-b)/k$ (see Sec. 5) and b is close to 1 (see Fig. 2), the exponent r in this case is small, implying that the velocity of fragments has a weak dependence on their mass: thus, there is an abrupt transition between the case when nearly all the fragments are slower than V_{esc} and the opposite one. The transition is less sharp in the probabilistic case, since in this case for every mass the fragment velocities have a significant chance variability — and it is much smoother when r (instead of k) is fixed to the value $1/6$, consistent with experimental data. In this case, even after a supercatastrophic impact a significant fraction of mass is always reaccumulated, because most of the mass resides in small fragments and is forced to move comparatively fast, so that conservation of energy causes the largest fragments to be slow and to reaccumulate. Fig. 7 shows what happens in the probabilistic case if the bodies in the small–size tail of the mass distribution (fragments smaller than about 20 m) are treated deterministically, as described in Sec. 5: both the curves become fairly similar to the ones corresponding to the fully deterministic case, because the tail fragments dominate the escaping mass distribution.

A physically important consequence of the different behaviors of the $k = 9/4$ and the $r = 1/6$ models is apparent from Figs. 8 and 9, which give, for the same models used to draw Fig. 6 and 7, the fraction of reaccumulated mass versus target size. Again, for the $k = 9/4$ models (but less for the probabilistic ones), there is a comparatively sudden transition from negligible to substantial reaccumulation, which occurs when the velocity of most fragments becomes lower than V_{esc}. This would imply comparatively short collisional lifetimes of small asteroids — say, in the 1 to 50 km diameter range. Also, in this case small asteroids would be unlikely to be *rubble piles*, but would mostly look like single, competent fragments. The opposite conclusion would follow from the $r = 1/6$ models, for which the figures show that even small bodies are always reaccumulated; recall, however, that when V_{esc} is small only a small number of massive fragments may become reaggregated.

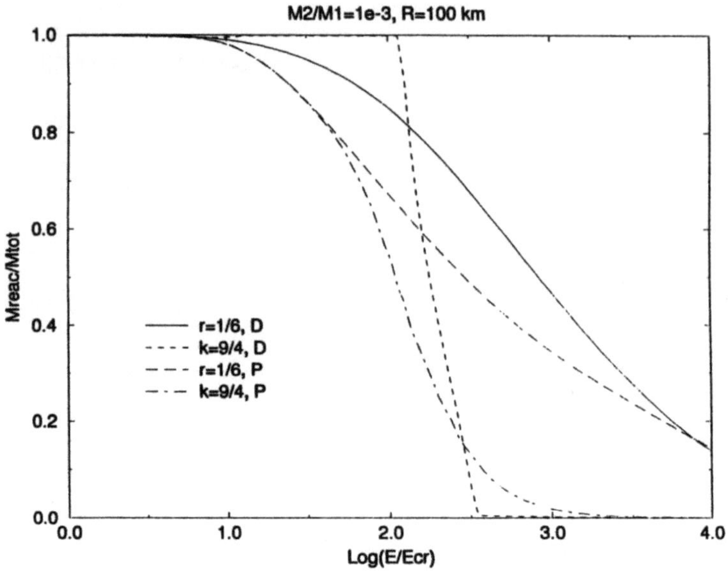

Fig. 6. The fraction of reaccumulated mass for an impact with $M_2/M_1 = 10^{-3}$ against a target of radius 100 km, versus the logarithm of the impact energy. The latter parameter is normalized to the minimum energy required for fragmentation. Different fragment velocity vs. mass relationships are used (see text), with D standing for "deterministic model" and P for "probabilistic model".

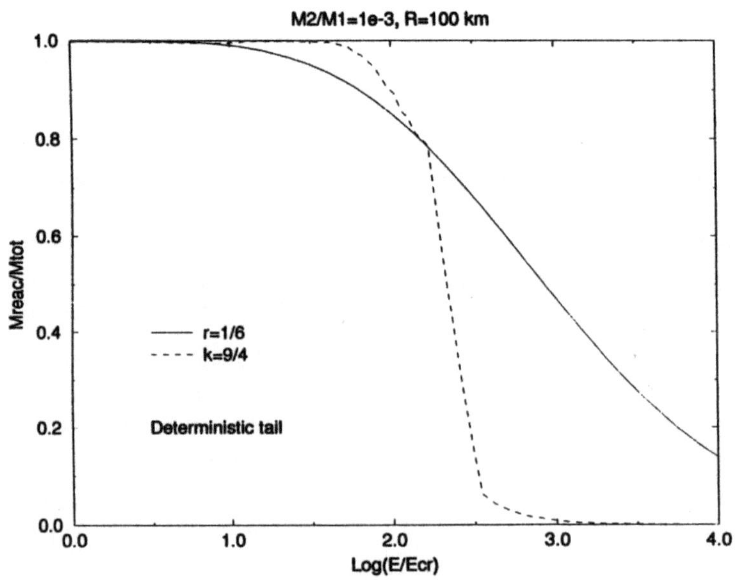

Fig. 7. The same as Fig. 6 for the probabilistic case, but treating always in a deterministic way the small–size tail of the fragment mass distribution.

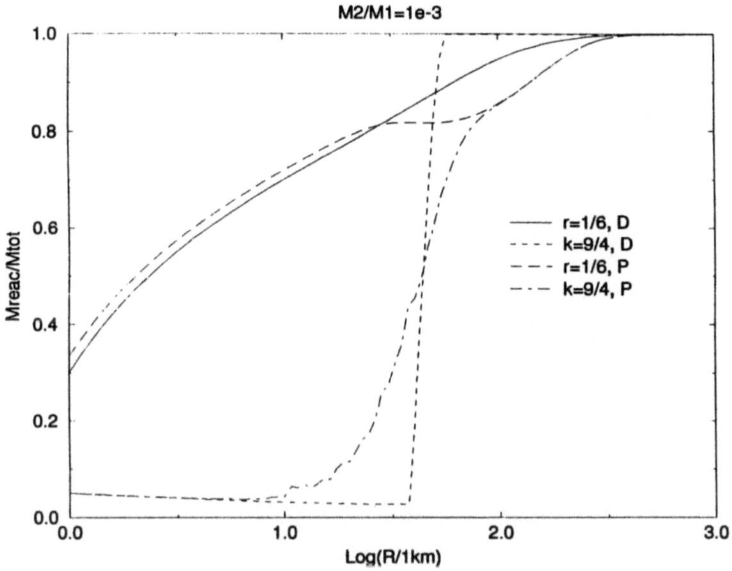

Fig. 8. The fraction of reaccumulated mass for a shattering impact with $M_2/M_1 = 10^{-3}$ versus the logarithm of the target radius. The same four models have been used as in Fig. 6.

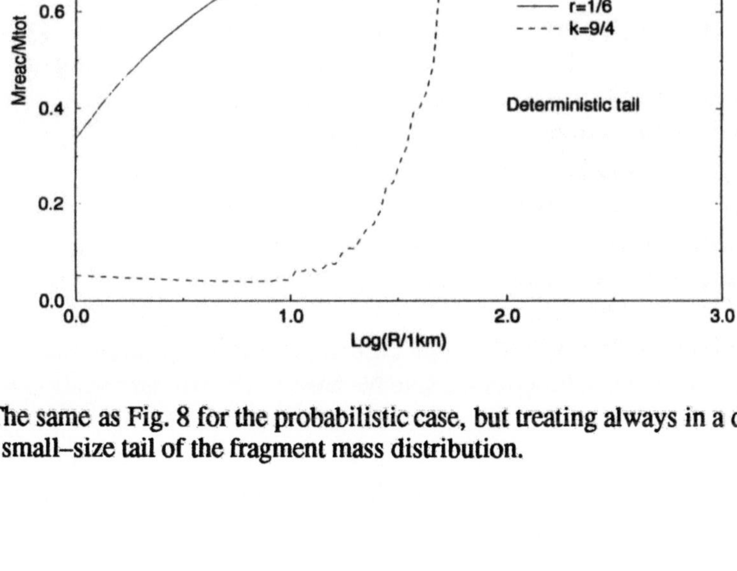

Fig. 9. The same as Fig. 8 for the probabilistic case, but treating always in a deterministic way the small–size tail of the fragment mass distribution.

This is possibly an effective mechanism to form km–sized contact or nearly–contact binary asteroids, such as recently observed by radar (Ostro *et al.*, 1990).

For a smaller projectile–to–target mass ratio (10^{-4}, still sufficient for fragmentation), Fig. 10 shows that both the $r = 1/6$ models and the $k = 9/4$ models display a transition from complete fragment escape (with the chance exception of some large fragments in the probabilistic cases, giving rise to irregular curves) to substantial reaccumulation. Of course here the transition occurs at smaller target sizes (less than 10 km radius), but there is a shift of about a factor 3 between the two types of models, due to the lower velocity of the big fragments in the $r = 1/6$ cases. For a barely shattering impact (Fig. 11), on the other hand, all the models yield similar results, with the transition occurring at a radius of ≈ 5 km (plus or minus a factor 2) and large fluctuations in the probabilistic cases.

When $M_2/M_1 = 10^{-5}$, we are in the cratering case (see Sec. 4), with Eqs.(14) yielding between 2% and 3% of excavated mass (see Fig. 12; the crater mass increases with target size because of the decreasing impact strength in this size range). Again the reaccumulation of crater ejecta is effective at smaller sizes in the $r = 1/6$ models, and is anyway practically complete for bodies exceeding 100 km diameter.

These results provide some hints to the collisional history of 951 Gaspra, the asteroid about 13 km in mean diameter recently approached and imaged by the Galileo probe (Belton *et al.*, 1992). Gaspra's global shape is strongly irregular, with two very large concavities (about 8 km across) evident on its limb. However, the general appearance of the surface is fairly smooth, with no sharp edges and a relative paucity of km–sized craters (while several linear "grooves" are apparent). To explain these features and the distribution of colors and albedo variations, which hints to the presence of surface regolith, Greenberg *et al.* (1992) have suggested that Gaspra's interior is a rubble pile, overlain by a substantial regolith layer; the large concavities would represent giant impact basins, formed by events close to the catastrophic break–up threshold. This interpretation is consistent with our models, in particular those including a significant ejecta velocity vs. mass dependence, as they show that even for bodies as small as Gaspra a large fraction of the material ejected after impacts can be reaccumulated by self–gravity both in the shattering and in the cratering regime (see Figs. 10–12). Thus Gaspra may have been originated in a shattering event followed by the reaggregation of a few large blocks; the subsequent formation of some giant craters may have produced the surface regolith and erased the small–scale topography.

Figs. 13 and 14 show another type of diagram, which should be compared with Fig. 5 of Davis *et al.* (1985) and Fig. 5 of Housen and Holsapple (1990). Here the lower (dashed) curve corresponds to the critical projectile–to–target mass ratio for fragmentation, which is a function of size because of the strength scaling relationship (1). The other curves give the threshold mass ratio for collisional disruption, with one half of the target mass dispersed to infinity. While for small asteroids, whose gravity is not efficient in reaccumulating fragments, the fragmentation and

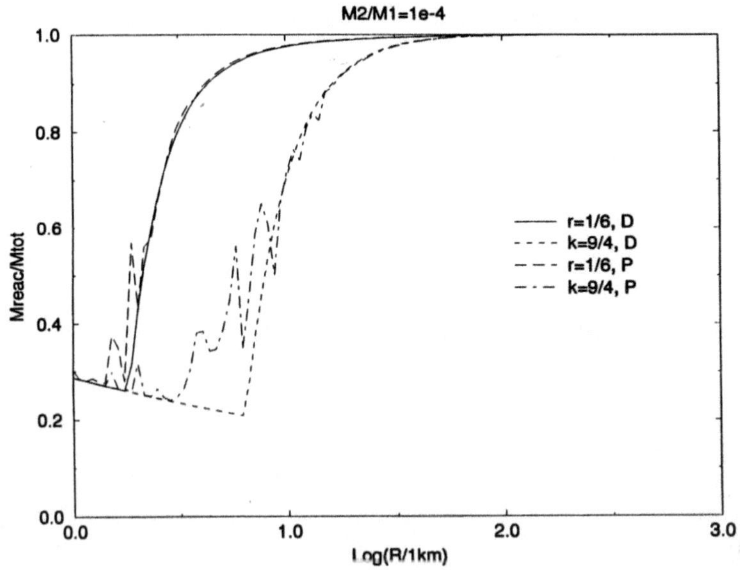

Fig. 10. The same as Fig. 8, but for $M_2/M_1 = 10^{-4}$.

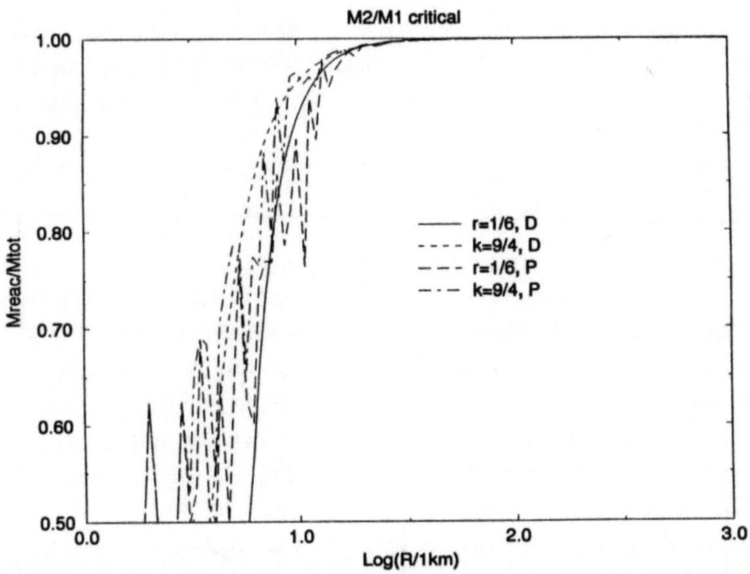

Fig. 11. The same as Fig. 8, but for M_2/M_1 just exceeding the threshold value for fragmentation.

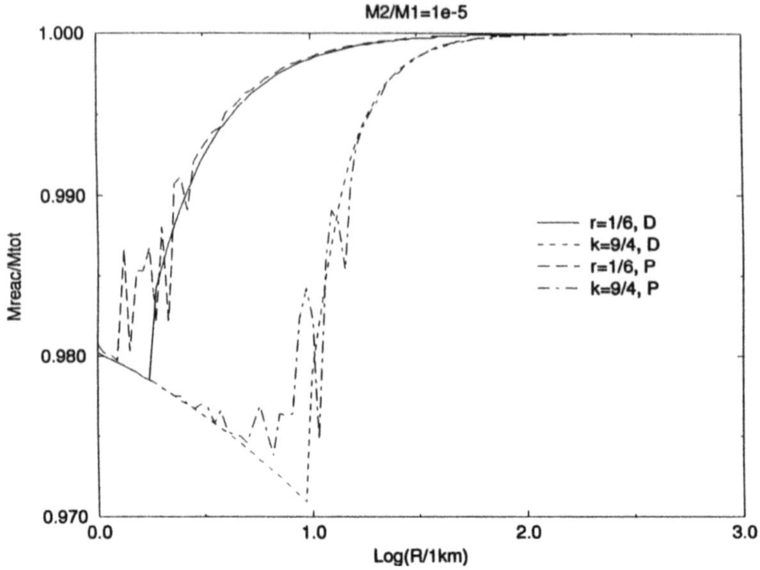

Fig. 12. The same as Fig. 8, but for $M_2/M_1 = 10^{-5}$. This ratio results in a cratering event.

disruption thresholds coincide, for asteroids $\approx 100\,km$ across there is a large gap between the two thresholds, due to the gravitational reaccumulation process. In this case, the curves show that the for the $r = 1/6$ models a sudden discontinuity appears when a critical size is exceeded, while for the $k = 9/4$ case there is a more gradual transition. This behaviour can be explained as follows: in the $k = 9/4$ case, when all the fragment velocities are close to each other, a modest specific energy (i.e., mass ratio) increase may be enough to exceed V_{esc}; on the other hand, in the $r = 1/6$ case, after self–gravity has become strong enough to reaccumulate the largest fragments, a moderate increase in the specific energy causes a mass shift to fragments of smaller sizes, which receive a larger fraction of the available energy, thus preventing the largest fragments from starting faster (and being able to escape).

This sudden "runaway reaccumulation" phenomenon in the $r = 1/6$ models is illustrated in Figs. 15 and 16. Here the reaccumulated mass fraction is plotted versus M_2/M_1 for two values of the target radius, close to each other but on the two sides of the critical value. While with the $k = 9/4$ model the reaccumulated mass always decreases in a nearly–monotonic way when the impact gets more energetic, the opposite happens with the $r = 1/6$ model: for the lower size, when the fragmentation threshold is exceeded by a small amount, there is still a small drop of the reaccumulated mass fraction to values $< 1/2$, but this drop disappears at the larger size, since in this case more energetic collisions result initially into more effective reaccumulation; only for much higher mass ratios the reaccumulated fraction decreases, falling again below $1/2$ only for values of M_2/M_1 exceeding the fragmentation threshold by almost two orders of magnitude.

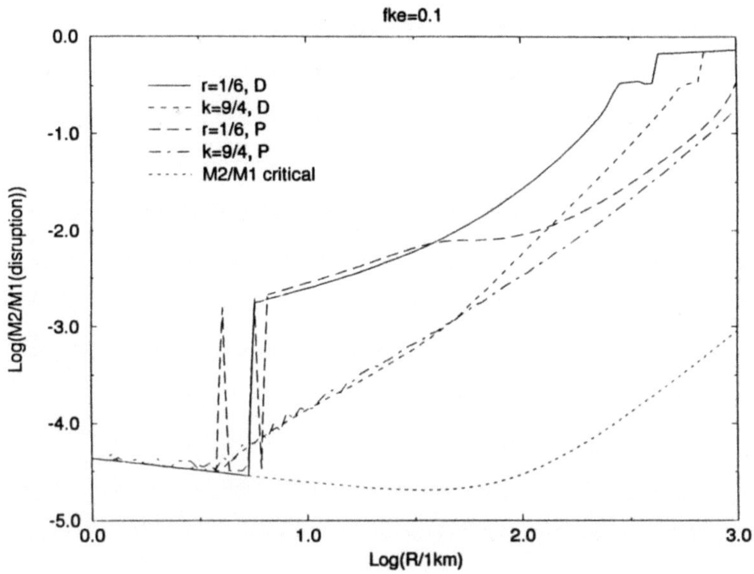

Fig. 13. The critical mass ratio for *disruption* versus the logarithm of the target size, for the four models used in the previous figures. The bottom (dashed) line is the threshold mass ratio for *fragmentation*.

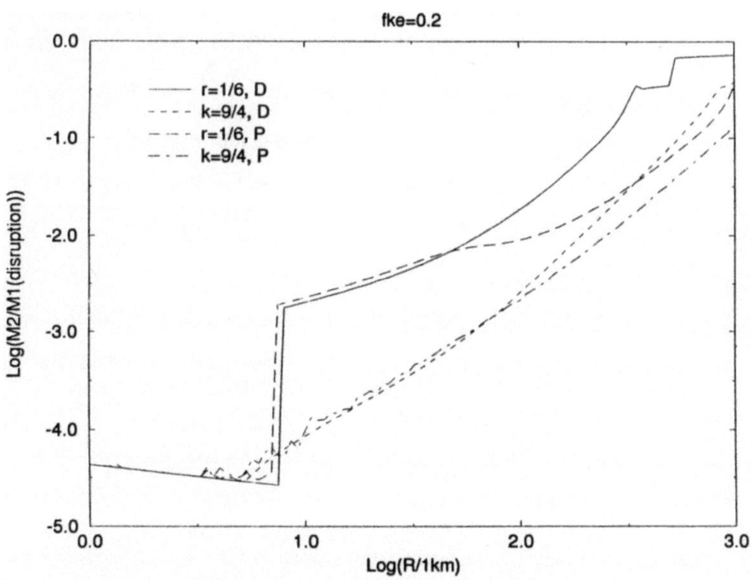

Fig. 14. The same as Fig. 13, but with $f_{ke} = 0.2$.

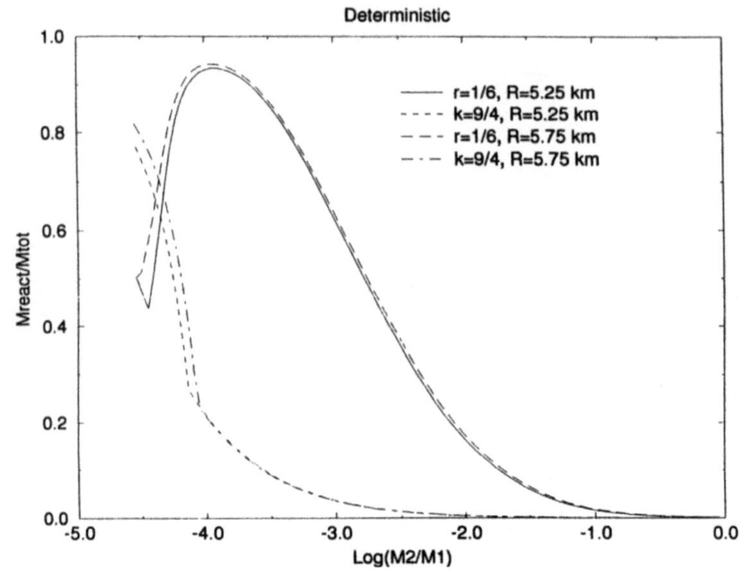

Fig. 15. The reaccumulated mass fraction versus M_2/M_1 for the deterministic $r = 1/6$ and $k = 9/4$ models and for two values of the target radius, close to each other but on the two sides of the critical value for which reaccumulation starts to be effective.

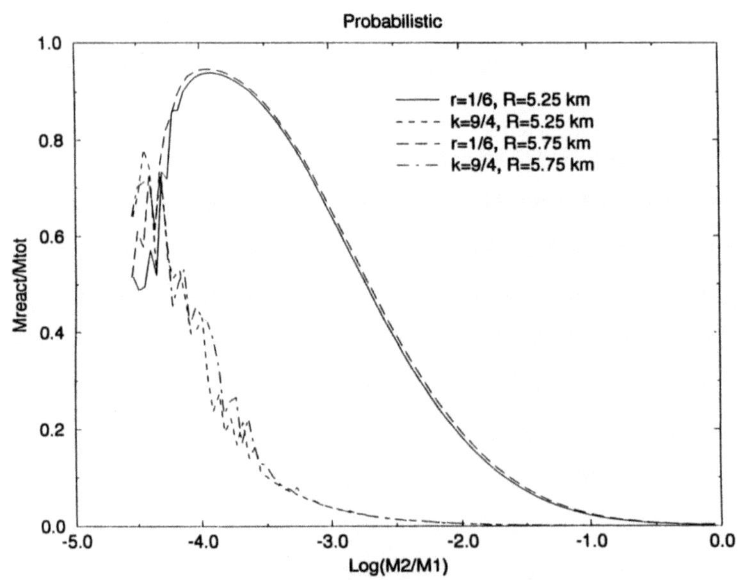

Fig. 16. The same as Fig. 15 but for the probabilistic models.

This is just an example of what we believe is an important finding of this work: for asteroidal–sized bodies, even keeping fixed the (poorly known) parameter f_{ke}, the predicted collisional outcomes depend in a sensitive way on the adopted fragment velocity vs. mass model. Take for instance the formation of an asteroid family from a target $\approx 200\,km$ across, which requires a collisional event capable of ejecting to infinity a substantial fraction of the mass of the parent body. Fig. 13 shows that, even keeping $f_{ke} = 0.1$ and limiting ourselves to the four models described earlier, the required projectile–to–target mass ratio ranges from about $1/300$ ($k = 9/4$, probabilistic) to $1/30$ ($r = 1/6$, deterministic), implying a projectile diameter between $30\,km$ and $65\,km$. This entails about a factor 4 of uncertainty in the available number of projectiles (≈ 1100 versus 300, respectively; see Fig. 3 of Cellino *et al.*, 1991), with a similar uncertainty in the collision lifetimes and therefore in the expected number of families. With the average collision rate derived by Farinella and Davis (1992; see Sec. 1) and the current population of main–belt asteroids, the expected number of disrupted $200\,km$–sized parent bodies may range between 1.5 and 5. The agreement with the results of family searches such as that of Zappalà *et al.* (1990) is good, but reducing the uncertainty would be very important.

7. Conclusions and Future Work

The algorithm described in this paper is simple enough to be applied in a quick and effective way to a variety of studies on the collisional evolution of small solar system bodies. In Sec. 6 we have presented some examples related to the evolution of asteroids and the origin of asteroid families, but similar analyses — with comparable resulting uncertainties — could be performed for other cases of catastrophic or nearly–catastrophic impacts occurred in the solar system. A list might include: the formation of Vesta's family from a giant crater preserving intact most of the basaltic crust of this asteroid (Davis *et al.*, 1985; Rosa *et al.*, 1992); the collisional origin of the dust bands discovered by IRAS (Sykes *et al.*, 1989); the formation of reaccumulated, rapidly spinning (or binary) asteroids (Farinella *et al.*, 1982, 1985; Weidenschilling *et al.*, 1989); the origin of very large craters and impact basins observed on many natural satellites (Chapman and McKinnon, 1986); the collisional disruption of Hyperion (Farinella *et al.*, 1983); the formation of planetary rings from disrupted satellites (Harris, 1984).

Of course, many input parameters of the collisional models — and also the validity of several empirical relationships used in Secs. 2 to to 6 — are still subject to considerable uncertainty. We are confident that this uncertainty will be reduced in the near future as a result of the combination between further laboratory experiments and complex numerical simulations of the collisional disruption process. We think that the algorithm presented here is simple and flexible enough to easily incorporate new theoretical and experimental results. Once this is done, the output of our code can be written as a three–dimensional matrix, specifying how many objects end up in the i–th mass bin following a collision between two bodies lying in the j-th and k–th mass bins. This matrix can then be read by a program aimed at simulating the time evolution of an initial asteroid (or planetesimal, or planetary ring particle) population by computing in any time step for every pair of mass bins the expected number of collisions, as in the work of Greenberg *et al.* (1978), Davis *et al.* (1979, 1985, 1989), and Farinella *et al.* (1992). We are going to present in a forthcoming paper some results on the sensitivity of the outcomes of the simulated asteroid collisional evolution (in terms of the observable size distribution), with respect to the fragment velocity vs. mass correlation models, as discussed in Sec. 6, and other poorly known input parameters. We also plan to develop in the future, along the same lines discussed here, a more comprehensive algorithm capable of computing impact outcomes, not only in terms of sizes but also including spins, i.e., angular momenta of rotation — for the physics involved in this problem see Davis *et al.* (1989), Cellino *et al.* (1990) and Farinella *et al.* (1992).

Acknowledgements

We are grateful to the reviewer R. Greenberg for pointing to us the recent hydrocode modelling work, and to A. Campo Bagatin, A. Cellino, D.R. Davis, G. Martelli, P. Paolicchi and V. Zappalà for useful discussions and comments. J.–M. Petit's stay in Pisa in 1991 was supported through the CNRS–CNR researcher exchange agreement. P. Farinella acknowledges financial support by the Observatoire de la Côte d'Azur and the EEC research grant no. SC1–0011–C(GDF).

References

Asphaug, E., Benz, W., and Melosh, H.J.: 1992, Asteroid impact studies with SPH 3D (abstract), *Bull. Amer. Astr. Soc.*, in press.

Belton, M.J.S., Veverka, J., Thomas, P., Helfenstein, P., Simonelli, D., Chapman, C., Davies, M.E., Greeley, R., Greenberg, R., Head, J., Murchie, S., Klaasen, K., Johnson, T.V., McEwen, A., Morrison, D., Neukum, G., Fanale, F., Anger, C., Carr, M., Pilcher, C.: 1992, Galileo encounter with 951 Gaspra: First pictures of an asteroid, *Science* **257**, 1647–1652.

Cellino, A., Zappalà, V., Davis, D.R., Farinella, P., and Paolicchi, P.: 1990, Asteroid collisional evolution, I. Angular momentum splash: Despinning asteroids through catastrophic collisions, *Icarus*, **87**, 391–402.

Cellino, A., Zappalà, V., and Farinella, P.: 1991, The size distribution of main–belt asteroids from IRAS data, *Mon. Not. R. astr. Soc.* **253**, 561–574.

Chapman, C.R., and McKinnon, W.B.: 1986, Cratering on planetary satellites, in *Satellites*, eds. J.A. Burns and M.S. Matthews (Univ. of Arizona Press), pp. 492–580.

Chapman, C.R., Paolicchi, P., Zappalà, V., Binzel, R.P, and Bell, J.F.: 1989, Asteroid families: Physical properties and evolution, in *Asteroids II*, eds. R.P. Binzel, T. Gehrels and M.S. Matthews (Univ. of Arizona Press), pp. 386–415.

Davis, D.R.: 1990, On social disintegration in the solar system: The disruption of families, presented at the Third International Workshop on Catastrophic Disruption of Small Solar System Bodies (Kyoto, Japan, September 1990).

Davis, D.R., Chapman, C.R., Greenberg, R., and Harris, A.W.: 1979, Collisional evolution of asteroids: Populations, rotations, and velocities, in *Asteroids*, ed. T. Gehrels (Univ. of Arizona Press), pp. 528–557.

Davis, D.R., Chapman, C.R., Weidenschilling, S.J., and Greenberg, R.: 1985, Collisional history of asteroids: Evidence from Vesta and the Hirayama families, *Icarus* **62**, 30–53.

Davis, D.R., Farinella, P., Paolicchi, P., and Zappalà, V. (eds.): 1986, *Catastrophic Disruption of Asteroids and Satellites*, *J. Ital. Astron. Soc.* **57**, n. 1.

Davis, D.R., Farinella, P., Paolicchi, P., Weidenschilling, S.J., and Binzel, R.P.: 1989, Asteroid collisional history: Effects on sizes and spins, in *Asteroids II*, eds. R.P. Binzel, T. Gehrels and M.S. Matthews (Univ. of Arizona Press), pp. 805–826.

Davis, D.R., and Ryan, E.V.: 1990, On collisional disruption: Experimental results and scaling laws, *Icarus* **83**, 156–182.

Dobrovolskis, A., and Burns, J.A.: 1984, Angular momentum drain: A mechanism for despinning asteroids, *Icarus* **57**, 464–476.

Farinella, P., Paolicchi, P., and Zappalà, V.: 1982, The asteroids as outcomes of catastrophic collisions, *Icarus* **52**, 409–433.

Farinella, P., Milani, A., Nobili, A.M., Paolicchi, P., and Zappalà, V.: 1983, Hyperion: Collisional disruption of a resonant satellite, *Icarus* **54**, 353–360.

Farinella, P., Paolicchi, P., and Zappalà, V.: 1985, Original abundance of asteroids from their present rotational properties, *Mon. Not. R. astr. Soc.* **216**, 565–570.

Farinella, P., Paolicchi, P., Cellino, A., and Zappalà, V.: 1988, Escape velocity for fragments of collisionally shattered solar system bodies. *Bull. Obs. Astron. Belgrade* **138**, 88–90.

Farinella, P., and Davis, D.R.: 1992, Collision rates and impact velocities in the main asteroid belt. *Icarus* **97**, 111–123.

Farinella, P., Davis, D.R., Paolicchi, P., Cellino, A., and Zappalà, V.: 1992, Asteroid collisional evolution: An integrated model for the evolution of asteroid rotation rates, *Astron. Astrophys.* **253**, 604–614.

Fujiwara, A., Kamimoto, G., and Tsukamoto, A.: 1977, Destruction of basaltic bodies by high–velocity impact. *Icarus* **31**, 277–288.

Fujiwara, A., and Tsukamoto, A.: 1980, Experimental study on the velocity of fragments in collisional breakup. *Icarus* **44**, 142–153.

Fujiwara, A., Cerroni, P., Davis, D.R., Ryan, E., Di Martino, M., Holsapple, K., and Housen, K.: 1989, Experiments and scaling laws on catastrophic collisions, in *Asteroids II*, eds. R.P. Binzel, T. Gehrels and M.S. Matthews (Univ. of Arizona Press), pp. 240–265.

Gault, D.E., Shoemaker, E.M., and Moore, H.J.: 1963, Spray ejected from the lunar surface by meteoroid impact, *NASA Tech. Note* **D–1767**.

Greenberg, R., Wacker, J.F., Hartmann, W.K., and Chapman, C.R.:1978, Planetesimals to planets: Numerical simulation of collisional evolution, *Icarus* **35**, 1–26.

Greenberg, R., Nolan, M.C., Bottke, W.F., Kolvoord, R.A., Veverka, J., and the Galileo imaging team: 1992, Collisional and dynamical evolution of Gaspra (abstract), *Bull. Amer. Astr. Soc.*, in press.

Harris, A.W.: 1984, The origin and evolution of planetary rings, in *Planetary Rings*, eds. R. Greenberg and A. Brahic (Univ. of Arizona Press), pp. 641–659.

Hartmann, W.K.: 1969, Terrestrial, lunar and interplanetary rock fragmentation, *Icarus* **10**, 201–213.

Hartmann, W.K.: 1988, Impact strength and energy partitioning in impacts into finite solid targets. *Lunar Planet. Sci.* **XIX**, 451–452.

Housen, K.R., and Holsapple, K.A.: 1990, On the fragmentation of asteroids and planetary satellites, *Icarus* **84**, 226-253.

Housen, K.R., Schmidt, R.M., and Holsappple, K.A.: 1991, Laboratory simulations of large scale fragmentation events, *Icarus* **94**, 180–190.

Kresàk, Ľ.: 1977, Mass content and mass distribution in the asteroid system, *Bull. Astron. Inst. Czech.* **28**, 65–82.

Nakamura, A., and Fujiwara, A.: 1991, Velocity distribution of fragments formed in a simulated collisional disruption, *Icarus* **92**, 132–146.

Nolan, M.C., Asphaug, E., and Greenberg, R.: 1992, Numerical simulation of impacts on small asteroids (abstract), *Bull. Amer. Astr. Soc.*, in press.

Öpik, E.J.: 1951, Collision probabilities with the planets and the distribution of interplanetary matter, *Proc. Royal Irish Acad.* **54A**, 164–199.

Ostro, S.J., Chandler, J.F., Hine, A.A., Rosema, K.D., Shapiro, L.L., and Yeomans, D.K.: 1990, Radar images of asteroid 1989 PB, *Science* **248**, 1523–1528.

Paolicchi, P., Cellino, A., Farinella, P., and Zappalà, V.: 1989, A semiempirical model of catastrophic breakup processes, *Icarus* **77**, 187–212.

Paolicchi, P., Verlicchi, A., and Cellino, A.: 1992, A new semiempirical model for catastrophic impact processes (abstract), *Bull. Amer. Astr. Soc.*, in press.

Piotrowski, S.I.: 1953, The collisions of asteroids, *Acta Astron. Ser. A* **6**, 115–138.

Rosa, F., Cellino, A., Zappalà, V., and Farinella, P.: 1992, Crater ejecta from Vesta, in preparation.

Ryan, E.V., Asphaug, E., and Melosh, H.J.: 1990, Hydrocode predictions of collisional outcomes: Effects of target size, *Lunar Planet. Sci.* **XXI**, 1155–1156.

Ryan, E.V., and Melosh, H.J.: 1992, Understanding asteroid collisions: The effect of target size on collisional outcome (abstract), *Bull. Amer. Astr. Soc.*, in press.

Stöffler, D., Gault, D.E., Wedekind, J., and Polkowski, G.: 1975, Experimental hypervelocity impacts into quartz sand: Distribution and shock metamorphism of ejecta. *J. Geophys. Res.* **80**, 4062–4077.

Sykes, M.V., Greenberg, R., Dermott, S.F., Nicholson, P.D., and Burns, J.A.: 1989, Dust bands in the asteroid belt, in *Asteroids II*, eds. R.P. Binzel, T.Gehrels and M.S.Matthews (Univ. of Arizona Press), pp.336–367.

Weidenschilling, S.J., Paolicchi, P., Zappalà, V.: 1989, Do asteroids have satellites?, in *Asteroids II*, eds. R.P.Binzel, T.Gehrels and M.S.Matthews (Univ. of Arizona Press), pp. 643–658.

Zappalà, V., Farinella, P., Knežević, Z., Paolicchi, P.: 1984, Collisional origin of asteroid families: Mass and velocity distributions, *Icarus* **59**, 261-285.

Zappalà, V., Cellino, A., Farinella, P., Kneževič, Z.: 1990, Asteroid families. I. Identification by hierarchical clustering and reliability assessment, *Astron. J.* **100**, 2030–2046.

A COMPARISON BETWEEN 2D AND 3D WAVELET ANALYSIS METHOD FOR ASTEROID FAMILY DYNAMICAL STUDY

PHILIPPE BENDJOYA
Observatoire de la Côte d'Azur, BP 229 06003 Nice cedex 4, France

Abstract. As found in previous studies (Bendjoya *et al.* 1991) the wavelet analysis is a reliable tool for the determination of asteroid families. A comparison between the 2D and a newly developed 3D wavelet analysis is performed on 14 "fictitious" families numerically generated the members of which are *a priori* known. It clearly appears that the 3D method brings improvements to the 2D approach. Indeed the 3D analysis appears well suited for all cases even for those corresponding to the very dense Flora region and for "exotic" families such as filamentar ones. With the 2D method, satisfying results are obtained for the simulations which correspond to 70% of the real asteroid families previously found. The aim of this paper is to show quantitatively the improvement of the 3D method versus the 2D one.

Key words: Asteroid – families of asteroids – wavelet analysis

1. Introduction

The basic, and today generally accepted idea for explaining the existence of asteroid families is due to Hirayama who identified, from 1918 , large concentrations in the distribution of the asteroid orbital elements. Families were formed by collisional events. They originated from the catastrophic disruption of single parent bodies, which were destroyed by high-velocity impacts with other objects of asteroidal size. Resulting from this collision, the family members should be fragments which were ejected with comparatively low relative velocities, so that their orbital elements did not change too much with respect to each other and to the original parent body.

Because of the gravitational perturbations of the planets (mainly Jupiter), the osculating elements are not well suited to the asteroid family search, and the proper elements (the "quasi" first integrals of the motion) appear to be the right quantities for such investigations. Practically, the asteroid families are searched in the 3D space of proper elements: a', e', and i' respectively the proper semi-major axis, proper eccentricity and proper inclination. Several improvements have been obtained in the last decades about the theory of proper elements (for a review see Lemaitre 1992).

The results about family determination appeared for a long time method dependent. The first encouraging results appeared when two objective different methods gave results with a non negligible overlapping of the results. This good agreement obtained by Zappala *et al.* 1990, and Bendjoya *et al.* 1991 occurred not only about the number of families but also for their memberships. The agreement appeared not only on the big well known families such as Eos, Themis and Koronis, but also for the less well identified smallest ones . The wavelet clustering method was used by Bendjoya *et al.* (1991) and had been performed on each of the three projected planes of the proper element space computed by Milani and Knežević (4.2 version 1991). Then, the families had been obtained by making the intersections of the sets

Celestial Mechanics and Dynamical Astronomy **57**: 29–36, 1993.

of asteroids belonging to the detected structures in each of these three planes (for more details see Bendjoya *et al.* 1991).

In order to identify asteroid families directly in the $3D$ proper element space, we developed (Bendjoya *et al.* 1992) a new wavelet analysis method. This method was applied on artificial families. These families were numerically generated assuming a break-up event model derived from laboratory experiments. The reliability and the limitations of this new method, have been described in Bendjoya *et al.* 1992 and a comparison with the hierarchical clustering method used by Zappala *et al.* has also been made.

In the present paper, we present a comparison between the $2D$ wavelet method and the $3D$ one on the same set of artificial families. As our aim is to test the efficiency of these two methods, the analyses have been only performed in the osculating element space. We will focus our attention on the results and will discuss the improvements of the $3D$ method. Our intention is, on one hand, to show that the $2D$ method gives good results for a given range of simulations and hence the family proposed in Bendjoya *et al.* 1991, corresponding to these simulations are reliable. On the other hand we hope to convince the reader that the $3D$ wavelet method is more performing and will bring more information with a greater level of confidence on a forthcoming asteroid family classification obtained from an analysis performed on a new set of improved proper elements (Milani and Knežević ,1992).

The reader will find all the informations about the methods and the model of family generation in the above mentioned papers and can find more details on the wavelet transform in Bendjoya and Slezak (1992) for a review of the applications of this method to dynamical systems.

2. The Analysis

The wavelet cluster analysis ($2D$ or $3D$) consists in computing at each node of a network superimposed to the studied set of data, a wavelet coefficient. This coefficient is an indicator of local surdensities in the vicinity of a given node and for a given scale. It is possible to give an arbitrary level of significance to the wavelet coefficient versus chance. This is done by comparing the values of the wavelet coefficients obtained from the analysis of the set of data to the ones derived from a study performed on a random distribution of proper elements conserving the local density of the initial set of data. For all the studied scales the significant wavelet coefficients are then selected and are used for modelling the cluster of asteroids responsible for the high values of these coefficients.

The main discrepancy between the $2D$ and $3D$ methods occurs in the modelling of these significant clusters. Within the $2D$ method frame the clusters are modelled in each plane by ellipses, the parameters of which are adjusted from the local set of significant coefficients and the local background. The structures in the $3D$ space are then obtained by intersections, as recalled in the introduction. Fig. 1 shows

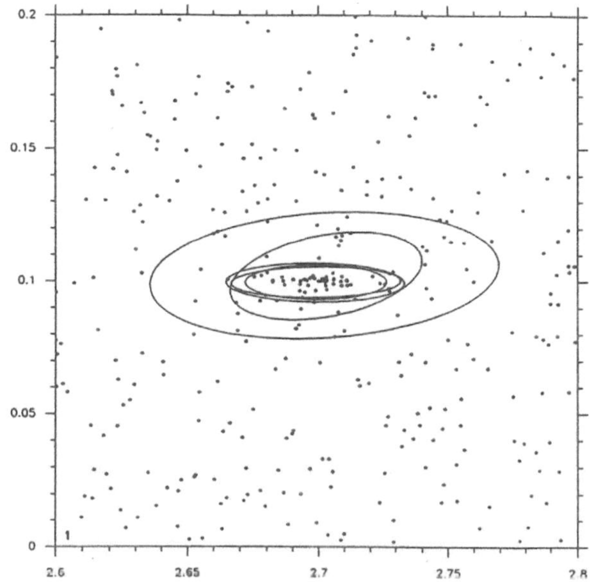

Fig. 1. The fit of ellipses around a family of 51 members in a background of 300 objects. The different ellipses correspond to different scales in the (a,e) plane.

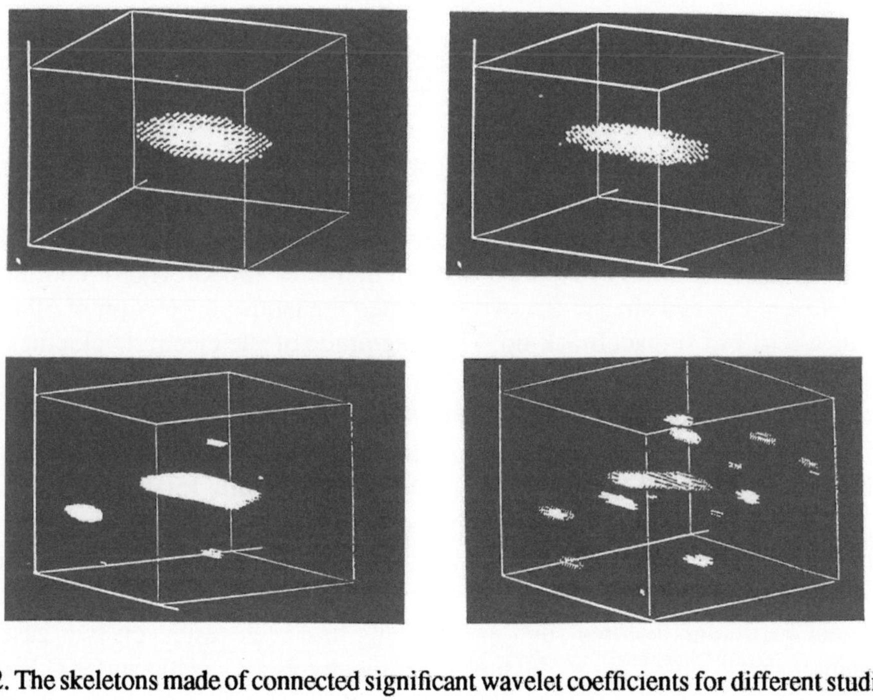

Fig. 2. The skeletons made of connected significant wavelet coefficients for different studied scales

TABLE I
Number densities for the different zones of the asteroid belt and for the three cases analyzed in the simulations.

	Density
Zone 2 ($2.065 \leq a < 2.300$ AU)	0.16
Zone 3 ($2.300 \leq a < 2.501$ AU)	0.08
Zone 4 ($2.501 \leq a < 2.825$ AU)	0.04
Zone 5 ($2.825 \leq a < 2.958$ AU)	0.03
Zone 6 ($2.958 \leq a < 3.030$ AU)	0.05
Zone 7 ($3.030 \leq a < 3.278$ AU)	0.04
Simulations Background 300	0.04
Simulations Background 600	0.08
Simulations Background 900	0.12

the fit of the ellipses in the (a,e) plane, for different scales and for a family of 50 members generated in a background of 300 objects.

For the $3D$ analysis the significant clusters of asteroids are directly modelled by the set of selected coefficients which form the skeleton of the cluster. The skeletons (that is to say the connected significant wavelet coefficients) are shown in fig. 2 for different scales and for the same example as mentioned above. According to the philosophy of the wavelet analysis the smaller is the studied scale, the thinner are the details extracted from the data set, and hence the greater is the number of skeletons.

The dynamical families have been generated in a volume defined in the osculating element space by: $2.6 \leq a \leq 2.8$ A.U; $0. \leq e \leq 0.2$ and $0. \leq \sin i \leq 0.2$ (where a, e, $\sin i$ are respectively the semi-major axis, the eccentricity and the sine of the inclination). For a parent body of diameter 100 km and of coordinates $a = 2.7$, $e = 0.1$ and $\sin i = 0.1$, we simulated the isotropic ejection of 50 members as a result of impact break-up. The magnitude of the ejection velocities was assumed to follow a truncated power law distribution (Farinella *et al.* 1992) in agreement with the data from laboratory experiments.

Four values of the minimum ejection velocity, V_{min}, which quantify the violence of the impact, were chosen: namely V_{min} = 50 m/s, 100 m/s 150 m/s and 200 m/s. Then we generated at random three different backgrounds (BG) having respectively 300, 600, and 900 asteroids. Each family has been plunged in each background, which means that we consider 12 different simulations. The ability of the methods to detect particular features such as filaments has been tested on two cases of filamentar shape with V_{min} = 150 m/s and respectively BG = 300 and 900. Table I (reproduced from Bendjoya *et al.* 1992) shows the different number densities.

In order to test the validity of the artificial families, the number density of the simulations has been compared to the number density of the suspected families

TABLE II
Summary of the simulations. N is the number of asteroids rediscovered in the family and I the number of interlopers.

BackGround	$V_{min}(m/s)$							
	50		100		150		200	
	N.	*I.*	**N.**	*I.*	**N.**	*I.*	**N.**	*I.*
	3D WAM							
300	**48**	*1*	**47**	*0*	**39**	*3*	**32**	*9*
600	**46**	*0*	**45**	*1*	**40**	*5*	**34**	*20*
900	**48**	*1*	**46**	*3*	**31**	*9*	**18**	*16*
filament 300	.				**48**	*2*	.	
filament 900	.				**48**	*8*	.	
	2D WAM							
300	**45**	*0*	**46**	*4*	**30**	*1*	**0**	*0*
600	**48**	*3*	**50**	*24*	**49**	*34*	**0**	*0*
900	**51**	*12*	**44**	*22*	**0**	*0*	**0**	*0*
filament 300	.				**39**	*4*	.	
filament 900	.				**0**	*0*	.	

in the 6 studied zones (see Zappala *et al.*, 1990) defined by the mean motion resonances with Jupiter in the main belt. The number density is defined by:

$$d = \frac{N_{tot}}{\text{Volume}}$$

where N_{tot} is the total number of asteroids in the considered zone, and the volume $(\Delta e \cdot \Delta \sin i \cdot \Delta a/a)$ is normalized to $(0.1 \cdot 0.1 \cdot 0.1/a)$.

It appears that, except for zone 2 which is the Flora region, most of the real families have to be compared to the simulated families of 50 members in a background of 300 asteroids.

3. The Results

The number of interlopers (i.e of asteroids found by the analysis as family members, but which were not generated by the simulated break-up) has been estimated for all considered cases. The greater is the number of genetical members found by the method and the lower is the number of interlopers, the more reliable are the results. Table II displays the results for both $2D$ and $3D$ analyses. A look at this Table permits to easily exhibit the robustness of the $3D$ method versus the $2D$ one.

First of all, one can see that the $2D$ method is unable to detect very spread families (V_{min} = 200 m/s). This is mainly due to the projection of the set of data on the three planes which smooths completely the contrast between the family and the background. This effect can be visualized on figs. 3a and 3b which show for

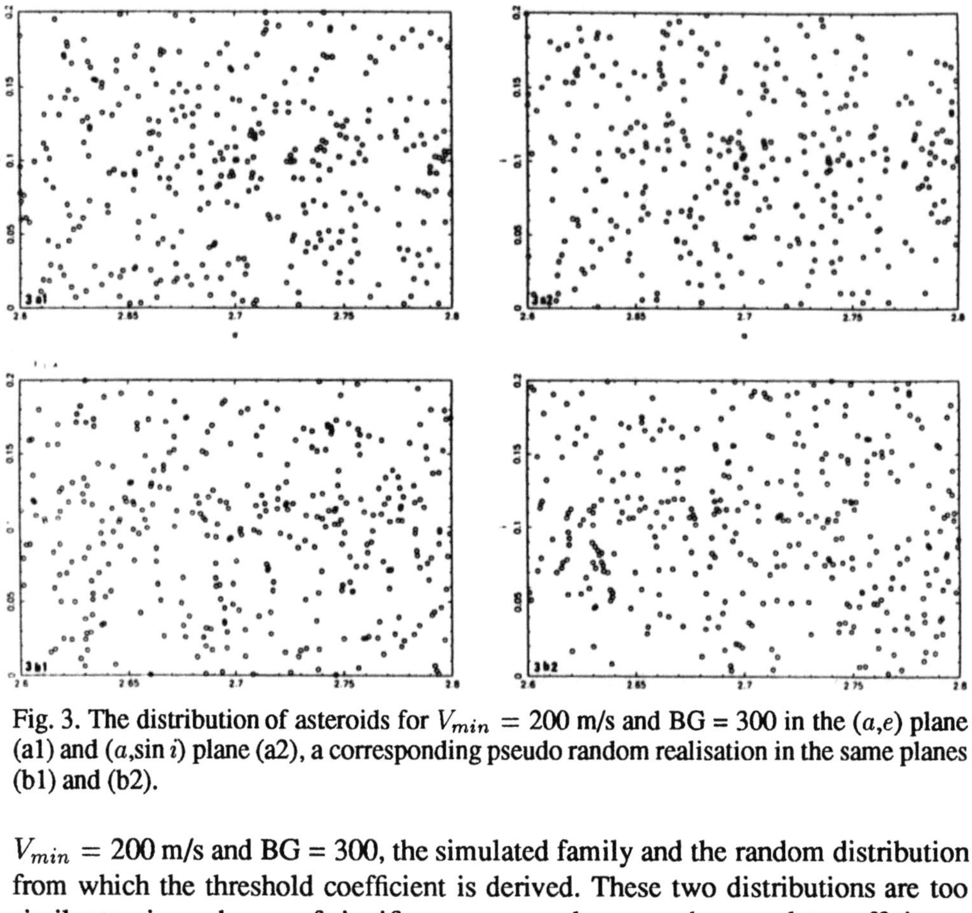

Fig. 3. The distribution of asteroids for $V_{min} = 200$ m/s and BG = 300 in the (a,e) plane (a1) and $(a,\sin i)$ plane (a2), a corresponding pseudo random realisation in the same planes (b1) and (b2).

$V_{min} = 200$ m/s and BG = 300, the simulated family and the random distribution from which the threshold coefficient is derived. These two distributions are too similar to give a degree of significance versus chance to the wavelet coefficients. The total inefficiency of the $2D$ method is pointed out for the simulations V_{min} = 200 m/s with BG = 300, 600 and 900 and for V_{min} =150 m/s with BG = 900. In the present study, the level of significance has been chosen equal to 5/1000. This means that the selected coefficient and hence, the associated structures of asteroids, have less than 5 over 1000 risks to be chance fluctuations. Even for a lower threshold of 1/100, the $2D$ method does not exhibit any significant structures for the above mentioned cases.

The $3D$ analysis, for its part, is able to extract a spread family even from a very dense background. Even if the number of interlopers is non negligible for all the cases with $V_{min} = 200$ m/s, the family is however detected. For the cases $V_{min} = 150$ m/s the number of interlopers stays reasonable.

The second failure of the $2D$ method appears in what concern the detection of the filament family with BG = 900. The explanation is now due to the quasi impossibility for the algorithm to fit ellipses above this particular structure. It clearly appears that such "exotic" features are very well isolated by the $3D$ method, with a few number of interlopers even in a dense background. This is mainly due to the modelling of the structures by means of skeletons which allows to follow any shape of surdensities.

Happily, for the $2D$ method and hence for the results published in (Bendjoya *et al.* 1991), such spread families do not seem to be realistic models for most of the asteroid families. They must rather be considered as academic cases the aim of which is to test the reliability of the methods, or may be the models of multi break-up events or of the mixing of different families. These two last hypotheses concerns a very little number of big asteroid families the edges of which are blurred (such as the Flora or Eunomia clusters). It appears that most of the asteroid families of zones 4, 5, 6 and 7 have to be compared to the simulations with BG= 300 and V_{min} = 50 m/s for the smallest ones as Maria, Dora, Veritas ., or with V_{min} = 100 m/s for the largest families like Eos, Koronis or Themis (for the nomenclature see Bendjoya *et al.* 1991). The $2D$ method gives in these cases good results, which permits to be confident in the results about these zones. The two methods are almost equivalent for these cases with a slight advantage for the $3D$ method concerning the number of interlopers.

Concerning the comparison between the simulations and the zones of the main belt, zone 3 could be modelled by a mixing of the cases BG = 600 with V_{min} = 50 m/s and V_{min} = 100 m/s. Indeed families of different expanse lie in this zone. The results obtained by the $2D$ analysis allows *a posteriori*, to be confident in the composition of the denser families of this zone namely: Vesta, Tinchen, Amalasuntha and Leonce. Indeed this method gives for the corresponding simulation 48 members and 3 interlopers. For the spreadest families Polona and 1969UN a non negligible ratio of interlopers can be expected (50 members with 24 interlopers, for the corresponding artificial family).

Once again the $3D$ investigation allows to believe in a better determination of the families in this zone with a lower ratio of interlopers, since for the corresponding fictitious families the results are respectively 46 members and no interloper and 45 members with 1 interloper.

Finally, the much debated zone, known as the Flora region, is zone 2. This zone corresponds to the simulations with BG = 900 (see Table I). The efficiency of the $2D$ wavelet method is limited to the sample V_{min} = 50 m/s for which 51 members are found with 12 interlopers, whereas for V_{min} = 100 m/s the number of interlopers represents 50 % of the found family members.

Even the dense families found in this zone must have many interlopers. Moreover the presence of these interlopers may bridge different families leading to the detection of one single big family (such as Berolina) rather than several small families. This zone lying between the secular resonance ν_6 and the mean motion resonance 7:2 with Jupiter is particularly dense. This large number of asteroids may imply a larger probability of collisions and consequently family creations but not necessary spread ones.

Since the number of interlopers is considerably reduced when the $3D$ analysis is applied on the corresponding simulated families, we can expect from it, in this zone, a sharper determination of families each of them having a common physical origin.

4. Conclusion

The direct $3D$ wavelet analysis appears to be more satisfactory for the search of asteroid families in the $3D$ proper element space than the $2D$ method.
As pointed out in section 3, the $3D$ approach allows to detect the majority of structures with a little number of interlopers. Even peculiar features like filaments can be extracted from a dense background. While such structures or spread ones cannot be detected by the $2D$ wavelet analysis either because of the impossibility to model the structure by an ellipse or because of the projection effects which smooth the contrast between the family and the background. Nevertheless, one can remark that, the families detected in (Bendjoya *et al.* 1991), by the $2D$ method have great chance to really exist since they have been found in spite of the decrease of their contrast. Finally the ability of the $3D$ wavelet analysis to follow the shape of the families can give us a way to discriminate and characterize different scenarii for the origin of families.

Acknowledgements

I would like to thanks Ch. Froeschlé, Cl. Froeschlé and R. Gonczi for their very helpful discussions comments and suggestions. I am also very grateful to J. Henrard for his constructive criticisms and enlightened advice.

References

Bendjoya Ph., Slezak E. and Froeschlé Cl.: 1991, "The wavelet transform: a new tool for asteroid family determination", *Astron. Astrophys.*, **251**, 312–330.

Bendjoya Ph., Cellino A., Froeschlé Cl. and Zappalà V.: 1992, "Asteroid Dynamical families: a reliability test for two identification methods", *Astron. Astrophys.*, accepted

Bendjoya Ph. and Slezak E.: 1992, "Wavelet analysis and applications to some dynamical system", *Celest. Mech.*, in press

Farinella P., Gonczi R.,Froeschlé Ch. and Froeschlé Cl.: 1992, "The injection of asteroid fragments into resonances", *Icarus*, in press

Hirayama K.: 1918, "Group of asteroids probably of common origin", *Proc. Phys.-Math. Soc. Japan*, **11**, 354–361

Lemaitre A.: 1992, "Proper elements: what are they?" *Celest. Mech.*, in press

Milani A. and Knežević, Z.: 1990, "Secular perturbation theory and computation of asteroid proper elements",*Celest. Mech.*, **49**, 247–411.

Milani A. and Knežević, Z.: 1992, "Asteroid proper elements and secular resonances", *Icarus*, in press

Zappala V., Cellino A., Farinella P. and Knežević, Z.: 1990, "Asteroid families: identification by hierarchical clustering and reliability assessment",*Astron J.*, **100**, 2030–2046

Zappalà V. and Cellino A.: 1992, "Asteroid families: Recent results and present scenario",*Celest. Mech.*, in press

ASTEROID "CLANS": SUPER-FAMILIES OR MULTIPLE EVENTS?

ALBERTO CELLINO and VINCENZO ZAPPALÀ
Osservatorio Astronomico di Torino

Abstract. Among the recently determined asteroid families (Zappala' et al., 1992) a number of very large clusterings are recognized. Some of them agglomerate few families previously identified as single groupings. In most cases, the new big families (called "clans") seem to be composed by different sub-clusterings connected each other by very narrow "bridges". In the present paper we analyze their possible origin and evolution: a primordial single event followed by subsequent collisions of the fragments, a super-catastrophic original impact or different collisional events overlapping by chance.

Key words: Asteroids – Asteroid Families

1. Introduction

During the last two years the general improvement of the techniques of asteroid proper elements computation and the use of new clustering analysis procedures has allowed to shed some light on the old-debated subject of asteroid dynamical families. As a matter of fact, a good agreement has been achieved between the results obtained by means of hierarchical clustering techniques of Multivariate Data Analysis (Zappalà et al., 1990) and by means of a Wavelet Cluster Analysis method (Bendjoya et al., 1991). Both techniques have been applied to the extensive data set of asteroid proper elements provided by Milani and Knežević (1990, version 4.2). The effectiveness of the family identification methods applied in the above papers has been recently tested by means of suitable numerical simulations by Bendjoya et al. (1992). The results of the tests show that both techniques give in general satisfactory results in cases bearing a general resemblance with the situations which are encountered in the study of the real asteroid belt. This allows to be confident on the overall reliability of the family lists presented in the above papers. The large discrepancy between the results obtained by both Zappalà et al. (1990) and Bendjoya et al. (1991) with respect to the results of the family searches carried out by Williams (1979, 1989), who found a much larger number of families, has been explained by Zappalà et al. (1991) as a consequence of the differences between the adopted identification methods (the Williams' method being more "liberal") more than as a consequence of the different data sets of asteroid proper elements used. Farinella et al. (1991) have pointed out that the family list of Zappalà et al. (1990), allows to recognize different kinds of asteroid "families": in particular, while the term "family" itself should be applied only to clusters of objects showing both a definite statistical significance and a consistency from the point of view of further physical evidence of a common genetic relationship (taxonomic consistency), the statistically significant clusterings found by Zappalà et al. (1990) can be generally separated in two main categories: well defined clusterings formed by a number of members very close in the space of proper elements, and with sharp boundaries

Celestial Mechanics and Dynamical Astronomy **57**: 37–47, 1993.

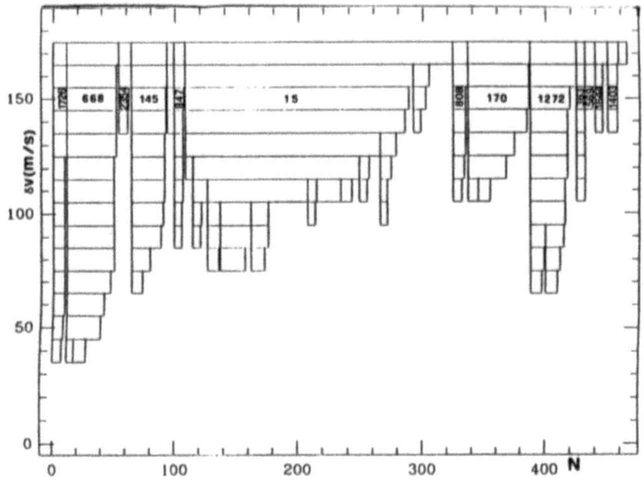

Fig. 1. A stalactite diagram showing the difference between typical "clusters" (sharp arms of stalactites, like 1726, 668, 145, 1272) and a "clan" (the large arm identified by no. 15)

(type II families or "clusters", following the Farinella et al. proposed nomenclature), and large groupings of objects with uncertain boundaries, such that the final membership of these groupings is strongly determined by the adopted choice of the critical distance level at which statistically significant clusterings are defined in different zones of the belt (type III families, or "clans"). This can be more easily understood by means of an inspection to the "stalactite diagrams" introduced in the Zappalà et al. (1990) paper (see Fig. 1). These diagrams show for different levels of the adopted distance in the proper elements space (for an explanation of the chosen notion of distance see Zappalà et al., 1990, and Milani et al., 1992) the different existing groupings of objects (identified by means of their lowest-numbered member); the statistically significant groupings are found on the basis of the derivation of a critical distance level (Quasi Random Level QRL, see again Zappalà et al., 1990, for further explanations). Thus, an inspection to Fig. 1 shows the most apparent differences between a typical "cluster", represented by a deep and sharp stalactite, and a typical "clan", which appears as a large stalactite arm which erodes and splits progressively for lower values of the distance. As pointed out by Zappalà and Cellino (1992), and Bendjoya et al. (1992), in the case of "clans" the choice of the QRL is critical, in the sense that a restrictive choice (i.e., a low value of the critical distance) can lead to identify only a minor fraction of the really existing family members, while a more "liberal" choice of the QRL leads to a larger fraction of family members, but with a largely increased amount of random interlopers, which can represent even a 50% of the derived clustering members, in some cases.

The existence of "clusters" and "clans" can be related to different causes: more or less dense background of non-family objects, different ejection velocities of the fragments, leading to a more or less pronounced spread of the family, etc. Among the possible explanations, one which cannot be ruled out is the superposition of different clusterings in a same portion of the proper elements space. This can be due either to the purely random chance to have two independent break-up events in a same cell of the proper elements space, involving completely unrelated parent bodies, but could be also due to multigeneration disruptions of bodies coming originally from the same parent. This fact is confirmed by a recent paper by Farinella and Davis (1992), which have computed the intrinsic collision probabilities for the entire set of asteroids for which an estimate of proper elements is currently available, and have shown that in some cases (like the asteroids belonging to the Koronis family) the probability of impact between members of the same family is noticeably higher with respect to the impact probability with the rest of the asteroid population.

Recently, a new enlarged data set of asteroid proper elements (version 5.7 by Milani and Knežević, 1992) has become available. This list comprehends also a significant amount of non-numbered objects for which a reliable estimate of orbital proper elements is anyhow available. The Hierarchical Clustering Method of Zappalà et al., (1990) has been applied to this enlarged data set (Zappalà et al., 1992), taking profit of the overall improvements of the adopted procedure of clustering identification which have been suggested on the basis of the experience achieved by means of the numerical tests performed by Bendjoya et al. (1992). While the results of this analysis will be published elsewhere, we want here to focus our attention on the differently identified "clans", analyzing in some detail their properties, in order to see whether some indication on their possible origin can be derived from the (limited) knowledge of the physical properties of their members (essentially albedo and taxonomic types), and from the observed features of these clusterings in the space of orbital proper elements. The "clans" identified are 5, and are listed, according with the name of their lowest-numbered member, as the Flora, Vesta, Eunomia, Nysa and Hygiea clans. Each of them will be separately analyzed in what follows.

2. Flora

Zappalà et al. (1990) did not identify a single clustering in the so-called Flora region in the inner belt; this was due to the relatively large errors of their adopted proper elements in this region close to the important ν_6 secular resonance, and to the restrictive procedure used for identifying families; on the other hand, Cellino et al. (1991) noticed that the size distribution of the asteroids in the Flora region can be described by a power law characterized by a steep slope, which is not typical of the "normal" asteroidal population, but is common in the case of the families identified in the Zappalà et al. paper. The results of the simulations performed by

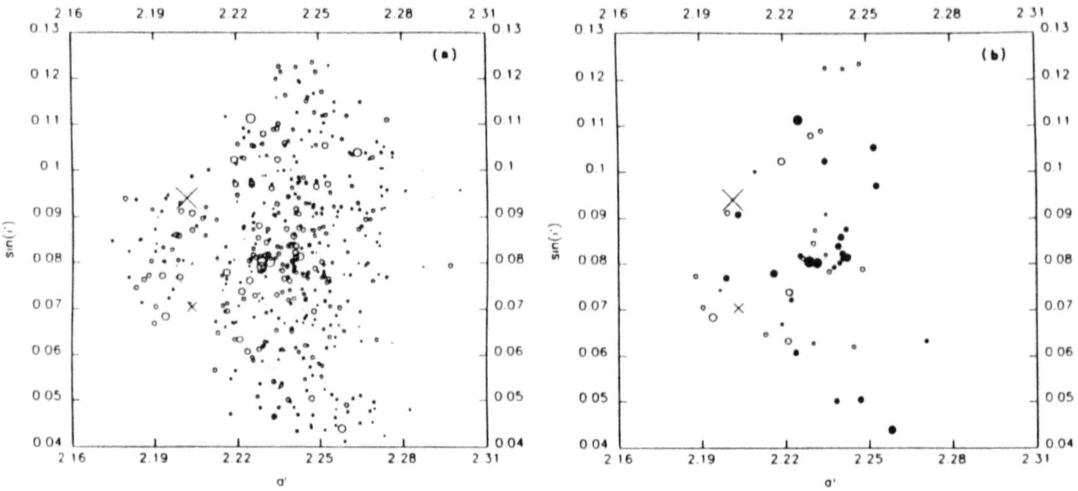

Fig. 2. (a) a'-sin i' plot for the Flora clan. The sizes of the symbols are proportional to the diameters of the corresponding objects; the larger objects 8 Flora and 43 Ariadne are indicated by crosses. (b) The same as (a), but only objects with a known albedo p_V are plotted (filled symbols: $p_V < 0.10$; open symbols: $p_V \geq 0.10$.

Bendjoya et al. (1992) suggested some improvements to the adopted procedure of families identification in cases in which very dense backgrounds like that found in the Flora region are concerned. As a result, Zappalà et al. (1992) find now a typical "clan" in the Flora region. Fig. 2(a) shows the plot a'-sin i' of this grouping: the members are plotted as circles whose size is proportional to the diameter of the object; 8 Flora and 43 Ariadne are represented by crosses having a reduced size with respect to the scale chosen for the rest of the clan, since both of them are much larger than the other members. We have to notice that the diameters were taken, when available, from the IRAS list of asteroid diameters; alternatively, they were computed on the basis of the known absolute V magnitude (H) and assuming the average albedo derived from the IRAS-observed objects of the clan (0.11). This can introduce in the case of the Flora clan some error due to the large dispersion of the IRAS albedos of this group (Fig. 3), but on the other hand the sizes of most of the Flora members are so small that the derived uncertainties on their derived diameters cannot be very high.

Fig. 2(a) shows an apparent sub-grouping located at $a' = 2.20$ approximately; 8 Flora and 43 Ariadne are located at the edge of this subgrouping. On the other hand, we know that 8 Flora is captured by the clan at high values of distance, so it could be actually an interloper. Since Fig. 3 shows that the Flora clan is characterized by a high non-homogeneity of albedo, it can be interesting to see whether objects with similar albedo tend to form some definite sub-cluster; Fig. 2(b) is the same as

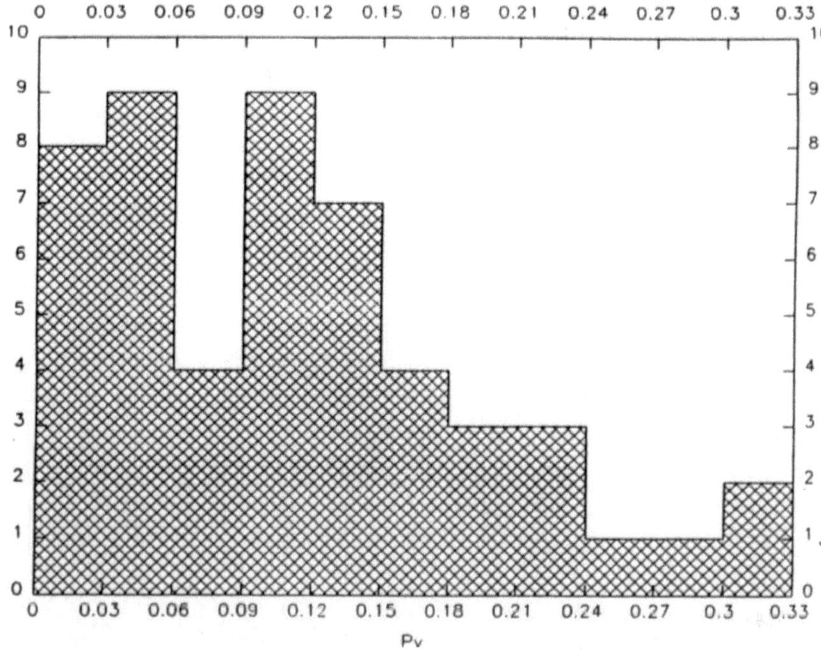

Fig. 3. Albedo histogram for the asteroids belonging to the Flora clan.

Fig. 2(a), but the fact that only objects for which a direct estimate of the albedo p_V is available are plotted, using different symbols for objects having $p_V < 0.1$ and $p_V \geq 0.1$. This Figure shows that no definite trend related to albedo is apparent. The subgrouping at $a' = 2.20$ is formed only by "bright" objects, but their number is too low for indicating a real separation from the rest of the clan.

As a conclusion, the Flora clan is one for which it is hard to infer any information about its origin; this is due to the lack of available data on albedos, and to the intrinsically complicated structure of this clan, which is probably strongly influenced also by the proximity to important mean motion and secular resonances.

3. Vesta

In their previous search, Zappalà et al. (1990) found two distinct families very close in the space of proper elements: one had as the lowest-numbered asteroid 4 Vesta, while the other was associated to 1933 Tinchen. The Vesta family is particularly important, since the unique taxonomic type of this object (V), due to its peculiar basaltic surface composition (see, e.g., Gaffey et al., 1989) makes it an ideal candidate for any check of the real collisional origin of a family based on an analysis of the taxonomic types of its members. For this reason, the fact that four asteroids belonging to the Vesta and the Tinchen families found by Zappalà et al. (1990) have been found to show a clear evidence of a basaltic composition (Binzel, private communication), must be considered of the highest importance.

Moreover, the Hierarchical Clustering method applied to the new extended data set of asteroid proper elements allows now to identify a single clan which incorporates the two Vesta and Tinchen groups previously found. These two groupings cannot be separated at the critical distance level QRL found in that zone of the belt, but they split into two different "arms of stalactite" at lower distance levels. The objects for which a basaltic composition has been inferred by the observations are equally shared out among the two sub-groupings. Fig. 4(a) shows the a'-e' plot of the clan; the position of the very big asteroid 4 Vesta (occultation observations by Millis et al., 1989, give a diameter of 561 ± 3 km) is indicated by a large asterisk. Two filled circles indicate two objects belonging to the C taxonomic type, which should be considered as probable interlopers; this conclusion is supported by the fact that both these objects are found at the edges of the clan. Four small asterisks indicate the position of the basaltic members of the clan. Fig. 4(b) is analogous to the previous one, but the asteroids belonging to the two subgroupings of Vesta and Tinchen are indicated by different symbols (the two probable C-type interlopers are not included in this plot). One can see that the two subgroupings are separated, with Vesta occupying a position just at the edge of both of them. The four basaltic objects are close to Vesta, two being found in each subgrouping. This fact suggests a common genetic relationship for the two subgroupings; on the other hand, it is not clear whether a single cratering event on Vesta can have originated simultaneously the whole clan, or different collisional events can explain what is observed. In this case, it is still not evident whether all these (at least two) events occurred on Vesta itself, or if some second generation disruption can have formed (at least partially) one of the two subgroupings. One observational constraint which should be taken into account is that polarimetric and photometric data indicate the presence of a large, hemispheric albedo spot on Vesta (see Cellino et al., 1987). This could be related to the presence of a single large crater, but the uncertainties are too big for drawing any definite conclusion.

The importance of the Vesta clan is special also taking into account that a very energetic cratering event on this large asteroid can probably explain the existence of a class of meteorites, the eucrites, which are known to bear a strict resemblance with the material which constitutes the surface of Vesta on the basis of the physical properties inferred from the optical observations; actually, some fragments of the surface could have reached the 3/1 mean motion resonance, and/or the ν_6 secular resonance, from where they could have been injected into Earth-crossing orbits (see, e.g., Wisdom, 1987, and Scholl et al., 1989). Recently, Cruikshank et al. (1991) have discovered some V-type objects having Earth-crossing orbits. These facts suggest a common origin for a variety of basaltic objects found in the inner solar system; this hypothesis will be tested quantitatively in a forthcoming paper (Rosa et al., 1993).

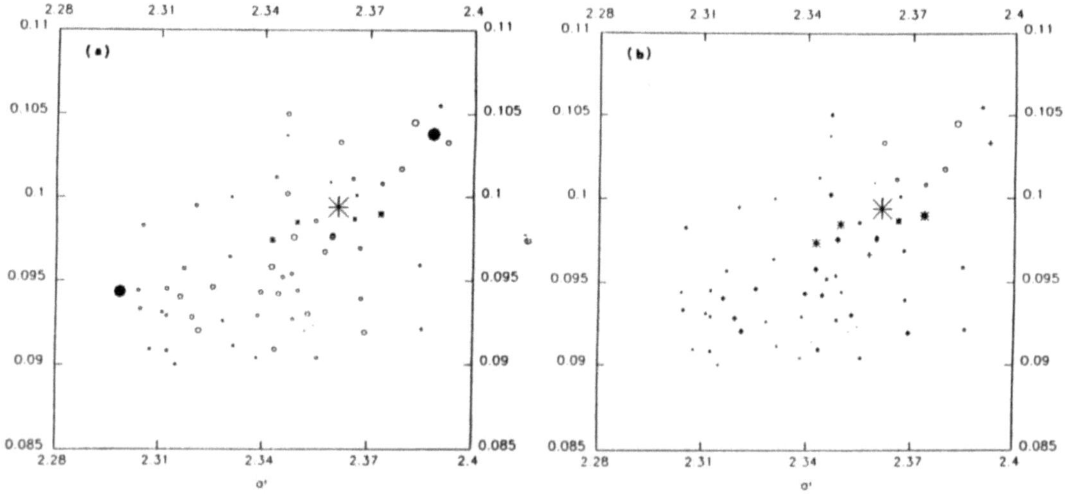

Fig. 4. (a) a'-e' plot for the Vesta clan. Symbol sizes proportional to the corresponding diameters, but for Vesta, indicated by a large asterisk. Filled symbols: two (probable interlopers) C objects. The four basaltic objects identified by observations have an asterisk superimposed to their symbol. (b) Same as (a), but the objects belonging to the Vesta and Tinchen sub-groupings are plotted with different symbols: open circles: Vesta sub-grouping; filled rhombs: Tinchen sub-grouping. Crosses indicate objects not belonging to the above subgroupings

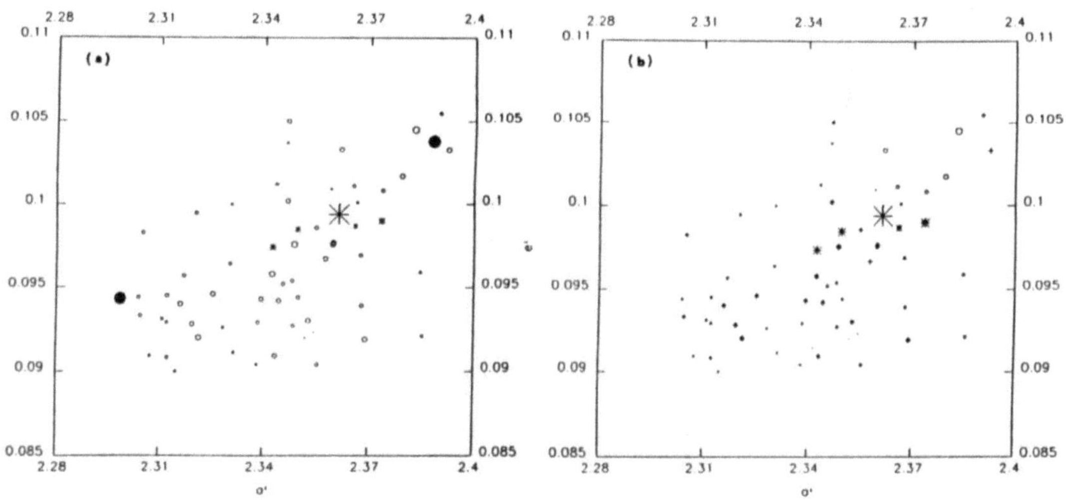

Fig. 5. (a) Same as Fig. 4(a), but for the Eunomia clan. Here the asterisks show the location of the objects 15 Eunomia and 85 Io. (b) Same as (a), but only objects with a known albedo are plotted. The meaning of the different symbols is the same as in Fig. 3(b).

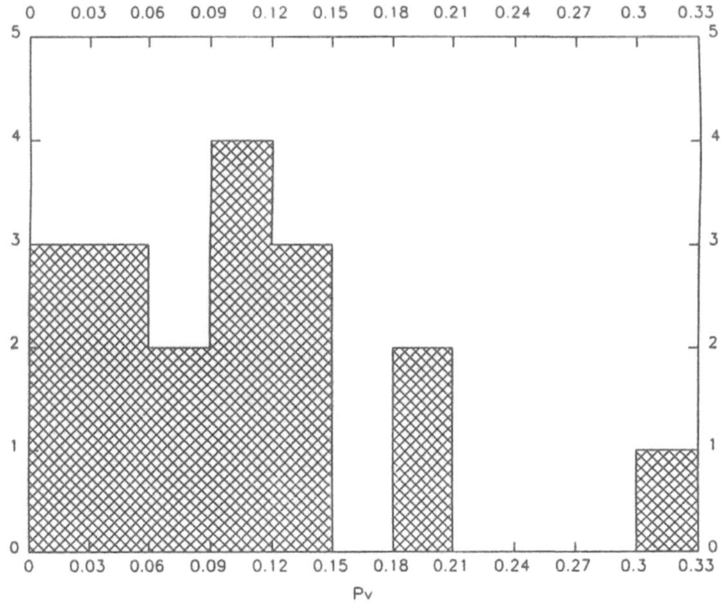

Fig. 6. Albedo histogram for the asteroids belonging to the Eunomia clan.

4. Eunomia

The Eunomia clan does not present any noticeable splitting in well defined sub-groupings. This fact is qualitatively confirmed by Fig. 5(a), which shows the a'-e' plot; also in this case, the size of each symbol representing the clan members is proportional to the size of each object, but for the case of the bigger asteroids 15 Eunomia and 85 Io, whose position is indicated by asterisks. Even in this case, the size of the non-IRAS observed members was derived on the basis of their absolute V-magnitude H and the average albedo of the clan, which turned out to be 0.11. Also in this case, however, the dispersion of the albedo for the objects for which this parameter is known is not negligible (Fig. 6), suggesting some surface heterogeneity among the members of the clan. Fig. 5(b) is analogous to Fig. 5(a), but the fact that only objects for which an albedo estimate is available are plotted, with different symbols for objects having p_V lesser or greater of 0.10. No definite conclusion can be drawn by this figure. The Eunomia clan could be either the product of a single break-up, or the result of two mutually overlapping families, possibly related to the two big objects 15 Eunomia and 85 Io, which are of different taxonomic types (S and C). On the other hand, a dozen of interlopers are expected for the Eunomia clan, based on the properties of the background in that zone of the belt (Zappalà et al., 1992), thus the observed albedo heterogeneity could be actually less substantial.

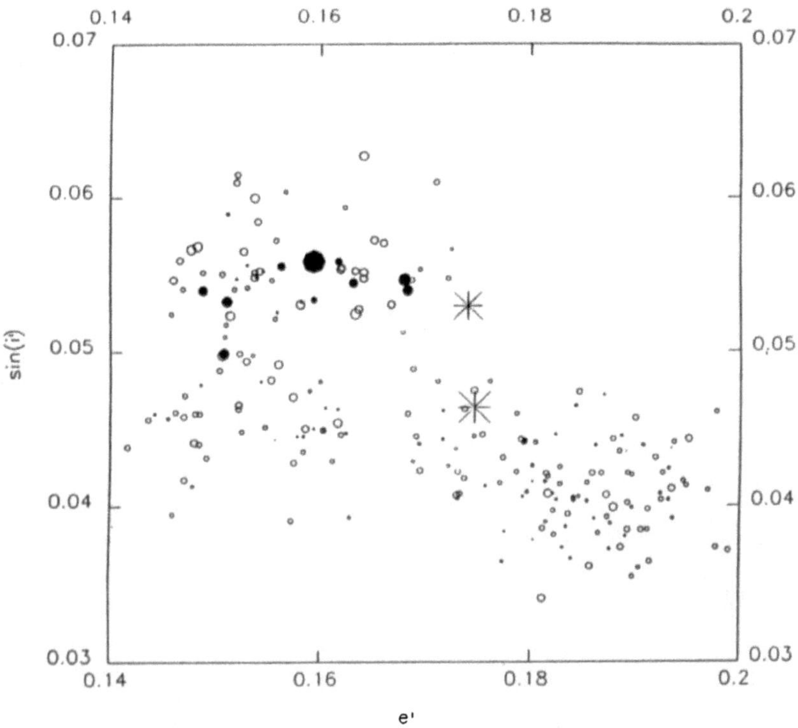

Fig. 7. e'-sin i' plot for the Nysa clan. Symbol sizes proportional to the objects' diameters, but for 44 Nysa and 135 Hertha, indicated by asterisks. The filled symbols refer to objects belonging to the F taxonomic type.

5. Nysa

Fig. 7 shows the e'-sin i' plot in the case of the Nysa clan. The two sizable objects 44 Nysa and 135 Hertha are indicated by asterisks. These two objects, classified, respectively, as E and M, may be chance interlopers of the clan, since they are clearly distinct with respect to the majority of the clan members, which are low-albedo objects, with a consistent fraction of asteroids classified as F (indicated by filled circles in the Figure). The fact that Nysa and Hertha can be interlopers is confirmed by an analysis of the observed distribution of asteroids in the immediate vicinities of the clan, which can be compared with the number of interlopers which is expected for this clan (Fig. 8). The presence of a significant amount of members belonging to the unusual taxonomic type F is obviously puzzling, and suggests a common origin for these objects, even if we have to notice that other F asteroids not belonging to any cluster exist in this zone of the belt. Fig. 7 suggests also a possible sub-splitting of the Nysa clan: in particular, a subgrouping having as the lowest-numbered object 878 Mildred, and formed mostly by small objects not including any F, is apparent in the bottom-right part of the plot. Hertha is located close to the edge of this sub-grouping. Since nothing is known about the albedo and taxonomic type of these objects, we cannot draw from this plot any definite

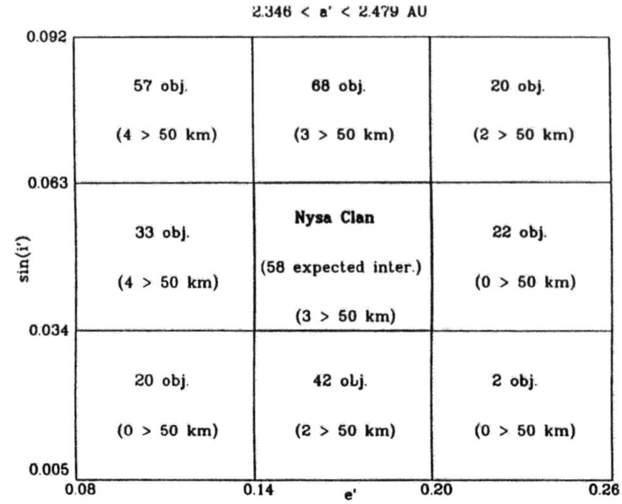

Fig. 8. Number of asteroids found in the immediate vicinity of the Nysa clan. The number of objects having diameter larger than 50 km is also indicated. These data suggest that 44 Nysa itself and 135 Hertha may be actually random interlopers of this clan.

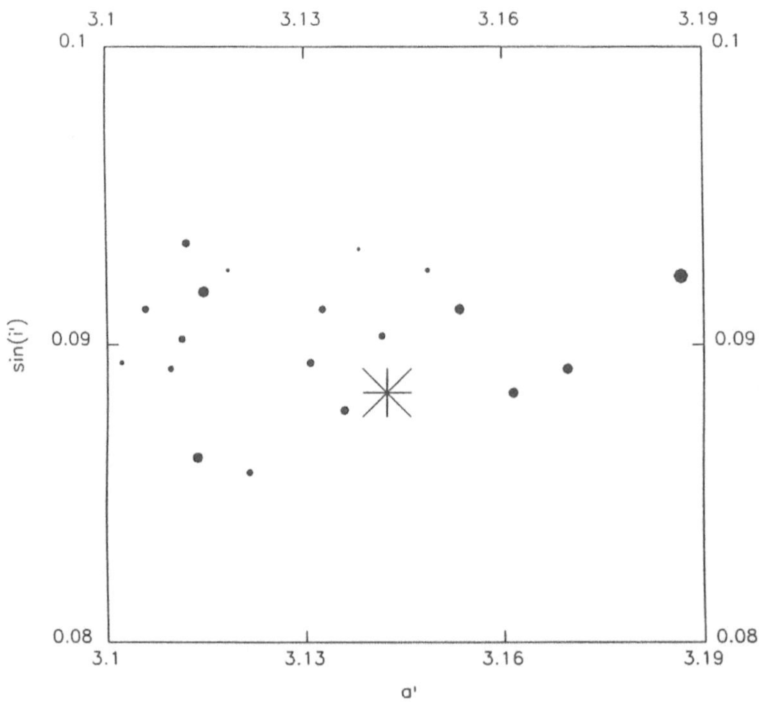

Fig. 9. a'-$\sin i'$ plot of the Hygiea clan. 10 Hygiea is indicated by an asterisk. The other members are indicated by symbols having sizes proportional to the corresponding diameters.

conclusion. Actually, we think that additional data on taxonomic types should be essential in order to have the possibility to discriminate between the possibilities of one or more collisional events as the origin of the presently observed clan.

6. Hygiea

Only a little information is available in the case of the Hygiea clan, which is located beyond 3 AU of heliocentric distance. In particular nothing is known about albedo and taxonomic type of the members of the clan, with the only exception of the large ($D = 429$ km) asteroid 10 Hygiea. The size of this asteroid could suggest an origin of the clan related to the occurrence of an energetic cratering event on Hygiea, in some way similar to the event which could be responsible of the existence of the Vesta clan. Fig. 12, which shows the a'-sin i' plot of this clan, does not contradict this hypothesis, due to the location of Hygiea along the lower edge of the clan in this projection. On the other hand, due to the lack of additional data, any conclusion seems premature.

References

Bendjoya Ph., Slezak E., Froeschlé C. (1991). *Astron. Astrophys.*, **251**, 312-330.

Bendjoya Ph., Cellino A., Froeschlé C., Zappalà V. (1992). *Astron. Astrophys*, in press.

Cellino A., Zappalà V., Di Martino M., Farinella P., Paolicchi P. (1987). *Icarus*, **70**, 546-565.

Cellino A., Zappalà V., Farinella P. (1991). *M.N.R.A.S.*, **253**, 561-574.

Cruikshank D.P., Tholen D.J., Hartmann W.K., Bell J.F., Brown R.H. (1991). *Icarus*, **89**, 1-13.

Farinella P., Davis D.R., Cellino A., Zappalà V. (1991). In *Asteroids Comets meteors 91*, in press, Flagstaff.

Farinella P., and Davis D.R. (1992). *Icarus*, in press.

Gaffey M.J., Bell J.F., Cruikshank D.P. (1989). In *Asteroids II*, (R.P. Binzel, T. Gehrels, M. S. Matthews, Eds.), pp. 98-127, University of Arizona Press, Tucson.

Milani A., and Knežević Z. (1990). *Celest. Mech.*, **49**, 247-411.

Milani A., Farinella P., Knežević Z. (1992). In *Comptes rendus de l'école de Goutelas*, (Cl. Froeschlé, and D. Benest, Eds.), Edition Frontière (in press).

Milani A., and Knežević Z. (1992). *Icarus*, in press.

Millis R.L., Franz O.G., Lebofsky L.A., Asphaug E., Hubbard W.B., Hunten D.M., A'Hearn M.F., Schnurr R., Klemola A.R., Osborn W., Vilas F., Potter A.E., Maley P.D., Manly P.L. (1989). *BAAS*, **21**(4), 1247.

Scholl H., Froeschlé Ch., Kinoshita H., Yoshikawa M., Williams J.G. (1989). In *Asteroids II*, (R.P. Binzel, T. Gehrels, M. S. Matthews, Eds.), pp. 845-861, University of Arizona Press, Tucson.

Rosa F., Zappalà V., Farinella P., Cellino A. (1993). In preparation.

Williams J.G. (1979). In *Asteroids*, (T. Gehrels, Ed.), pp. 1040-1063, University of Arizona Press, Tucson.

Williams J.G. (1989). In *Asteroids II*, (R.P. Binzel, T. Gehrels, M. S. Matthews, Eds.), pp. 1034-1072, University of Arizona Press, Tucson.

Wisdom J. (1987). *Icarus*, **72**, 241-275.

Zappalà V., Cellino A., Farinella P., Knežević Z. (1990). *Astron. J.*, **100**, 2030-2046.

Zappalà V., Cellino A., Farinella P. (1991). In *Asteroids Comets meteors 91*, in press, Flagstaff.

Zappalà V., and Cellino A. (1992). *Celest. Mech.*, **54**, 207-227.

Zappalà V., Cellino A., Farinella P., Milani A. (1992). Submitted to *Astron. J.*

CATASTROPHIC FRAGMENTATION AND FORMATION OF FAMILIES: PRELIMINARY RESULTS FROM A NEW NUMERICAL MODEL

PAOLO PAOLICCHI and ANDREA VERLICCHI
Dip. di Fisica, Università di Pisa
Piazza Torricelli 2, Pisa, Italy

and

ALBERTO CELLINO
Osservatorio Astronomico di Torino
Pino Torinese, Italy

Abstract. Preliminary results of an improved version of the semiempirical model for catastrophic break up processes developed by Paolicchi et al., (1989) are presented. Among the several changes with respect to the old version, the most important seem to be related to the new treatment of gravitational effects, including self-compression and reaccumulation of fragments. In particular, the new model is able to analyze processes involving both cm-sized objects, like those studied by means of laboratory experiments, as well as much larger bodies, for which self-gravitational effects are dominant; moreover, in this latter case the model seems in principle adequate to describe with the same physics very different phenomena, like the formation of plausible asteroid families and the creation of single, rapidly spinning, objects. This fact, if confirmed by refined analyses, may be of high importance for our general understanding of asteroid collisional evolution.

Key words: Asteroids–Asteroid Families–Catastrophic Fragmentation

1. Introduction

In a previous paper (Paolicchi et al., 1989, hereafter referred to as Paper I), a numerical model for the analysis of catastrophic fragmentation processes was developed. The model was a semiempirical one, assuming that all the main physical properties of the fragments (i.e. size, shape, rotation, velocity) are determined by an effective post-impact velocity field. This was assumed to be the sum of an uniform rotation plus a "breakup field", irradiating from a point ("irradiation point") located along the radius connecting the impact point and the centre of the target, and having a stronger intensity in directions closer to the impact point.

The present model, whose details will be given in a forthcoming paper, is much more refined, and aims to be more realistic from different points of view. The main change to the physics of the model with respect to that used in Paper I is to include gravitational effects which are important when studying the disruption of bodies larger than several tens of km in size. First, gravitational reaccumulation of fragments is taken into account by recapturing fragments whose velocity is less than escape velocity onto the largest fragment. Second, gravity produces a compressional strengthening throughout the interior of a large body such that the collisional energy density required for fragmentation increases with depth beneath the surface. Neither of the above effects was included in the model described in

Celestial Mechanics and Dynamical Astronomy **57**: 49–56, 1993.

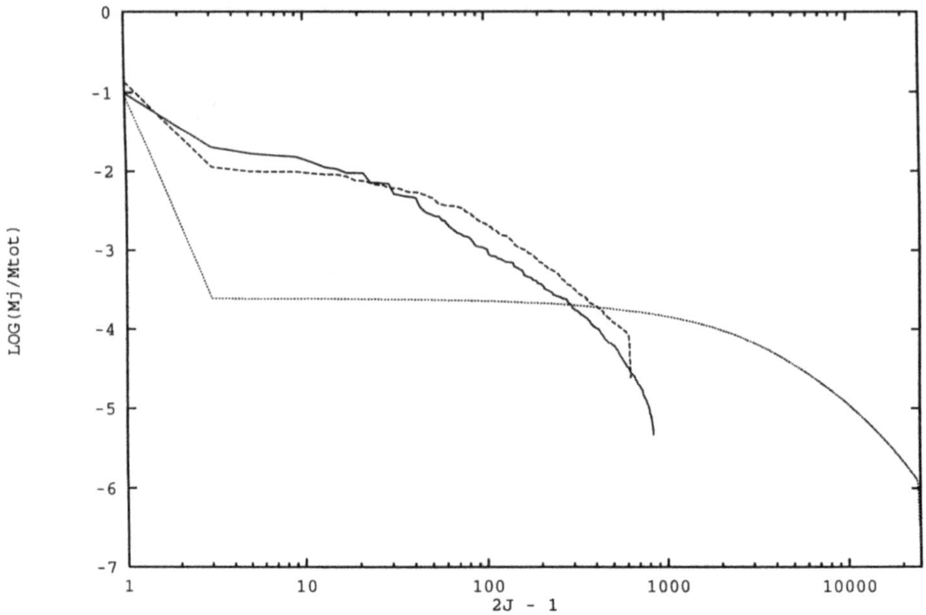

Fig. 1. Plots M(J) ("Kresak plots"; see for comparison Fujiwara et al., 1989; Chapman et al., 1989) for three synthetic "families" with a similar largest remnant but with different physical assumptions: no self-compressional effect (dotted line), self compression taking the mean value (dashed line); local self compression (solid line).

paper I, since that was mainly aimed to reproduce the outcomes of laboratory experiments, where the targets are a few cm in size. Moreover, the resulting fragments are now derived by means of an improved algorithm, which allows to avoid any mutual overlapping. This is another improvement with respect to Paper I, in which the main physical properties of the fragments were derived only in a purely statistical way.

2. Results

Among the many cases which we have analyzed, we decided to focus our attention on the problem of the catastrophic fragmentation of a 200 km sized target, and we analyzed this problem within three physical schemes, in order to analyze in more details the effect that self compression can have on the overall results:

(1). An "old" scenario, where there is not any self-compressional effect. We have to notice that even with the present version of the model, as well as in the case of Paper I, we have not a direct input value for the so-called *impact strength* S, an important parameter defined as the energy density in the body needed to produce a barely catastrophic outcome. In particular, the usual definition of S comes from

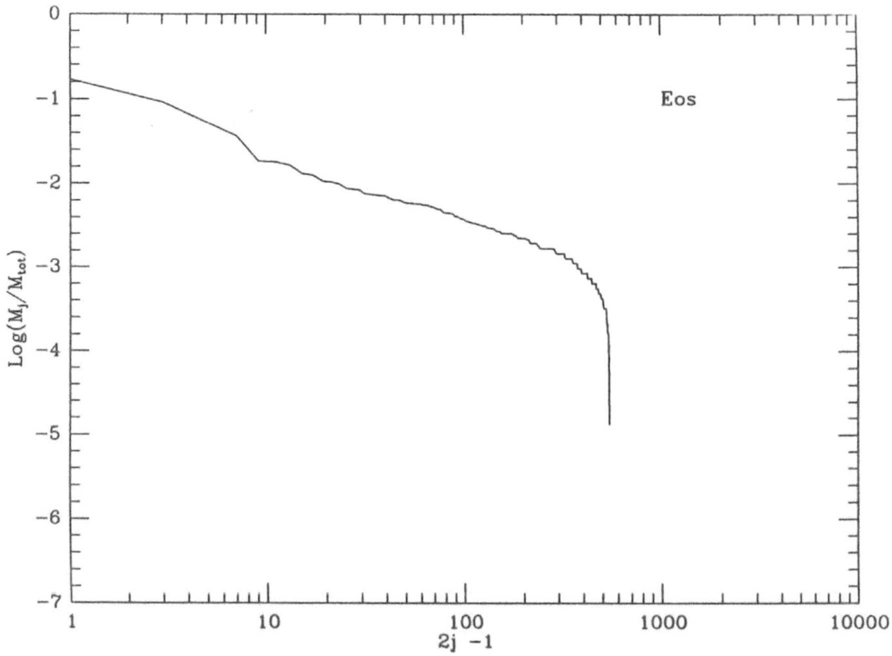

Fig. 2. Same as Figure 1, but for the case of the real Eos family, according with Zappalà et al., 1992. The sharp drop of the curve at smaller sizes is clearly mostly due to observational incompleteness of the data

the formula giving f_l, the mass of the largest fragment of a catastrophic impact, normalized to the mass of the original body (Fujiwara et al., 1977):

$$f_l = 0.5 \left(\frac{SM}{\rho E/2} \right)^{1.24}$$

where M and ρ are, respectively, the mass and density of the target body, E is the kinetic energy of the projectile, and S is the energy density value needed to have $f_l = 0.5$.

What we do in the framework of the model based on the notion of a velocity field, is to define a "breakup energy density" which has to be compared with the squared velocity differences within the target (this is explained in details in Paper I). According to the overall scenario arising from the laboratory experiments with "asteroid–like" targets (e.g., basalt), we took typical velocity values of the order of 10 m/s, and the resulting critical energy density for breakup was 10^6 erg/cm^3. This is an order of magnitude below the usual figures for S; that corresponds essentially to take into account that only a fraction of the original impact energy is converted into kinetic energy of the fragments; in the literature, this is usually accounted for by multiplying the impact energy times a coefficient f_{KE} which is, according with the laboratory experiments, of the order of 0.1.

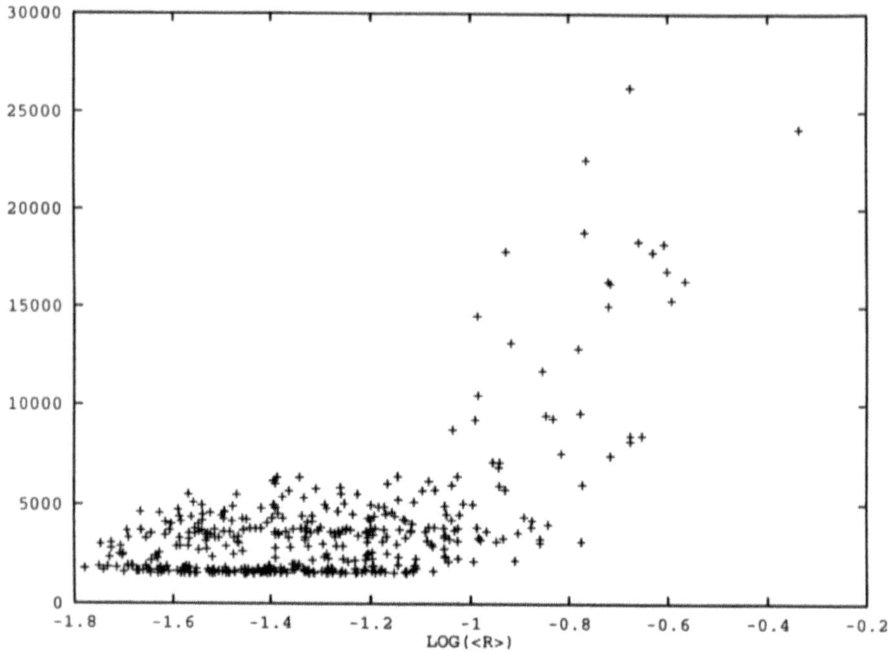

Fig. 3. The relation period- size for the fragments obtained in case (3). The size is given in units of the original target radius.

This physics, as can be shown also by Fig. 1, and in agreement with the conclusions by Davis et al. (1985), is not consistent with the formation of the observed families of asteroids (see, for a comparison, Fig. 2 which shows the case of the well known Eos family, according with the results of Zappalà et al., 1992). In particular, the results of the model show that, in addition to a largest remnant which can be fairly massive, due also to reaccumulation processes, only very small fragments can be formed.

(2). A second case includes in the model the "self compressional effect" defined by Davis et al. (1985), intended as a correction to the impact strength depending on the diameter D:

$$S = S_0 + A(\pi/15)G(\rho D)^2$$

where G is the gravitational constant, and S_0 here means the "uncompressed" value for S. While the (empirical) parameter A was assumed to be 1 in the Davis et al. (1985) paper, and values up to 3 are suggested by various laboratory experiments, a recent paper by Housen et al. (1991) suggests values possibly higher than 10; in our model there is the further problem of the –eventual– reduction of the self-compressional term when we consider our break-up energy density, since this is practically defined as the product $S \cdot f_{KE}$, and nothing is known about possible variations of f_{KE} in high–pressure cases.

One could guess, intuitively, that the self compression will be released as a spring; this should mean a larger f_{KE} in high–pressure regimes, up to 1, but this is only -at least for the moment- a purely speculative idea.

Between the most conservative assumption (A of the order unity, f_{KE} always the same, resulting in self-compressional factor even less than .1 especially in the case of strain-rate dependent models) and the most optimistic estimate (A of the order ten, f_{KE} of the order unity, what gives a self-compressional factor of the order ten), we decided to assume the value $A = 3$ also in the energy-density formula (where obviously the term has to be divided by the density). As shown in Fig.1, the resulting mass distribution is far more similar to that observed in asteroids families, in the sense that there is not an enormous gap between the mass of the largest remnant and the other fragments.

(3). Finally, we took into account the case in which the self-compressional term is considered as a "local" effect:

$$S = S_0 + 2/3\pi G\rho^2(R^2 - r^2)$$

where R is the radius and r is the distance from the centre. In this way the core is stronger and the peripheral parts are softer; this assumption seems to be the most correct from a physical point of view. Under this assumption, the model was able to create a fairly good "synthetic family". In particular, we got a largest remnant having a mass about ten times smaller of the original 200 km target (e.g. a 90 km body), and about 50 fragments larger than 20 km; the mass distribution, represented in Fig. 1 in a "Kresak plot" (so as to get straight lines for the power-law relationships, see Kresak, 1977) supports this conclusion; the comparison shows, qualitatively, a dramatic improvement with respect to case (1) and a –less spectacular– one with respect to case (2), in the sense that the initial drop of the mass of fragments is even less sharp. A more quantitative analysis of these results is clearly beyond the scope of the present work, and is postponed to a forthcoming paper.

In Table I we list the most important physical properties which can be derived by means of our model for the main "members of the family" obtained in case (3), i.e., the fragments having masses larger than 1/1000 of the original target; the period distribution for these objects is represented also in Fig. 3, which shows a generally faster rotation for smaller fragments, in a general agreement with the (few) available data for the major asteroid families. It has to be pointed out that the largest remnant is rotating slightly slower with respect to its post impact period, due to the effect of recapturing of fragments having ejection velocities smaller than the escape velocity ("splash effect", see Cellino et al., 1990). It has also to be remarked that the masses of the fragments were computed assuming an ellipsoidal shape; consequently, only half (about) of the mass is recovered in the computed fragments; this fact may suggest to increase the number of fragments or/and – more reliably– to increase their real mass. The resulting best fit exponent of the power-law describing the mass distribution varies from almost two (if we assume

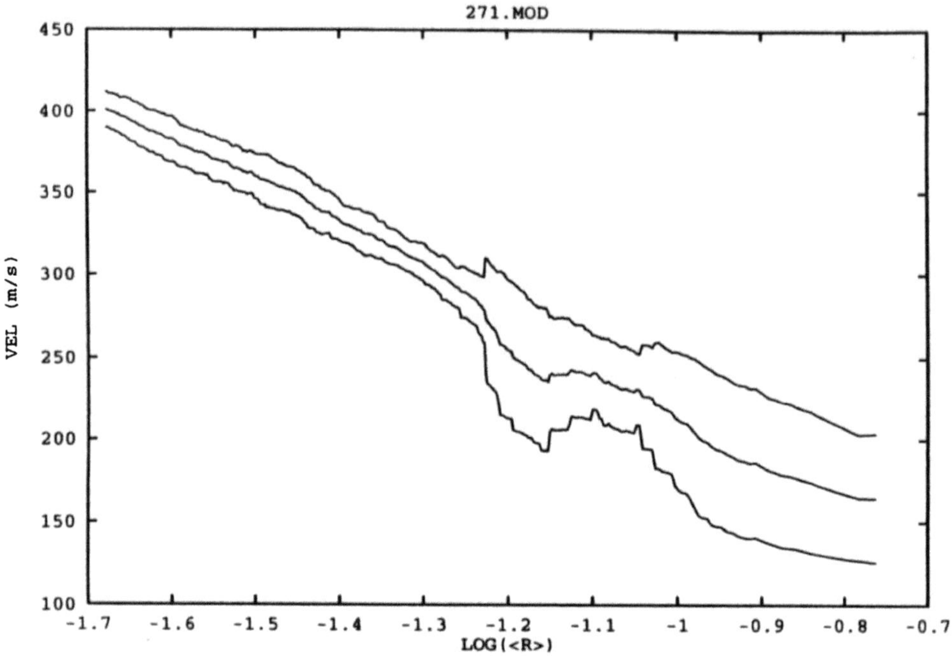

Fig. 4. "Running box" plot showing the relation ejection-velocity vs. size for all the 424 fragments obtained in case (3). The upper and lower curves represent the limits defined by a statistical $1 - \sigma$ deviation. Also in this case, the sizes are given in units of the original target radius.

that the lacking mass is going into debris) to more moderate values (of the order of 1.5).

Fig. 4, showing the relation ejection-velocity vs. size for all the fragments computed with the simulation, indicates that the smaller fragments are typically faster; a result which is of interest for the study of catastrophic fragmentation processes.

3. Conclusions

Even if at a preliminary stage, the above discussed results are of some interest for the physical studies concerning the formation of asteroids families. The "synthetic family" obtained in case (3) looks not too different from some of the observed asteroids families, for what concerns the mass distribution, the ejection velocities, and also the rotation (even with some overestimate of the spin rate for the small fragments). There is another point to be remarked: if the "self compressional" term is so effective, we can obtain families without the assumption of an unusual material fracture strength, which should be difficult to reconcile with experiments and theoretical considerations; moreover, the hypothesis of an increasing f_{KE} (the

TABLE I

The "synthetic family" described in the text: list of the fragments more massive than 1/1000 of the parent body; the columns show, respectively: (1-2) shape (b/a and c/a); (3) mass in units of the parent body; (4) velocity; (5-6) rotational velocity, both original (ω) and after relaxation (ω_{fin}) (see Paper I for an explanation); (7-8) logarithm of the corresponding ratios between the rotational and translational energy; (9) mean radius (in units of the original target radius).

b/a	c/a	m %	V (m/s)	ω	ω_{fin}	E_R/E_T	E'_R/E_T	$< r >$
.989	.937	9.781	61.44	.261E-03	.260E-03	-1.7996	-1.8006	.4608
.849	.621	2.018	160.84	.597E-03	.385E-03	-2.4599	-2.6496	.2722
.706	.529	1.669	155.18	.491E-03	.411E-03	-2.5026	-2.5783	.2555
.744	.521	1.569	158.71	.683E-03	.373E-03	-2.4339	-2.6939	.2504
.666	.489	1.510	155.06	.352E-03	.345E-03	-2.7255	-2.7343	.2472
.692	.471	1.287	157.81	.572E-03	.353E-03	-2.5717	-2.7649	.2343
.727	.455	1.111	160.70	.129E-02	.746E-03	-1.9438	-2.1633	.2232
.593	.398	1.064	152.44	.768E-03	.343E-03	-2.4365	-2.7622	.2199
.614	.402	.9490	153.25	.339E-03	.240E-03	-2.9857	-3.1145	.2117
.755	.515	.9490	177.75	.805E-03	.771E-03	-2.2899	-2.3036	.2117
.741	.476	.9457	175.98	.109E-02	.746E-03	-2.1602	-2.3022	.2115
.667	.399	.7163	151.90	.107E-02	.847E-03	-2.0209	-2.0933	.1928
.585	.367	.7162	147.06	.497E-03	.388E-03	-2.6334	-2.7126	.1928
.626	.383	.6988	148.63	.623E-03	.419E-03	-2.5456	-2.6799	.1912
.563	.355	.6967	147.63	.787E-03	.386E-03	-2.4609	-2.7171	.1910
.641	.415	.5096	185.16	.384E 03	.280E-03	-3.2179	-3.3371	.1721
.601	.399	.4993	181.59	.342E-03	.335E-03	-3.1461	-3.1553	.1709
.620	.364	.4802	140.90	.130E-02	.105E-02	-1.8647	-1.9297	.1687
.653	.375	.4740	142.72	.203E-02	.201E-02	-1.3851	-1.3875	.1680
.657	.399	.4678	186.34	.919E-03	.658E-03	-2.5067	-2.6133	.1672
.540	.328	.4553	134.18	.157E-02	.489E-03	-2.0196	-2.5242	.1657
.676	.396	.3571	199.63	.191E-02	.835E-03	-2.1886	-2.5431	.1529
.589	.351	.3189	178.67	.817E-03	.676E-03	-2.5414	-2.6220	.1472
.568	.333	.2982	119.87	.184E-02	.158E-02	-1.4757	-1.5399	.1439
.650	.393	.2881	205.89	.922E-03	.665E-03	-2.7239	-2.8269	.1423
.547	.330	.2750	179.27	.806E-03	.537E-03	-2.7302	-2.8443	.1401
.584	.338	.2660	113.28	.226E-02	.193E-02	-1.2884	-1.3556	.1386
.550	.324	.2659	111.35	.206E-02	.204E-02	-1.2722	-1.2739	.1385
.542	.321	.2389	134.81	.238E-02	.150E-02	-1.5873	-1.7369	.1337
.633	.362	.2354	195.12	.224E-02	.163E-02	-1.9289	-2.0334	.1330
.555	.329	.2115	182.64	.145E-02	.144E-02	-2.0773	-2.0825	.1284
.616	.349	.1952	148.71	.292E-02	.176E-02	-1.5969	-1.7725	.1250
.599	.342	.1864	99.74	.300E-02	.292E-02	-0.9841	-0.9949	.1231
.545	.336	.1758	206.14	.148E-02	.478E-03	-2.7121	-3.1999	.1207
.605	.371	.1648	215.98	.448E-03	.353E-03	-3.4677	-3.5602	.1181
.604	.354	.1629	216.83	.118E-02	.109E-02	-2.5396	-2.5701	.1177
.570	.332	.1582	185.69	.226E-02	.191E-02	-1.8654	-1.9370	.1165
.655	.384	.1506	228.88	.119E-02	.885E-03	-2.7588	-2.8512	.1146
.547	.324	.1499	77.05	.137E-02	.106E-02	-1.6239	-1.6898	.1144
.578	.343	.1482	215.46	.111E-02	.917E-03	-2.6562	-2.7336	.1140
.643	.366	.1437	229.40	.212E-02	.201E-02	-2.1233	-2.1385	.1129
.656	.382	.1362	232.89	.150E-02	.884E-03	-2.7074	-2.8951	.1109
.617	.352	.1253	77.52	.228E-02	.172E-02	-1.2764	-1.3574	.1078
.605	.347	.1148	226.52	.200E-02	.196E-02	-2.1905	-2.1987	.1047
.621	.350	.1114	219.70	.249E-02	.191E-02	-2.1257	-2.2075	.1037
.552	.332	.1107	221.95	.619E-03	.600E-03	-3.1902	-3.1992	.1035
.558	.341	.1098	221.22	.438E-03	.433E-03	-3.4868	-3.4908	.1032
.538	.324	.1092	219.59	.169E-02	.164E-02	-2.3031	-2.3123	.1030
.618	.369	.1068	231.45	.897E-03	.681E-03	-3.0856	-3.1734	.1022
.599	.342	.1038	201.56	.304E-02	.296E-02	-1.7523	-1.7627	.1013
.549	.323	.1033	209.05	.255E-02	.126E-02	-2.2614	-2.5156	.1011

fraction of impact energy converted into kinetic energy of the fragments) in the high–pressure regime, quoted in the preceding Section, appears as an interesting one, which should be more deeply analyzed in the forthcoming papers. Since the pressure depends on the position within the target, we could obtain a jump in the required energy (and in the characteristics of the event) between impacts shattering a good part of the target, saving the stronger core, and others completely shattering the body. The former impact should be characterized by the reaccumulation of fragments, and a moderate f_{KE} (consistent with the values found in the laboratory experiments); the outcome in this case might be often a fast rotator (such as the triaxial ellipsoids discussed in Farinella et al., 1981; see also Farinella et al., 1982). The other kind of impact should lead to the formation of a family. It is well known that the coexistence, with the same physics and within the same size range, of fast rotators and families is one of the main difficulties for drawing a consistent scenario of the asteroid collisional evolution (see the Fig. 1 in Davis et al., 1989, and the related discussion).

At this preliminary stage we are not claiming to have solved the puzzle, but we think that this may be a fruitful direction for more accurate investigations.

Acknowledgements

We thank D. R. Davis, P. Farinella and V. Zappalà for useful comments and suggestions.

References

Cellino A., Zappalà V., Davis D.R., Farinella P., Paolicchi P. (1990). *Icarus*, **87**, 391.
Chapman C.R., Paolicchi P., Zappalà V., Binzel R.P., Bell J.F., (1989). In *Asteroids II*, (R.P. Binzel, T. Gehrels, M. S. Matthews, Eds.), pp. 386-415, University of Arizona Press, Tucson.
Davis D.R., Chapman C.R., Weidenschilling S.J., Greenberg R. (1985). *Icarus*, **62**, 30.
Davis D.R., Weidenschilling S.J., Farinella P., Paolicchi P., Binzel R.P. (1989). In *Asteroids II*, (R.P. Binzel, T. Gehrels, M. S. Matthews, Eds.), pp. 805-826, University of Arizona Press, Tucson.
Farinella P:, Paolicchi P.,Tedesco E., Zappalà V. (1981). *Icarus*, **46**, 113.
Farinella P:, Paolicchi P., Zappalà V. (1982). *Icarus*, **52**, 409.
Fujiwara A., Kamimoto G., Tsukamoto A. (1977). *Icarus*, **31**, 277.
Fujiwara A., Cerroni P., Davis D.R., Ryan E., Di Martino M., Holsapple K., Housen K. (1989). In *Asteroids II*, (R.P. Binzel, T. Gehrels, M. S. Matthews, Eds.), pp. 240-265, University of Arizona Press, Tucson.
Housen K.R., Schmidt R.M., Holsapple K.A. (1991). *Icarus*, **94**, 189.
Kresak L. (1977). *Bull. Astron. Inst. Czech.*, **28**, 65.
Paolicchi P., Cellino A:, Farinella P., Zappalà V. (1989). *Icarus*, **77**, 187.
Zappalà V., Cellino A., Farinella P., Milani A. (1992). *A. J.*, submitted.

STRUCTURE OF THE ASTEROID BELT FOR SMALL MEMBERS

TSUKO NAKAMURA
National Astronomical Observatory,
Osawa, Mitaka, Tokyo 181 Japan

Widely accepted recognition of the belt asteroids is that they are part of a proto-planet that could not grow up to the planetary size probably due to the gravitational perturbations by Jupiter, and asteroids are fragments produced as a result of mutual collisions (Wetherill, 1989). This view has been more and more supported through recent laboratory experiments of collision (Fujiwara, 1989), the related theoretical studies (Wetherill, 1989), and physical observations of asteroids (Binzel *et al.*, 1989). If this collisional view is correct, it is natural to expect that, in general, smaller asteroids or photometrically fainter asteroids, are products from more frequent and more recent collisions. Then, a question comes into mind about what is the spatial distribution for such small members in the belt, especially compared with the distribution for large and medium-sized asteroids. With such a viewpoint in mind, we investigated in this report the distribution of semi-major axis (a) of asteroids in the asteroid belt as a function of their diameter. Although the most abundant and uniform database of asteroid diameter is the one observed by the IRAS satellite (Tedesco, 1989), we found that, for our purpose, the entry number of the IRAS data of about 1800 is insufficient particularly regarding small asteroids. Therefore, we estimated the diameters for about 4500 numbered asteroids from their H-magnitudes, using empirical relations between H-mag and diameter (D) obtained for the IRAS asteroids. Those relations are: log D(km) = 3.5 - H/5 for S-type, and log D(km) = 3.8 - H/5 for C-type. As most of the asteroids whose taxonomic type, or equivalently, (B-V) and (U-V) colors are measured fall, when plotted on the log D vs. H diagram, between the above two formulae, a maximum error of our diameter estimate will be about a factor of two. As for the asteroids of unknown taxonomic type, the S-type formula is applied to the inner belt asteroids whereas the C-type one to the outer belt asteroids. >From the log D vs. a plot for about 4500 numbered asteroids, it is found that the inner and outer edges of the belt are more swelled toward the smaller diameter, and at the 3/1, 5/2 and 7/3 resonant gaps their widths are narrowed toward the smaller diameter. Since these swellings (called "bulges" here) are seen fairly smoothly ranging from 100 km through a few km, the bulges are not due to the error in diameter estimate. And since the bulges are seen similarly both for the osculating semi-major axis and for the proper semi-major axis whose recent version was kindly provided by Dr. A. Milani, the bulges are not due to the orbital variation of asteroids. Moreover, a simplified analysis based on a size distribution of asteroids in the belt and the lifetime of asteroids in the resonant gaps showed that the probability that the bulges took place by chance as a result of accidental fluctuation in small sample statistics is negligibly small. Therefore, the existence of the bulges in the distribution of log D vs. a of asteroids

Celestial Mechanics and Dynamical Astronomy **57**: 57–58, 1993.

seems real, though more refined studies are obviously necessary to establish the reality of the bulges. The resonant gaps and probably the belt edges are generally supposed to be a result of the secular perturbations by Jupiter together with the help of some inner planets. However, there is no reason at all that the action of the gravitational perturbations is dependent on the size of asteroids. Hence, it is inferred that the bulges were formed by the mutual collisions between asteroids. According to high-velocity impact experiments of rocks in the laboratory, it is generally observed that smaller fragments after collisions show a tendency to gain larger velocity relative to the target rock (Nakamura and Fujiwara, 1991). If this fact is scaled up to the size of actual asteroids, increased swelling of the bulges toward the smaller diameter can easily be understood at least qualitatively. Perhaps because the time scale of the fragment supply due to collision is shorter for smaller asteroids than the gravitational depletion time scale by Jupiter, small asteroids can be thought to "ooze" out of the Jovian gravitational barrier at the belt edges and at the gaps. From the shapes of the bulges, we tried to make a crude estimate of the relative velocity of small asteroids after collision. Using an energy conservation relation before and after a collision, the bulge width expressed by the unit of the semi-major axis as a function of diameter was converted to the total velocity change. It is found that a relative velocity of fragments with D = 20-5 km in every bulge is roughly 100 - 200 m/s under the assumption of ejection from the edges of the belt or the gaps. These relative velocities well compare, within a factor of two, with the ones calculated for members of the typical Hirayama families of asteroids (Fujiwara, 19892). If the bulge structure found here is extrapolated to even smaller diameter, it is possible that the Kirkwood gap widths are much more narrowed and perhaps filled with asteroids for the diameter of less than, say, 5 km. We imagine that the bulges are maintained by the balance between collisional supply of fragments and the gravitational depletion by Jupiter.

References

Wetherill, G.W.: 1989, in *Asteroids II*, R.P. Binzel, T. Gehrels and M.S. Matthews (eds), University of Arizona Press, 661–680.

Fujiwara, A. et al.:1989, in *Asteroids II*, R.P.Binzel, T.Gehrels and M.S. Matthews (eds), University of Arizona Press, 240–265.

Binzel, R.P., Farinella, P., Zappala, V. and Cellino, A.: 1989, in *Asteroids II*, R.P. Binzel, T. Gehrels and M.S. Matthews (eds), University of Arizona Press, 416–441.

Tedesco, E.F.: 1989, in *Asteroids II*, R.P. Binzel, T. Gehrels and M.S. Matthews (eds), University of Arizona Press,1090–1138.

Nakamura, A. and Fujiwara, A.: 1991, *Icarus*, **92**, 132.

Fujiwara, A.: 1982, *Icarus*, **52**, 440.

THE TROJAN ASTEROID BELT:
PROPER ELEMENTS, STABILITY, CHAOS AND FAMILIES

ANDREA MILANI
Department of Mathematics, Pisa University,
Via Buonarroti 2, I-56100 Pisa, Italy
and
Groupe E.U.R.O.P.A., Observatoire de Paris/Meudon
E-mail milani@dm.unipi.it

Abstract. I have computed proper elements for 174 asteroids in the 1 : 1 resonance with Jupiter, that is for all the reliable orbits available (numbered and multi–opposition). The procedure requires numerical integration, under the perturbations by the four major planets, for 1,000,000 years; the output is digitally filtered and compressed into a "synthetic theory" (as defined within the *LONGSTOP* project). The proper modes of oscillation of the variables related to eccentricity, perihelion, inclination and node define proper elements. A third proper element is defined as the amplitude of the oscillation of the semimajor axis associated with the libration period; because of the strong nonlinearity of the problem, this component cannot be determined by a simple Fourier transform to the frequency domain. I therefore give another definition, which results in very good stability with time. For 87% of the computed orbits, the stability of the proper elements –at least over 1 $M\,yr$– is within the following bounds: 0.001 AU in semimajor axis, 0.0025 in eccentricity and sine of inclination. Half of the cases with degraded stability of the proper elements are found to be chaotic, with e–folding times between $16,000$ and $660,000\,yr$; in some other cases, chaotic behaviour does not result in a significantly decreased stability of the proper elements (stable chaos). The accuracy and stability of these proper elements is good enough to allow a search for asteroid families; however, the dynamical structure of the Trojan belt is very different from the one of the main belt, and collisional events among Trojans can result in a distribution of fragments difficult to identify. The occurrence of couples of Trojans with very close proper elements is proven not to be statistically significant in almost all cases. As the only exception, the couple 1583 Antilochus – 3801 Thrasimedes is significant; however, it is not easy to account for it by a conventional collisional theory. The Menelaus group is confirmed as a strong candidate collisional family; Teucer and Sarpedon could be considered as significant clusters. A number of other clumps are detected (by the same automated clustering method used for the main belt by Zappalà et al., 1990,1992), but the total number of Trojans with reliable orbits is not large enough to detect many significant candidate families.

Key words: asteroids–libration–proper elements–stability–chaos–families

1. The Other Asteroid Belt

The motion of the asteroids of the so–called *main belt* has been for a very long time an important source of dynamical problems to be handled by Celestial Mechanics. Out there there is another asteroid belt, the *Trojan belt*, which was believed, by some, to be less relevant to the history and dynamical structure of the Solar System only because the larger distance and the lower albedo biased the rate of asteroid discovery; now the number of Trojans is believed to be of the same order as that of the main belt asteroids (Schoemaker et al., 1992).

The problem of the 1:1 libration orbits has always fascinated the specialists of Celestial Mechanics, to the point that to collect a complete list of references is an impossible task. The theory I found most useful is the one by Érdi (1988; and references therein), which is based on a 3–D model with Jupiter in a fixed

Celestial Mechanics and Dynamical Astronomy **57**: 59–94 1993.

elliptic orbit, and which I shall often quote later. However, very few people have tried to work on the orbits of the real objects according to a realistic model of the perturbations by all the major planets; the most important work on this has been done by Schubart and Bien (Bien and Schubart, 1987; Schubart and Bien, 1987; and references therein). This might be due to the intrinsic difficulty of the problem: the very high inclination of many Trojans makes it very difficult to use any theory truncated with respect to the powers of inclination and eccentricity; the singularity of collision occurring for values of the critical argument just outside the libration region makes both analytical and semianalytical theories much more difficult.

However, the problem is there for us to solve, and the fact that the already available tools are not suitable to tackle the problem is not a good excuse not to try. Therefore I decided to collect as much information as possible on the behaviour of Trojan orbits and on the dynamical structure of the Trojan belt, by means of a large set of numerical integrations of orbits, in the framework of a realistic model of the gravitational perturbations by the major planets.

Now that CPU time is not any more a scarce resource (how many personal workstations are really used during the night?) the main problem with numerical integrations is to understand what the results mean. Unless spectacular instabilities occur, very little information is contained in the final state, even after a very long integration. The main challenge is then to compress the information –contained in hundreds of megabytes of orbital elements time series– in a theory comparable in form and readability to a simple analytical theory. Such a problem has been solved for non–resonant planetary orbits with the *synthetic theory* method (Carpino et al., 1987), and some techniques (such as digital filtering) can be copied from that case.

However, it turns out that there are some features of the orbits of the real Trojans which cannot be accounted for by a synthetic theory of the same form of the one successfully used to represent the orbits of the major outer planets. This arises from one of the same reasons discussed above, namely from the fact that the three *small parameters* occurring in a theory of the Trojans, the sine of the libration amplitude, the proper eccentricity, and the sine of the proper inclination, can have values as high as $0.58, 0.22, 0.61$ (see later, e.g. Tables 1 and 2, and Figure 10). Thus a theory which essentially forces a linearisation of the results through linear spectral analysis cannot represent well the orbit over a long period; this even for perfectly regular, quasiperiodic orbits. Here I propose a modified synthetic theory technique, which can be shown to represent the dynamical behaviour of Trojan orbits over a long time span (of the order of millions of years) and with satisfactory accuracy (typically within a thousandth of AU in semimajor axis and a few thousandths in $e, \sin I$), and this for most orbits of real Trojans.

Two pieces of information can be extracted by this technique from the output of a number of numerical integrations of Trojan orbits. *Proper elements*, that is three characteristic orbital parameters which, unlike the osculating elements, are almost constant (within a narrow range of variations) for a very long time, are computed for all the Trojans observed for long enough to provide reliable osculating orbital elements. *Resonances* and *chaotic behaviour* can be studied from the few difficult cases in which the stability of the proper elements is poor.

The proper elements, being almost constant, allow to search for *families* of objects which have been on very close orbits a long time ago, and which could be interpreted as the remains of a catastrophic collision. Information on resonances and chaos, especially when combined with the data on the time invariance of the proper elements, allows to discuss the stability of the Trojan belt. In this paper these consequences of my computations are only tentatively explored, and a long list of open problems is given. However, I hope to have provided –to myself as well as to others– a large enough set of constraints which should allow to found a future dynamical model of the Trojan belt on a solid basis.

This paper is organized as follows. Section 2 contains the definitions of the proper elements, and information on how they have been computed for 174 Trojan asteroids with well known orbits; the proper elements themselves are given in two Tables at the end of the paper. Section 3 discusses the accuracy of the proper elements, by means of a stability test using the running box method; also the results on the estimation of the maximum Lyapounov characteristic exponent are given, and the –more or less understood– relationship between stability and chaos is briefly discussed. Section 4 presents a preliminary search for asteroid families in the Trojan belt; although the method used is consistent with the most rigorous ones used for the main belt (Zappalà et al., 1990, 1992), the results are limited, simply because the number of Trojans observed so far is not large enough. Section 5 gives a list of problems, which I believe are relevant to understand the dynamics and the history of the Trojan belt. For these, either I have not been able to find a solution, or (worse) the problem looks more difficult now than when I started studying the dynamics of Trojans; a significant amount of future work, including possibly my own, needs to be done.

2. Computation of Proper Elements

The critical argument of the 1:1 resonance is $\lambda - \lambda'$, where λ is the mean longitude of the Trojan and λ' the mean longitude of Jupiter; since this argument cannot circulate, but librates around either $\pi/3$ or $-\pi/3$, it can be expressed as:

$$\lambda - \lambda' = \chi + D\cos(\theta) + \dots \tag{2.1}$$

where θ is the argument of libration, $\chi = \pm\pi/3 + \dots$, D is the libration amplitude, and the little dots stand for anything else, which is supposed to be "higher order". As a result of the libration, the semimajor axis a of the Trojan also oscillates $\pi/2$ out of phase:

$$a = a' + d\sin(\theta) + \dots \tag{2.2}$$

around a mean value given by the semimajor axis a' of Jupiter. In the simplest possible approximation, D and d are constant and θ is a linear function of time, with a fixed frequency f: $\theta = f(t - t_\circ)$. This approximation is not too bad, if the dynamics of the Trojan is studied within a simplified model, in which Jupiter is on a fixed elliptic orbit (restricted, 3–D, 3–body model): as shown by Érdi (1988), the little dots are indeed of order 2 with respect to a small parameter, which is of the same order of D and d; moreover, the ratio d/D is given by:

$$\frac{d}{D} = \sqrt{3\mu}a' + \ldots \simeq 0.278 \tag{2.3}$$

where μ is the ratio (mass of Jupiter)/(mass of the Sun + mass of Jupiter), d is measured in AU, D in radians. However, the theory by Érdi does not account for the perturbation from all the other planets. When the perturbation of the outer planets (mostly the one by Saturn) is added in, two effects arise. First, Saturn (and the other planets) perturb directly the orbit of the asteroid, inducing both short periodic and long periodic changes in the orbital elements. Second, the orbital elements of Jupiter are not constant any more; this indirectly affects the orbit of the Trojans, and makes it hard to use a 3–body type theory as an intermediary orbit, because the secular changes in the eccentricities and inclinations are slow but by no means small.

For the orbits which are not locked in a resonance, an approximate representation is available for the solution of the complete N–body problem: the classical linear theory of secular perturbations. The nonsingular orbital elements $h = e\sin(\varpi), k = e\cos(\varpi)$ (ϖ is the longitude of the perihelion) and $p = \sin(I)\sin(\Omega), \sin(I)\cos(\Omega)$ are represented as a Fourier series:

$$\begin{aligned} h &= e_P \sin(gt+\phi) + \textstyle\sum_k \xi_k \sin(g_k t + \phi_k) + \ldots \\ k &= e_P \cos(gt+\phi) + \textstyle\sum_k \xi_k \sin(g_k t + \phi_k) + \ldots \\ p &= \sin(I_P)\sin(st+\psi) + \textstyle\sum_k \eta_k \sin(s_k t + \psi_k) + \ldots \\ q &= \sin(I_P)\cos(st+\psi) + \textstyle\sum_k \eta_k \sin(s_k t + \psi_k) + \ldots \end{aligned} \tag{2.4}$$

with twice as many *secular fundamental frequencies* g_k, s_k as there are perturbing planets, and two *proper frequencies* g, s; the forced amplitudes ξ_k, η_k are proportional to the proper eccentricities e_k and to the sine of the proper inclinations I_k of the perturbing planets, while the *linear proper eccentricity and inclination* e_P, I_P are function of the initial conditions of the asteroid. In (2.4) the little dots stand for the terms which are of higher order, with respect to the small parameters e_P, e_k, I_P, I_k; some higher order terms can also be computed, to improve the accuracy in the representation of the solution, and to compute nonlinear *proper eccentricity and inclination* more accurately constant (Milani and Knežević, 1990, 1992). There are two main difficulties in applying a similar procedure to the Trojan case. First, the coefficients of an analytic theory (e.g., the ratios $\xi_k/e_k, \eta_k/\sin(I_k)$) are not simple combinations of Laplace coefficients, but more complicated functions of the orbital elements for which no explicit expression is known. Second, the little dots standing for the higher order terms are not negligible at all when the proper inclinations are as high as it is typical of the Trojan belt; even an higher

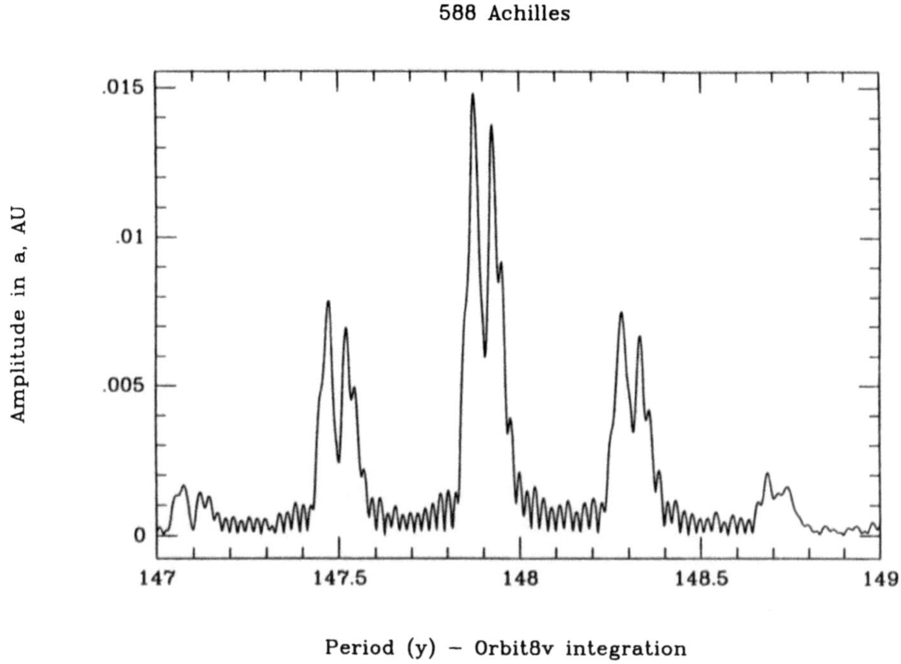

Fig. 1. Spectrum of the changes in the semimajor axis of *588 Achilles*: the region around the frequency f of the libration of the critical argument is shown. The two main features on the sides of the main line correspond to the frequencies $f \pm (g_5 - g_6)$. The finer structure inside each line cannot be interpreted because the data time span of 1 Myr is not long enough.

order analytical theory (analogous to Milani and Knežević), for the very fact of using an expansion in powers of the inclination, could not give accurate results.

All these difficulties notwithstanding, some theory will one day be manufactured to describe the full complexity of the dynamical behaviour of the Trojans; the semianalytical theories (see Williams, 1969; Morbidelli, 1991) are, in principle, the most promising. However, I always found useful, before constructing a theory, to know what the results are for a representative set of examples; numerical integration can of course provide us with such a set. The problem is how to represent the solution given by the numerical integration.

The output of any numerical integration is a time series for the osculating orbital elements. A well established technique (Carpino et al., 1987; Schubart and Bien, 1984) allows to generate a mean elements time series by digital filtering; the problem of the short periodic perturbations by Saturn et al. can therefore be handled in a comparatively easy way. As for the long periodic perturbations, the technique of the *synthetic theory*, used to transform the output time series in a Fourier series, and developed for *Project LONGSTOP*, accounts for higher and higher degree terms by assuming a linear superposition of harmonics, formed by integer combinations of the secular fundamental frequencies (Carpino et al., 1987;

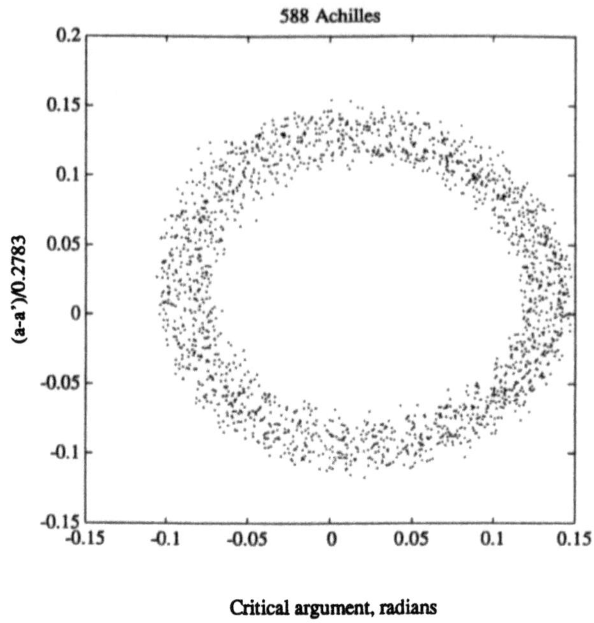

Fig. 2. Critical argument $\lambda - \lambda' - \pi/3$, in radians, and renormalized difference in semimajor axis between the Trojan and Jupiter, that is $(a - a')/(0.2783\ AU)$, for *588 Achilles*; the librations are not exactly circular, nor the center is at the origin of the coordinate axis, but the average amplitude of libration is well defined.

Nobili et al., 1989). The analogous of the *LONGSTOP* technique, applied to the Trojan case, would be to represent the little dots of equations (2.1), (2.2), (2.4) as trigonometric terms with frequencies obtained from integer combinations of f, g, s and g_k, s_k. The reason why this would not work is shown in Figure 1, where I have plotted a portion of the Fourier transform of the semimajor axis of *588 Achilles*, corresponding to frequencies near the libration frequency f. All the spectral lines in this figure can be interpreted as combination frequencies, e.g. the separation between the bunch of lines around 147.9 y and the one around 147.5 y corresponds to the frequency $g_6 - g_5$. However, the synthetic theory technique works only if the series of higher and higher degree combination frequencies converges, and quickly enough; as it is seen in Figure 1, the combination lines decrease too slowly to allow an accurate representation of the solution with a Fourier series of reasonable length.

To understand what is happening to the semimajor axis of a Trojan asteroid, I have computed the argument of libration θ from (2.1) and (2.2), neglecting all the little dots and taking also into account (2.3); that is, θ is the polar angle of the point with cartesian coordinates $(0.2783 \cdot (\lambda - \lambda' - \chi), a - a')$. Since the libration is not a circle –centered at the origin– in the plane of the above defined variables (see Figure 2), an oscillation with period equal to the libration period is superimposed to the long term trend of θ; moreover, there are short periodic perturbations induced by Saturn et al.. The time series for the angle θ is then digitally filtered, in such a way that all the oscillations with periods between 2.5 y and 250 y are removed; the main short periodic perturbation has a period of $\simeq 20\,y$, and the libration period is $\simeq 150\,y$, thus both the short periodic perturbations and the oscillation with period equal to the libration period are removed.

Since the filter has coefficients with sum 1 and even symmetry, linear trends are not distorted. This applies provided the angular variables are handled as real numbers (that is, the angular variables are continuous functions of time, unlike the principal values which jump by 2π at each revolution). Then the filtered time series for θ can be used to estimate a libration period (e.g. by linear fit) with short periodic perturbations removed. The result of this computation for *588 Achilles* and *1868 Thersites* are shown in Figure 3; the libration period changes over a long period (with arguments such as $(g_5 - g_6)t$) and the extremes of this oscillation correspond to the main side lines of the spectrum of Figure 1.

The standard representation for an oscillation with variable period is of the form $A\cos(ft + \tau)$ where τ oscillates slowly; the attempt to represent the changes in a of a Trojan as a Fourier series with arguments linear functions of time fails because the amplitude of the oscillation of τ is large, as shown (again for *588 Achilles*, and for *2594*) in Figure 4. However, by some kind of dumb luck the amplitude A is not changing in a long periodic way as τ does, and this both for the semimajor axis and for the critical argument; that is both D and d are approximately constant for a long time. This can be understood in an intuitive way, by using again the formula (2.3); the ratio of the amplitudes in the two conjugate variables depends only weakly upon the amplitudes themselves: the little dots in (2.3) are actually second order with respect to D (Érdi, 1988, eq. (8) and (4)). Therefore the area of the libration curve is approximately proportional to the square of any one of the two amplitudes, and the amplitudes d and D are, to a good approximation, functions of the adiabatic invariant obtained by representing the problem as a slowly varying restricted elliptic problem.

The proper element related to the libration amplitude can therefore be defined as the amplitude of the oscillation of either a or $\lambda - \lambda'$ with argument θ; for the latter variable, the filtered time series is used, while the variables a, a', λ, λ' are represented by a time series passed through a different filter, which removes the short periodic perturbations (periods between 2.5 y and 50 y) but not the periods close to 150 y. The only tricky point is that the Fourier component with argument θ of either a or $\lambda - \lambda'$ needs to be computed in line with the numerical integration, because the frequency f is comparatively fast and a time series for 1 My would be too large to be stored, if sampled well enough to compute a component with a period $\simeq 150\,y$. For this purpose I have modified my own numerical integration program by adding two parallel filters (with dark bands for periods up to 50 and 250 y respectively, designed according to Carpino et al., 1987), and the computation of θ as defined above; the sums to be used to compute the Fourier transforms with

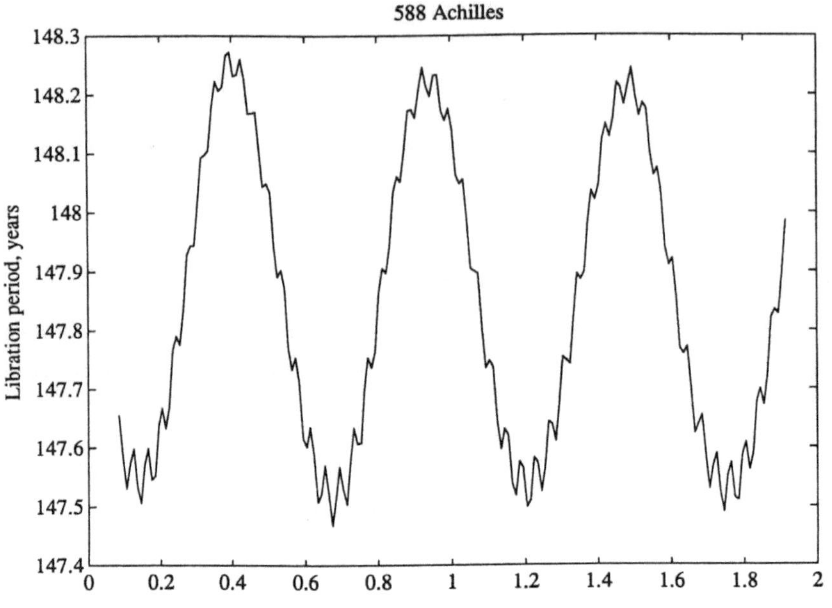

(a)

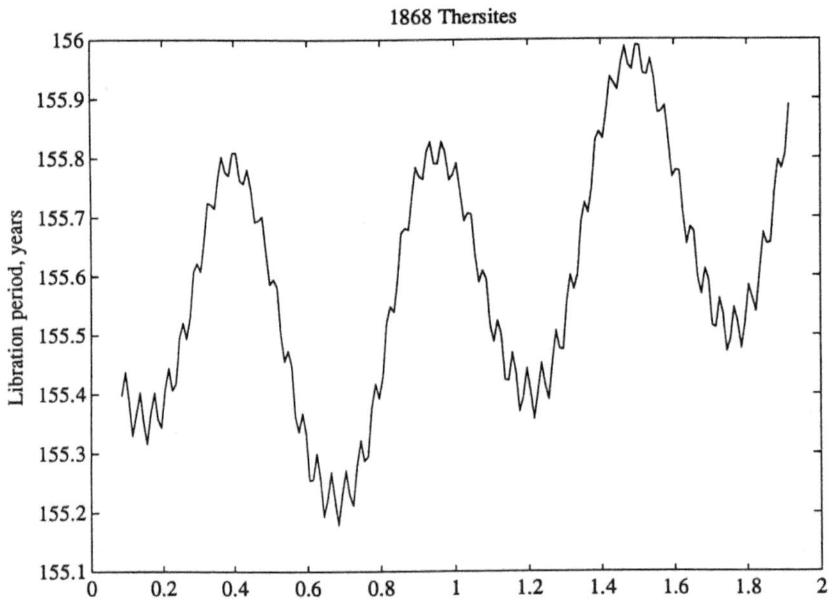

(b)

Fig. 3. Changes in the period of libration, in years, as a result of longer period perturbations: (a) for *588 Achilles* the main changes are correlated with the changes in the eccentricity of Jupiter, that is with the frequency $g_5 - g_6$; (b) for *1868 Thersites*, with larger amplitude of libration, a more complex behaviour appears.

respect to θ are accumulated and written to disk at long intervals. This procedure I have used to compute the proper elements d and D, and the frequency f, for all the Trojan asteroids for which a good enough orbit is available from the observations.

The computation started from a list of osculating elements for Trojan asteroids, including 85 numbered asteroids, 86 observed at two oppositions, and 3 single opposition with a very long arc ($> 90\, days$); the orbital elements were for the same reference epoch (December 1991). The catalogue of orbits was provided to me by E. Bowell of the Lowell Observatory, as a byproduct of his continuing effort to maintain and improve a large catalogue of asteroid orbital data. For the unnumbered asteroids I am using in this paper an arbitrary numbering (with numbers above *6000*); to actually identify the indicated asteroid, the readers must refer to Table 2 where international designators and survey catalogue numbers are given.

In the same catalogue there were 97 other single opposition orbits, but I have not used them because the osculating orbital elements are too inaccurate. Table II of Van Houten et al. (1991) shows that the orbital elements of Trojans, derived from single opposition observations during the main surveys specifically targeted at the Trojan belt, are too poor to be useful, having an uncertainity an order of magnitude larger than the one introduced by my computation of proper elements (see Section 3 and Table 3). It is somewhat surprising that the surveys, which have been conducted with the specific purpose of discovering Trojans, have used observations spread over an arc of only 16 days, shorter by a factor of 2 than the arc length used in the Palomar–Leiden survey for main belt asteroids; this when it is well known that a longer arc is needed to determine with the same accuracy a Trojan orbit.

For each one of the 174 Trojans I have used, an orbit was numerically computed for a time span of 1 $Million\ years$, within a rather complete model, fully accounting for the perturbations by the four major outer planets; the accuracy of the integration was more than enough for our purposes (Milani and Nobili, 1988). The amount of computational resources used was not very significant; my portable workstation was running background jobs for a few weeks. To reduce the effect of the unmodelled perturbations by the inner planets, the initial conditions of both the outer planets and the asteroids have been referred to the center of mass of the inner solar system; this technique (essentially equivalent to the use of jacobian coordinates) sharply reduces the difference in the fundamental secular frequencies g_k, s_k between computations including and not including the inner planets; e.g. the value of the frequency g_6 of the perihelion of Saturn, as found by numerical integrations for $> 5\ My$, is 28.164 $arcsec/y$ without barycentric correction (Milani and Knežević, 1992), and 28.236 with barycentric correction; the best value, obtained by the *LONGSTOP* project with a complicated procedure to remove the effect of the inner planets on the initial conditions, is 28.246 (Nobili et al., 1989).

The values of the amplitudes d and D, and of the libration frequency f, are listed in the first 3 columns in Table 1 (for numbered asteroids) and Table 2 (for

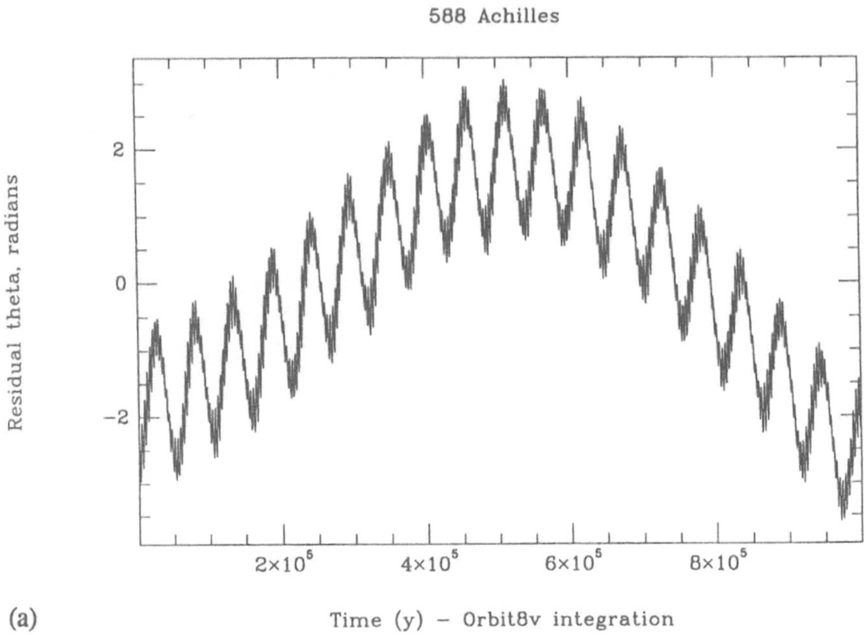

(a)

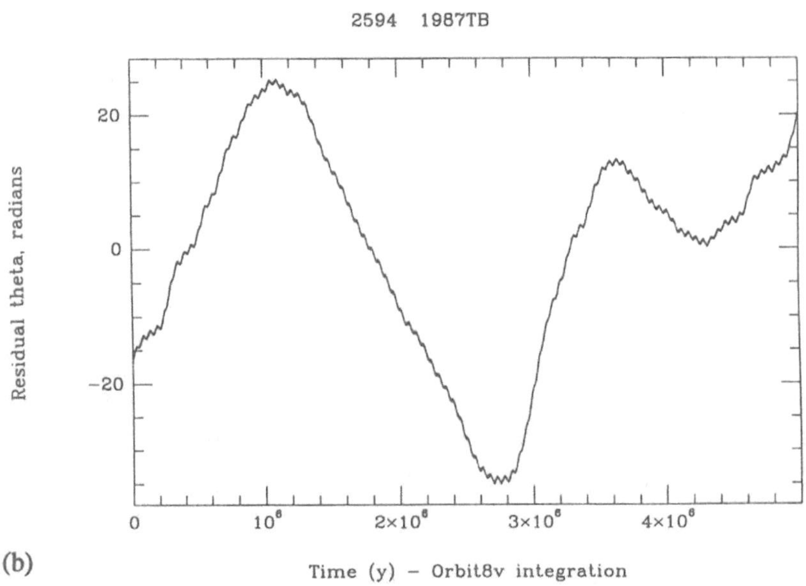

(b)

Fig. 4. The libration argument θ has been fitted to a straight line (thus computing the frequency f) and the residuals, in radians, plotted over the entire integration time span: (a) for *588 Achilles*, over 1 Myr, a regular pattern appears; (b) for *2594 1977 RR*, over 5 Myr, the irregular behaviour of θ suggests chaos.

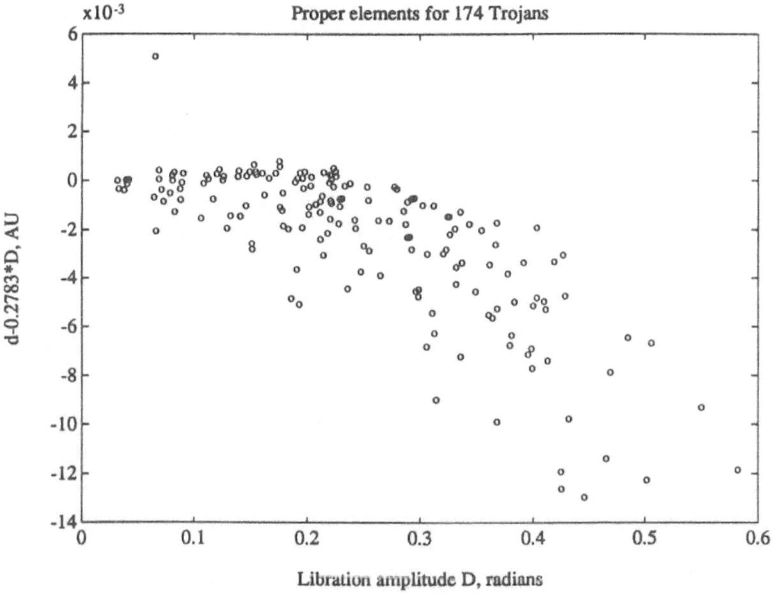

Fig. 5. The amplitude of libration d of the critical argument, in radians, and the amplitude of oscillation D of the semimajor axis, after the simple linear trend of eq. (2.3) has been removed. The trend corresponds to the higher order terms in Érdi (1988) theory, but there is also a wide scatter.

unumbered two opposition orbits). These data can also be used to check a posteriori the accuracy of the approximation (2.3) used to compute θ; Figure 5 shows the residuals of d with respect to the simple law $d = 0.2783 \cdot D$; there is a visible trend corresponding to the higher order terms (see Érdi, 1988, eq. (8)), but also some scatter, indicating the limitations of a theory based upon a very simplified model.

For the computation of the other two proper elements, the orbital elements h, k, p, q of the asteroid are taken from the digitally filtered time series; the perturbations with periods between 2.5 y and 250 y are removed almost perfectly (the attenuation factor is $< 10^{-3}$) by the filter, and the filtered time series is essentially equivalent to a set of mean elements. For the eccentricity related variables h, k the proper eccentricity is essentially defined by (2.4): the components with the known frequencies g_5, g_6, g_7 are determined by Fourier analysis and removed from the data; the largest nonlinear forced term, the one with frequency $2g_6 - g_5$, is also removed (this term is important, see Milani and Knežević, 1992). The residuals after this removal of the forced terms are analysed to find the fundamental frequency g; to this purpose I compute the angle $gt + \phi$ as a time series, again accounting for the number of revolutions (this can be done provided the time series for h, k is sampled at time intervals shorter than π/g); a linear fit to the angle gives the best estimate of the frequency g and the amplitude of the Fourier component with frequency g is used as proper eccentricity e_P. The values of e_P and g computed in this way are given in columns 4 and 5 of the Tables 1 and 2.

The above procedure is somewhat simplistic, in that the highly nonlinear character of the Trojan dynamics results in the presence of important nonlinear coupled modes; however, the combination frequencies containing f should correspond to periods shorter than 250 y, which are removed by the digital filter. The main problem arises from the coupled modes containing both g and s, and also possibly some of the g_k, s_k; they result in the splitting of the spectral line with frequency g into a complicated multiplet. However, for most Trojans the combination lines are not very strong and the determination of the proper mode with frequency g is not significantly impaired, at the level of accuracy of this work. Figure 6 shows the spectrum of the variable h for frequencies close to g, after removal of both the forced terms g_5, g_6, g_7 and $2g_6 - g_5$ and of the proper mode g. For *588 Achilles*, the multiplet surrounding g (made visible by the removal of the dominant line) has the highest amplitude line at the frequency $g - 2s$. The frequency $g - 2s$ corresponds to the behaviour described by Bien and Schubart (1987) and by Érdi (1988) with the geometric construction of an ellipse rotating with frequency s. It would be possible to compute proper elements in a more refined way by adding the $g - 2s$ line to the theory; the improvement would not be very significant, since there are many more spectral lines not accounted for by Érdi's theory. Even in a case with large inclination, such as *2363 Cebriones*, the $g - 2s$ frequency corresponds to the largest, but by no means dominant, spectral line in the residuals; although the residuals after fitting the forced and proper modes are larger, the proper eccentricity is still stable enough (see Table 3).

The computation of the proper inclination was somewhat more problematic. As a matter of principle, (2.4) could be used again, and the forced terms could be identified and removed before proceeding to the determination of s and I_P. However, in this case the forced modes with frequencies s_6 and s_7 are much smaller than the proper s mode; if they are determined before removing the proper mode, the relative accuracy is often $> 10\%$. The only case in which the forced modes are large is when they are near resonant, and this means that the separation of the forced mode, from the proper mode with nearby frequency, is difficult. As for the s_8 mode, it is larger than the s_7 mode, but its period is $1,871,000\, y$ and there is no way to compute it accurately with only 1 My of data. As a result, I have taken the simplistic approach of computing the proper angle $st + \psi$ as the polar angle of (q, p) (accounting for the number of revolutions) and then fitting s; of course p, q have been computed with inclination and node with respect to the invariable plane of the outer solar system (as determined by *LONGSTOP*). The sine of the proper inclination is then taken as the amplitude of the Fourier component with frequency s. The accuracy of this procedure is not really limited by the "noise" left by the forced terms, but by the time span. I have used the method by Ferraz-Mello (1981) to correctly separate the target frequency from the constant (or longer period) terms, but this is not always enough. The period of the node

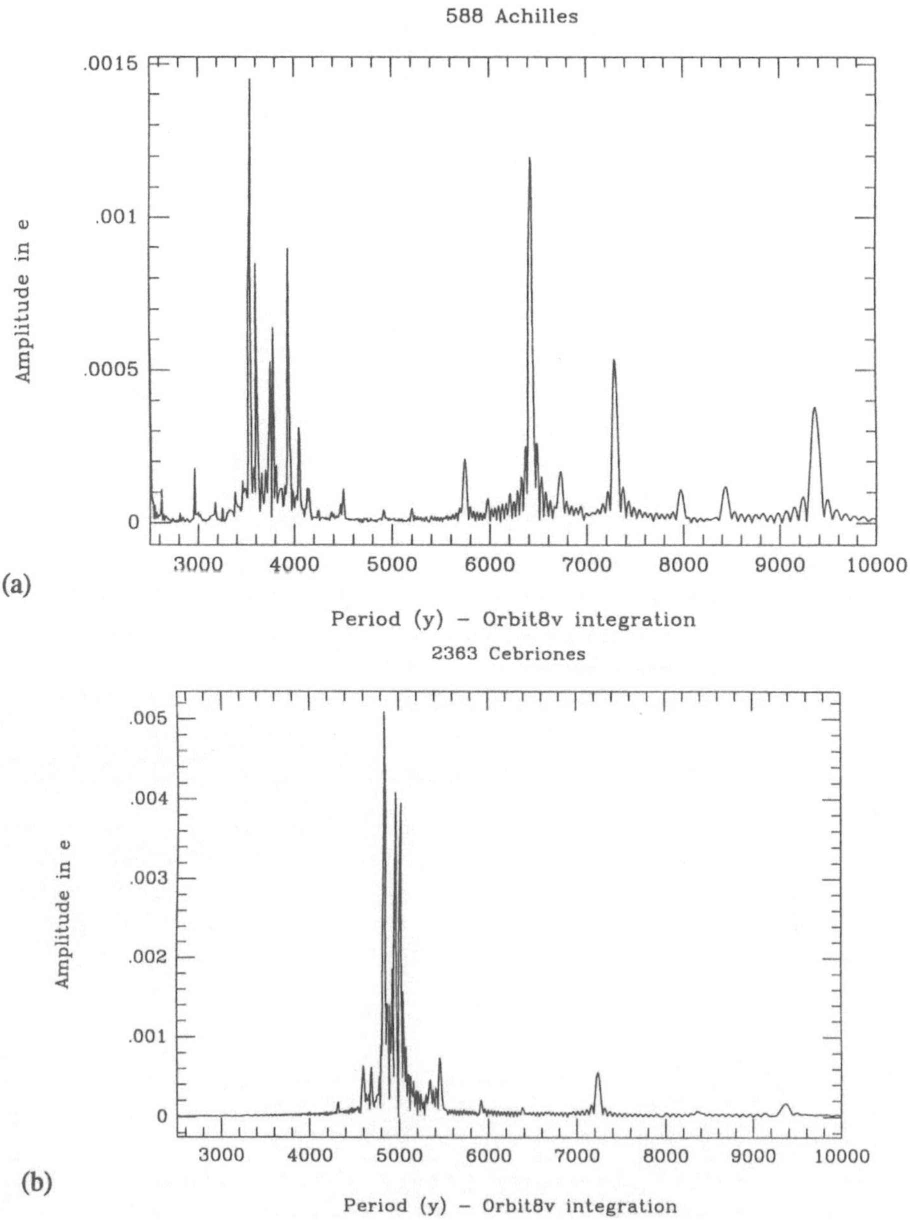

(a)

(b)

Fig. 6. The spectrum of the mean element $h = e \sin \varpi$ after removal of the main lines, namely the ones with frequencies $g, g_5, g_6, g_7, 2g_6 - g_5$: (a) for *588 Achilles*, the residuals contain a multiplet around the removed proper mode g (corresponding to a period of $3,763\,yr$), but also other lines not accounted for by any existing theory; (b) for *2363 Cebriones*, with a proper inclination of 32.7°, the $g - 2s$ line predicted by existing theories is the largest, but there are many more forming a complex multiplet around the removed g mode (at $4,990\,yr$).

of a Trojan can range between $\simeq 2,700,000\,y$ (for *1208 Troilus*) and $\simeq 38,000\,y$ (for *2594* and *6031*). When it is comparable to the integration time span, neither the frequency nor the amplitude are well determined. In 4 such cases, I decided to extend the numerical integration to 5 My; this was done for *1208, 2146, 4833* and *6057* because of their long nodal periods, and also for 3 more asteroids for another reason which will be apparent from Section 3. For all the 174 Trojans $\sin I_P$ and s are given in columns 6 and 7 of Tables 1 and 2; the last two columns specify whether the libration is around L_4 (*Greek*, or leading Trojan) or L_5 (*Trojan*, or trailing Trojan), and the integration time span.

The Tables of this paper, as well as some auxiliary information, are available in computer readable form from the `ftp anonymous server` of my department (*INTERNET* address `gauss.dm.unipi.it`, that is 131.114.6.55; look in the `pub/propel/trojans` directory).

3. Accuracy, Stability, Chaos

The level of accuracy in the proper elements, which is actually required, is limited to the physically meaningful range, and the latter is limited both by the observational accuracy of the astrometry and by the impossibility to study collisional processes between too small asteroids that far. A level of accuracy corresponding to about 0.001 (rms) in both proper e and $\sin I$, to about 0.01 AU in a (that is, in *d*), and to about 20 m/s in relative velocity after escape of the fragments of a catastrophic disruption, can be considered fully satisfactory, at the present state of the art. The question is whether I have achieved this level of accuracy.

I have always been very diffident toward internal consistencies and formal covariances as a way to assess the accuracy of a result; what is needed is a test which directly verifies the property of the proper elements which makes them useful, namely their stability over a long time span. This is comparatively easy to do for main belt asteroids, if the proper elements are computed by an analytical formula: the time series of osculating elements obtained as output of an accurate numerical integration is processed to give the corresponding time series for the proper elements; the changes of the supposedly constant proper elements directly measure their accuracy (Milani and Knežević, 1990; 1992). For the Trojans, such an analytic theory does not exist, and the proper elements as defined in this paper are themselves derived from the output of a long numerical integration. The only external check I could think of, consists of a change in the time span over which the output of the numerical integration is analysed. I have taken a running box of length $500,000\,y$, and moved it by steps of $50,000\,y$, thus obtaining 11 separate computations of my proper elements (for the 7 orbits which have been computed for 5 My, the box was 2.5 My long and the shifts were by $250,000\,y$). The stability of the proper elements and of the proper frequencies is defined as the difference between the maximum and the minimum of these 11 values. To compare both with the physically meaningful levels discussed above, and with the values given by

Milani and Knežević (1990; 1992), you should take into account that the value (maximum–minimum) generally corresponds to $\simeq 2\sqrt{2}$ times the rms value of the changes with respect to the mean value.

The results of this test have been encouraging; as an example, the values for *588 Achilles* are given in the first line of Table 3; they are so small, that the statement –given in Section 2– that a more refined theory is not needed appears fully justified in this case. Of course there are also some asteroids for which the results are much less satisfying. Table 3 lists all the cases in which either $\Delta d > 0.001\ AU$ or $\Delta e_P > 0.0025$ or $\Delta(\sin I_P) > 0.0025$: there are 7, 10 and 14 such cases respectively, a total of 22 asteroids with "degraded" proper elements. This proportion of difficult cases is not very different from the onc occurring in the main belt, and does not inhibit the use of a proper elements catalogue to search for asteroid families (see Milani and Knežević, 1992; Zappalà et al., 1992). However, we would like to understand why some proper elements are less proper than others.

Beside the example of *588*, Table 3 contains also some other interesting cases, including all the 7 orbits computed for 5 My. The main indicators available to detect irregular dynamical behaviour are the *Lyapounov characteristic exponents (LCE)*. An orbit is by definition *chaotic* when some Lyapounov exponents are positive; all the Lyapounov exponents are zero when the solution is a conditionally periodic orbit of the kind described by the Kolmogorov–Arnold–Moser theorem; it is not known whether other cases can occur with a non vanishing probability.

For each one of the 174 Trojan orbits computed numerically, I have simultaneously solved the *equation of variations*, that is the linearized differential equation for relative motion of two nearby orbits. The initial conditions for the variational equations were randomly chosen displacement vectors $v(0) = v_0$. It is possible to obtain an indication on the value of the maximum Lyapounov exponent by monitoring the function

$$\gamma(t) = \log \frac{|v(t)|}{|v_0|} \tag{3.1}$$

where $v(t)$ is the solution of the variational equations. With probability 1 (with respect to a random choice of the initial conditions v_0) the maximum Lyapounov exponent is: $\chi = \lim_{t\to+\infty} \gamma(t)/t$. Any calculation being finite, only an estimate can be obtained by means of the function $\gamma(t)$. I have used a linear fit to $\gamma(t)$ over the available time span to estimate the maximum LCE; the constant absorbs most of the initial increase of the function $\gamma(t)$ due to the linear divergence of nearby orbits occurring even in a regular orbit, and this allows the detection of a slower linear trend (see Figure 7). In this way a positive LCE χ can generally be detected with an integration time span of between 6 and 7 times $1/\chi$ ($1/\chi$ is the *Lyapounov time*). With a time span of 1 My all the cases in which the fit to $\gamma(t)$ gives a slope $\simeq 6 \times 10^{-6}\ y^{-1}$ or larger can be interpreted as positive detections of a positive LCE; of course there are always marginal cases, in which visual inspection of the plot of $\gamma(t)$ is recommended before drawing any conclusion; thus a number of

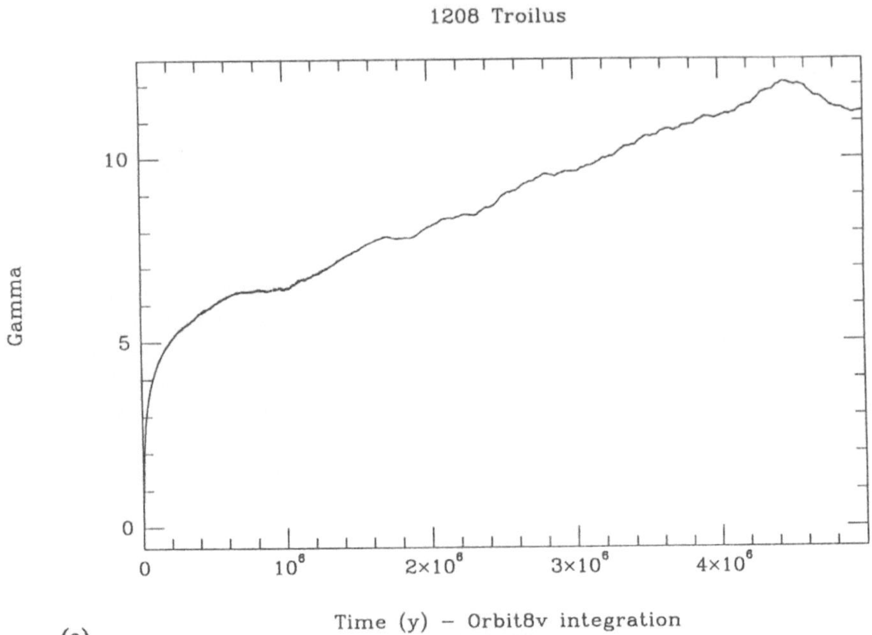

(a)

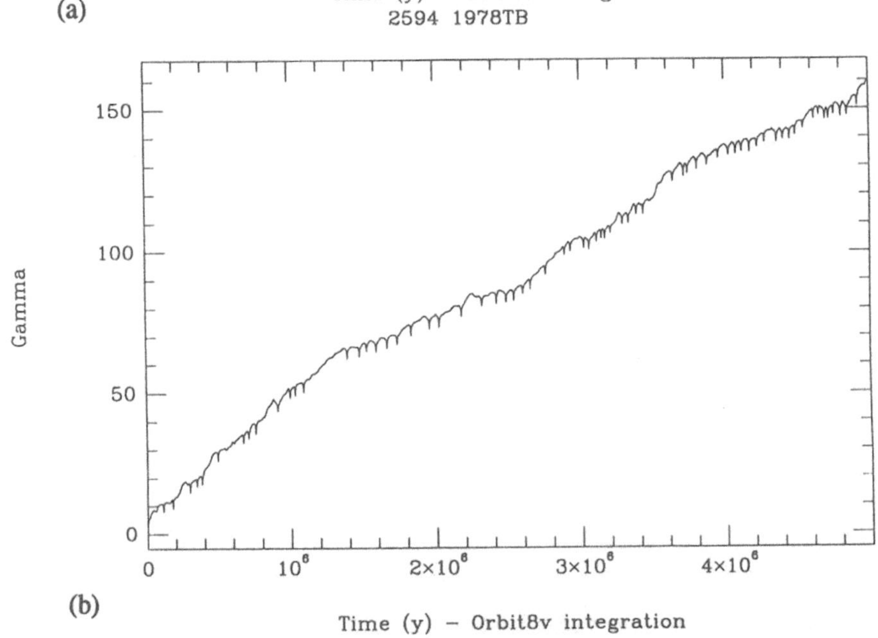

(b)

Fig. 7. The divergence ratio γ (see eq. (3.1)) as a function of time. The presence of a linear trend indicates a positive Lyapounov exponent. (a) for *1208 Troilus*: although the time span is too short for strong chaotic effects, the best fit value of $1.5 \times 10^{-6}\, yr^{-1}$ indicates the orbit is indeed chaotic. The use of the classical formula $\gamma(t)/t$ at the end of the integration time span would give a wrong value and would not allow a reliable detection of chaos. (b) for *2594* the presence of chaos is quite obvious; the Lyapounov time is only $\simeq 35,000\, yr$.

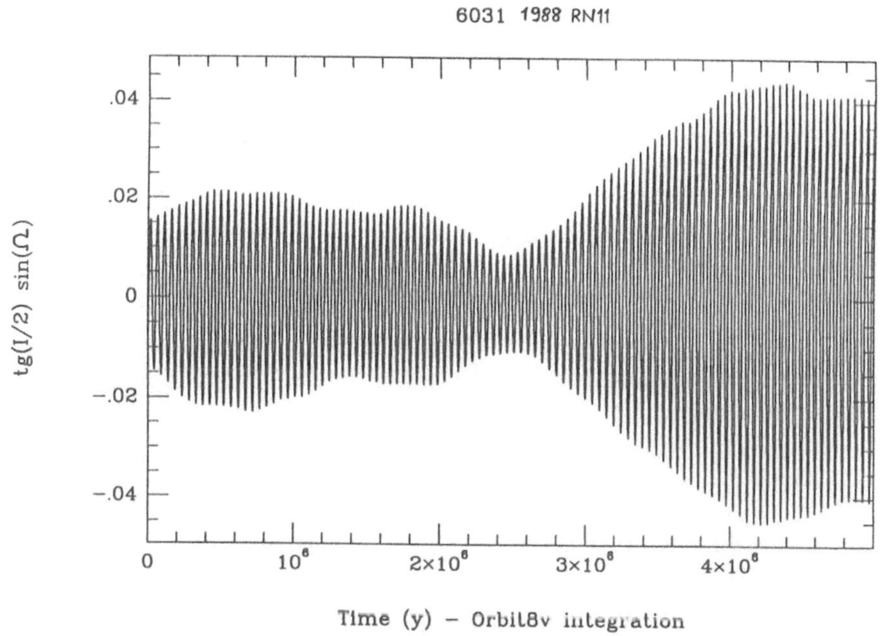

Fig. 8. The worst case of unstable chaos: *6031 88RN11* undergoes a shift in the orbital behaviour after $\simeq 2\,Myr$ which results in a change in the proper elements by a factor $\simeq 2$. As an example the element $tg(I/2)\sin\Omega$ shows a wild change in the amplitude of oscillation.

cases with a result of the fit in the range 4 to $5 \times 10^{-6}\,y^{-1}$ are likely to be chaotic, but a longer integration is needed for a positive detection.

The values of the maximum LCE, computed with the method discussed above, are given in the last column of Table 3; all the orbits for which the value was $\geq 6 \times 10^{-6}$ have been included in the Table. 24 such cases have been found among the 174 orbits computed. For the orbits which have been computed for 5 My, even a smaller value of the maximum LCE can be considered as a positive detection of chaos: the value 1.5×10^{-6} for *1208 Troilus* is indeed a positive detection, and indicates a chaotic behaviour with a Lyapounov time longer than any known example in the asteroid belt (Figure 7a). Some other Trojans have remarkably large LCE, such a *2594* (Figure 7b); 3 of these strongly chaotic cases (*2594, 6031, 6067*) have also been computed for 5 My to look for large scale instability. Figure 8 shows the most striking case of large scale instability, occurring for asteroid *6031*, with I_P changing by a factor $\simeq 2$.

For each entry, the instabilities of the proper elements and frequencies as derived from the running box test are also given, in the same order as in the Tables 1 and 2. For all the other orbits I have computed, neither chaotic behaviour nor instability of the proper elements (above the levels discussed above) was detected. For the asteroids which are listed in Table 3, the main problem is to understand the origin and long term dynamical effect of both instability and chaoticity. I cannot claim I have solved all the problems posed by Table 3; on the contrary, I understand only a few of these dynamical and computational problems, and a list of unsolved problems is given in Section 5. My experience with the study of the dynamics of a large population of objects indicates that it is very important to find a right balance between quantity and quality; in other words, extending the study to a larger sample and investigating in depth a few peculiar cases are both useful, and it would be wrong to give up any of the two. Therefore I will try to give at least a first approximation classification of all the "exceptional" cases of Table 3, to be able to point out the cases worthy of further study, without giving up my approach based upon large scale computations and the study of general dynamical properties of large populations.

All the confirmed chaotic orbits and orbits with degraded proper elements, a total of 37 cases, belong to one of the following classes:

(A) chaotic orbits, with an instability of the proper elements d and D of significant size, although not very large in absolute terms: *1437, 3451, 4709, 4835, 4946.* Asteroid *4835* is also remarkable for its very high eccentricity (however it is not unique, since the recently discovered *1991XX* has a comparable eccentricity). In some of these cases there is some instability in the proper element I_P, but not in e_P; this might be a clue to the critical arguments involved in the chaotic behaviour, but I do not have an explanation for the occurrence of chaos.

(B) stable chaotic orbits, for which a positive LCE does not correspond to a significantly increased instability of the proper elements and frequencies: *1404, 1868, 1869, 2207, 2797, 4489, 4543, 4708, 4754, 4827, 6001, 6049, 6056, 6079.* Stable chaos is a poorly understood phenomenon, but it is now clear that it does occur in many orbits of real solar system bodies, sometimes in very extreme forms (Milani and Nobili, 1992). The resonances involved are likely to be of high order; on the other hand, with a ratio f/g always above $\simeq$ 18, a ratio g/s above $\simeq$ 12, and a ratio n/f always above $\simeq$ 12, low order resonances are hard to find in the Trojan belt.

(C) unstable chaos, with both positive LCE and large scale instabilities: *1173, 2594, 6031, 6067, 6073.* The only resonances which are known to occur in the Trojan belt are secular resonances involving the node, mainly $s - s_6$, which is certainly responsible for the spectacular instability of *6031* shown in Figure 8, and is also implied in the behaviour of *1173* and *6073*. This resonance has already been studied (see Nakai, and Kinoshita, 1985; Bien and Schubart, 1986), but a wild behavior such as that of *6031* has never been suspected. Another puzzle arises from the fact that *1871* is also close to the $s - s_6$ resonance but it shows no indications of chaos. It is somewhat surprising to find a large scale instability, as in the case of *6031*, for an object for which there are no a priori reasons to suspect a dynamical age shorter than that of most asteroids; on the other hand, there are no indications of a transition to a *comet* class orbit, that is to a Jupiter crossing state. For *2594, 6067* the origin of the chaotic behaviour is unknown: these are large libration amplitude and low inclination orbits, and this of course implies stronger perturbation by Jupiter; however, the Trojan with the largest libration amplitude is *6075* with $D = 33.3°$, and it does not show any indication of chaotic behaviour. I have not been able to find any resonance of a "reasonable" order which could be responsible for their behavior. The only possible explanation is that some prejudices on what is a "reasonable" resonance need to be abandoned.

(D) orbits with large I_P, and very slow s: *1208, 2146, 2363, 3317, 4833, 4834, 4867, 6047, 6057*. The main problem is to understand whether the difficulty is computational (that is, the proper elements have not been computed accurately because of the need to solve for effects with a period close to the data time span) or dynamical (that is, there are secular resonances involved). In the cases of *4833* and *6057*, whose node precesses with a period of $1,026,000$ and $699,000\ y$ respectively, the "cure" consisting in an increased integration time span has worked reasonably well; not so for *1208 Troilus* and *2146 Stentor*, for which there are very complicated long periodic perturbations. *1208* has $s = -0.15\ arcsec/y$ and it is not clear whether the time span of 5 My is enough; on the other hand the influence of some secular resonance cannot be excluded. For *2146* $s = +3.28\ arcsec/y$ and there is the possibility that the secular small divisor $2s - 2g_5$ plays a significant rôle, as proposed by Bien and Schubart (1987); the finding of a (very slow) chaotic behaviour might be rated as a confirmation of this hypothesis. All the cases in this class have periods of the node above $280,000\ y$, and it could be argued that for the 5 cases in which the orbit has been computed for 1 My and the running box was only $500,000\ y$ long, the proper element I_P computed with half data is likely to be much less accurate of the one computed with all the data, thus the instability estimate of Table 3 should be pessimistic. However, not all the cases with slow s show the same degradation.

(E) cases in which the proper eccentricity has a degraded stability, without any other sign of either chaos or instability: *884, 4057, 6023, 6050*. None of these cases has a high inclination, and the occurrence of complicated combination lines containing both g and s is not likely to be responsible; some other mechanism, which I do not understand, must be involved. Of course it is possible to investigate these cases with degraded e_P one by one, but this is not a solution. An algorithm for the computation of proper elements needs to be automatic and objective, to allow the processing of large and ever increasing catalogues of orbits; to find a specific fix for each case would not help.

As a conclusion, I am rather satisfied that the proper elements I have computed are as accurate as they should be. Out of the 22 cases which are not stable according to the specifications, 11 are chaotic; in these cases, it is already a remarkable result that proper elements can, in some cases, be only moderately unstable, that is the finding of various degrees of stable chaos, with very small average changes to proper elements over one Lyapounov time. Of the other 11 cases, 6 (may be 7) can be explained by the presence of very long periods which should be handled with even longer integrations, and only 4 to 5 seem to point to some dynamical phenomenon not accounted for in the algorithm used to compute proper elements. It is certainly possible to improve upon this result, but these improvements would be relevant only for a small minority of Trojans, and therefore I believe the present version of my proper elements is good enough to be used, e.g. to search for Trojan families.

4. Search for Families

The outline of the procedure to identify asteroid families is discussed e.g. in Milani et al. (1992); once proper elements are available (together with some information on their stability), we need to define a distance between each couple of objects in the same region. There is unfortunately no unique optimal choice of a metric; Zappalà et al. (1990; 1992) have used two metrics which appear to be good compromise solution, and compared the results obtained with both, to increase the reliability of the detected families. Their metric d_1 (see Zappalà et al., 1992, eq. (2.1)) has the property of overweighing the differences in semimajor axis, metric d_2 overweighs the difference in inclination. For the Trojan belt the difficulty of the choice of a metric is compounded by the very peculiar geometry of the Trojan belt; it is indeed not a belt, but more like a couple of towers, with an height (in the direction perpendicular to the plane of the orbit of Jupiter) of more than 6 AU. When seen in the velocity space –were the arguments about dispersion of fragments from a catastrophic collision should apply– the shape is even more peculiar. A change of velocity, in the along track direction, of about 200 m/s, can result in the expulsion of a Trojan from the libration region, while a change of velocity of 7 Km/s in the direction perpendicular to the orbital plane can change a stable Trojan orbit in another equally stable (what happens with the radial direction, that is with the range in e_P, is less clear; see Section 5). As a result, the use of a metric which overweights the velocity difference in the along track direction by a factor 3, as the d_1 of Zappalà et al., changes the shape of the Trojan region in a very significant way; d_2 would on the contrary stretch the already very large separations in I_P. I am not claiming to have found an ideal solution to this problem; as a first attempt, I have used the "trivial" metric:

$$d_3 = \sqrt{\frac{1}{4}(\frac{\delta d}{a})^2 + 2(\delta e_P)^2 + 2(\delta \sin I_P)^2} \tag{4.1}$$

which is obtained by ignoring the off diagonal terms in the equations linking the velocity difference components and the changes in the proper elements.

The next step is the classification, for which I have used the algorithm of *hierarchical clustering*, consistently with Zappalà et al. (1990; 1992). In short, the

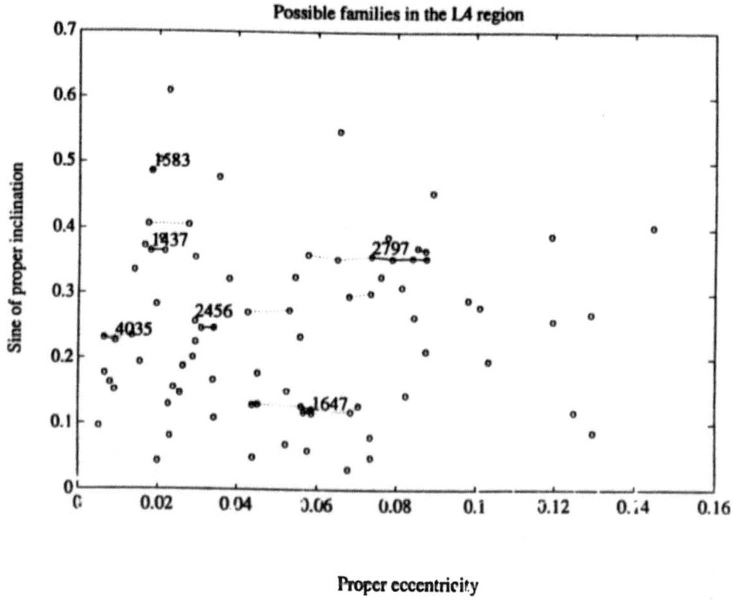

(a)

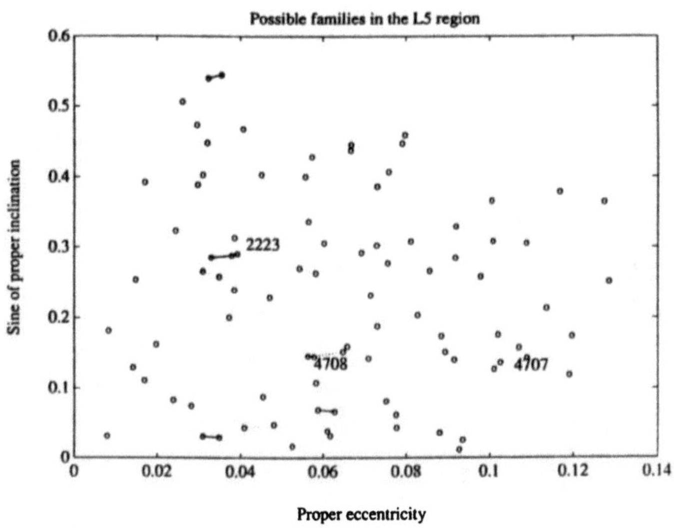

(b)

Fig. 9. Proper elements e_P and $\sin I_P$: the third proper element d matters much less, because its contribution to the relative velocities is minor; however, the distances have been computed in 3 dimensions by eq. (4.1). The segments indicate a distance corresponding to less than 208 m/s (dotted line), and to less than 130 m/s (full line). (a) in the L4 swarm, the most remarkable features are the *1583–3801* couple and the clusters headed by *1647 Menelaus* and by *2797 Teucer*; (b) in L5 there is the *2223 Sarpedon* cluster. The others clumps may or may not be significant.

distances between all the couples of Trojans are sorted in ascending order; for each value of the cutoff Q, the objects with a distance $d_3 < Q$ are "clustered" together. This can be represented either as a graph, with the objects at distance less than Q joined by a line as in Figure 9, or as a "stalactite" table such as Table 4, were the rows correspond to a cutoff value (the unit is 10^{-2}) and list the objects joining the cluster at that cutoff, the columns correspond to the different clusters. The real difficulty is in deciding were to stop; the procedure has no natural termination, not until all the objects are clumped together at a very large cutoff.

I have not copied the procedure used by Zappalà et al. (1990) to decide were to place the cutoff (what they call the *Quasi Random Level*), because it would not work; being of a statistical nature, that procedure requires large numbers of asteroids to work reliably, and the number of Trojans is not large enough (the more complicated procedure used by Zappalà et al., 1992, could work). In preparing Table 4 and Figures 9a–9b I have assumed that a cutoff $Q = 0.016$ (which corresponds, multiplying by na, to $\simeq 208\ m/s$ in relative velocity; the dotted lines in the Figures and the top 3 lines in the Table) give results to be rated from marginal to unreliable; the results in which we can have some confidence should be obtained with a cutoff not above 0.01 ($130\ m/s$; the full lines in the Figures). I acknowledge that more work should be done to compute the correct value of the Quasi Random Level; this is just a first attempt, mostly with the purpose of deciding whether the presently available number of (good) Trojan orbits is enough to begin working more in depth on the family problem.

I am still left with the problem of deciding, for a given cutoff level, how many members a cluster must have to be rated significant. For this purpose, there is an argument I can apply to the Trojans which was not available for the main belt. There are two regions, around L4 and around L5, which are physically disjoint (no Greek ever gets as close as 3 AU from a Trojan); nevertheless, the proper elements, as defined here, are in the same range. It is enough to omit any check on the L4/L5 flag of Tables 1 and 2, when computing the distance between two Trojans, to find –in the list of distances sorted in increasing order– the spurious distances between an L4 object and an L5 one, together with the meaningful distances between objects in the same region. E.g., if I chose as cutoff $Q = 0.01$ ($\simeq 130\ m/s$), there are 13 couples at a lower distance in the L4 swarm, 6 in the L5 one, and 18 spurious couples formed with one in L4 and one in L5. Since the spurious couples have no possible physical interpretation, they are a measure of how many couples would be within such a distance by chance only: the number of chance occurrences in L4 and in L5 together should be statistically almost equal to the number of spurious ones, and indeed the numbers are close. I also would like to mention that a computation done with Poisson statistics, assuming a more or less uniform distribution, would predict much less than 18 spurious couples; this is a good example of the pitfalls of Poisson statistics when applied to a clumpy distribution. Therefore I do not use Poisson statistics at all, and I conclude from this test that the couples at distances $d_3 \leq 0.01$ are not statistically significant, and that at least 3 objects must belong to a cluster before it can be proposed as a family. I have not proven that 3 are enough, and I need to insist on the cautionary statement that I am listing "candidate", or "possible", families, to be confirmed with other independent information.

I need to stress, also for the purpose of comparing my work with that of Schoemaker et al. (1989; 1992), that the above statement does not imply that two Trojan asteroids with very close proper elements cannot be collisionally related; I am only saying that, with one exception (see below), there is no way to prove such a relationship on the basis of the statistical analysis of the dynamical information available. Two Trojans may well be fragments of the same parent body, but until we either find at least a third one near enough, or find some other evidence independent from the proper elements, we cannot prove it. Out of the many "Trojan couples" proposed by Schoemaker et al., some are here rejected (because my better computation of proper elements result in a larger distance), many are confirmed as couples but not as candidate families, and a few are confirmed as families because I have found at least one more neighbour; this is the case of Teucer in L4 and Sarpedon in L5.

The most striking result apparent from Table 4 is the couple *1583 Antilochus – 3801 Thrasimedes*; the actual distance is $d_3 = 0.0007$, which corresponds to less than $10\, m/s$. As a matter of fact, the differences in both e_P and $\sin I_P$ are of a few 10^{-4}; although the stability of the proper elements of *1583* and *3801* is very good, the difference is consistent with zero. The libration amplitude difference of $1.2°$ (corresponding to 0.006 AU in d) is significant and accounts for most of the distance. No other known Trojan is nearby (within $d_3 \leq 0.02$). Although I cannot use a Poisson statistics argument (not to contradict myself), it is clear that such a distance is unlikely to occur by chance (the nearest spurious L4/L5 couple is at a distance larger by a factor 3). The only conclusion I can draw so far is that this couple has some physical meaning and requires some dynamical explanation. The problem is, such a small relative velocity is hard to account even with a standard collisional model, taking into account that the escape velocity from *1583 Antilochus* is $\simeq 65\, m/s$. P. Farinella has proposed a non standard collisional model, involving binary asteroids; if we can confirm it, you will hear from us.

The other interesting results are the Menelaus family, which appears reliable, with between 4 and 8 members, depending from the cutoff (this family was also proposed in Schoemaker et al.), and the two less reliable but probably significant families of Teucer (4 to 6 members, but with visible chaining effect) and Sarpedon (3 members only, but very compact). The other six columns in Table 4 correspond to what I might call *clumps*, by using the Zappalà et al. (1992) terminology slightly out of context: some of them, but probably not all of them, could be real collisional families; I have not done enough work to be able to give a quantitative assessment of the probabilities.

These results may appear limited, and indeed they are; however, I need to stress that the problem of identifying families among the Trojans is intrinsically more difficult than in the main belt. The size, in the velocity space, of the Trojan belt, is much smaller than the size of the main belt; on the other hand from the collisional processes in the Trojan belt, the relative velocities being comparable, the volume in the velocity space occupied by a family should be comparable. The most extreme case of this difficulty is when a fragment from a catastrophic collision ends up outside the stable libration region, and experiences a close approach to Jupiter; since ejection velocities of $\simeq 200\ m/s$ for fragments are possible (as it can be seen in the main belt families), a Trojan family can have as a member an object in a cometary orbit, in a meteorite type orbit, and (more likely) in an hyperbolic orbit of escape from the solar system. Further problems could result from the internal dynamical structure of the Trojan belt. E.g. by comparing with Table 3, it is found that 5 of the asteroids appearing in Table 4 have chaotic orbits; since 3 of them display stable chaos (case (B) of Section 3), the other 2 only moderately unstable chaos (case (A)), the results on possible families still stand, although it is hard to predict what could happen over a time span much longer than 1 Myr.

The main conclusion I can draw from this initial search for Trojan families is that the number of Trojans with good enough orbits is marginal for a reliable detection of families. Not only I have been able to use only 84 Trojans in L4 and 90 in L5, but the size distribution is quite limited, because most of the orbits I have used have been discovered by chance while looking for main belt asteroids. The data from the Trojan surveys are mostly useless for the reasons discussed in Section 2, and this has not allowed to use the data on enough faint asteroids, which have a considerable probability of being fragments from the larger ones. Thus the most urgent task, to be able to go forward with the families of Trojans, is either the discovery of a significant number of faint Trojans or at least the recovery of many of the 97 lost objects, observed at a single opposition.

5. Open Problems

The availability of such a large set of data on the dynamics of Trojans solves some problems, and raises many more we were unable to see before. I would like to list the main ones:

(1) Figure 10 shows the distribution of "chaotic" orbits, defined here as positive identifications of Lyapounov exponents $\geq 6 \times 10^{-6}\ y^{-1}$, and "ordered" orbits, for which a positive LCE was not confirmed; they are marked with a + and an o respectively. Figure 10a, displaying libration amplitude D versus proper inclination I_P, definitely shows some pattern: chaos occurs more often for large libration amplitudes, and also more often (but not exclusively) at low inclination. The strength of the most intense perturbation by Jupiter increases with the libration amplitude; on the contrary the increase in inclination decreases the average perturbation, but not the maximum (even a very inclined orbit can have a close approach when the node is in the right position); thus the concentration of crosses and circles in different portions makes some sense, although one would like a quantitative

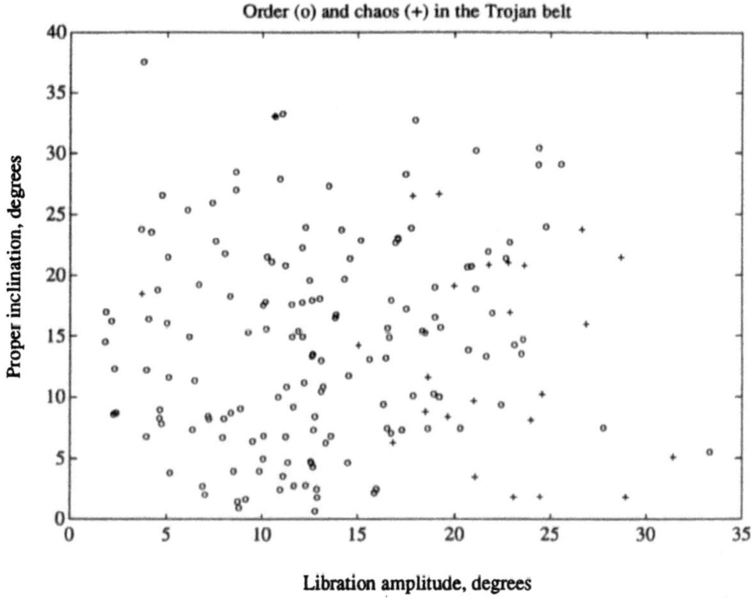

(a)

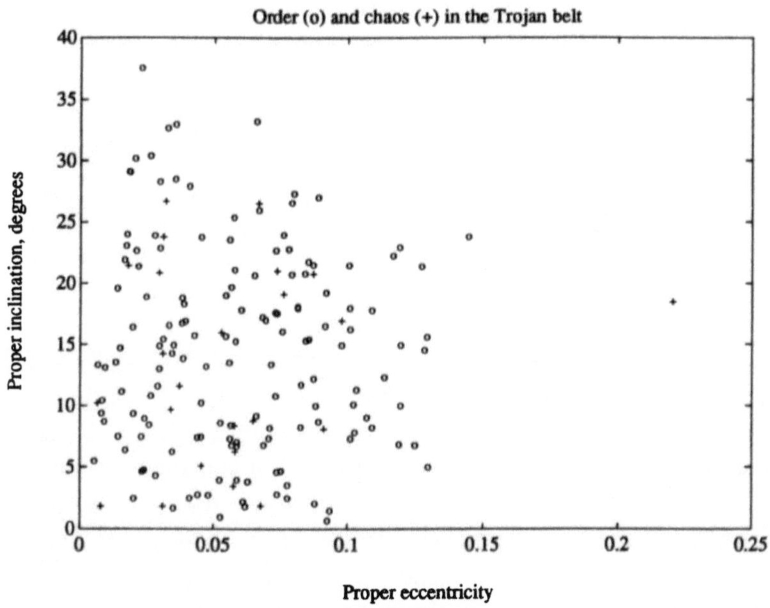

(b)

Fig. 10. Proper elements for all the 174 Trojans with well known orbits; the L4 and L5 swarms are superimposed. Circles indicate "regular" orbits, crosses are chaotic; the threshold for positive detection of chaos has been set at $6 \times 10^{-6}\, yr^{-1}$. (a) I_P and D (both in degrees). A large libration amplitude corresponds often, but by no means always, to chaotic orbits. (b) e_P and I_P (in degrees); apart the strange situation of *4835*, there is no obvious pattern.

model. However, there is no large region where all the orbits are chaotic. On the contrary I cannot see any special pattern in Figure 10b, which shows the proper eccentricity e_P versus proper inclination I_P. Then, why are some orbits chaotic, and others (with similar libration amplitude) either regular or at least with smaller LCE? Which are the resonances responsible for the chaos? As already pointed out in Section 3, it is not easy to find a resonance when the basic frequencies are so far apart; the distribution of the chaotic orbits in the frequency space (Figure 11) does not give any clue I can recognize. Of all the chaotic cases, I have an explanation in terms of resonances for only 3, namely *1173*, *6073* and *6031*, for which the $s - s_6$ resonance is involved.

(2) I have listed in Section 3 a total of 5 strongly unstable, 5 moderately unstable, and 14 stable chaotic orbits (again *1208* is excluded from this count because its chaos is much slower). Why are some chaotic orbits unstable, others stable? Of course when the resonance responsible is a major one, the corresponding chaotic layer is thicker, and, in a region of strong perturbations, resonance overlap can occur. However, excluding again the cases in the $s - s_6$ secular resonance, all the resonances involved should be of rather high order, given the ratios of the frequencies (Figure 11); why then is some of them more important than others? As an example, it is true that no Trojan has a ratio f/n smaller than $1/12$, but I cannot understand why $1/12$ is more important than $1/13$.

(3) Figure 10 also shows the overall shape of the Trojan region in the proper elements space (I have superimposed the L4 and the L5 swarms). Are the boundaries of that region stability boundaries? As an example, can we assume that no Trojan can have a larger libration amplitude than *6075*? And why there are no Trojans with $0.15 < e_P < 0.22$? Should we think that *4835* is a "comet", that is an object with a dynamical age significantly less than the age of the solar system? Although it is chaotic, it is only mildly so, much less than other moderate eccentricity Trojans. At least over 1 My, *4835* only shows a mild instability of the libration amplitude (which by the way could be the result of some long periodic effect); the hypothesis that it might be a temporary Trojan has no supporting evidence. Since another Trojan with comparable eccentricity has been found, it is difficult to invoke some exceptional protection mechanism. As a result of my computation of the orbit of *4835*, the problem of the high e boundary of the Trojan belt looks more difficult than before. An even more difficult problem is raised by the wild behaviour of *6031*; either there can be "comets" in the middle of the Trojan region, or a Trojan orbit can be unstable on a large scale such as in Figure 8 and still remain in the libration region for the age of the solar system: both hypothesis are somewhat paradoxical.

(4) The proper elements as defined in this paper are stable enough for many purposes, including the search for Trojan families. However, a proper element should be something more than an arbitrary function which changes little with

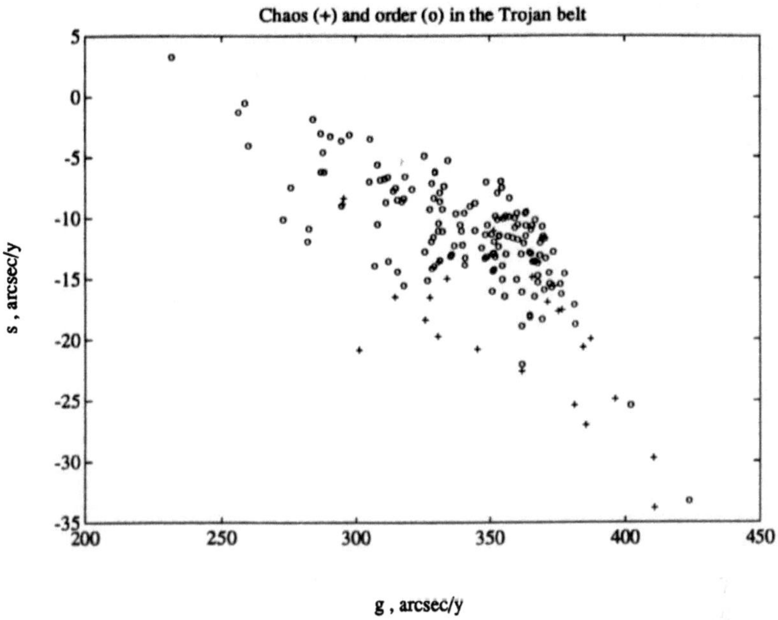

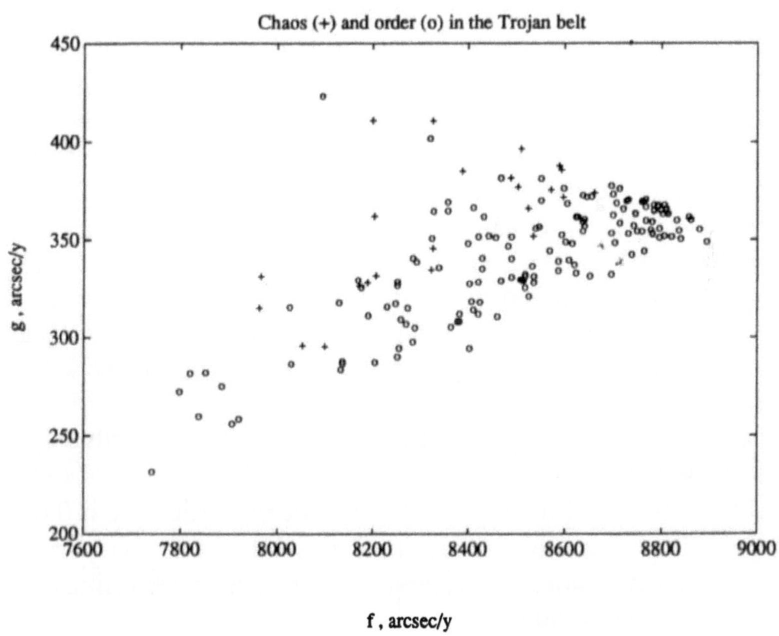

Fig. 11. The distribution of 174 Trojans in the frequency space; the units are $arcsec/yr$; crosses indicate positive detections of chaos. (a) perihelion frequency g versus node frequency s; more frequent chaos for shorter periods of the node, which occur for more strongly perturbed orbits. (b) libration frequency f versus g.

time over a particular solution. Even a large scale chaotic system can have a quasi–constant quantity, e.g. an average value of some quantity changing wildly with time but in a statistically regular way. The most chaotic systems, such as Anosov systems, are ergodic and have long term averages which tend to a limiting value, hence the averages over long time spans are quasi–integrals. A proper element should be the action variable of an approximating integrable system; that is, the set of proper elements should define a conditionally periodic orbit which approximates the exact solution for a very long time.

This is one reason why I did not follow the same procedure defined by Bien and Schubart (1987) to compute proper elements for Trojans. Bien and Schubart define the proper elements D and e_P in a way different, but not entirely so, from the one used here (warning: for Schubart and Bien, D is the difference between maximum and minimum of $\lambda - \lambda' - \chi$, that is twice the value reported in Tables 1 and 2). For them, the proper element is not the amplitude of a specific spectral line with a single frequency, but the average amplitude of the oscillation obtained by superimposing all the Fourier components with a frequency close to the proper one. E.g. the proper element d (in their notation, A) is the amplitude of the oscillation obtained by superimposing all the spectral lines shown in Figure 1. In my approach, the proper elements d and D are the amplitudes of a single Fourier component, only with an argument θ which is not linear in time. A similar difference occurs in the definition of e_P: I use the amplitude (in h, k) of the line with frequency g, Bien and Schubart use the superposition of all the spectral lines with frequencies close to g. The goal underlying my choice was to have a proper element which exactly defines an approximate solution, and is not only a statistical quantity. It is quite remarkable that the proper elements defined in a different way by Bien and Schubart and by me have almost the same values. The differences, for the 40 Trojans whose proper elements had been computed by Bien and Schubart, have a maximum of 0.92°, an rms of 0.33° in their D (printed in their Table 1 as an integer number of degrees). For e_P, a large discrepancy occurs for *1208* and *2594*, and this is understandable; for the others, the maximum difference is 0.0026 and the rms difference is 0.0007 (less than the value of the last digit printed by Bien and Schubart). For I_P, given by Bien and Schubart in degrees with only one digit after the decimal point, there are discrepancies up to 0.92° for *2594* and *1437*, for the others the maximum difference is 0.13° and the rms difference is 0.04°. Thus I have succeeded in giving a definition compatible with the one used previously, and which gives more information on the orbit of the Trojans, by referring explicitly to a conditionally periodic solution.

However, I am not very happy with the size of the residuals remaining after the removal of the forced and the proper modes, especially for h, k (see Figure 6). An unmodelled component with a typical rms of a few parts in 10^{-3} remains; this means that the theoretical goal of representing the solution as a sum of known harmonics has been achieved only roughly. For the inclination variables p, q a similar problem arises because I have not removed the forced terms (see Section

2); I have actually computed the forced terms, apart from s_8, but the accuracy is not very good. For the libration amplitude, the residuals are small but the use of the variable θ, whose time dependence is not known, also amounts to give up an explicit representation as a Fourier series in some known argument. I have to conclude that a real proper elements theory, which also implies an explicitly known approximate solution, has been achieved only with poor accuracy. The problem is how to obtain it, either with a synthetic method or otherwise.

(5) Secular resonances for Trojan orbits are not really understood, although quite a large numerical evidence has been accumulated (see e.g. Bien and Schubart 1986, Nakai and Kinoshita 1985). It would be helpful to be able to compute the width of resonances such as $s - s_6, s - s_7, s - s_8$ and $2s - 2g_5$, $2s - 2g_7$; it is somewhat of a paradox that the resonance widths appear (numerically) to be small, even for $s - s_6$, and then such a large scale instability as *6031* (Figure 8) can occur.

(6) To really understand the frequency of occurrence of Trojan families, we would need to develop a theory of the collisions between Trojans. Collision probabilities between main belt asteroids can be computed quite accurately (see e.g. Farinella and Davis, 1992), but the corresponding computations for Trojans are more difficult, because it is not possible to assume that the mean longitudes of two Trojans are not correlated. A new theory should be developed taking into account the model of the motion of the Trojans implied by the proper elements.

I am not listing the problems relative to the classification of Trojans into families, some of which have been mentioned in Section 4. I am also not sure I have really exhausted the list of open problems, but I believe there is enough work for us to do.

Acknowledgements

I thank the people who encouraged this work and gave some good advice: A. Barucci, E. Bowell, M. Carpino, P. Farinella, P. Goldreich, J. Henrard, A. Nobili, J. Schubart, G. Valsecchi, E. Zappalà. I did this research while on leave from the University of Pisa and visiting the Observatoire de Paris/Meudon; I thank the Observatoire de Meudon, and in particular B. Sicardy. The European Space Agency supported my stay in Meudon with the "G. Colombo" fellowship for 1991/92.

References

Bien, R. and Schubart, J.: 1986, "Two Trojans near a secular resonance: (1173) Anchises and (1871) Astyanax, in *Asteroids Comets Meteors II*, Lagerkvist C.–I. and Rickmann, H. eds., Uppsala University, pp. 157–160.

Bien, R. and Schubart, J.: 1987, "Three characteristic parameters for the Trojan group of asteroids", *Astron. Astrophys.* **175**, 292–298.

Carpino, M., Milani, A. and Nobili, A.M.: 1987. "Long–term numerical integrations and synthetic theories for the motion of the outer planets", *Astron. Astrophys.* **181**, 182–194.

Érdi, B.: 1988, "Long periodic perturbations of Trojan asteroids", *Celest. Mech.* **43**, 303–308.

Farinella, P. and Davis, D.: 1992, "Collision rates and impact velocities in the main asteroid belt", *Icarus*, in press.

Ferraz-Mello, S.: 1981, "Estimation of periods from unequally spaced observations", *Astron. J.* **86**, 619–624.

Milani, A., and Knežević, Z.: 1990, "Secular perturbation theory and computation of asteroid proper elements". *Celestial Mechanics* **49**, 247–411.

Milani, A., and Knežević, Z.: 1992, "Asteroid proper elements and secular resonances." *Icarus* **98**, 211–232.

Milani, A. and Nobili, A.M.: 1988, "Integration error over a very long time span" *Celestial Mechanics* **43**, 1–34.

Milani,A. and Nobili, A.M.: 1992, "An example of stable chaos in the Solar System" *Nature* **357**, 569–571.

Milani, A., Farinella, P. and Knežević, Z.: 1992,"On the search for asteroid families", in *Physics and Dynamics of Minor Bodies of the Solar System*, Benest, D. and Froeschlé, C. eds., Editions Frontière, in press.

Morbidelli, A.: 1991, "Perturbation methods and asteroid dynamics", Ph.D. thesis, Univ. Namur, Belgium.

Nakai, H. and Kinoshita, H.: 1985, "Secular perturbations of asteroids in secular resonance", *Celestial Mechanics* **36**, 391–407.

Nobili, A.M., Milani, A. and Carpino, M.: 1989, "Fundamental frequencies and small divisors in the orbits of the outer planets" *Astron. Astrophys.* **210**, 313–336.

Schoemaker, E. M., Schoemaker, C. S. and Wolfe, R.F.: 1992, "Systematic survey for bright Jupiter Trojans", in *Proceedings of ACM 91*, Bowell, E. ed., Lunar and Planetary Institute, in press.

Schoemaker, E. M., Schoemaker, C. S. and Wolfe, R.F.: 1989, "Trojan asteroids: population, dynamical structure and origin of the L4 and L5 swarms", in *Asteroids II*, Binzel. R. and Mattews, M. eds., Arizona University press, pp. 487–523.

Schubart, J. and Bien, R.: 1984, "An application of Labrouste's method to quasi–periodic asteroidal motion", *Celestial Mechanics* **34**, 443–452.

Schubart, J. and Bien, R.: 1987, "Trojan asteroids: relations between dynamical parameters", *Astron. Astrophys.* **175**, 299–302.

Van Houten, C.J., Van Houten–Groeneveld, I., Wisse-Schouten, M., Bardwell, C., Green, D.W.E., and Gehrels, T.: 1991, 'The second Palomar–Leiden Trojan survey", *Icarus* **91**, 326–333.

Williams, J.G.: 1969, "Secular perturbations in the Solar System" Ph.D. Thesis, Univ. California Los Angeles.

Zappalà, V., Cellino, A., Farinella, P. and Knežević, Z.: 1990, "Asteroid families I: identification by hierarchical clustering and reliability assessment", *Astron. J.* **100**, 2030–2046.

Zappalà, V., Cellino, A., Farinella, P. and Milani, A.: 1992, "Asteroid families II: extension to unumbered multi opposition asteroids", preprint.

TABLE 1: PROPER ELEMENTS AND FREQUENCIES

Numbered Trojan asteroids

No.	Name	d AU	D deg	f deg/y	e_P	g "/y	sin I_P	s "/y	L4 L5	M y
588	Achilles	.0314	6.45	2.434	.1032	344.44	.1967	-11.00	4	1
617	Patroclus	.0236	5.02	2.350	.1005	310.63	.3662	-6.76	5	1
624	Hektor	.0880	18.99	2.316	.0543	335.72	.3259	-12.95	4	1
659	Nestor	.0495	10.03	2.466	.1297	355.44	.0870	-16.40	4	1
884	Priamus	.0525	10.82	2.425	.0883	353.25	.1739	-12.31	5	1
911	Agamemnon	.0778	16.95	2.291	.0207	317.22	.3857	-8.59	4	1
1143	Odysseus	.0481	9.84	2.440	.0521	365.40	.0689	-10.90	4	1
1172	Aneas	.0481	10.15	2.366	.0602	331.45	.3056	-7.89	5	1
1173	Anchises	.1132	23.99	2.358	.0914	381.51	.1404	-25.40	5	1
1208	Troilus	.0468	10.63	2.200	.0354	258.55	.5446	-0.50	5	5
1404	Ajax	.0925	19.98	2.312	.0761	334.28	.3270	-15.00	4	1
1437	Diomedes	.1273	28.73	2.213	.0179	330.86	.3653	-19.74	4	1
1583	Antilochus	.1064	24.36	2.181	.0183	282.16	.4858	-10.84	4	1
1647	Menelaus	.0387	7.93	2.439	.0587	359.32	.1168	-9.96	4	1
1749	Telamon	.0660	13.61	2.422	.0686	365.95	.1185	-13.55	4	1
1867	Deiphobus	.0782	17.50	2.230	.0294	286.67	.4738	-6.22	5	1
1868	Thersites	.1060	22.88	2.313	.0979	345.35	.2906	-20.81	4	1
1869	Philoctetes	.1005	21.04	2.386	.0576	387.56	.0596	-19.97	4	1
1870	Glaukos	.0462	9.49	2.429	.0169	363.51	.1114	-9.48	5	1
1871	Astyanax	.1284	27.76	2.311	.0142	401.84	.1299	-25.37	5	1
1872	Helenos	.1091	23.55	2.313	.0148	364.69	.2538	-17.98	5	1
1873	Agenor	.0563	12.08	2.327	.1168	308.12	.3791	-10.48	5	1
2146	Stentor	.0162	3.76	2.150	.0226	231.51	.6093	3.28	4	5
2148	Epeios	.0224	4.61	2.430	.0239	354.56	.1553	-7.46	4	1
2207	Antenor	.0812	16.86	2.406	.0582	373.98	.1082	-15.61	5	1
2223	Sarpedon	.0655	13.88	2.358	.0379	340.41	.2882	-9.54	5	1
2241	79WM	.0596	12.59	2.365	.1008	329.03	.3081	-11.54	5	1
2260	Neoptolemus	.0192	4.03	2.385	.0195	334.39	.2828	-5.25	4	1
2357	Phereclos	.0337	6.88	2.448	.0480	363.27	.0478	-9.59	5	1
2363	Cebriones	.0784	17.99	2.177	.0323	259.73	.5397	-4.06	5	1
2456	Palamedes	.0785	16.63	2.358	.0293	351.53	.2566	-11.92	4	1
2594	78TB	.1433	31.42	2.278	.0454	410.95	.0879	-33.88	5	5
2674	Pandarus	.0345	7.01	2.461	.0879	360.41	.0361	-11.72	5	1
2759	Idomeneus	.0478	10.22	2.336	.0872	313.99	.3670	-7.76	4	1
2797	Teucer	.1039	22.81	2.275	.0737	327.72	.3579	-16.54	4	1
2893	Peiroos	.0781	16.44	2.372	.0471	356.11	.2287	-12.92	5	1
2895	Memnon	.0339	7.38	2.293	.0666	294.43	.4377	-3.64	5	1
2920	Automedon	.0986	21.69	2.271	.0164	325.69	.3733	-12.71	4	1
3063	Makhaon	.0605	12.64	2.389	.0557	349.14	.2339	-10.55	4	1
3240	Laocoon	.0541	11.07	2.444	.0775	365.30	.0622	-12.91	5	1
3317	Paris	.0216	4.71	2.292	.0790	290.29	.4475	-3.28	5	1

No.	Name	d AU	D deg	f deg/y	e_P	g "/y	sin I_P	s "/y	L4 L5	M y
3391	Sinon	.0903	19.28	2.339	.0426	351.56	.2711	-14.18	4	1
3451	84HA1	.1181	26.66	2.212	.0309	314.77	.4027	-16.52	5	1
3540	Protesilaos	.0786	17.10	2.297	.1192	306.90	.3898	-13.88	4	1
3548	Eurybates	.0890	18.62	2.388	.0438	376.54	.1290	-16.25	4	1
3564	Talthybius	.0444	9.26	2.394	.0844	337.31	.2645	-9.62	4	1
3596	Meriones	.0809	17.77	2.275	.0275	311.14	.4056	-8.68	4	1
3708	74FV1	.0087	1.79	1.802	.1283	330.78	.2511	-11.01	5	1
3709	Polypoites	.0584	12.46	2.339	.0138	328.30	.3353	-7.12	4	1
3793	Leonteus	.0489	10.47	2.335	.0576	318.40	.3603	-6.58	4	1
3794	Sthenelos	.0194	3.91	2.470	.1250	348.95	.1177	-13.12	4	1
3801	Thrasymedes	.1112	25.56	2.172	.0181	281.74	.4861	-11.89	4	1
4007	Euryalos	.0585	12.17	2.400	.0154	357.13	.1937	-9.85	4	1
4035	86WD	.0742	15.61	2.374	.0092	356.71	.2268	-11.46	4	1
4057	85TQ	.0599	12.26	2.440	.0736	367.80	.0487	-13.65	4	1
4060	Deipylos	.0490	10.19	2.403	.1292	331.51	.2690	-13.44	4	1
4063	Euforbo	.0477	10.04	2.371	.0735	331.42	.3011	-8.58	4	1
4068	Menestheus	.0821	17.52	2.341	.0680	340.79	.2963	-13.25	4	1
4086	85VK2	.0955	20.82	2.292	.0789	326.60	.3541	-15.10	4	1
4138	Kalchas	.0567	11.64	2.434	.0439	369.46	.0486	-11.66	4	1
4348	Poulydamas	.0408	8.36	2.439	.0893	353.43	.1516	-11.45	5	1
4489	88AK	.0994	21.77	2.280	.0293	331.34	.3551	-13.62	4	1
4501	Eurypylos	.0781	16.33	2.390	.0079	368.69	.1631	-13.04	4	1
4543	Phoinix	.1227	26.88	2.279	.0529	361.87	.2746	-22.61	4	1
4707	Khryses	.0392	7.99	2.450	.1091	351.63	.1429	-12.85	5	1
4708	Polydoros	.0937	19.65	2.381	.0578	375.54	.1450	-17.71	5	1
4709	Ennomos	.0862	19.23	2.237	.0319	295.53	.4484	-8.39	5	1
4715	89TS1	.0970	21.05	2.301	.0244	340.64	.3234	-13.82	5	1
4722	Agelaos	.0231	4.69	2.455	.1027	350.63	.1362	-11.29	5	1
4754	Panthoos	.0885	18.67	2.368	.0372	365.96	.2004	-14.92	5	1
4791	Iphidamas	.0610	13.47	2.260	.0797	287.89	.4591	-6.21	5	1
4792	Lykaon	.0617	12.75	2.416	.0563	362.61	.1461	-12.04	5	1
4805	Asteropaios	.0604	12.60	2.393	.0714	348.25	.2315	-11.34	5	1
4827	Dares	.0886	18.54	2.388	.0648	371.57	.1522	-16.95	5	1
4828	Misenus	.0566	11.85	2.385	.0856	339.28	.2660	-11.02	5	1
4829	Sergestius	.1054	22.39	2.352	.0198	381.70	.1625	-18.73	5	1
4832	Palinurus	.0573	12.06	2.371	.1090	328.13	.3054	-11.90	5	1
4833	Meges	.0485	11.03	2.196	.0656	256.04	.5477	-1.26	4	5
4834	Thoas	.0393	8.62	2.279	.0890	287.42	.4545	-4.60	4	1
4835	89BQ	.0231	3.71	1.500	.2207	301.21	.3158	-20.88	4	1
4836	Medon	.0614	13.00	2.358	.0813	331.02	.3103	-10.40	4	1
4867	89SZ	.0493	10.90	2.260	.0406	286.63	.4678	-3.04	5	1
4902	89AN2	.0115	2.36	2.433	.0090	354.27	.1515	-6.95	4	1
4946	88BW1	.1075	23.65	2.270	.0875	326.08	.3542	-18.39	4	1
5012	9507PL	.0613	12.57	2.435	.0735	366.91	.0803	-13.49	4	1

TABLE 2: PROPER ELEMENTS AND FREQUENCIES

Unnumbered, multi opposition Trojan asteroids

No.	Name	d AU	D deg	f deg/y	e_P	g "/y	sin I_P	s "/y	L4 L5	M y
6001	73SY	.0715	15.05	2.371	.0309	351.60	.2455	-11.06	4	1
6002	73SD1	.0830	17.29	2.399	.0560	372.99	.1270	-15.69	4	1
6003	73SH1	.0532	10.93	2.433	.0199	369.57	.0429	-10.68	4	1
6004	73SM1	.1090	23.46	2.321	.0133	369.37	.2345	-18.31	4	1
6005	73SO1	.0225	4.59	2.446	.0825	352.11	.1437	-9.83	4	1
6006	73SQ1	.0545	11.20	2.429	.0566	363.33	.1180	-11.44	4	1
6007	73SR1	.0349	7.18	2.428	.0256	357.37	.1464	-8.33	4	1
6008	73SW1	.0900	18.93	2.375	.0451	370.12	.1777	-15.90	4	1
6009	73SA2	.0701	14.48	2.416	.0230	373.55	.0811	-12.74	4	1
6010	85TG3	.0624	13.06	2.387	.0294	352.74	.2250	-10.10	4	1
6011	86TR6	.1010	21.58	2.336	.0064	366.61	.2308	-16.43	4	1
6012	86TS6	.0246	5.08	2.417	.0288	348.67	.2014	-7.04	4	1
6013	86VG1	.0353	7.56	2.328	.0778	308.01	.3879	-5.60	4	1
6014	87YU1	.0376	8.04	2.339	.0853	311.97	.3715	-6.61	4	1
6015	88BX1	.0924	21.06	2.190	.0203	275.33	.5030	-7.47	4	1
6016	88BY1	.0524	11.18	2.340	.0840	318.01	.3551	-8.37	4	1
6017	88QY	.1058	24.38	2.166	.0260	272.58	.5062	-10.10	5	1
6018	88RO	.0239	4.99	2.395	.0755	333.02	.2770	-7.38	5	1
6019	88RA1	.0086	1.84	2.363	.0692	329.55	.2922	-6.24	5	1
6020	88RF1	.1033	22.85	2.258	.0730	317.77	.3865	-15.52	5	1
6021	88RG1	.0782	17.08	2.286	.0170	315.49	.3924	-8.45	5	1
6022	88RG10	.0615	12.65	2.425	.0281	370.54	.0749	-11.68	5	1
6023	88RL10	.0628	12.85	2.442	.0776	367.80	.0433	-14.74	5	1
6024	88RN10	.0780	16.53	2.356	.0543	346.92	.2701	-12.41	5	1
6025	88RO10	.0697	15.15	2.298	.0296	314.86	.3890	-7.49	5	1
6026	88RR10	.0887	18.99	2.333	.0329	348.14	.2851	-13.26	5	1
6027	88RS10	.0627	12.87	2.435	.0617	370.75	.0315	-13.27	5	1
6028	88RY10	.0431	8.80	2.447	.0525	365.81	.0170	-10.62	5	1
6029	88RH11	.0768	15.86	2.420	.0610	376.24	.0385	-15.41	5	1
6030	88RM11	.0773	15.98	2.415	.0408	377.80	.0435	-14.58	5	1
6031	88RN11	.1102	23.08	2.387	.0309	385.58	.0308	-27.06	5	5
6032	88RY11	.0564	12.24	2.302	.0758	304.90	.4065	-6.99	5	1
6033	88RD12	.0608	12.53	2.424	.0238	369.93	.0830	-11.46	5	1
6034	88RE12	.0579	12.09	2.391	.0978	339.86	.2580	-12.19	5	1
6035	88RH12	.0433	8.86	2.443	.1071	351.09	.1575	-12.91	5	1
6036	88RP12	.0253	5.15	2.453	.0627	360.07	.0667	-9.58	5	1
6037	88RS12	.0553	11.31	2.440	.0751	364.80	.0815	-12.79	5	1
6038	88RT12	.0494	10.05	2.454	.1190	354.72	.1189	-15.05	5	1
6039	88RH13	.0870	18.49	2.349	.0581	351.11	.2629	-14.36	5	1
6040	88RL13	.0549	11.52	2.380	.0348	344.46	.2579	-8.73	5	1
6041	88SW1	.0656	13.83	2.370	.0917	336.70	.2845	-12.25	5	1
6042	88SK2	.0430	8.72	2.460	.0935	361.72	.0256	-12.92	5	1
6043	88SP2	.0109	2.27	2.303	.1137	338.91	.2132	-10.55	5	1

No.	Name	d AU	D deg	f deg/y	e_P	g "/y	sin I_P	s "/y	L4 L5	M y
6044	88SA3	.0785	16.74	2.341	.0812	335.25	.3079	-13.09	5	1
6045	88SG3	.0394	8.32	2.364	.0385	329.52	.3135	-6.19	5	1
6046	88SJ3	.0668	14.30	2.334	.0564	327.56	.3371	-9.26	5	1
6047	88SL3	.0193	4.15	2.323	.0557	305.29	.4000	-3.47	5	1
6048	88TH1	.0543	11.24	2.415	.0730	353.56	.1879	-11.40	5	1
6049	88TZ1	.0806	17.88	2.250	.0667	295.21	.4456	-8.85	5	1
6050	88TA3	.0857	17.85	2.396	.1021	361.73	.1755	-18.86	5	1
6051	89AU1	.0806	16.74	2.404	.0586	372.21	.1226	-15.38	4	1
6052	89AV2	.0947	20.63	2.292	.0650	328.35	.3531	-14.12	4	1
6053	89BL	.0310	6.34	2.443	.0706	355.93	.1278	-9.84	4	1
6054	89BW	.0099	2.12	2.352	.1010	329.05	.2798	-8.33	4	1
6055	89BB1	.0299	6.18	2.415	.1196	332.31	.2580	-11.06	4	1
6056	89CW1	.1146	24.57	2.330	.0066	384.78	.1767	-20.68	4	1
6057	89CH2	.0391	8.63	2.259	.0353	283.67	.4770	-1.85	4	5
6058	89CK2	.0213	4.49	2.366	.0379	325.59	.3225	-4.87	4	1
6059	89DJ	.1028	22.63	2.269	.0215	329.20	.3649	-13.92	4	1
6060	89EO11	.1105	24.76	2.229	.0173	315.46	.4068	-14.39	4	1
6061	89SC7	.1015	21.92	2.312	.0392	350.79	.2904	-15.99	5	1
6062	89TS2	.0627	13.06	2.400	.0082	360.45	.1812	-10.50	5	1
6063	89TU5	.0625	12.76	2.446	.0926	367.78	.0121	-15.29	5	1
6064	89TO11	.0970	20.68	2.342	.0384	361.73	.2396	-16.06	5	1
6065	89UC5	.0920	19.20	2.395	.1197	361.82	.1734	-21.99	5	1
6066	89UO5	.0861	18.34	2.345	.0308	351.99	.2658	-13.12	5	1
6067	89UX5	.1340	28.96	2.313	.0079	410.59	.0310	-29.76	5	5
6068	90DK	.0108	2.22	2.439	.0524	352.86	.1498	-7.90	4	1
6069	90TV12	.0279	6.06	2.301	.0573	297.58	.4285	-3.15	5	1
6070	90VU1	.0680	14.59	2.328	.1273	312.01	.3647	-13.50	5	1
6071	91EL	.0171	3.66	2.334	.1445	294.44	.4036	-8.97	4	1
6072	91GX1	.0633	13.18	2.399	.0263	359.28	.1876	-10.73	4	1
6073	4523PL	.1158	24.47	2.364	.0678	396.39	.0314	-24.91	4	1
6074	6541PL	.0964	20.26	2.375	.0227	381.36	.1295	-17.13	4	1
6075	6581PL	.1499	33.30	2.248	.0052	423.39	.0960	-33.26	4	1
6076	6591PL	.0794	16.52	2.401	.0452	371.97	.1296	-14.50	4	1
6077	9602PL	.0645	13.32	2.418	.0343	368.78	.1088	-12.00	4	1
6078	5030T2	.0191	3.92	2.427	.0874	342.67	.2118	-9.01	4	1
6079	5187T2	.0993	20.98	2.362	.0340	376.99	.1669	-17.60	4	1
6080	5493T2	.1073	23.08	2.321	.0341	364.87	.2463	-18.13	4	1
6081	2035T3	.0618	12.67	2.435	.1011	359.93	.1269	-15.05	5	1
6082	3104T3	.0317	6.68	2.368	.0919	321.07	.3295	-7.60	5	1
6083	3108T3	.0447	9.14	2.442	.0348	366.98	.0290	-10.14	5	1
6084	4035T3	.0545	11.50	2.366	.0729	332.32	.3023	-9.25	5	1
6085	4101T3	.0650	14.15	2.294	.0450	309.13	.4031	-6.87	5	1
6086	4179T3	.0417	8.51	2.445	.0587	363.29	.0693	-10.56	5	1
6087	4317T3	.0561	11.59	2.420	.0658	358.56	.1594	-11.61	5	1
6088	4369T3	.0352	7.21	2.438	.0709	355.19	.1425	-10.00	5	1
6089	5191T3	.0698	14.53	2.399	.0827	354.63	.2036	-13.87	5	1

TABLE 3: STABILITY OF PROPER ELEMENTS

No.	Name	Δd AU	ΔD deg	Δf deg/y	Δe_P	Δg "/y	$\Delta sin I_P$	Δs "/y	LCE $1/(10^5 y)$
588	Achilles	.00001	.002	.0010	.00025	.025	.00061	.025	0.3
884	Priamus	8	17	6	362	246	144	292	0.2
1173	Anchises	17	33	4	30	652	975	59	0.9
1208	Troilus	11	28	8	333	397	896	14	0.15
1404	Ajax	52	112	1	30	103	86	131	0.9
1437	Diomedes	204	519	57	168	264	237	593	1.7
1868	Thersites	52	125	17	46	92	33	144	1.1
1869	Philoctetes	24	52	7	15	123	160	63	1.5
2146	Stentor	49	115	4	507	198	136	4	0.06
2207	Antenor	12	29	8	44	45	49	40	0.6
2363	Cebriones	6	14	2	44	511	483	46	0.4
2594	78TB	50	133	25	1486	500	913	324	2.90
2797	Teucer	10	23	8	20	54	39	38	0.7
3317	Paris	3	7	3	176	406	252	38	0.2
3451	84HA1	131	326	27	73	129	257	420	1.9
3540	Protesilaos	70	165	17	31	149	63	116	0.4
4057	85TQ	1	11	35	574	843	66	532	0.0
4489	88AK	33	81	13	9	108	37	77	1.1
4543	Phoinix	45	125	19	35	119	55	128	1.1
4708	Polydoros	67	155	18	12	369	81	147	0.8
4709	Ennomos	95	221	14	36	31	410	105	0.7
4754	Panthoos	48	110	15	16	273	81	112	0.8
4827	Dares	46	108	15	13	264	110	95	1.0
4833	Meges	32	75	3	155	78	379	23	0.08
4834	Thoas	6	11	11	165	162	289	16	0.3
4835	89BQ	191	416	2317	43	177	48	45	0.6
4867	89SZ	15	38	14	100	174	305	5	0.4
4946	88BW1	166	397	32	52	304	150	432	0.7
6001	73SY	22	44	7	7	81	83	46	0.6
6023	88RL10	15	14	49	819	1.328	133	781	0.2
6031	88RN11	176	382	57	8658	2.882	3612	731	1.99
6044	88SA3	82	174	2	10	187	29	170	0.3
6047	88SL3	5	2	5	79	293	325	32	0.3
6049	88TZ1	13	33	8	41	16	51	27	0.7
6050	88TA3	18	29	16	297	431	91	321	0.3
6056	89CW1	12	36	6	5	121	70	55	0.9
6057	89CH2	9	21	3	134	165	99	16	0.08
6067	89UX5	140	352	39	3539	1.117	1159	500	6.38
6071	91EL	5	18	10	234	256	217	216	0.3
6073	4523PL	120	283	20	255	718	1169	640	2.7
6079	5187T2	34	71	5	11	232	90	113	1.2

TABLE 4: POSSIBLE TROJAN FAMILIES

L4 region (Greeks)

Family	Antilochus	Menelaus	Teucer	Diomedes	4035	Palamedes
1.6		1749 3548 6053 6076	3793 6052			2456
1.4						
1.2				2920	6004	
1.0			2797 4086		4035 6011	
0.8		6002 6051	4946 6016			
0.6				1437 6059		6001 6080
0.4		1647 6006				
0.2	1583 3801					

L5 region (Trojans)

Family	Sarpedon	88RA1	Polydoros	Kryses
1.6		6019 6044	4827	
1.4				4707 4722 6081
1.2				
1.0	6026	6084		
0.8				
0.6	2223 6061		4708 4792	

LONG-TERM EVOLUTION OF ASTEROIDS NEAR A RESONANCE

JOHN D. HADJIDEMETRIOU AND G. VOYATZIS
Department of Theoretical Mechanics
University of Thessaloniki, GR-540 06 Thessaloniki, Greece

We report here recent work on the long term evolution of an asteroid near the 3:1 resonance with Jupiter, for planar motion, by making use of a mapping model (Hadjidemetriou, 1993). Some additional results are also included. The mapping is four dimensional and is a model for the Poincaré mapping, on a surface of section, of the elliptic restricted three body problem, near the 3:1 resonance, defined at the times $t_n = nT$, where T is the period of Jupiter in its orbit. The mapping is equivalent to a symplectic integrator, obtained from the generating function

$$\begin{aligned} W &= \sigma_n S_{n+1} + \nu_n N_{n+1} + T[H_0(S_{n+1}, N_{n+1}) + \\ &\quad \mu H_1(S_{n+1}, \sigma_n, N_{n+1}) + \mu e' H_2(S_{n+1}, \sigma_n, N_{n+1}, \nu_n)]. \end{aligned} \qquad (1)$$

through the equations

$$\sigma_{n+1} = \partial W/\partial S_{n+1}, S_n = \partial W/\partial \sigma_n, \nu_{n+1} = \partial W/\partial N_{n+1}, N_n = \partial W/\partial \nu_n. \qquad (2)$$

The function $H_0 + \mu H_1 + \mu e' H_2$ is the usual averaged Hamiltonian valid near the 3:1 resonance, where correction terms have been added to H_1 and H_2, in order to include the higher order resonances missing in the original form and $\mu = 0.00095387535,\ e' = 0.048$.

For $e' = 0$ we have N =*constant* and the equations (1),(2) for S, σ are decoupled from the other two. The mapping is now two dimensional, and the asteroid is trapped in a fixed plane N =*constant*, as shown in Fig.1a. The evolution of the eccentricity undergoes oscillations with small amplitude and looks regular. This picture changes completely near the resonance region when $e' \neq 0$. Then N is variable and the motion from one N =*constant* plane to the next is now possible. If we start near a critical curve of the $e' = 0$ case, chaotic motion develops, as

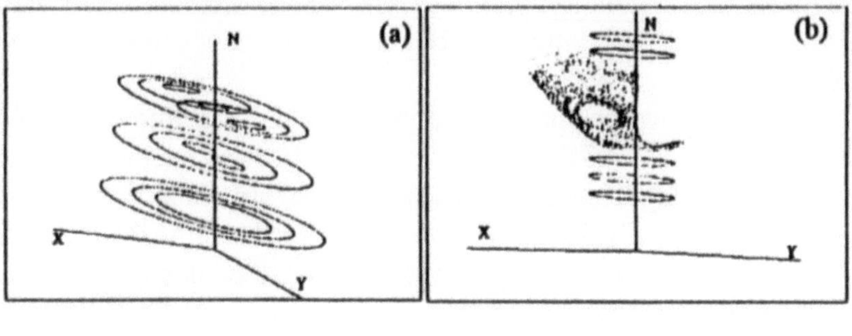

Fig. 1. The evolution in the three dimensional space X, Y, N (a)$e' = 0$, invariant curves for three different N levels. (b)$e' = 0.048$, initial conditions $S = 0.0116$, $\sigma = 0$, $\nu = \pi/2$ and $N = 1.350,\ 1.365,\ 1.380,\ 1.400$ (chaotic), $1.480,\ 1.485$

Celestial Mechanics and Dynamical Astronomy **57**: 95–96, 1993.

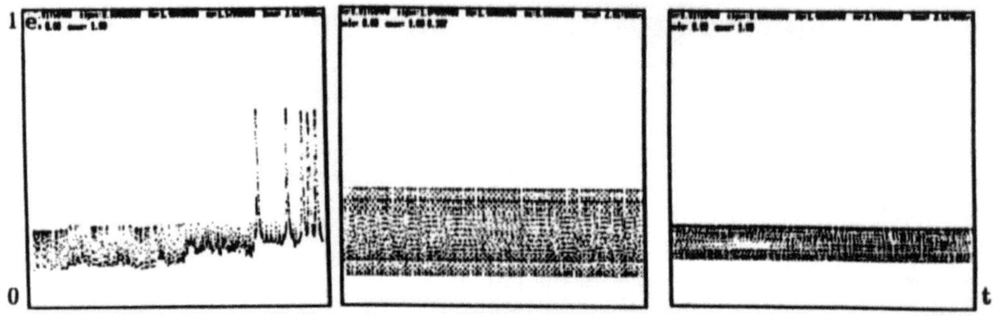

Fig. 2. Evolution of the eccentricity near the 3:1 resonance for $S = 0.0116,\ N = 1.4$ and: (a) $\sigma = 0,\ \nu = \pi/2$, (b) $\sigma = 1,\ \nu = 0$ and (c) $\sigma = 0,\ \nu = \pi$

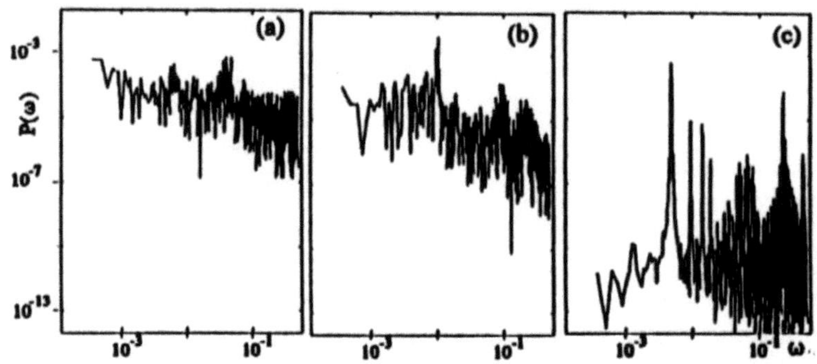

Fig. 3. The power spectrum of the time evolution of the eccentricity

shown in Fig.1b. In this latter case the evolution of the eccentricity of the asteroid with respect to the time is shown in Fig. 2a: The eccentricity remains small for a long time interval (in the present case it is 3×10^6 years) and then it jumps to high values, due to chaotic motion. This is not however the only type of evolution. The evolution depends on the phase also (i.e. on the initial values of σ and ν) and we may have oscillations of the eccentricity with amplitude either higher than 0.3 (the critical value for the asteroid to become Mars crosser), as in Fig.2b, or with small amplitude as in Fig.2c. This latter case however corresponds to a small region of phase space.

In Figs.3a,b we show the spectral analysis (FFT) of the time series of Fig.2a, for a short time interval (Fig3a) and for a long time interval, including the jump in the eccentricity (Fig.3b). They both show the typical structure of chaotic motion, i.e. the power spectrum density does not converge to zero. It is interesting to note that the chaotic property is clearly present well before the jump in the eccentricity. In Fig.3c we present the power spectrum of the time series of Fig.2c. We note that it goes exponentially to zero as the frequency goes to zero.

References

Hadjidemetriou, J.D.:1993, *Asteroid motion near the 3:1 resonance*, Celest. Mech. (to appear)

THE STABILITY OF THE PLANETARY THREE-BODY PROBLEM : INFLUENCE OF THE SECULAR RESONANCES

PHILIPPE ROBUTEL
Astronomie et Systèmes Dynamiques, Bureau des Longitudes, 77 Avenue Denfert-Rochereau, 75014 Paris, France

Key words: Planetary Theory - K.A.M. Theory - Resonances - Stability - Frequency Analysis

At first sight, K.A.M. theory seems to be very useful for obtaining results of global stability in planetary problems. In fact, it describes the topology of nondegenerate quasi–integrable elliptical problems and provides result of stability for all time. In 1963 Arnold (Arnold, 1963) demonstrates a theorem of conservation of invariant tori in degenerated problems, that he apply to the planar planetary three-body problem to prove its stability. This result of stability was extended in 1992 to the spatial case (Robutel, 1992). But the proof of these results is only valid for very small values of the planetary masses, eccentricities and inclinations, which are largely inferior to these met in our solar system. In order to overcome this difficulty, we are going to use Laskar's numerical method of frequency analysis (Laskar, 1992a) which enable us to have a global vision of the secular phase space for two planets in the neighbouring conditions of those of Jupiter and Saturn. The study of such a problem, which is more realistic than those based upon the K.A.M. theory, will lead us to give results of stability available in finite but very long time.

To compute the hamiltonian of the problem, we chose canonical heliocentric formalism (Laskar, 1992b) which enable us to make the reduction of the center of mass. This reduction leads to the study of the motion of two planets around a fixed sun. As the angular momentum is a constant of the motion, we can eliminate both the longitude of the ascending nodes and the inclination. These two consecutive reduction lead to a more simple problem with only four degrees of freedom.

In order to compute the secular hamiltonian, we expand it in Taylor series up to order six in Poincaré variables. Then, the Lindstedt–Poincaré normalization methode is applied up to order two in the planetary masses, due to the existence of small denominators because of the vicinity of the (2:-5) mean motion resonance. This provides us with a quasi–integrable hamiltonian system with two degrees of freedom. A last transformation consists in the diagonalization of the quadratic part of the secular hamiltonian. We finally obtain the following expression of the hamiltonian :

$$H(x,\bar{x};\alpha) = \nu_1(\alpha)|x_1|^2 + \nu_2(\alpha)|x_2|^2 + H_4(x,\bar{x};\alpha) + H_6(x,\bar{x};\alpha)$$

where α is the ratio of the semi–major axis, x_j and $\bar{x}_j$ are the images of the mean Poincaré canonical variables by the previous linear map, H_j are homogeneous polynomials of degree j in $(x,\bar{x})$ which coefficients depend on α, and the purely

Celestial Mechanics and Dynamical Astronomy **57**: 97–98, 1993.

imaginary quantities ν_j are the two eigenfrequencies of the linearized secular problem.

Because of the denominators, the quantities H_4 and especially, H_6 tend to infinity when α reaches the value α_c which corresponds to the separatrix of the (2:-5) mean motion resonance. This reflect the fact that the process of normalization is not convergent in the libration area of the $2\lambda_1 - 5\lambda_2$ argument. On the contrary, if the system stays in the circulation area, the secular hamiltonian of order two provides a good approximation of the real average motion. Thus, if α stays in the interval defined by its actual value α_a and a maximal value α_m lower than α_c, the quantities H_4 and H_6 increase as α approaches α_m, which increases the width of the chaotic areas existing in the neighbourhood of the secular resonances.

For these reasons, several phase space have been studied for different values of the parameter α, from the actual value of α to its maximal one α_m.

The frequency analysis reveals that all the trajectories appear as really regular, at least over a time span of 40 Mya, except perhaps for $\alpha = \alpha_m$ in the neighbourhood of the (1:-3) secular resonance where the methode frequency analysis indicates the destruction of tori. Indeed, the phase space contains here the alternation of hyperbolic points and very small islands of libration around elliptic points. But the width of these islands is so small (lower than 10^{-4}) that the expected chaotic regio neighbouring hyperbolic points is of negligible size, an then have no significant dynamical effect.

Therefore, even if the K.A.M. theory is not directly applicable to this problem, its stability is insure for very long time span in almost all the phase space. Then, we can conclude that, the secular resonances are not generators of unstability in the secular problem of two planets like Jupiter and Saturn.

References

Arnold, V.: 1963, 'Small denominators', *Rus. Math. Surv.* **18**, 85–118.

Laskar, J.: 1992a, 'Analytical framework in Poincaré variables for the motion of the solar system', *Long-term dynamic behaviour of natural and artificial N-body systems* A.Roy (ed), Kluwer Publ. 93–114.

Laskar, J. 1992b, 'Frequency Analysis for Multi–Dimensional Systems. Global Dynamics and Diffusion', *submitted to Physica D*.

Robutel, P.:1992, In preparation.

THE MAIN MEAN MOTION COMMENSURABILITIES IN THE PLANAR CIRCULAR AND ELLIPTIC PROBLEM

MICHÈLE MOONS AND ALESSANDRO MORBIDELLI

Département de mathématique FUNDP
8, Rempart de la Vierge, B-5000 Namur, Belgique

Abstract. We study the motion of asteroids in the main mean motion commensurabilities in the frame of the planar restricted three–body problem. No assumption is made about the size of the eccentricity of the asteroid. At small to moderate eccentricity, we recover existing results (shape of the phase space and location of secondary resonances). We also provide global pictures of the dynamics in the region of secondary resonances. At high eccentricity, the phase space portraits of the integrable part of the Hamiltonian show new families of stable orbits for the 3:2 and 2:1 cases and the secular resonances ν_5 and ν_6 are located.

Key words: mean motion resonances, secondary resonances, secular resonances, planar circular and elliptic restricted problem, action–angle variables, surfaces of section

1. Introduction

The motion of asteroids in a mean motion commensurability with Jupiter has been studied many times these last years and extensive references to that subject can be found in Yoshikawa (1989) and in Henrard (1988). In this work, we revisit the dynamics in the main mean motion commensurabilities (3:2, 2:1, 7:3, 5:2, 3:1 and 4:1) in the frame of the planar restricted three–body problem Sun–Jupiter–Asteroid.

Our perturbation method makes use of Arnold action–angle variables, which we introduce and manipulate in a semi–numerical way, following Henrard (1990). Moreover, we compute the averaged Hamiltonian and its derivatives in closed form, namely avoiding expansions in the eccentricity of the asteroid. The variables we introduce are non singular at $e = 0$ (following Ferraz–Mello and Sato, 1989). In this way we are able to recover the existing results on the location of the main secondary resonances at moderate eccentricity (Lemaître and Henrard, 1990), as well as to locate the secular resonances ν_5 and ν_6 at large eccentricity; these play a fundamental role as soon as the perturbation of Saturn on Jupiter's orbit is taken into account and they are the subject of a specific paper (Morbidelli and Moons, 1993). Furthermore, we compute Poincaré sections of the averaged elliptic problem (in closed form with respect to e and e'), in order to investigate the chaotic zone originated by the secondary resonances at moderate eccentricity. The existence of this (bounded) chaotic layer has been indicated first by Murray (1986), using a low–order truncated model, and later by Wisdom (1987), computing Lyapunov exponents of pure numerical integrations in the planar elliptic problem. Here, we provide global pictures of the dynamics in this region.

Celestial Mechanics and Dynamical Astronomy **57**: 99–108, 1993.

2. The Hamiltonian

To study the mean motion resonances between asteroids and Jupiter, we take the well–known Hamiltonian of the restricted three–body problem (see, for instance, Szebehely, 1967)

$$\mathcal{H} = L' - \frac{1-\mu}{2a} - \mu\left(\frac{1}{|\mathbf{r}-\mathbf{r}'|} - \frac{\mathbf{r}|\mathbf{r}'}{r'^3}\right) \tag{1}$$

and introduce canonical variables appropriate to the resonance involved. For the $(p+q)/p$ resonance, these variables are (Schubart, 1968)

$$\begin{aligned} \sigma &= \frac{p+q}{q}\lambda' - \frac{p}{q}\lambda - \varpi \quad , & S &= L - G \\ -\nu &= \frac{p+q}{q}\lambda' - \frac{p}{q}\lambda - \varpi' \quad , & N &= \frac{p+q}{p}L - G \\ \lambda' & \quad , & \Lambda' &= L' - \frac{p+q}{p}L \end{aligned} \tag{2}$$

with $L = \sqrt{(1-\mu)a}$ and $G = L\sqrt{1-e^2}$ and the usual notations λ (resp. λ'), ϖ (resp. ϖ'), a (resp. a') and e (resp. e') for the mean longitude, the longitude of pericenter, the semi–major axis and the eccentricity of the asteroid (resp. Jupiter). We choose as units the universal gravitational constant, the total mass of the Sun–Jupiter system and the semi–major axis of Jupiter. The Hamitonian is averaged with respect to the fast frequencies and takes the form

$$H(\sigma, S, \nu, N) = -\frac{1-\mu}{2a} - \frac{p+q}{p}\sqrt{(1-\mu)a} - \mu\left(\frac{1}{|\mathbf{r}-\mathbf{r}'|} - \frac{\mathbf{r}|\mathbf{r}'}{r'^3}\right) \tag{3}$$

This Hamiltonian is composed of an integrable part $H_0(\sigma, S, N)$ and a perturbation proportional to the eccentricity of Jupiter.

Having in mind a study of the dynamics at high eccentricity, we do not perform the classical developments in power series of the eccentricity nor in Fourier series of the resonance variables. Following Ferraz–Mello and Sato (1989), we compute the value of the averaged Hamiltonian (3) using Poincaré–type variables which are non–singular for small eccentricities. The explicit form of this Hamiltonian, however, is not known, but its value can be computed with accuracy for any given value of the phase space variables (σ, S, ν, N).

3. The Circular Problem

The integrable approximation $H_0(\sigma, S, N)$ of the Hamiltonian (3) can be considered as a one–degree of freedom Hamiltonian in (σ, S) depending on a parameter N. The phase space portraits of such an Hamiltonian are well–known for small

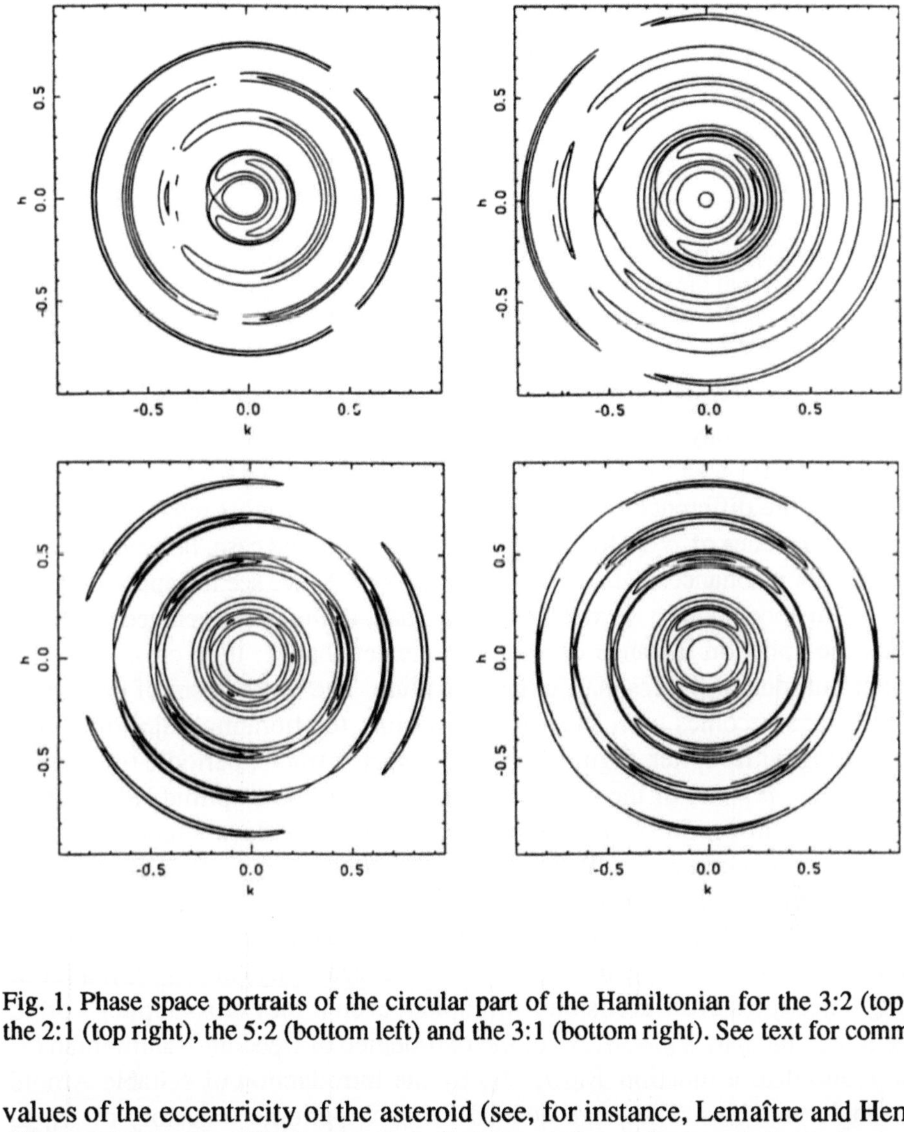

Fig. 1. Phase space portraits of the circular part of the Hamiltonian for the 3:2 (top left), the 2:1 (top right), the 5:2 (bottom left) and the 3:1 (bottom right). See text for comments.

values of the eccentricity of the asteroid (see, for instance, Lemaître and Henrard, 1990, and Lemaître, 1984). We show in figure 1 the evolution of these phase space portraits when the eccentricity increases. Each picture is composite and represents the phase space in Poincaré–type variables ($k = e\cos\sigma, h = e\sin\sigma$) for four different values of the parameter N. The central zone of these target–like pictures corresponds to the smallest value of N and we recognize there the usual three different topological regions: internal counterclockwise circulation, libration and external clockwise circulation for the angle σ. From the center to the edge of the targets, we can follow the evolution of the libration zone when the parameter N increases: the mean value of the eccentricity in the libration zone also increases, but the maximal amplitude of its variation decreases and the libration zone becomes

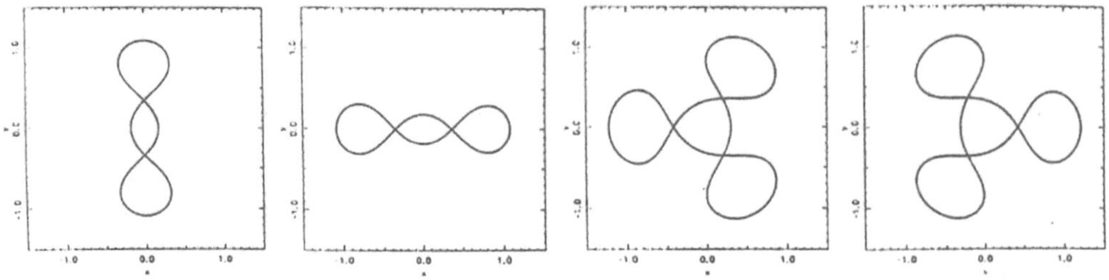

Fig. 2. Periodic trajectories of the non–averaged circular problem in rotating coordinates: Jupiter is at (1,0). From the left to the right: periodic orbit at $\sigma = 0$ in the 2:1 case, periodic orbit at $\sigma = \pi$ in the 2:1 case, periodic orbit at $\sigma = 0$ in the 3:2 case and periodic orbit at $\sigma = \pi$ in the 3:2 case.

thinner and thinner. Conversely, the libration amplitude, measured in semi–major axis, increases with large N; this gives to the resonance the well–known V–shaped form in a diagram (a, e).

Excluding the difference of width, the phase space of most resonances is the same for high values of N as for small ones. This is not the case, however, for the 3:2 and the 2:1 resonances, where, for high values of N, we see the apparition of a new libration zone about $\sigma = \pi$. This corresponds, in the non–averaged problem, to orbits the aphelion distance of which is greater than 1: they cross the orbit of Jupiter but, due to the resonance, never collide. The mechanism of protection from collisions becomes more clear if we consider the librating trajectories in a frame rotating with Jupiter. Figure 2 shows, in such a frame, periodic trajectories at $\sigma = 0$ (resp. $\sigma = \pi$) of the non–averaged problem; the amplitude of libration is zero and the direction of pericenter (resp. apocenter) stays aligned with the direction of conjunction. For librating trajectories about $\sigma = 0$ (resp. $\sigma = \pi$) with non–zero amplitude of libration, the direction of pericenter (resp. apocenter) oscillates about the direction of conjunction with a maximal amplitude determined by the initial conditions in semi–major axis and eccentricity; in this way, a lot of librating trajectories avoid collision with Jupiter, even if they cross Jupiter's orbit.

The integrable part $H_0(\sigma, S, N)$ of the two-degree of freedom Hamiltonian (3) can be reduced to a function $K_0(J_1, J_2)$ by the introduction of suitable Arnold action–angle variables. These are defined for each periodic trajectory of H_0 as (Henrard, 1990)

$$\begin{aligned} \psi_1 &= \frac{2\pi}{T_1} t \quad , \quad J_1 = \frac{1}{2\pi} \oint S \, d\sigma \\ \psi_2 &= \nu - \rho(\psi_1, J_1, J_2) \quad , \quad J_2 = N \end{aligned} \tag{4}$$

where $\oint$ denotes the integration along the considered trajectory, t being the time, T_1 the period and $\rho(\psi_1, J_1, J_2)$ a periodic function. Let us recall that, as the topology of the phase space associated to H_0 is not uniform, there is a discontinuity in the

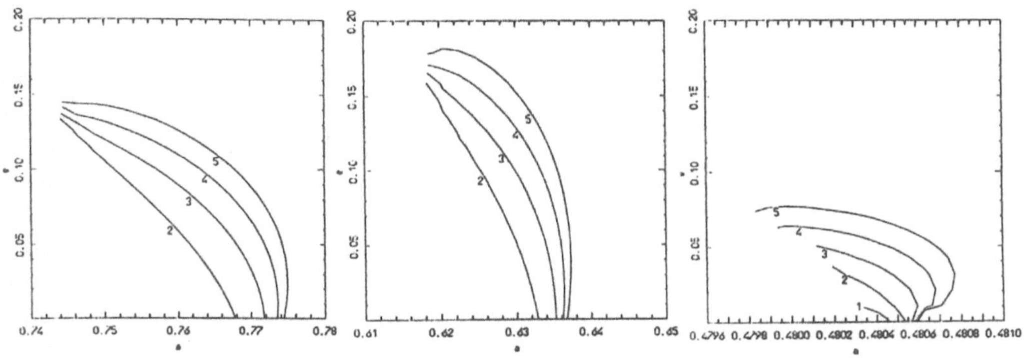

Fig. 3. Secondary resonances $\omega_1/\omega_2 = 1, 2, 3, 4, 5$ (denoted by 1, 2, 3, 4, 5) in the 3:2 (on the left), the 2:1 (at the centre) and the 3:1 (on the right) cases.

definition of the action–angle variables when crossing the separatrix of the first degree of freedom. The canonical transformation (4), however, can be considered as a global transformation of the phase space as it conserves its topological structure.

4. The Elliptic Problem

When the action–angle variables (4) are introduced into the Hamiltonian (3), this latest takes the form

$$K = K_0(J_1, J_2) + \sum_{n\geq 1} e'^n K_n(\psi_1, J_1, \psi_2, J_2) \tag{5}$$

where each term K_n admits the Fourier decomposition

$$K_n(\psi_1, J_1, \psi_2, J_2) = \sum_m c_{n,m}(J_1, J_2, e') \cos(m\psi_1 + n\psi_2) \,. \tag{6}$$

Therefore, any commensurability $m\dot{\psi}_1 + n\dot{\psi}_2 = 0$ is associated to a secondary resonance the strength of which is proportional to e'^n. In order to locate these secondary resonances, we compute the unperturbed frequencies (Henrard, 1990)

$$\omega_1 = \dot{\psi}_1 = \frac{\partial K_0}{\partial J_1} = \frac{2\pi}{T_1} \quad , \quad \omega_2 = \dot{\psi}_2 = \frac{\partial K_0}{\partial J_2} = \frac{1}{T_1}\int_0^{T_1} \frac{\partial H_0}{\partial N}\, dt \tag{7}$$

by numerically integrating along the periodic trajectories of the integrable Hamiltonian $H_0(\sigma, S, N)$.

We have plotted in figure 3 the location of the main secondary resonances ($\omega_1/\omega_2 = 2, 3, 4, 5$) as functions of the initial conditions of integration $(a, e, \sigma = \sigma^*)$ where $\sigma^* = \pi/2$ for the 3:1 and $\sigma^* = 0$ for the other cases; for the set of initial conditions we have chosen, there are no colliding trajectories with Jupiter. For the 7:3, the 5:2 and the 4:1 cases these secondary resonances are very close to the separatrix of the first degree of freedom. Similar results have been found

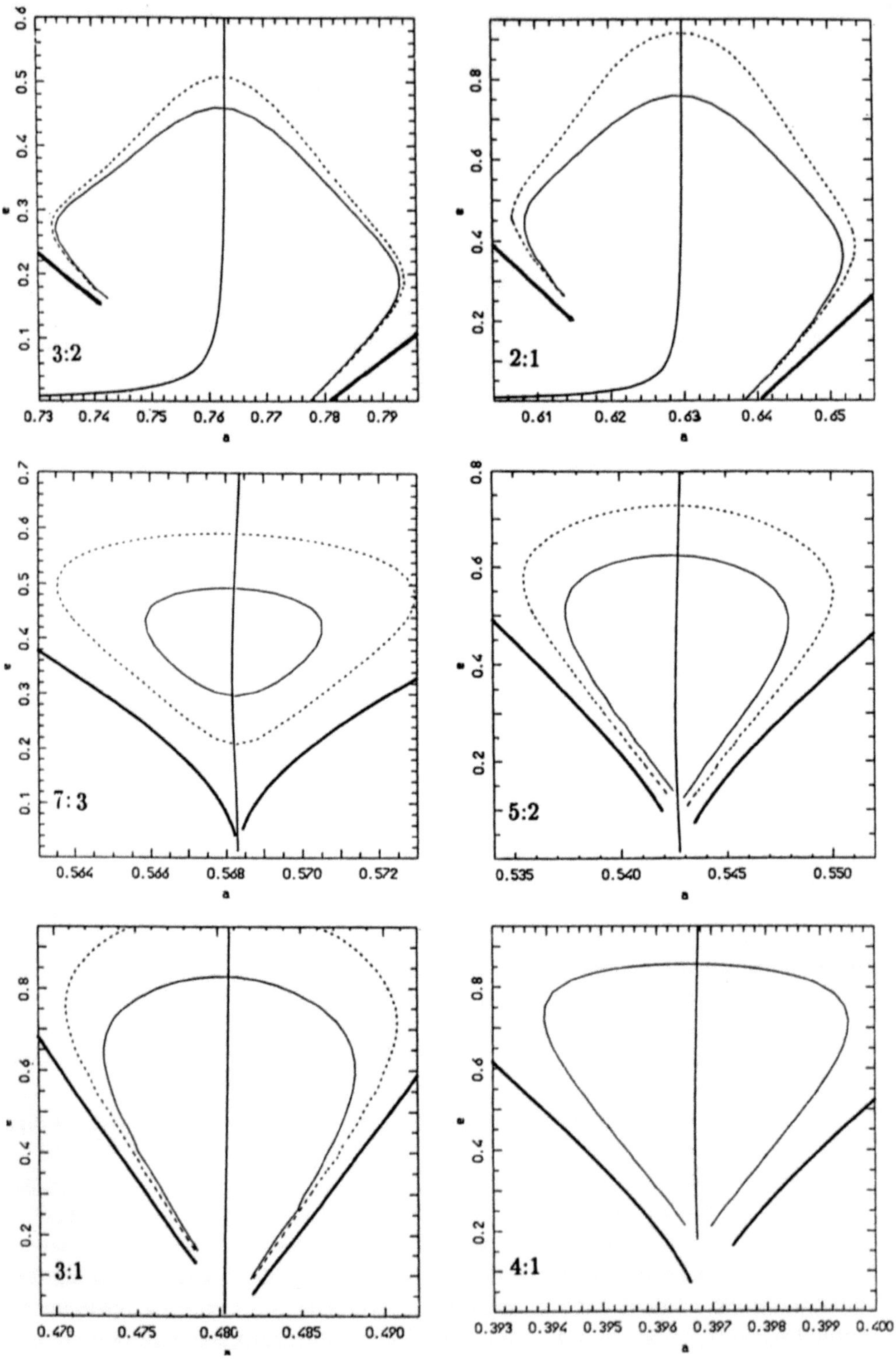

Fig. 4. Secular resonances ν_5 (solid line) and ν_6 (dashed line).

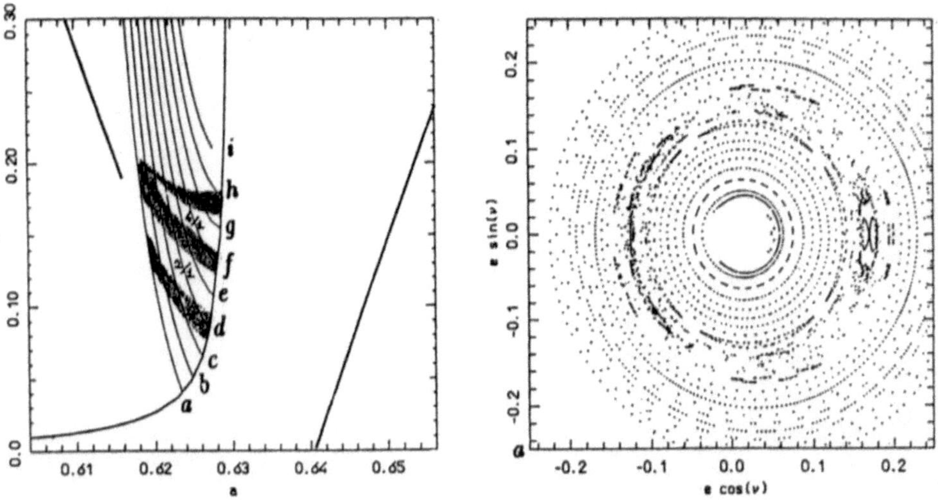

Fig. 5. On the left: the surfaces H=constant where we have computed Poincaré sections ($\sigma = \nu = 0$); the dashed arcas denote the location of the chaotic layers at $\nu = 0$. On the right: the Poincaré section on level a.

by Lemaître and Henrard for the 2:1 and the 3:2 cases (Lemaître and Henrard, 1988 and 1990) and by Henrard and Caranicolas (1990) for the 3:1 case, using an analytic model truncated at degree 8 in eccentricity (degree 2 for the 3:1); the agreement is very good at low eccentricity for the 2:1 commensurability.

Moreover, since our model is in closed form with respect to the eccentricity, we can look for resonances at large e. As a matter of fact, the resonance with $m = 0$ and $n = 1$ can be found, and this can be expected to be a relevant one, since it is proportional to e'. This resonance corresponds to the corotation of the asteroid longitude of perihelion ϖ with that of Jupiter ϖ'. If we consider that the rotation of ϖ' can be described by the superimposition of two modes, with frequencies g_5 and g_6 (respectively 4.257 and 28.246 arcsec/year) induced by Saturn's perturbation, this resonance splits into the two secular resonances ν_5 ($\omega_2 \equiv -\langle\dot{\varpi}\rangle + g_5$) and ν_6 ($\omega_2 \equiv -\langle\dot{\varpi}\rangle + g_6$). The location of these two secular resonances is given in figure 4, and their dynamical effect is the subject of a specific paper (Morbidelli and Moons, 1993) where the secular resonance ν_{16} between the nodes is moreover taken into account at non–zero inclination. In each picture of figure 4, the two thick lines and the central line denote respectively the separatrices and the family of stable equilibrium points of the averaged circular problem H_0.

5. Chaotic Motion at Moderate Eccentricity

In 1986, Murray, on the basis of a truncated model at degree 2 in eccentricity, pointed out the existence of a chaotic layer at moderate eccentricity in the 2:1 commensurability. Henrard and Lemaître (1986) with additions in (Lemaître and

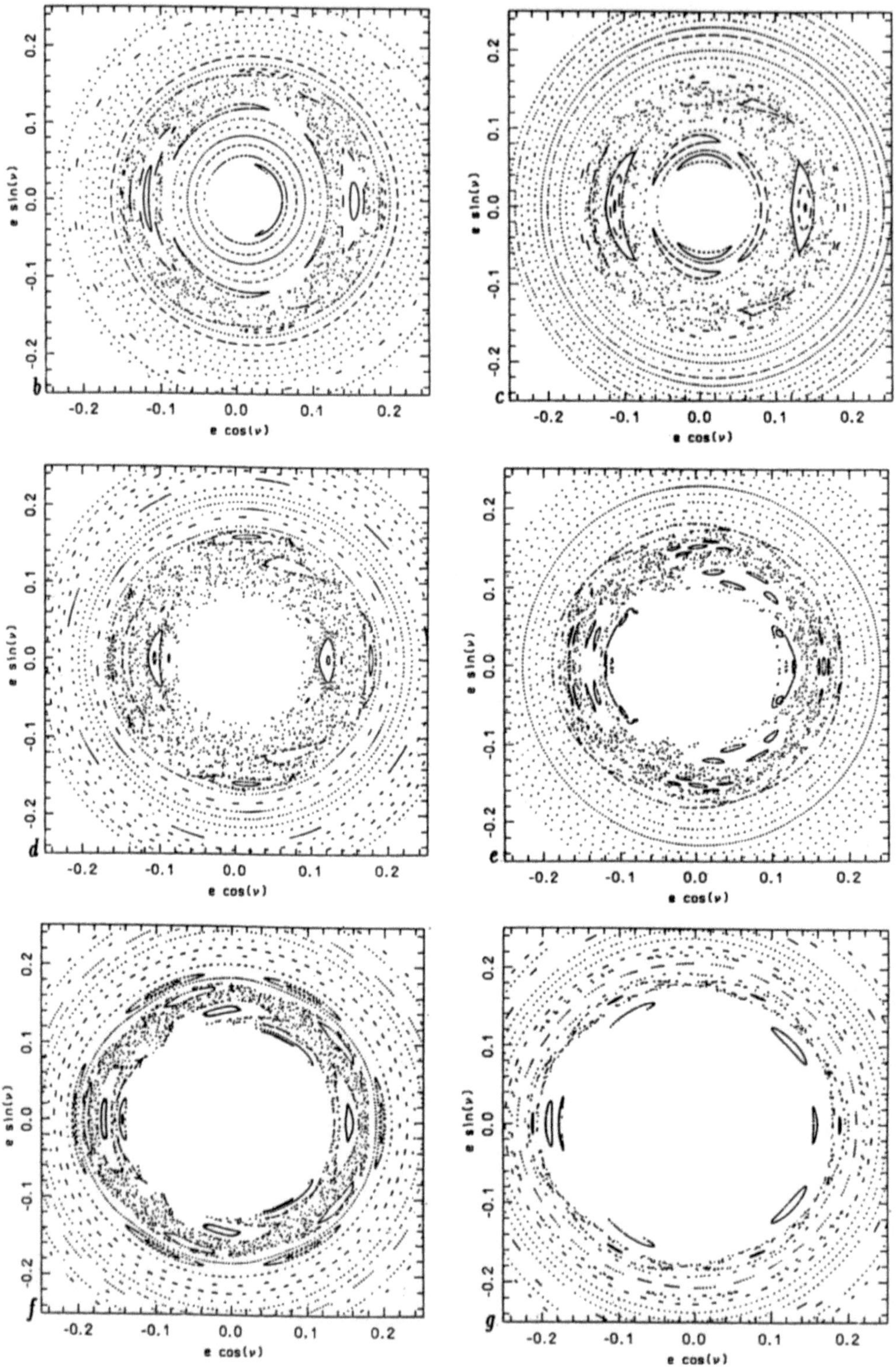

Fig. 6. Poincaré sections on levels *b*, *c*, *d*, *e*, *f*, *g*.

Henrard, 1990) have pointed out that such a chaotic layer is in the region where the main secondary resonances are present (see figure 3). Finally, Wisdom (1987) determined in the (a, e) plane at $\sigma = 0, \nu = 0$ the shape of the chaotic layer, by computing LCEs on numerical integrations in the planar elliptic three–body problem. However, a global picture of the dynamics, showing regular islands and chaotic regions produced by these resonances, has never been obtained, up to now. This is due to the fact that the averaged models (which, being two–dimensional, are the only ones which allow to compute Poincaré sections) were all truncated in the eccentricity, and therefore not completely reliable (see Lemaître and Henrard, 1990, for a complete discussion of this problem).

The recent work by Ferraz–Mello and Sato (1989), allows to compute the averaged Hamiltonian in closed form with respect to e; however, since the model is expanded up to degree 2 in e', it is not suitable to investigate secondary resonances of higher order like the ones in figure 3. Moons (1993) extended their work in order to obtain an averaged Hamiltonian in closed form with respect to e' also, which is thus perfectly suitable to study numerically secondary resonances of any order. This is what we use here for the computation of Poincaré sections. In short, we have an Hamiltonian $H(\sigma, S, \nu, N, e')$ with $e' = 0.0485$, and we compute the section at $\sigma = 0, \dot{\sigma} < 0$.

The picture on the left in figure 5 shows the 9 energy surfaces we have investigated, on a a, e diagram at $\sigma = 0$ and $\nu = 0$, the thick lines denote the stable family of equilibrium points and the separatrices of the averaged circular problem H_0. The picture on the right in figure 5 and those in figure 6 are the Poincaré sections on the levels *a, b, c, d, e, f, g* respectively (on levels *h* and *i* the dynamics turns out to be perfectly regular). One can easily recognize the regular islands corresponding to the main secondary resonances.

In the left picture of figure 5 we have qualitatively reported the location of the chaotic layer as measured on the axis $\nu = 0$ in the different Poincaré sections. The two "holes" are given by the regular islands of the 2/1 and 4/1 secondary resonances. This result is in very good agreement with that of Wisdom (1987).

References

Ferraz–Mello, S. and Sato, M.: 1989, 'The very–high–eccentricity asymmetric expansion of the disturbing function near resonances at any order', *Astron. Astrophys.* **225**, 541–547.

Henrard, J.: 1988, 'Resonances in the planar elliptic restricted problem', *Long-Term Dynamical Behaviour of Natural and Artificial N-Body Systems (A. E. Roy ed.), Kluwer A.P.*, 405–425.

Henrard, J.: 1990, 'A semi–numerical perturbation method for separable Hamiltonian systems', *Celest. Mech.* **49**, 43–68.

Henrard, J. and Caranicolas, N.: 1990, 'Motion near the 3/1 resonance of the planar elliptic restricted three body problem', *Celest. Mech* **47**, 99–121.

Henrard, J. and Lemaître, A.: 1986, 'A perturbation method for problems with two critical arguments', *Celest. Mech.* **39**, 213–238.

Lemaître, A.: 1984, 'High–order resonances in the restricted three–body problem', *Celest. Mech.* **32**, 109–126.

Lemaître, A. and Henrard, J.: 1988, 'The 3/2 resonance', *Celest. Mech.* **43**, 91–98.

Lemaître, A. and Henrard, J.: 1990, 'On the Origin of Chaotic Behavior in the 2/1 Kirkwood Gap', *Icarus* **83**, 391–409.
Moons, M.: 1993, 'On the resonant Hamiltonian in the restricted three–body problem', *Internal report, Department of Mathematics*, FUNDP, Namur.
Morbidelli, A., and Moons, M.: 1993, 'Secular resonances inside mean motion commensurabilities: the 2/1 and 3/2 cases', *Icarus*, in press.
Murray, C.D.: 1986, 'Structure of the 2:1 and 3:2 Jovian resonances', *Icarus* **65**, 70–82.
Poincaré, H.: 1902, 'Sur les planètes du type d'Hécube', *Bull. Astron.* **19**, 289–310.
Schubart, J.: 1968, 'Long-period effects in nearly commensurable cases of the restricted three–body problem', *Astron. J.* **73**, 99–103.
Szebehely, V.: 1967, 'Theory of orbits', *Academic Press*, New York.
Wisdom, J.: 1987, 'Urey Prize Lecture: Chaotic dynamics in the Solar System', *Icarus* **72**, 241–275.
Yoshikawa, M: 1989, 'A survey of the motions of asteroids in the commensurabilities with Jupiter', *Astron. Astrophys.* **213**, 436–458.

STOCHASTICITY OF TWO COMETS IN RESONANCE WITH JUPITER

D.BENEST and R.GONCZI
O.C.A. Observatoire de Nice, B.P. 229, F-06304 NICE Cedex 4, FRANCE

Abstract. The orbital stochasticity of comets P/Ciffréo (1985 XVI) and P/Maury (1985 VI), at the present time near the 5/3 and 4/3 resonances with Jupiter, is investigated using Lyapunov Characteristic Indicators. First results indicate a strong stochastic behaviour for the two comets, mainly induced by encounters with Jupiter, which looks roughly like the behaviour of the group of comets in 1/1 resonance with Jupiter.

Key words: Comets - Orbits - Stochasticity

1. Introduction

The two comets P/Ciffréo (1985 XVI) and P/Maury (1985 VI) were discovered by astronomers from Observatoire de la Côte d'Azur (O.C.A.), respectively with the 1.5 meter Schmidt Telescope of Observatoire de Calern (O.C.A.) and during the Second Palomar Sky Survey (see details in Benest, Ciffréo, Heudier, 1990, and Benest, Gonczi, Maury, 1992, hereafter called Papers I and II).

The present orbits of both comets P/Ciffréo and P/Maury cross the orbit of Jupiter and approach the orbit of Mars (see Figure 3 in Papers I and II). Their orbital periods are at present near the 5/3 and 4/3 resonances with Jupiter: in a rotating frame, the comets move around the Sun along fairly regular curves with 5 loops (for P/Ciffréo) and 4 loops (for P/Maury), described during about 3 heliocentric revolutions of Jupiter (see Figure 2 in Papers I and II). These 5- and 4-loops curves rotate slowly around the Sun, in a motion classically called circulation; as the aphelion distance of the comets are slightly greater than the mean jovian heliocentric distance, the most distant (from the Sun) part of each loop can sometimes, during the circulation, reach Jupiter's region; if this coincides with a passage of the comet through the jovian orbital plane, a more or less close encounter between the comet and Jupiter may occur, which is known to induce a more or less drastic change in the cometary orbital elements. As is now well-known, such close encounter with a planet introduces generally stochasticity in the long-term dynamical behaviour of the cometary orbit: i.e. two orbits with very close initial conditions diverge exponentially and have very different ultimate fates.

We plan to study the stochasticity of the orbits of our two comets P/Ciffréo and P/Maury. For this purpose, we compute quantities related to the theory of Lyapunov Characteristic Exponents.

Two orbits initially close diverge either linearly or exponentially depending on whether the initial points lie in an integrable or in a stochastic region of the phase space. This classical property has been extensively used as an indicator of stochasticity, and theoretically developed by Lyapunov, whose Characteristic Exponents (hereafter called LCE) indicate quantitatively how fast nearby orbits diverge and thus the degree of unpredictability of such orbits. But as the LCE's are

Celestial Mechanics and Dynamical Astronomy **57**: 109–112, 1993.

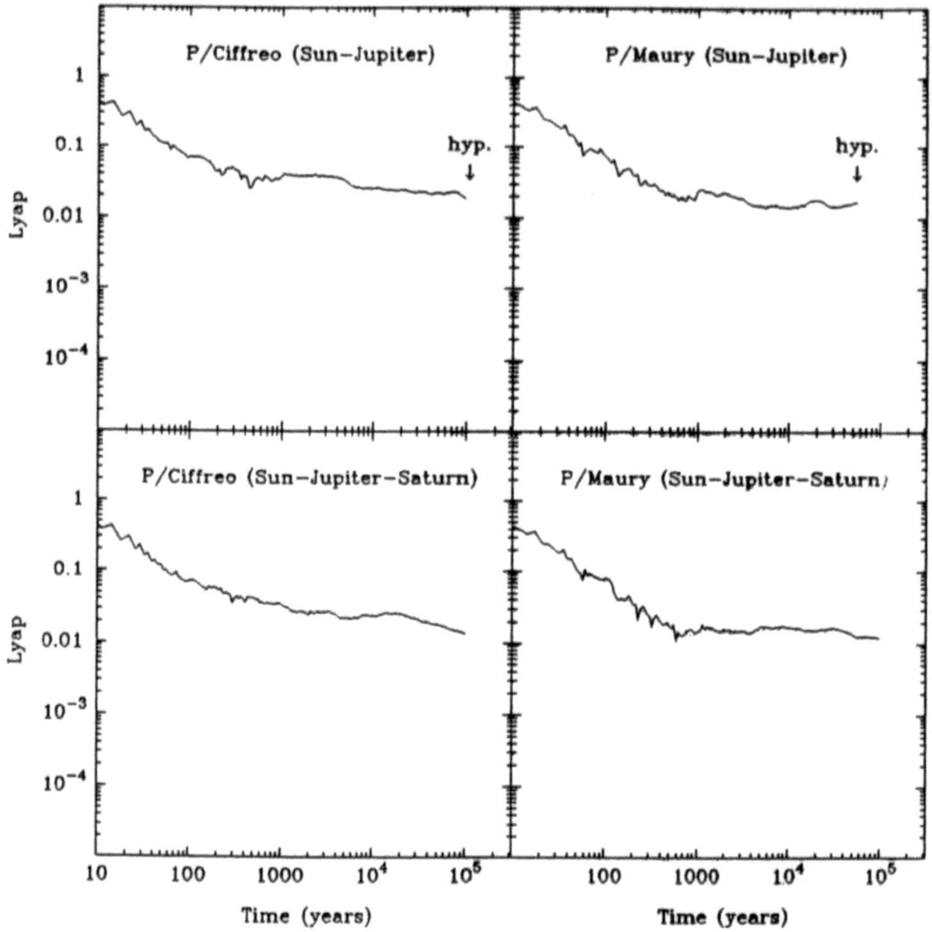

Fig. 1. Evolution versus time of the main Lyapunov Characteristic Indicator (whose limit is identified with the main Lyapunov Characteristic Exponent) for comets P/Ciffréo (left) and P/Maury (right) in the restricted three-body Sun-Jupiter-comet (top) and four-body Sun-Jupiter-Saturn-comet (bottom) models; "hyp"=hyperbolic escape.

stricto sensu limiting values at $t \to \infty$, they are evidently impossible to compute practically. We then define the Lyapunov Characteristic Indicators (LCI's) as the truncated values of the LCE's for a finite time; these LCI's provide however a fairly good quantitative measure of the stochasticity for a given orbit (for more details, see e.g. Gonczi and Froeschlé, 1981).

Our dynamical model is the classical elliptic three-dimensional restricted three-body problem (Sun - Jupiter - comet); we use also a restricted four-body model where we take into account the influence of Saturn. The equations of motion are integrated during 10^5 years, using a Bulirsch-Stoer method with variable step size. LCI's are computed every three steps. We present here our first results.

2. Results

Figure 1 shows the evolution versus time of the main LCI (whose limit at $t = 10^5$ years may be identified with the main LCE) for comets P/Ciffréo (left) and P/Maury (right), computed in the classical restricted three-body model Sun-Jupiter-comet (top) and in a restricted four-body model Sun-Jupiter-Saturn-comet (bottom). In all cases, the value of the Lyapunov Indicator does not decrease under 0.01, which is much higher than the limiting value ($< 10^{-4}$) found at $t > 10^5$ years for regular (i.e. non stochastic) orbits (see e.g. Benest, Gonczi, 1992a). P/Ciffréo and P/Maury thus show a strong stochastic behaviour; as there is no significant change when Saturn is added in the computation, we can infer that the stochasticity is mainly due to encounters with Jupiter. We may remark that the two comets are linked to their resonance with Jupiter only during a few thousands of years backward and forward (see Papers I and II), so that they probably leave the resonance long before their fatal encounter with Jupiter.

Besides our two comets, we study the stochasticity of the three comets now known to be at present at or near the 1/1 resonance with Jupiter, namely P/Slaughter-Burnham (1981 XVIII), P/Boethin (1986 I) and P/Ge-Wang (1988 VIII) (Benest, 1990). In this study, we compute the LCI's also for fictitious orbits where we vary the initial elements of the comets, essentially the mean anomaly, the eccentricity and the inclination. First results for P/Slaughter-Burnham (Benest, Gonczi, 1992a) and for P/Boethin (Benest, Gonczi, 1992b) indicate that, although the LCI's for the actual comets show a strong stochasticity as for P/Ciffréo and P/Maury, some values of these initial elements correspond to regular orbits.

Therefore, we have to evaluate also for P/Ciffréo and P/Maury the respective influences of the initial mean anomaly, eccentricity and inclination of the cometary orbit on the stochasticity of the orbits. Finally, we will undertake more precise studies, which could take into account other perturbations (e.g. from the other planets, or even the so-called non-gravitational forces known to act often on comets).

Acknowledgements

We are grateful to Dr Hans Rickman for his fruitful suggestions.

References

Benest, D.: 1990, ' P/Ge-Wang joins P/Slaughter-Burnham and P/Boethin in the club of comets in 1/1 resonance with Jupiter', *Celest. Mech.* **47**, 361-374

Benest D., Ciffréo J., Heudier J.L.: 1990, ' A study of the prediscovery motion of comet P/Ciffréo and its future', *Astron. Astrophys.* **228**, 539-545

Benest D., Gonczi R.: 1992a, in A.Harris & E.Bowell, ed(s)., *Asteroids, Comets, Meteors 1991*, Lunar and Planetary Institute Publication Services, in press

Benest D., Gonczi R.: 1992b, in J.Stohl & I.P.Williams, ed(s)., *Meteoroids and their Parent Bodies*, Astronomical Institute of the Slovak Academy of Sciences, in press

Benest D., Gonczi R., Maury A.: 1992, ' Dynamics of comet P/Maury', *Astron. Astrophys.*, in press

Gonczi, R., Froeschlé, C.: 1981, ' The Lyapunov Characteristic Exponents as indicators of stochasticity in the Three Body Restricted Problem', *Celest. Mech.* **25**, 271-280

ORBITAL EVOLUTION OF GIANT COMET-LIKE OBJECTS

TSUKO NAKAMURA
National Astronomical Observatory, Mitaka, Tokyo 181 Japan
and
MAKOTO YOSHIKAWA
Communications Research Laboratory, Kashima, Ibaraki 314 Japan

Abstract. We investigated by numerical integrations the long-term orbital evolution of four giant comets or comet-like objects. They are Chiron, P/Schwassmann- Wachmann 1 (SW1), Hidalgo, and 1992AD (5145), and their orbits were traced for 100–200 thousand years (kyr) toward both the past and the future. For each object, 13 orbits were calculated, one for the nominal orbital elements and other 12 with slightly modified elements based on the rms residual of the orbit determination and on the number of observations. As past studies indicate, their orbital evolution is found to be very chaotic, and thus can be described only in terms of probability. Plots of the semi-major axis (a) and perihelion distance (q) of the objects treated here seem to cross each other frequently, suggesting a possibility of their common evolutionary paths. About a half of all the calculated orbits showed q- or a-decreasing evolution. This indicates that, at least on the time scale in question, the giant comet-like objects are possibly on a dynamical track that can lead to capture from the outer solar system. We could hardly find the orbits with perihelia far outside the orbit of Saturn ($q > 15$ AU). This is perhaps because the evolution of the orbits beyond Saturn is so slow that substantial orbital changes do not take place within 100–200 kyr.

Key words: Cometary orbital evolution, giant comets, numerical integrations.

1. Introduction

The origin of short-period (SP) comets is not yet fully understood. In addition to the classical Oort cloud, a few replenishment sources of SP comets from the outer solar system have so far been proposed, such as the inner Oort cloud (Hills, 1981) and the trans-Neptunian (or Kuiper) comet belt (e.g. Fernández, 1980). However, it seems still not clear about which source is the main supplier of comets into SP orbits.

We calculated the orbital evolution of about 160 actual SP comets ($P < 200$ yr), 3400 years and 1000 years toward the past and the future respectively based on a realistic dynamical model (Nakamara and Yoshikawa, 1991; hereafter referred to as Cosmo-DICE). In Nakamura and Yoshikawa (1992), we could make much clearer the capture mechanism of SP comets by Jupiter than that studied first by Everhart (1972) in a simplified dynamical model. Therein we also found several cases of multi-stage capture in which comets evolved from near-Saturn orbits and then were relayed under the control of Jupiter.

Modern observation techniques succeeded in discovering unusual comet-like objects like Chiron and 1992AD (Possibly 1992 QB1 is also one of such objects). They are characterized by their estimated size, at least one order of magnitude larger than the nuclei of typical comets. Such objects, including P/Schwassmann-Wachmann 1 and Hidalgo (944), are tentatively called here "Giant Comet-like Objects" (GCO) or "Chiron-type Objects" for convenience. Although cometary

Celestial Mechanics and Dynamical Astronomy **57**: 113–121, 1993.

activity has not yet been identified in 1992AD and Hidalgo, their physical observations strongly suggest that they are of cometary rather than asteroidal nature (*IAU Circular* 5434; Weissman *et al.*, 1989).

Another point worthy of consideration is concerning the size of the GCO. Recent formation theories of comets regarded as planetesimals in the proto-solar nebula predict that giant comets of more than several tens km diameter are rare (Greenberg *et al.*, 1984; Yamamoto and Kozasa, 1988). Hence it is interesting to pursue dynamically where the GCO came from.

Since their current orbits are under the influence of both Jupiter and Saturn in terms of a and q, it is natural to ask whether or not the orbits of the GCO are on an intermediary phase between SP orbits and their source orbits in relation to the multi-stage capture cases found in Cosmo-DICE. The aim of this paper is, therefore, to examine, in the framework of a realistic dynamical model of the planetary system, if there is any evolutionary relation between the four GCOs, and how the GCOs have an orbital link both to SP comets and to their source orbits. In section 2, a unified criterion is given as for the variation range of starting orbital elements. Section 3 describes the long-term orbital behavior of each object. Section 4 discusses whether the GCO as a group showing a similar dynamical behavior is justified or not.

2. Method of Numerical Integrations

2.1. A Unified Criterion for the Variation of Orbital Elements

The orbits of most periodic comets are so unstable and chaotic that two orbits with slightly different initial conditions often deviate substantially after several thousand years' orbital integration. This implies that it is practically impossible to calculate the actual orbits of real comets beyond, say, 10 thousand years, even if the best available orbital elements are used.

As a compromise with the difficulty, Oikawa and Everhart (1979) followed Chiron's long-term orbital behavior by calculating many orbits of it whose starting orbital elements are changed from its nominal values within a reasonably small range. From those orbits, they inferred Chiron's orbital evolution statistically in terms of probability. Lacking other effective replacements, subsequent researchers followed the same approach (e.g., Scholl, 1979; Hahn and Bailey, 1990). We also do similarly.

If reliable errors of the Keplerian elements for an object obtained from orbit determination are always given in sources like *MPC* and *IAUC*, it would be most reasonable to calculate the collective orbits by changing the orbital elements within the error range. However, such cases are very rare. Because of it, as the variation range, some adopted one unit of the last decimal place of the published orbital elements, or others used the difference between the values from two orbit determinations. But these approaches are somewhat arbitrary and seem to lack the

TABLE I
Relevant data for the observations and the orbits of our four objects

	Chiron	SW1	Hidalgo	1992AD
Element source	MPC13999	MPC18255	MPC19480	MPC19850
P(arcsec)	0."9	1."12	1."24	0."75
N	71	661	210	59
Time span	1895-1988	1902-1991	1920-1991	1977-1992

theoretical basis for the variation.

We, instead, adopt here a unified criterion for selecting the variation range of orbital elements. If the orbit of an object is determined from its well-distributed observations of a number ($N \gg 1$) based on a sound least- squares procedure, the error of each orbital element should be estimated from the rms residual after orbital improvement. The Gaussian statistics of errors teaches us that the standard deviation (SD) of the error in an orbital element (E_j, $j = 1$–6) can be expressed as

$$\mathrm{SD}(E_j) = \sigma(T_{jj}/N)^{1/2}, \tag{1}$$

where σ is the rms residual per observation (in arcsec) and T_{jj} the diagonal element of the inverse of the variance-covariance matrix in the solution. This means that the error dE_j in an orbital element after N observations is improved by a factor of $N^{-1/2}$. Therefore, $\sigma(T_{jj})^{1/2}$ can be considered to correspond to the error per observation ΔE_j ; that is,

$$dE_j = \Delta E_j/N^{1/2}. \tag{2}$$

Since all the objects treated here lie far outside the Earth's orbit, we can expect approximately $\sigma \cong \Delta\omega \cong \Delta\Omega \cong \Delta i \cong \Delta M$, and $\Delta a \cong \sigma a$. And if we assume a relation

$$\Delta a \cong \Delta q, \tag{3}$$

Δe is calculated as

$$\Delta e = (1 - e)\Delta a/a - \Delta q/a \cong -e\Delta a/a. \tag{4}$$

In Table I, the relevant data about the orbits of our four objects are summarized. As each N in the table is sufficiently large, it is expected that the error estimate of orbital elements discussed above is valid for the four objects. We admit that the above estimated dE_j is sometimes only theoretical and will not always reflect the true error of orbital elements, because error propagation often takes place in a more complicated way in actual orbit determinations (see the examples in *MPC*). However, we still believe that our dE_j is a more objective measure of the errors in orbital elements than the variation ranges used by previous researchers.

Then, we calculate 13 orbits for each object, one for the nominal orbital elements taken from the sources given in Table I (a_0, e_0, i_0, ω_0, Ω_0, M_0) and other 12 modifications with $a_0 \pm da$, $e_0 \pm de$, ... and $M_0 \pm dM$.

2.2. Numerical Integration

The dynamical model adopted here is a seven-body problem with planets running from Jupiter through Pluto. Starting position and velocity vectors for the planets were taken from the JPL DE200. We used an extrapolation-type integrator (Bulirsch and Stoer, 1966) with an automatic step-size control for planetary close-approaches. The relative truncation error was set to 1×10^{-14}. Our maximum time interval of integration, 200 kyr, was selected as a reasonable trade-off between the mean step-size of integration (CPU time) and accumulation of the round-off error (e.g. Brouwer, 1937).

3. Orbital Evolution of Each Object

A rough sketch of the orbital evolution of a comet will be that it first enters the planetary region from very far, either from the Oort cloud or the Kuiper belt, stays there for a while and eventually leaves as the result of close encounters with planets. As this dynamical lifetime will perhaps be the order of million years (My), short-term orbital integration may not distinguish whether a comet in question is on a capture phase or on an ejection phase. For this reason, we integrated the orbit of each object toward both the past and the future for 100–200 kyr, and sometimes more up to 1 My. Longer integrations, on the other hand, may severely deteriorate even the reliability of collective orbits because of accumulation of the round-off error. Since the orbital behavior of comets is of very chaotic nature, its interpretation can be different depending on what time scale we are concerned with. It is possible that a comet is on an ejection phase in the long run, while it looks like on a capture phase into SP orbits for a time scale of several kyr. Hence we consider here two kinds of the time scale of evolution, that is, ± 10 kyr and ± 200 kyr.

3.1. 1992 AD

This object was discovered by the Spacewatch telescope (T. Gehrels) as a slow-moving asteroid and marked a record of q-distance (= 8.67 AU). Subsequent physical observations showed that this has an unusually red color and a low albedo, suggesting the nature of cometary nuclei. Its diameter was obtained to be about 140 km as a minimum estimate from IR observations (*IAUC* 5434, 5449). Although 1992 AD much resembles Chiron in terms of q and size, there have been no reports of coma detection yet.

The time evolution of a and q of 1992AD is shown for ± 200 kyr in the top of Figure 1. We plot the calculated 13 orbits *on the same graph*, in an attempt

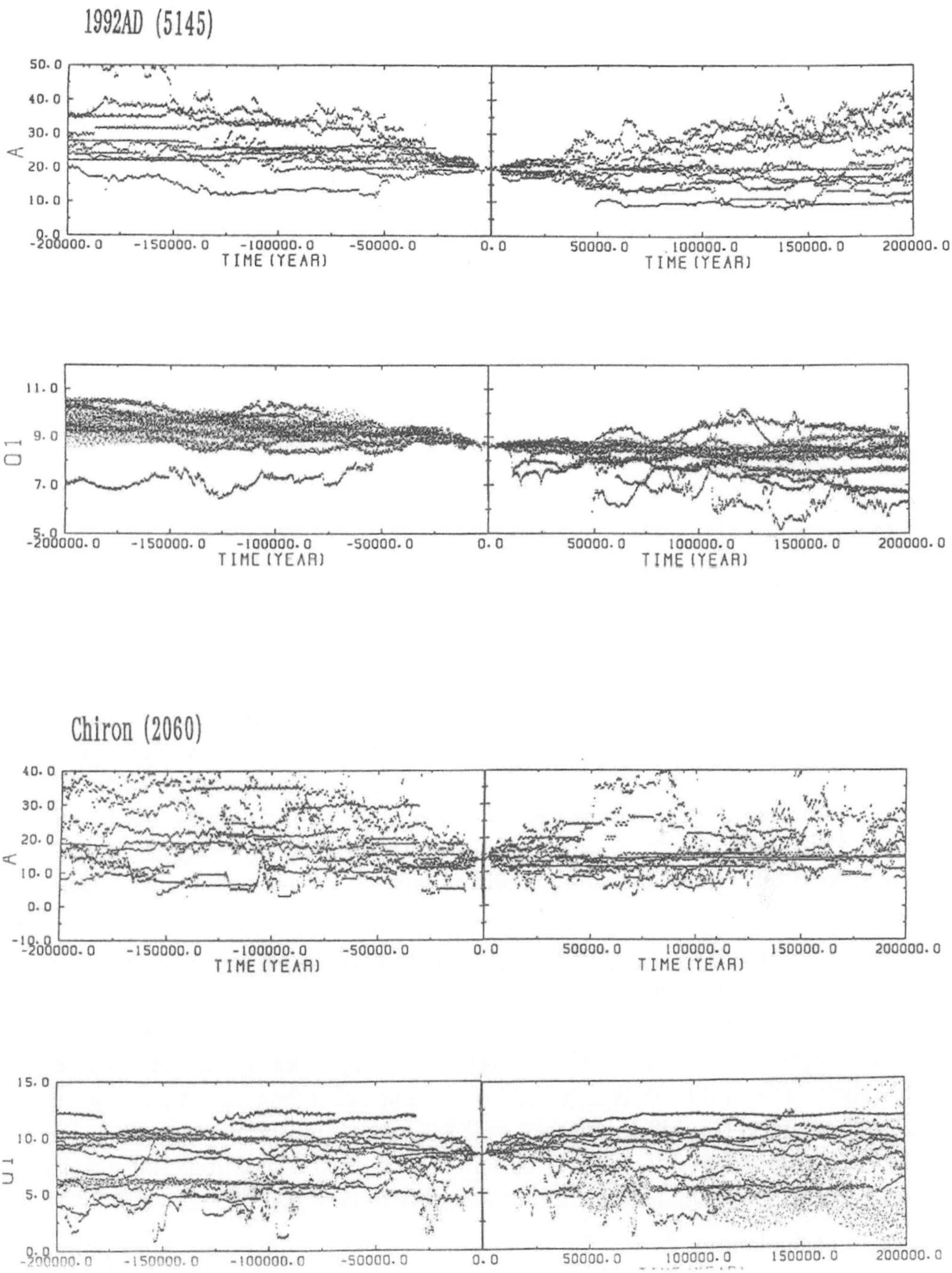

Fig. 1. Long-term (200 kyr) orbital evolutions of 1992 AD and Chiron for the semi-major axis (A) and perihelion distance (Q1). The 13 orbits explained in section 2 are plotted on the same graph, in order to give a visual impression that a denser set of curves in the graph corresponds to a more probable orbit.

to show that a denser set of points in the graph corresponds to a more probable orbit. In the plot of ± 10 kyr not shown here, it is seen that the orbital behavior of 1992AD is reliably predictable 6000 years and 4000 years toward the past and the future respectively; actually the orbit of 1992 AD is found to be the least chaotic among the four objects. According to Figure 1, the inward evolution of a and q is highly probable in the past, because the past a and q are smaller than the present values only in one case. Toward the future, q still continues to decrease with a considerable probability, while a seems to have a nearly equal chance for decreasing and increasing. These results thus suggest that 1992 AD has a more gentle and slower evolution compared with the other three objects.

3.2. Chiron

Because coma activity has recently been detected in Chiron (*IAUC* 4770) and its estimated diameter is as large as about 200 km (Bus *et al.*, 1989), this object is really worth being called a giant comet, so that its dynamical origin is of special interest.

Hahn and Bailey (1990) have made a detailed investigation on the orbital history of Chiron. The differences between their calculations and ours are as follows: We used a more refined set of orbital elements (see Table I) than theirs. Our variation range of elements given by equation (2) is about ten times smaller than theirs. They traced 83 orbits whereas our number of orbits is 13. This is because our belief is that if a more reliable set of orbital elements is available for analysis, a smaller number of orbits around the nominal orbit should be enough to get a likely orbital path.

For the past 200 kyr (see the bottom of Figure 1), about 75% of a- behaviors were decreasing from 20–40 AU to the present value; in particular, three cases were of direct capture via a single planetary encounter (within 5 gravitational radii of the planet) from a = 100–400 AU, 100–160 kyr ago. During the future 200 kyr, only about 30% of the orbits acquired lower values of a than the present one. The rest of the orbits ($\sim$ 70%) increased to 20–30 AU from the present value. Four cases out of the 70% were of ejection to a = 230–2500 AU, within 100–130 kyr. The estimated half-lifetime of Chiron to stay in orbits of $a < 50$ AU was 0.54 My.

In our calculations, the past/future asymmetry of the SP phase of Chiron pointed out by Hahn and Bailey (1990) was not so clearly seen. For both the past and the future, the evolution of q seems more complicated than that of a. About a half of q-orbits stayed between 12 AU and the present value, while another half showed very chaotic variations between the present value and 1–2 AU. Although capture into orbits of q =1–2 AU is not rare, it is temporary and seldom exceeds 2–3 kyr.

3.3. P/SCHWASSMANN-WACHMANN 1

For the time scale of ±10 kyr, a of SW1 for the past showed an inward evolution with a high probability from 8–9 AU to the present value and that for the future indicated an outward evolution also with a considerable probability to a = 10–15 AU. The future evolution of q has a decreasing trend, though the past q has a nearly equal chance of decreasing and increasing.

The evolution of SW1 for the time scale of ±200 kyr is very diverse (see the top of Figure 2). The past behavior of a is of definitely decreasing nature from a very wide range of the initial a. Four cases were direct captures from nearly parabolic or hyperbolic orbits. The future evolution of a is much more violent. Only 4 or 5 orbits were found to stay between 5–20 AU. About 65% orbits were ejected to nearly parabolic orbits including hyperbolic ones within 200 kyr. This is probably due to strong interactions of SW1 with Jupiter and thus the estimated half-lifetime was as short as 310 kyr. As for the past 200 kyr, q has a nearly constant probability of being anywhere in the range between 3 AU and 10 AU. But the future q is fairly likely to be less than the present value. Therefore it is likely for SW1 to appear perhaps as a brilliant comet sometime in the future.

3.4. HIDALGO

In spite of the very comet-like characteristic of the present orbit of Hidalgo, detection of its non-gravitational acceleration has not yet been successful (Yeomans, 1991). The q behavior for the ±10 kyr time scale is remarkable in that, with a high reliability, q decreases from 2.5 AU ($\sim$ 0 AD) through the present value, attains a minimum of about 1 AU around 5500 AD, and again increases back near the present q. Such a behavior had already been predicted by Kozai (1979) with the secular perturbation theory. Thus, if Hidalgo is a dormant comet with little experience in the inner solar system region, it may ignite cometary activity within a few thousand years in the future.

The a evolution of Hidalgo during ±200 kyr is similar to that of SW1 (the bottom of Figure 2). The ejection probability for the future is three times higher than that for the past. Although the q evolution for the same period is very chaotic, there is a vague trend of q-decreasing throughout the past and the future. One peculiar point about Hidalgo's future orbits is that about 25% from the calculated 30 orbits showed close approaches to the sun within 0.1 AU (Three cases were considerably less than one solar radius as close as to the sun). In this respect, it is interesting to note that Arnold (1965) had found that about 20% of Hidalgo-like orbits calculated by him became sun-approachers in a simplified Monte Carlo simulation. Thus, if Hidalgo-like giant comets are not a rare phenomenon, this mode of evolution may be a new source of SP comets through, perhaps, a splitting mechanism as suggested by the Kreutz group.

P/Schwassmann-Wachmann 1

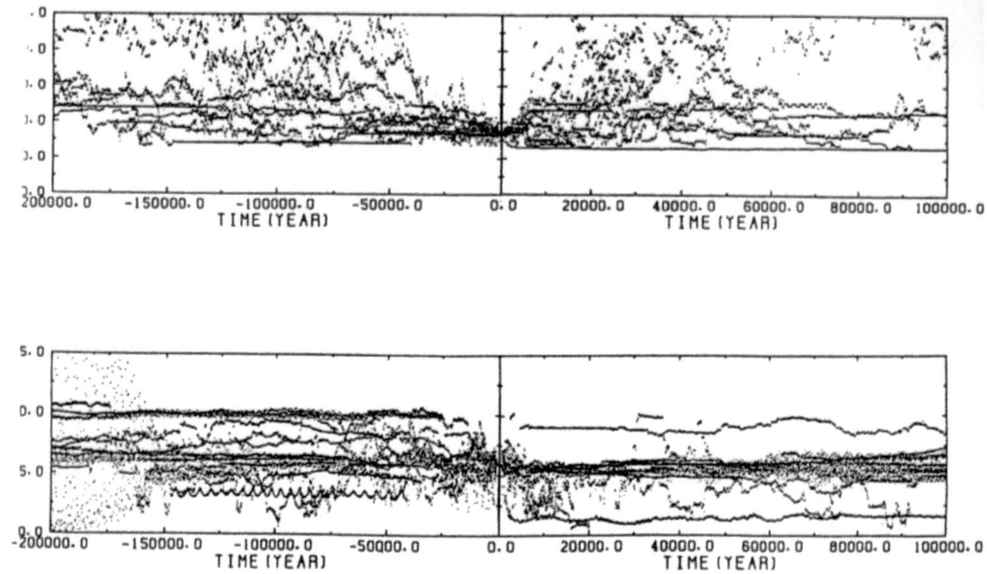

Hidalgo (944)

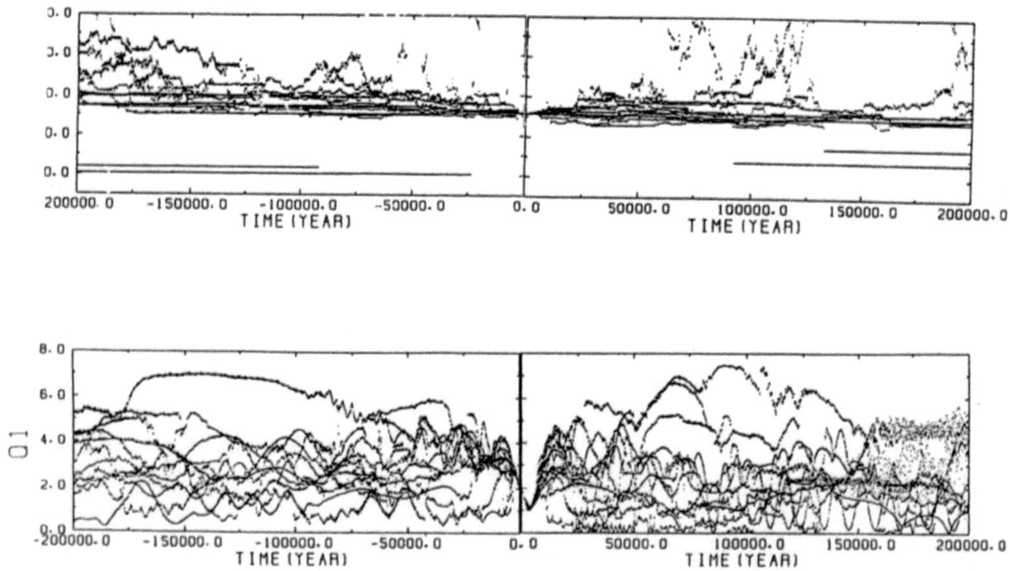

Fig. 2. Long-term (200 kyr) orbital evolution of P/Schwassmann-Wachmann 1. See the caption of Fig.1 for other explanations.

4. Discussion

Now we are in a position to examine whether or not the four GCOs treated here are in a dynamically related group. Firstly, about a half of all the calculated orbits showed q- or a-decreasing evolution as a general tendency. In particular, if the evolution is limited only to the past 200 kyr, more than 80% orbits were found to evolve from orbits with values of a of more than 20–30 AU. This indicates that the GCO orbits are possibly on the capture track from the outer solar system. Secondly, comparison with Figure 1 and Figure 2 reveals that the a- and q-plots of the four GCOs frequently cross each other, with a less degree for 1992AD. These facts, therefore, suggest that the GCOs, possibly except for 1992AD, are likely to have a common evolutionary path. This preliminary result, of course, needs further confirmation, because the behavior of other elements such as inclination is not yet fully analyzed.

We could not find the orbits with perihelia beyond 15 AU toward both the past and the future, even for 1992AD whose present orbit lies outermost among the four GCOs. This is perhaps because the q evolution of orbits outside Saturn's orbit is so slow that substantial orbital changes do not take place during ±200 kyr. Alternatively, there is no reason to exclude a possibility that these GCOs were brought about directly from the Oort cloud or the Kuiper belt to the present orbits, without experiencing the multi-stage capture by Uranus and/or Neptune.

References

Arnold, J.R.: 1965, *Astrophys. J.* **141**, 1536.

Brouwer, D.: 1937, *Astron. J.* **46**, 149.

Bulirsch, R., and Stoer, J.: 1966, *Numerische Mathematik* **8**, 1.

Bus, S.J., Bowell, E., Harris, A.W., and Hewitt, A.V.: 1989, *Icarus* **77**, 233.

Everhart, E.: 1972, *Astrophys. Lett.* **10**, 131.

Fernández, J.A.: 1980, *Mon. Not. R. Astron. Soc.* **192**, 481.

Greenberg, R., Weidenschilling, S.J., Chapman, C.R., and Davis, D.R.: 1984, *Icarus* **59**, 87.

Hahn, G., and Bailey, M.E.: 1990, *Nature* **348**, 132.

Hills, J.G.: 1981, *Astron. J.* **86**, 1730.

Kozai, Y.: 1979, in Dynamics of the Solar System, ed(s)., *R.I. Duncombe*, D. Reidel, Dordrecht, p. 231.

Nakamura, T., and Yoshikawa, M.: 1991, *Publ. Natl. Astron. Obs. Japan* **2**, 293.

Nakamura, T., and Yoshikawa, M.: 1992, in Asteroids,Comets, Meteors 1991, ed(s)., , The Univ. of Arizona Press, Tucson, in the press.

Oikawa, S., and Everhart, E.: 1979, *Astron. J.* **84**, 134.

Scholl, H.: 1979, *Icarus* **40**, 345.

Weissman, P.R., A'Hearn, M.F., McFadden, L.A., and Rickman, H.: 1989, in Asteroids II, ed(s)., *R.P. Binzel, T.Gehrels, and M.S. Matthews*, The Univ. of Arizona Press, Tucson, p. 880.

Yamamoto, T., and Kozasa,T.: 1988, *Icarus* **75**, 540.

Yeomans, D.K.: 1991, *Astron. J.* **101**, 1920.

ON A TEMPORARY CONFINEMENT OF CHAOTIC ORBITS OF FOUR DIMENSIONAL SYMPLECTIC MAPPING: A TEST ON THE VALIDITY OF THE SYNTHETIC APPROACH

C. FROESCHLÉ and J.M. PETIT

Observatoire de Nice, B.P. 229, Nice Cedex 4, France

Abstract. Poincaré maps for Hamiltonian systems with 3 degrees of freedom lead to the study of four dimensional symplectic mappings. As a test for the validity of a synthetic mapping of order 3 using gradient informations, we study the evolution with time of Liapounov Indicators in the case of the four dimensional standard map with chaotic and stable zones. Both Liapounov Indicators show the same behaviour for the real and synthetic mappings. They reveal exploding diffusion phenomena for temporarily confined chaotic orbits. The distribution of the time of explosion fits well with a Poisson law for the real mapping, but not for the synthetic one. However the mean time of explosion is essentially the same in both cases.

Key words: Hamiltonian – mapping – Liapounov

1. A Synthetic Approach

Poincaré maps are now of common use for studying the qualitative behaviour of differential equations (see Hénon, 1981). Moreover, in order to study stability problems, many authors have sought explicit algebraic mappings which approximate, at least qualitatively, the Poincaré maps obtained from the original Newton equations. Froeschlé and Petit (1990, paper I) have reviewed some of these mappings and showed that they are reliable only as long as one remains within the domain of validity of the approximations made in order to isolate either — in the case of deterministic mappings — an integrable part and some instantaneous perturbations, or — for stochastic mappings — a source of endogenous/exogenous stochasticity (see Froeschlé and Rickman, 1988). All these mappings are *ad hoc* and reliable only in some region of the phase space and for some specific purpose. In paper I we built a mapping valid everywhere in the phase space, following an idea already used by Varosi et al. (1987) but in the framework of non–Hamiltonian systems (i.e., systems where attractors do exist). The method consists in coarse–graining the phase–space surface of section and then interpolating the value of the image of a point. Linear interpolation requires a rather fine graining of the phase space, hence it is necessary to compute a lot of points on the grid. However, Taylor expansions of order 3 and 5 (denoted by T3 and T5) can provide very good results as long as symmetrical interpolation formulae are applied, for which it is necessary to use an extended grid. Since there are cases where one cannot cross a given limit, asymmetrical interpolation formulae have been tested, but their accuracy was found to be inferior. Therefore Petit and Froeschlé (1992, paper II) have developed another type of interpolation (denoted by $\tau3$), where the information, including that on the gradients, is stored to the same level of accuracy only for the

Celestial Mechanics and Dynamical Astronomy **57**: 123–130, 1993.

nearest–neighbouring vertices. Thus, not only images of vertices are computed, but also tangential mappings at each vertex.

There are in any case two key parameters: the number of bins in each direction N = (total number of cells)$^{1/D}$, where D (= 2 and 4 in papers I and II) is the dimension of the surface of section, and M the order of the Taylor expansion. In order to explore the validity of the synthetic approach, we applied our method in two cases (paper I and II):

1. An algebraic area–preserving mapping for which the computation of orbits is very fast. This allows to follow a large number of orbits and to carry out enough iterations for a meaningful comparison.
2. A special case of the restricted three–body problem, already studied by Duncan et al. (1989).

We continue here this study in the case of Hamiltonian systems with 3 degrees of freedom. In this case the surface of section has 4 dimensions and interpretations of the results are more difficult. Again, using a symplectic four-dimensional mapping appears suitable as a model problem in order to test both the different synthetic mappings and the hash function method. Furthermore a four-dimensional surface of section is populated by points and not by a continuous trajectory. Therefore, if we perform other sections, the probability of a point falling on an arbitrarily chosen surface vanishes. Therefore, we must take not a true section but rather a *slice* having a small but finite thickness in order to catch some points. The following four dimensional symplectic mapping already used by Froeschlé (1972) has been taken as a model problem:

$$
\begin{aligned}
x^{(n+1)} &= x^{(n)} + a_1 \sin(x^{(n)} + y^{(n)}) \\
&\quad + b \sin(x^{(n)} + y^{(n)} + z^{(n)} + w^{(n)}), \\
y^{(n+1)} &= x^{(n)} + y^{(n)}, \\
z^{(n+1)} &= z^{(n)} + a_2 \sin(z^{(n)} + w^{(n)}) \\
&\quad + b \sin(x^{(n)} + y^{(n)} + z^{(n)} + w^{(n)}), \\
w^{(n+1)} &= z^{(n)} + w^{(n)}.
\end{aligned}
\qquad (\bmod 2\pi)
$$

A trajectory is a set of points in the (x, y, z, w) space. It is first projected on the three-dimensional (x, y, z) space; then a series of nine slices are taken, defined by $|z - z_0| < 0.01$ with nine regularly spaced values for z_0. Parameter values are $a_1 = -1.3$, $a_2 = -1$ for both figures; $b = 0.075$ for fig. 1 and $b = 0.275$ for fig 2. In the first case, the points fall on well defined curves in each slice; this suggests that they fall on a two-dimensional surface in (x, y, z) space. In the second case, the situation is rather different: the points appear to fill a three-dimensional region in (x, y, z) space. Both figures have been obtained by the tangential Taylor 3 ($\tau 3$) synthetic mapping. Figures for the point-set Taylor 3 (T3) mapping are identical. For figures of the real mapping, see Froeschlé (1972), figs. 11.d and 11.f. But again the figures are identical at least in the integrable case.

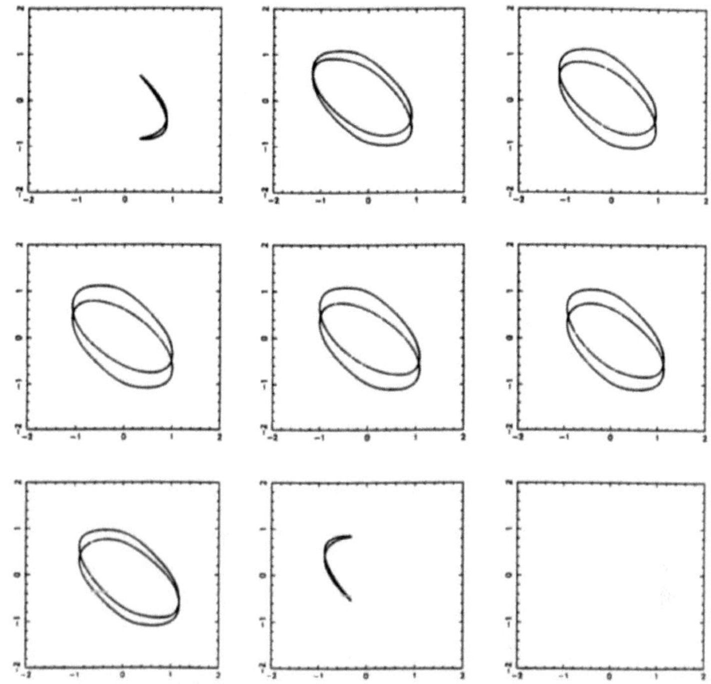

Fig. 1. Slices of the subspace (x, y, z) for the tangential Taylor 3 interpolation of the four-dimensional standard map, for $a_1 = -1.3$, $a_2 = -1.$ and $b = 0.075$, the starting point being $P = (1, 0, 0.5, 0.5)$. Then slices are defined by $|z - z_0| < 0.01$ with $z_0 = -1.5 + i/3$, where i varies from 1 to 9 from left to right and top to bottom.

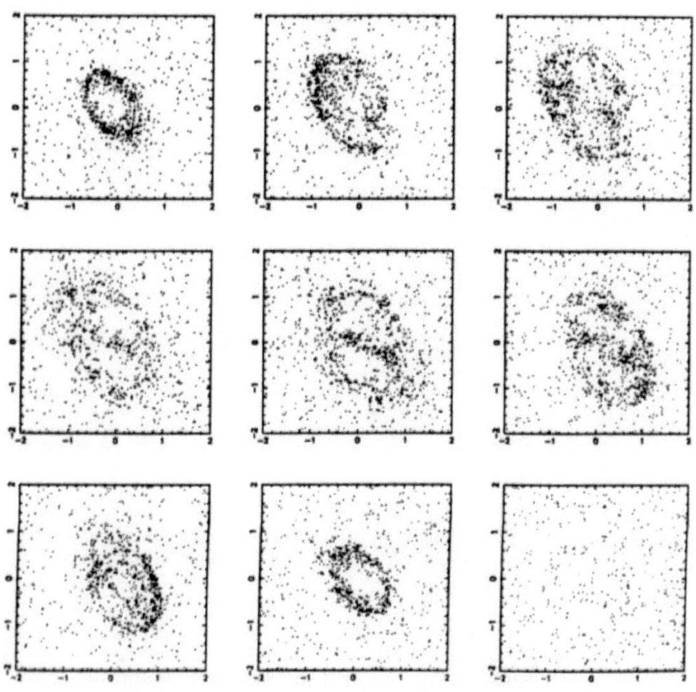

Fig. 2. Same as fig. 1, but with $b = 0.275$.

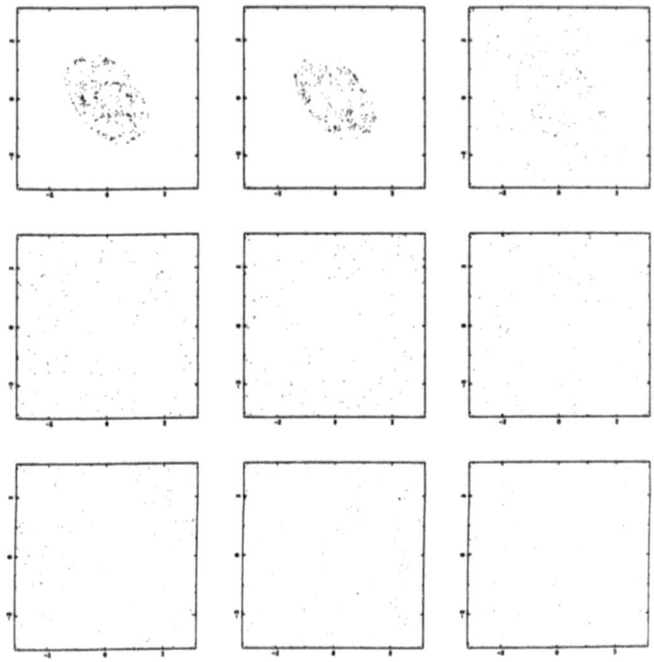

Fig. 3. (a) Median slice ($|z| < 0.01$) for the orbit of fig. 1 with $b = 0.275$ for the real mapping. In each square are plotted the points of 40000 successive iterations.

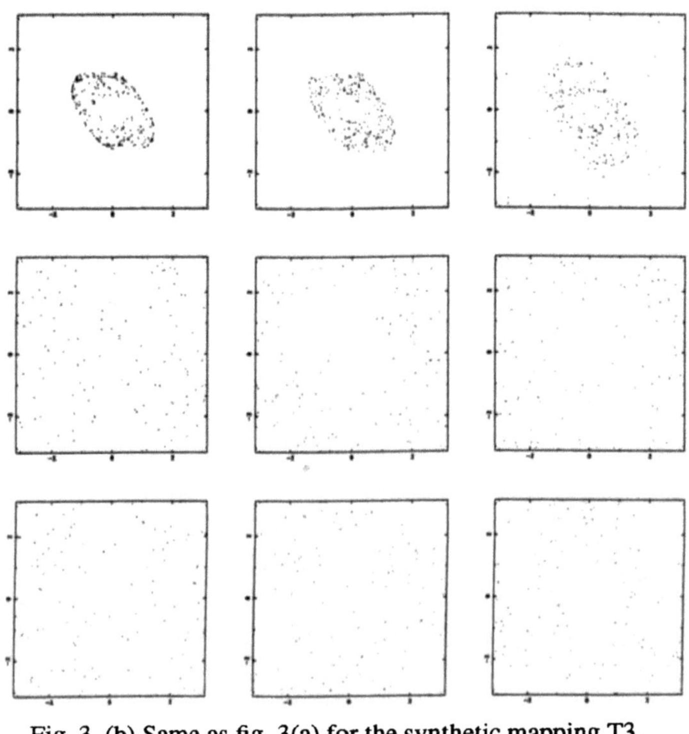

Fig. 3. (b) Same as fig. 3(a) for the synthetic mapping T3.

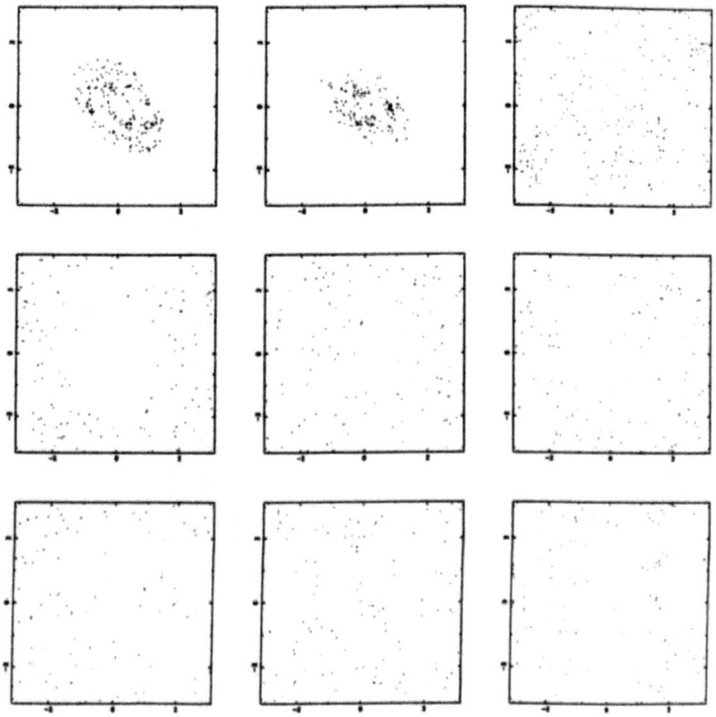

Fig. 3. (c) Same as fig. 3(a) for the synthetic mapping $\tau 3$.

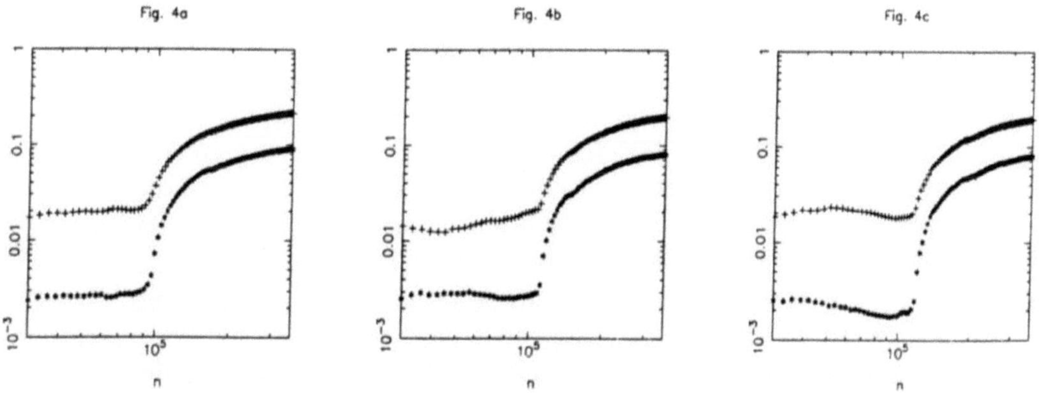

Fig. 4. (a) Positive Liapounov characteristic numbers for the same orbit than in fig. 3a, for the real mapping. (b) Same as fig. 4a for the synthetic mapping T3. (c) Same as fig. 4a for the synthetic mapping $\tau 3$.

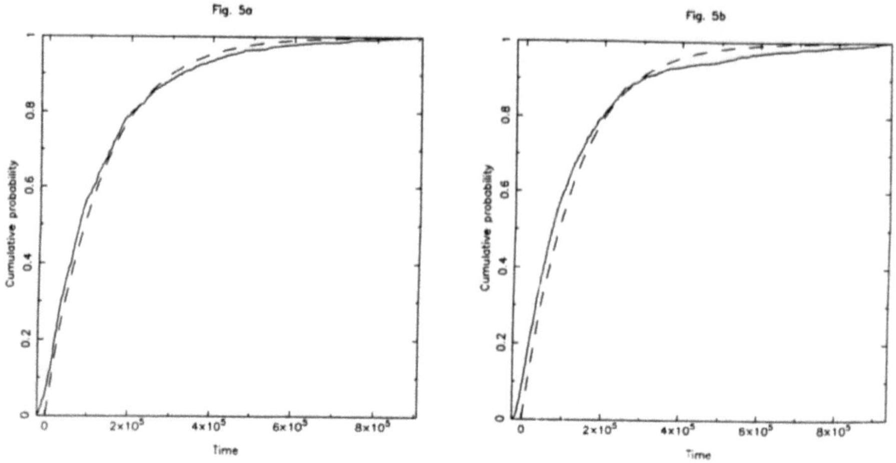

Fig. 5. (a) Distribution of the explosion time for a set of 1000 orbits randomly distributed in a cube of size 10^{-8} around the same orbit than in fig. 3a, for the real mapping (solid line). Dashed line: Poisson distribution with the same mean value. (b) Same as fig. 5a for the synthetic mapping $\tau 3$.

A rather different technique of study consists in determining numerically the Liapounov characteristic exponents of the trajectories. This allows one to distinguish between quasi-periodic and chaotic orbits and to map the regular and chaotic regions in phase space. The question raised by Froeschlé (1970) on the number of quasi-integrals for a one parameter set of orbits of the non planar restricted three-body problem has been solved more easily, using the LCEs, by Gonczi and Froeschlé (1981). The combination of the two methods (slice cutting and LCEs) has been very useful to discover and illustrate a sticking-exploding phenomenon. This phenomenon has been visualized plotting on figs. 3a-c the median slices ($|z| < 0.01$) of the projections along the w axes of the points P_n for $(i-1) \times 40000 \leq n < i \times 40000$, with i varying from 1 to 9 from left to right and top to bottom, and $b = 0.275$, for the real, T3 and $\tau 3$ mappings respectively. Figs. 4a-c show on a log-log scale, the behaviour of the two positive Liapounov characteristic indicators γ_i^n (the limits of which are identified with the Liapounov characteristic numbers) as a function of time (i.e. the number n of iterations). Note that one should not pay too much attention to the fact that the times of explosion coincide. Actually, we found slightly worse agreement for the explosion time when using the real mapping on two different computers that handle the last bit in floating point operations differently, and that put the operations in different order due to differences in the compilers.

In order to study more precisely the question of the exploding time we have computed this time for 1000 different initial conditions taken at random in a small box of size $r = 10^{-8}$ centered at $P = (1., 0., 0.5, 0.5)$. For the real mapping the distribution (solid line) fits well with a Poisson distribution (dashed line) (see fig.

5a). This may be due to the fact that the confined part is already chaotic since the Liapounov indicators do not decrease to zero. Of course such studies are more costly using the synthetic maps (for this particular mapping). The results shown on fig. 5b have been obtained with the tangential synthetic mapping $\tau 3$. There is no discrepancy between the mapping distribution and the Poisson distribution of the same mean value. In addition, the value of the mean and variance for this distribution are very close to that of the real case. These values are 129 650 and 150 900 iterations respectively for the real mapping and 126 120 and 164 960 for the synthetic mapping. Actually, for both distributions, only orbits escaping within a time span of 10^6 iterations have been taken into account. In 1000 computed orbits, 21 did not escape for the real mapping and 18 for the synthetic mapping $\tau 3$.

2. Conclusions

Synthetic maps appear to be valuable tools for celestial mechanics. Even for problems with more than two degrees of freedom where the situation is less straightforward, the use of a hash function has allowed to overcome the difficulties (memory management; see paper II). Results of probabilistic essence are also recovered even if, of course, individual orbits are different.

Acknowledgements

We would like to thank the Institut Mediterranéen de Technologie for providing us with an account on a CRAY-YMP. This allowed us to carry out the computation regarding the distribution of exploding time in the case of the $\tau 3$ synthetic mapping. The use of that computer speeds up the computation by a factor of 21 compared to an IBM-Risc 6000 320H, and by a factor of about 150 compared to a SUN SLC. The computing time was about 420 seconds for one orbit with 500 000 iterations.

References

Duncan, M., Quinn, T., Tremaine S.: 1989, "The long term evolution of orbits in the solar system. A mapping approach." *Icarus* **82**, 402

Froeschlé, C.: 1970, "A numerical study of the stochasticity of dynamical systems with two degrees of freedom." *Astron. Astrophys.* **9**, 15

Froeschlé, C.: 1972, "Numerical study of a four dimensional mapping." *Astron. Astrophys.* **16**, 172

Froeschlé, C., Petit, J.M.: 1990, "Polynomial approximation of Poincaré maps for Hamiltonian systems." *Astron. Astrophys.* **238**, 413

Froeschlé, C., Rickman, H.: 1988, "Monte Carlo modelling of cometary dynamics." *Celestial Mech.* **43**, 265

Gonczi, R., Froeschlé, C.: 1981, "The Liapounov characteristic exponents as indicators of stochasticity in the three body restricted problem." *Celest. Mech.* **25**, 271

Hénon, M.: 1981, "Numerical exploration of Hamiltonian systems." *Cours des Houches* **XXXVI**, North Holland, Amsterdam, 57

Petit, J.M., Froeschlé, C.: 1992, "Polynomial approximation of Poincaré maps for Hamiltonian systems, II." *Astron. Astrophys.* submitted

Varosi, F., Grebogi, C., Yorke, J.A.: 1987, "Simplicial approximation of Poincaré maps of differential equations." *Phys. Lett. A* **124**, 59

A SIMPLE MAPPING FOR COMETS IN RESONANCE

J.E. CHAMBERS

Department of Astronomy, The University, Manchester M13 9PL, UK.
Harvard-Smithsonian Center for Astrophysics, 60 Garden St., Cambridge, MA 02138, USA.

Abstract. The dynamical evolution of a low inclination, intermediate period comet in mean-motion resonance with a giant planet can be described by a simple mapping giving the comet's heliocentric energy and ecliptic longitude at successive aphelia. Provided that the energy oscillations produced by the resonance are not too large, this mapping is linear. When the amplitude of the oscillations is close to the maximum possible within the resonance, the mapping can be improved by adding non-linear terms.

Key words: Comets – short-period comets – dynamics – mean-motion resonance

1. Introduction

Look closely at the long-term evolution of the orbit of a comet and you will see numerous examples of mean-motion resonances which occur whenever the comet's orbital period P becomes approximately equal to a simple fraction of that of a major planet. A comet whose mean-motion n $(= 2\pi/P)$ is roughly p/q times that of a planet (where p and q are integers) is said to be in the $p : q$ mean-motion resonance.

Marsden (1970) noted that short-period comets can be influenced by resonances, for example periodic comet Neujmin 1 currently librating about the 2:3 resonance with Jupiter. Everhart (1973) found many cases in which theoretical asteroids between Jupiter and Saturn became temporarily captured into a resonance with one or the other planet. There also appears to be some concentration of Halley-type comets towards resonances with Jupiter (Carusi *et al.* 1987).

Although complete numerical integrations of the equations of motion can be used to investigate the evolution of a particular comet in a particular resonance, a remarkably accurate description of the motion can be obtained using a simple mapping which gives the comet's orbital elements at successive aphelia. The mapping (qualitatively similar to that used by Chirikov & Vecheslavov (1989) to study comet Halley) is considerably faster than integrations but retains much of their fine detail. It may therefore be useful for making a comprehensive study of comets in resonance.

2. A Simple Mapping for Librations

All the essential elements of mean-motion resonance can be seen in the circular, planar, restricted three-body problem (CPRT). Here the sun and a single planet travel on fixed circular orbits about their barycentre while a comet of negligible mass moves in their combined gravitational field. The system can be completely described by four variables: the comet's osculating eccentricity e, true anomaly f, longitude relative to the planet $\gamma \equiv f + \varpi - f_J$ (where f_J is the planet's

Celestial Mechanics and Dynamical Astronomy **57**: 131–136, 1993.

true anomaly and ϖ is the comet's longitude of perihelion), and dimensionless heliocentric energy $\alpha \equiv 1 - a_{\rm res}/a$ (where a is the comet's semi-major axis and $a_{\rm res}$ is its value at exact resonance). The system evolves along a trajectory in phase-space which is confined to a three-dimensional hypersurface by the Jacobi integral $C(\alpha, e, \gamma, f) = 2H + 2Gm_\odot[(1 + m_J/m_\odot)a(1 - e^2)/a_J^3]^{1/2}$ =const where m_J and a_J are the planet's mass and semi-major axis, G is the gravitational constant and H is the system's Hamiltonian.

Consider a two-dimensional surface cutting this hypersurface. The trajectory will cross this "surface of section" at a set of points whose distribution depends on the initial conditions. Many trajectories produce points scattered across the surface apparently at random. In this case the evolution is highly sensitive to the initial conditions. On some parts of the surface however, the points are confined to closed curves by a second isolating integral. These essentially predictable "quasi-periodic" regions surround mean-motion resonances.

In both regimes we can imagine a mapping which gives the coordinates of a point on the surface in terms of those of the previous point. For example, using $f = \pi$ (i.e. the comet at aphelion) as the surface of section, the mapping could be

$$\alpha_1 = F_1(\alpha_0, \gamma_0, C)$$
$$\gamma_1 = F_2(\alpha_0, \gamma_0, C)$$

where e_1 can be found using $C(\alpha_1, e_1, \gamma_1, \pi) = C(\alpha_0, e_0, \gamma_0, \pi)$. Unfortunately, exact analytic expressions for F_1 and F_2 are unknown.

However, we can get an approximate mapping by replacing the planetary perturbations accumulated by the comet over one revolution (both direct and indirect) with a single impulse applied at perihelion. That is, the comet starts at aphelion with elements (α_0, γ_0), moves on a heliocentric Keplerian orbit to perihelion, receives an impulse changing its energy from α_0 to α_1, and returns to aphelion on a new Keplerian orbit with elements (α_1, γ_1). This approximation is most appropriate for comets with highly eccentric orbits and large aphelion distances, such as Halley.

Using the definition given above, the change in γ over one revolution is $(f_1 - f_0) + (\varpi_1 - \varpi_0) - (f_{J1} - f_{J0})$. Applying the impulse approximation, neglecting the small change in ϖ and recalling that $f_1 = f_0 = \pi$, gives

$$\gamma_1 = \gamma_0 - n_J(P_0/2 + P_1/2)$$

where n_J is the planet's mean motion, and $P = a^{3/2}$ is the comet's orbital period (in years when a is in AU). Within the quasiperiodic region surrounding a resonance, $|\alpha|$ is typically less than about 0.05, and to a very good approximation $a^{3/2} = a_{\rm res}^{3/2}(1 + 3\alpha/2)$. Hence

$$\gamma_1 = \gamma_0 - \frac{3}{4}k_2(\alpha_0 + \alpha_1) - k_2 \tag{1}$$

where $k_2 = n_J a_{\rm res}^{3/2} = q \times 2\pi$ for the $1 : q$ resonance, when a is expressed in AU and n_J in radians per year.

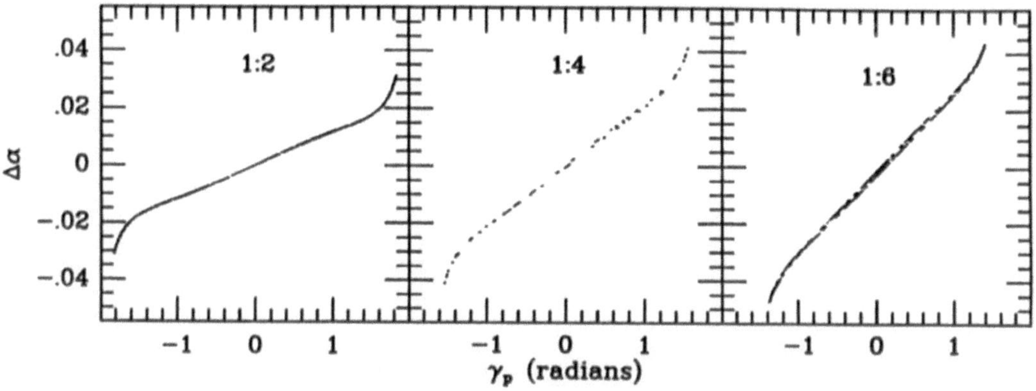

Fig. 1. The change in α per revolution versus γ_P for 1000 revolutions of a comet with $C = 1.5$ in the 1:2, 1:4 and 1:6 resonances with Jupiter.

It is useful at this point to define a new quantity $\gamma_P \equiv \gamma - f + n_J(M - 2\pi)/n$ where $0 < M < 2\pi$ is the comet's mean anomaly and n is its mean motion. γ_P is the value that γ would have if the planet and comet progressed along their osculating orbits until the latter next came to perihelion.

Fig. 1 shows $\Delta\alpha = \alpha_1 - \alpha_0$ versus γ_P for 1000 revolutions of a comet with $C = 1.5$ in the 1:2, 1:4 and 1:6 resonances with Jupiter. The values were calculated using full integrations of the equations of motion for the CPRT, with the initial values of α and γ chosen to place the comet at the extreme edge of the quasi-periodic resonance region. In each case the points lie on a curve which is approximately linear over most of its range. Hence provided that γ_P does not become too large:

$$\alpha_1 = \alpha_0 + k_1 \gamma_{P0} \tag{2}$$

where $\gamma_{P0} \equiv \gamma_0 - f_0 + n_J(M_0 - 2\pi)/n_0 = \gamma_0 - \pi(1 + n_J/n_0)$ and the empirically determined values of k_1 are 0.0116, 0.0212 and 0.0296 for the 1:2, 1:4 and 1:6 resonances respectively, obtained by fitting straight lines to the linear portions of the curves in Fig. 1.

Using the definition of γ_P, Kepler's third law $n = 2\pi a^{-3/2}$ and the surface of section condition $f_0 = M_0 = \pi$ leads to

$$\begin{aligned} F_1 \equiv \alpha_1 &= \alpha_0 + k_1(\gamma_0 - \pi) - k_1 n_J a_0^{3/2}/2 \\ &= \alpha_0(1 - 3k_1k_2/4) + k_1(\gamma_0 - \pi - k_2/2) \end{aligned}$$

where the last step uses the approximation $a^{3/2} = a_{\text{res}}^{3/2}(1 + 3\alpha/2)$.

For a resonance of the form $1 : q$, the above expression simplifies to

$$\begin{aligned} \text{For } q \text{ odd: } \alpha_1 &= \alpha_0(1 - 3k_1k_2/4) + k_1\gamma_0 \\ \text{For } q \text{ even: } \alpha_1 &= \alpha_0(1 - 3k_1k_2/4) + k_1(\gamma_0 - \pi) \end{aligned} \tag{3}$$

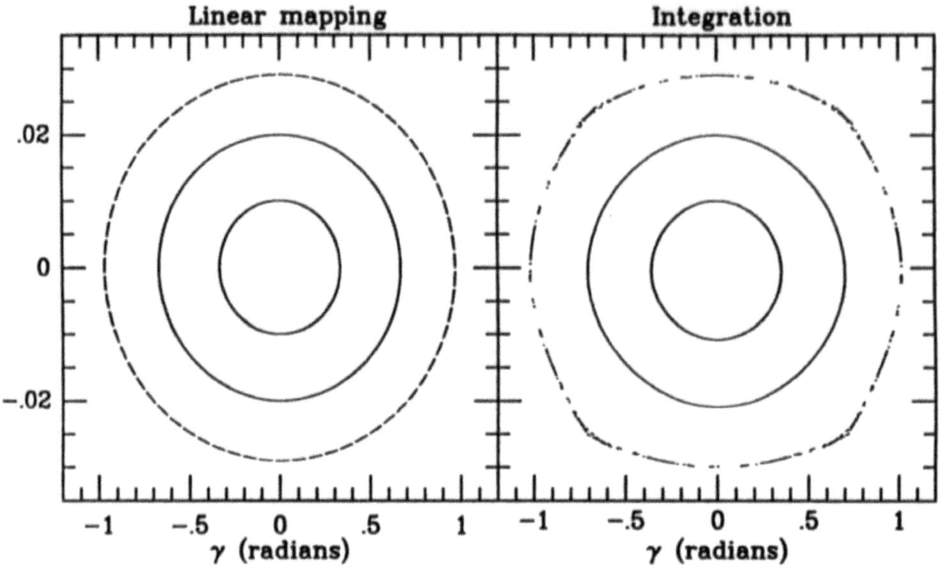

Fig. 2. Surfaces of section for a comet with $C = 1.5$ in the 1:6 resonance with Jupiter produced by the linear mapping (left) and a full integration of the equations of motion (right).

When q is odd, Eqns. 1 and 3 can be combined to give a linear mapping of the form

$$\begin{pmatrix} \alpha_1 \\ \gamma_1 \end{pmatrix} = \begin{pmatrix} k_3 & k_1 \\ (k_3^2 - 1)/k_1 & k_3 \end{pmatrix} \begin{pmatrix} \alpha_0 \\ \gamma_0 \end{pmatrix}$$

where $k_3 = 1 - \frac{3}{4}k_1k_2$.

The mapping matrix is such that successive points lie on an ellipse whose major and minor axes, $a_{\rm ell}$ and $b_{\rm ell}$, lie parallel to the γ and α axes, and are related by $a_{\rm ell}/b_{\rm ell} = (1-k_3^2)^{1/2}/k_1$. The angular progression of successive points around the ellipse is $\delta = \cos^{-1} k_3$. An equivalent mapping can be derived for the case when q is even by defining a new quantity $\gamma\prime \equiv \gamma - \pi$. The mapping for α and $\gamma\prime$ is then identical to the one given above.

Fig. 2 compares the evolution at the 1:6 resonance predicted by the linear mapping ($k_1 = 0.0296$ and $k_2 = 6 \times 2\pi \simeq 37.7$) with a full integration of the CPRT, for three initial values of α and γ. The integration does indeed produce roughly elliptical curves, even at the extreme edge of the quasi-periodic region.

Using the linear mapping, initial conditions further from exact resonance continue to produce larger, but otherwise identical, ellipses. In practice the quasi-periodic resonance region is finite in extent. This discrepancy results from the increasing non-linearity of the graphs in Fig. 1 as γ_P becomes large.

TABLE I
Angular progression (degrees) of successive points around the ellipses of Figs. 2 and 3.

	Integration	Linear Map	Quintic Map
$\alpha = 0.1$	81.6	80.6	80.5
$\alpha = 0.2$	80.8	80.6	80.8
$\alpha = 0.29$	84.7	80.6	83.1

3. Nonlinear Librations

The mapping can be improved for large initial values of α and γ by fitting a polynomial to the data in Fig. 1 instead of a straight line. A first step is to add a cubic term (the quadratic term is negligible) to Eqn. 2 to give

$$\alpha_1 = \alpha_0 + k_{11}\gamma_{\mathrm{PO}} + k_{13}\gamma^3_{\mathrm{PO}} \tag{4}$$

Combined with Eqn. 1 this gives a new mapping for α and γ. For a comet with $C = 1.5$ in the 1:6 resonance with Jupiter, $k_{11} = 2.66 \times 10^{-2}$ and $k_{13} = 3.14 \times 10^{-3}$. The mapping can be further improved by adding a term in γ^5_{PO}. In this case, the mapping constants are $k_{11} = 3.01 \times 10^{-2}$, $k_{13} = -4.20 \times 10^{-3}$ and $k_{15} = 3.04 \times 10^{-3}$. Fig. 3 shows surfaces of section for the 1:6 resonance produced using these cubic and quintic mappings which can be compared with Fig. 2.

One quantitative comparison which can be made between the integrations and the mappings is the angular progression δ of successive points around the "ellipse". Table 1 gives δ for the three sets of initial conditions used in Figs.2 and 3 for the integrations and the linear and quintic mappings. The agreement is generally good. In particular the quintic map correctly predicts an increase in δ near the edge of the quasi-periodic resonance region.

Adding non-linear terms produces several new phenomena which are qualitatively similar to those seen in full integrations. The non-linearity first appears as a distortion of the libration ellipses, which become box shaped. For larger initial values of α and γ the ellipses may fragment into "island chains"—secondary librations, produced when the orbital and libration periods are themselves commensurable. Finally, when the initial conditions are sufficiently large, the comet makes very close approaches to the planet, and the trajectory becomes chaotic.

4. A Quick Look at Reality

In the CPRT, a comet initially in a resonance remains there indefinitely. In reality this doesn't happen. However, over a short time interval a low inclination comet in

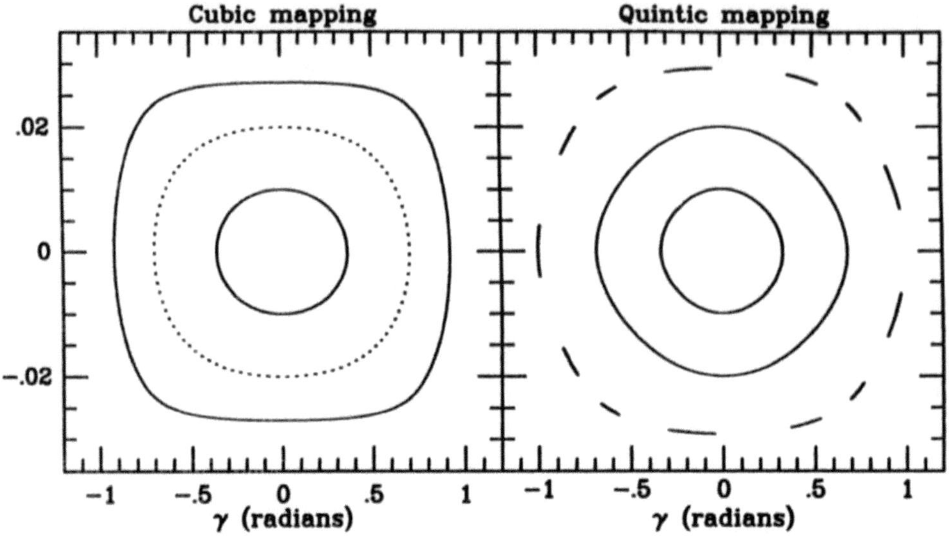

Fig. 3. Surfaces of section for a comet with $C = 1.5$ in the 1:6 resonance with Jupiter produced by the cubic (left) and quintic (right) mappings.

the real solar system will still approximately follow the quasi-periodic "ellipses" of the CPRT since this is an excellent first approximation. Over longer time spans, the inadequacies of the model become apparent, and the comet wanders off its starting ellipse. The long-term evolution can be pictured as a drifting between the quasi-periodic curves, eventually leading to expulsion from the resonance into the chaotic sea surrounding it. However such is the abundance of resonances, that it will not be long before the comet encounters another.

Acknowledgements

I would like to thank M. E. Bailey, J. D. Hadjidemetriou, P. J. Message and many of the participants at the Symposium in Pléneuf-Val-André for their helpful comments and suggestions during the preparation and presentation of this paper.

This work was supported by the SERC.

References

Carusi, A., Kresák, L., Perozzi, E. & Valsecchi, G.B. (1987) Astron. Astrophys. **187,** 899–905.
Chirikov, B.V. & Vecheslavov, V.V. (1989) Astron. Astrophys. **221,** 146–154.
Everhart, E. (1973) Astron. J. **78,** 316–328.
Marsden, B.G. (1970) Astron. J. **75,** 206–217.

GEOPOTENTIAL FROM SPACE TECHNIQUES

P. EXERTIER
Observatoire de la Côte d'Azur, Département CERGA
Avenue Copernic, 06130 Grasse, France

Abstract. The monitoring of the perturbed motion of artificial satellites since the 1960's has provided a valuable resource for improved activities in geodesy. These observations and the improved geodetic techniques they fostered have provided an unparalleled means for studying both the gravity field and the Earth's shape. In this paper, we will review the various determinations of the Earth's gravity field produced at many research centers over the years from space techniques. The specific methods which have been used to measure our planet, such as satellite altimetry and satellite orbital perturbation analyses (over short and long periods of time), will be compared.

Key words: Artificial satellite – Earth gravity field – Earth shape

1. Introduction

Most simplistically, the determination of the figure of the Earth is basically a geometrical problem. The result of these efforts should be a catalog of cartesian coordinates of fundamental points in a unique earth-fixed coordinate frame. From these reference values and from relative positioning measurements (and successive interpolations) it would be possible to determine the location of any point on the surface of the Earth. On the over hand, when we introduce the notion of force, which is based on the Newton universal gravitational law and on the fundamental principles of the dynamics, the shape of the Earth maps into a complex gravitational field containing both gravity and centrifugal forces. The determination of the surface gravitational acceleration can be directly made using a gravimeter. But it is a purely static method and the observations of local gravity at any geometrical point of the surface of the Earth are delicate to carry out and complicated to interpret given environmental and topographic settings. The launching of artificial satellites in the 1960's onward provided a new tool for improved activities in geodesy. It has permitted a transcendence of the local methods which were applied to restricted regional areas and has given us access to the global field. However, the Earth gravity field knowledge is very important to construct a geodetic reference system. Now, the geoid is that equipotential surface of the geopotential that corresponds to mean sea level. The role of the geoid is fundamental in the determination and the representation of the figure of the Earth.

Satellite geodesy has traditionally been divided into two techniques; one based on geometric methods, and the other on dynamic methods. The geometric method utilizes the satellite simply as a high-flying target. Knowledge of the satellite's orbit is only required to locate the satellite during tracking. This method do generally require quasi-simultaneous measurements at different observatories. The dynamic method (perturbation methods due to H. Poincaré) treats the satellite as a sensor as it passes through the earth's gravitational field. Both methods have their merits and difficulties, but, as it is explained below, most of the information for geophysics

Celestial Mechanics and Dynamical Astronomy **57**: 137–153, 1993.

from satellite data analysis is based on the dynamic method. The publication of Kaula's *Theory of Satellite Geodesy* [Kaula, 1966] was one of the advances in satellite geodesy towards the end of the 1960's. His systematic development of close Earth satellite orbital theory quickly became the basis for many subsequent analyses of orbit perturbations. However, dynamic methods for developing Earth gravity field models from satellite tracking data require some assumptions about the geocentric positions of the tracking stations, and if the terrestrial coordinate frame is not perfectly known, the tracking station positions have to be determined with the dynamical parameters. Then, the derivation of a reference frame is intimately tied to the determination of satellite orbits through the dynamical equations of motion. Thus, there is a strong interrelationship between the concepts of the Earth's shape (including its orientation) and gravity field. Thusly, we will not try to separate these concepts. We will focus on the fundamental manifestations of the Earth gravity field, as seen using the dynamic methods of satellite geodesy, and we will illustrate the determination of the field over the years.

Improvements within geodetic techniques [Kovalevsky, 1971] have provided a unique means for studying the gravity field, the Earth's shape and related phenomena. In the early 1970's, for the first time, the space geodesy applied to Earth's Physics and Oceanography has been re-defined in its objectives. Consequently, numerous Research Programs have emerged.

2. Geopotential Improvements in the 1980's

The emergence of satellite geodesy as a useful tool for geophysical research has been stimulated by the development of high-accuracy tracking techniques in order to determine both satellite and station positions. The most advanced and accurate of these techniques, in the optical domain, is Satellite Laser Ranging (SLR). In fact, during just ten years, between the 1970's and the 1980's, the quality and quantity of SLR data have been improved several-fold due to the development of "second" and "third" generation laser ranging systems, as witnessed by their emergence as an increasingly important data set for Geodynamics studies [Degnan, 1985].

The primary goals of the SLR system development and demonstration were to directly measure tectonic plate motion at about the 0.01 m/yr level, to determine Earth orientation and accurate ephemerides for the satellite. Laser ranging activities have concentrated on satellites designed especially for geodetic laser operations, namely Lageos (launched by the U.S. in 1976), Starlette (launched by the French CNES in 1975) and Ajisai (launched by Japan in 1986). In all cases, the combination of satellite design and orbit (mass, shape and altitude) has provided a quasi-stable reference frame for studying the evolution of any time-dependent phenomena related to changes in the size, shape, and orientation of the Earth in addition to the long-wavelength component of the gravity field including its temporal variations. Most of the information for Geodynamics from laser ranging is from tracking these satellites. However, Lageos is the best target satellite for geodetic purposes, due to its small area-to-mass ratio, which reduces difficult-to-model non-gravitational (or surface) forces, and its high altitude (one Earth radius) which attenuates the effect of the short-wavelength parts of the Earth's gravity field. In the early 1980's, the possibility of computing orbit trajectories of sufficient accuracy to resolve station motions due to plate tectonics was realized [Christodoulidis et al., 1985; Tapley et al., 1985]. However, a minimum of 5-6 years of observations was needed in order to obtain unambiguous estimates of the overall tectonic velocities. In recent studies [Smith et al., 1990; Biancale et al., 1991], horizontal motions at many laser tracking station sites have been determined from at least 5 years of Lageos data. These motions are compatible, within a few mm/yr, with those predicted by geological models for stations located far enough from plate boundaries. Moreover, agreement between SLR and VLBI station positioning is now globally at the ± 2 cm level [Ray et al., 1991].

Improved ground tracking systems (e.g., laser ranging) along with other more conventional tracking systems (Doppler) have provided highly accurate means of sensing the perturbed motion of artificial satellites. However, satellite orbital data represent the long-wavelength part of the gravity spectrum; the short-wavelength parts are smoothed out by the attenuation of the gravity field with the altitude and by tracking coverage. Thus, from these measurements (tracking data), the broad features of the geopotential have been determined. Over the last fifteen years, various solutions for geodetic and geodynamic parameters have been produced at the National Aeronautics and Space Administration Goddard Space Flight Center (NASA/GSFC), the University of Texas at Austin Center for Space Research (UTCSR), and the joint Groupe de Recherches de Géodésie Spatiale - Deutsche Geodaetische Forchungs Institut (GRGS/DGFI). The geopotential models have increased in size and accuracy with the inclusion of newer tracking data, better non-gravitational force modeling, and more orbits over a wide range of inclinations and mean motions [Lerch et al., 1979; Marsh et al., 1988, 1990; Reigber et al., 1984; Schwintzer, Biancale et al., 1991]. As it is true, detailed models of the Earth's gravity field are important both for direct analysis of geophysical phenomena and for determination of satellite orbits. For one example, the ability to monitor tectonic motions through laser tracking depends on the accuracy of orbits which in turn depend fundamentally on the accuracy of the gravity model. Furthermore some of the models produced entirely from satellite tracking data (so called "satellite-only" models) have greatly contributed to improved modeling of the long-wavelength geoid. In itself it has extensive geophysical and oceanographic applications, for example, in understanding the slope of the sea surface across large ocean basins. However, before introducing several years of Lageos tracking data in the solutions, errors due to uncertainty in the values of low-order terms were the limiting factors in these and other investigations.

To resolve the long-wavelength geoid, some models have increasingly relied on subdecimeter laser ranging accuracies and special spacecraft, such as Lageos, which have easier to represent non-gravitational force modeling. Thus, Lageos tracking data have been instrumental in considerably improving the accuracy of low-order terms (i.e., degree and order 4) of the geopotential [Lerch et al., 1982; Reigber et al., 1985]. Consequently, the capabilities of monitoring tectonic motions using satellite laser ranging were also improved. The Earth's orientation parameters estimate contributed to the stability of harmonic coefficient solutions and improved positioning accuracy. These models have apparently achieved high accuracies of low degree and order geopotential coefficients ($l < 8$). The calibration of their long-wavelength part has been accomplished notably with independent satellite data (e.g., 24-Hour satellite accelerations), but the verification of the coefficient accuracies remains a difficult problem.

The short-wavelength features of the geoid, which have considerably less effect on satellite position than the longer ones, have, until the 1970's, been studied using surface gravimetry [Rapp, 1969]. This type of data, however, suffers from three failings: (i) very inhomogeneous and uneven accuracy, (ii) incomplete coverage, and (iii) significant uncertainty about how many of the data in use are based, at least in part, on some kind of geophysical estimation or correlation. The Smithsonian Astrophysical Observatory's (SAO) comprehensive gravity field model complete to degree and order 16 [Gaposchkin and Lambeck, 1971] was one in which, for the first time, the introduction of surface gravity observations led to an improved representation of the satellite motion. Further iterations were published by Gaposchkin such as the SE-III model which used the Isagex laser data [Gaposchkin, 1973]. However, by about the mid 1970's the solutions appeared to be approaching the limits of the methods of analyzing perturbation in the satellite motion as deduced from ground-based tracking data, whether camera, electronic or laser. The launching in 1975 of the U.S. satellite Geos-3 represented a next important step in measuring the Earth's gravity field. Then, a third source of information of the Earth's geoid has been provided by direct measurements of the sea surface. Through the use of space-borne altimetry measuring the height of the ocean surface, the short-wavelength features of the field have been estimated. Preliminary global solutions incorporating these observations have been published by several authors — the discussion of altimetry appears on Section 3.

These geopotential models, in fact, belong in the group of "general purpose" models intended for both terrestrial use and satellite orbit computations. On the other hand, special models have also been developed for the computation of satellite orbits or more specifically "tailored" for the orbit of specific satellites such as Geos-3, Seasat, Starlette, or Spot-2. The purpose of these developments was not to create a general geopotential model, but a specific one for the computation of precise ephemerides for a given satellite. For one example, the Starlette satellite has provided an important means for deriving geodetic and geodynamic parameters from near-Earth satellite laser ranging data [Marsh et al., 1985]. The PGS-1331 gravity model, while specialized for Starlette, reflects the substantial sensitivity of this satellite to the geopotential (constant and variable parts) and, in particular, to the higher degree and order coefficients. From the point of view of geodetic modeling, Starlette has provided a means to verify the tracking station locations obtained independently from Lageos. In case of satellite altimetry, because an accurate determination of the satellite position (radial) is required to separate satellite height variations from ocean surface height variations, such "tailored" gravity field models have also been developed [Lerch et al., 1982(b)].

3. Principal Contributions of Satellite Altimetry

The most productive program contributing to the improvement of global gravity coverage was satellite altimetry. The sea surface information provided by satellite altimetry contributed dramatically to the knowledge of the geoid (Fig. 1), the gravity anomalies over the oceans and in turn to the improvement of the spherical harmonic geopotential models. In the 1970's, radar altimeters have been carried out on Skylab, Geos-3 and Seasat. The Geos-3 (0.50 m - accuracy) and Seasat (0.10 m - accuracy) altimeter data have been analyzed by a number of investigators [Marsh, 1983] for detailed regional (e.g., [Barlier et al., 1982]) as well as global mean sea surface computations [Marsh, 1986]. But as it was mentioned earlier, the disadvantage of satellite altimetry from the point of view of geopotential modeling, is that the data coverage, due to its very nature, is restricted only to the ocean areas. Consequently, this requires the combination with the over types of data. By combining altimeter data with terrestrial gravity and satellite perturbation analysis, comprehensive models of the Earth's gravity field have been developed (in the form of spherical harmonic coefficients) to describe both the complex shape of the geoid and its resulting effects on near-Earth orbiting objects [Gaposchkin, 1980; Lerch et al., 1981; Reigber et al., 1984]. Many data sources of varying quality have contributed to these global solutions. The use of altimeter data resulted in significant improvements in the 1980's gravity field models. This is particularly true with the use of Geosat (0.20 m - accuracy) altimeter measurements since 1985. The dense coverage provided over the ocean areas has been particularly important for the definition of the coefficients above degree and order 20 in the models [Tapley et al., 1988; Marsh et al., 1990; Rapp and Pavlis, 1990; Reigber et al., 1991]. Extensive tests have been made to determine the improvements in terms of radial orbit and geoid accuracies. Orbit determination accuracy for Seasat in 1978 was about 2 m rms for the radial position of the satellite, whereas the present computations for the Geosat orbit are approaching 0.20 m rms.

Altimeter profiles and global mean sea surfaces have formed the basis for numerous geophysical studies in ocean areas where surface data have not previously been available [Douglas et al., 1987]. The outstanding example is undoubtedly the use of satellite altimetry for the study of the marine gravity field [Rapp, 1982, 1983, 1989]. Estimates of the mean sea surface and the deflections of the gravity intensity vector (vertical) based on altimetric measurements have been especially useful. Moreover this surface can also be used for the detection of bathymetric fea-

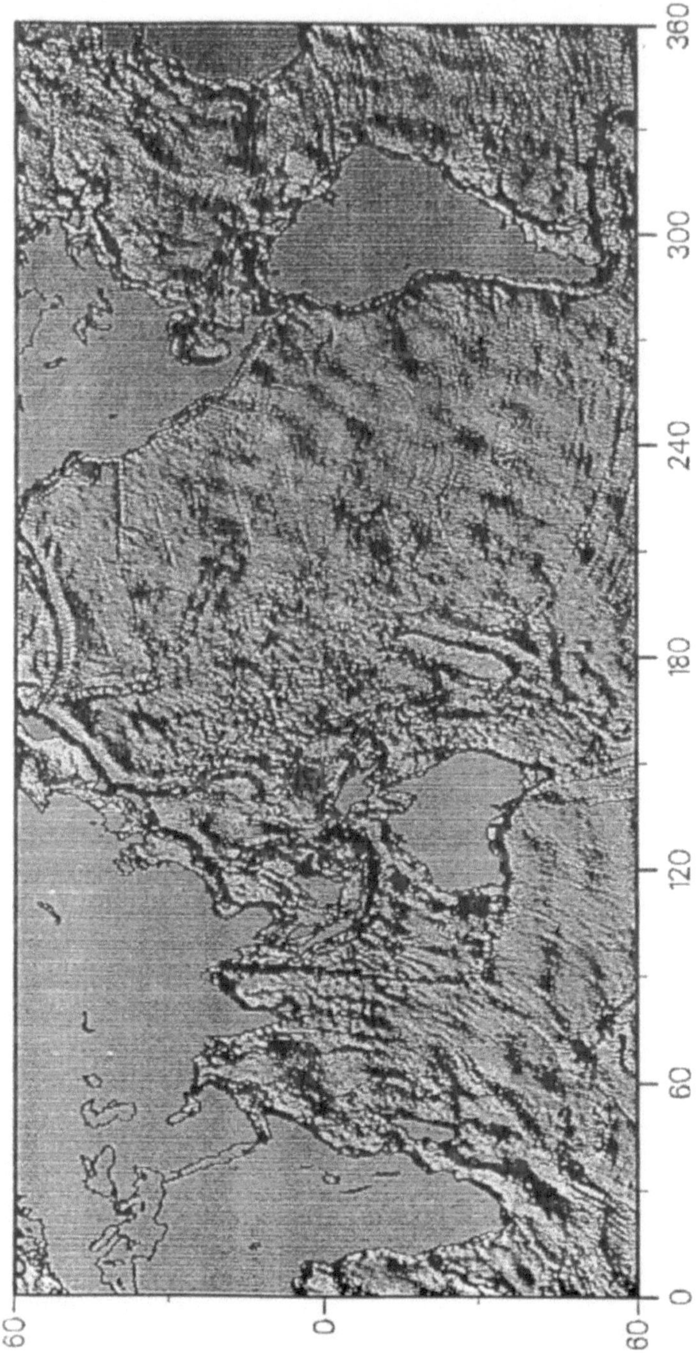

Fig. 1. The 0.125°x0.125° gridded sea surface height obtained from the mean profiles (Geosat data) after removal of the GEM-T2 (50x50) geoid heights (after [Houry et al., 1992].

tures. Mean sea surface undulations represent geoid anomalies which reflect lateral variations in the lithosphere and mantle mass distribution. Altimetry data have provided medium- and short-wavelength geoid observations over the whole oceanic domain. Interpretation of these data over tectonic features such as seamounts, fracture zones, or mid-ocean ridges has greatly improved our understanding of the thermal and mechanical structure of the lithosphere and of its interaction with the underlying mantle [Cazenave, 1992].

Of particular importance is the need of a fine spatial sampling of the marine geoid. Continued high-quality altimetric measurements will improve both the horizontal resolution of the geoid and its accuracy. In addition, this knowledge should help relax the repeat constraint for oceanographic missions. In this context, the 176-day exact repeat mission of ERS-1 (0.10 m - accuracy) is anticipated with great interest. In addition to the geodetic aspects which were the primary objectives of the analyses of altimeter data, sea surface height measurements substantially contributed to Oceanography in the field of: mesoscale current detection and variability, ocean tides, winds and waves, etc. As demonstrated over the past two decades first with Seasat in 1978, then with the Geosat Exact Repeat Mission in 1985, and more recently with the launch of ERS-1 in July 1991, satellite altimetry is a powerful technique for providing global information on the ocean circulation and ice sheet volume. In order to provide useful information about the climate system and improve predictability of the oceans and ice sheets, altimetric observations need to be carried out over several years. In this context of global change research, the joint U.S. and French Topex/Poseidon (T/P) mission (0.02 m - accuracy) will be sent into space (August 1992) to accurately measure the ocean topography and circulation for the next five years [Koblinsky et al., 1992].

But, the problem of measuring directly the broad permanent sea surface topography from altimetry is difficult primarily because both the low-degree geoid and the orbit of the altimeter satellite have significant errors at long wavelengths. This is particularly true, in the present instance, with the requirements of both ERS-1 and T/P missions in sensing directly the dynamic effects of ocean circulation at the sub-decimeter level. The use of precise and voluminous altimetry without adequate (precise) modeling of oceanographic phenomena appears to compromise the long-wavelength parts of the field. Likewise, gravimetry, while less sensitive because the data are weaker at long wavelength, also contains regional errors which might corrupt the solutions. In fact, it is important in the attempt to "see" the broad sea surface topography which is referenced to a specific geoid, and in this regard, satellite-only geopotential solutions wich are free from sea surface measurements are of special interest [Wagner, 1983]. The geoid undulations are two orders of magnitude larger than surface variations caused by the ocean circulation. In this manner, an accurate determination of the geoid is prerequisite to an estimate of the dynamic sea surface. In addition, the 0.10 m radial orbit accuracy requirement of T/P, driven by the radar altimeter system, is at least a factor of three beyond the capability of gravity models existing in 1985. These was the principal requirements for recovering a new generation of gravity field models based on satellite observations towards the end of the 1980's. However, the simultaneous solutions for the quasi-stationary sea surface topography (φ), the gravity field, the satellite orbits, and other geophysical parameters provided a substantially improved estimate for the gravity field and the long-wavelength signal of the general ocean circulation [Tapley et al., 1988; Denker and Rapp, 1990; Nerem et al., 1990; Marsh et al., 1990(b)]. While these results were encouraging, the large calibrated errors (0.20 m) in the solution (φ) were still relatively large.

4. Recent Gravity Field Models

Today, American (Goddard, Univ. of Texas, Ohio State Univ.) and European (DGFI/GRGS) models are being improved (1) to contribute no more than 0.10 m to the radial orbit error budget of T/P, and (2) to precisely define the long-wavelength geoid to support many oceanographic applications. At the same time, thanks to improvements in worldwide gravity data, they are being extended to high degrees, such as 180 and 360, primarily for geodetic and geophysical applications. The GEM-T1 gravity model [Marsh et al., 1988] has been the first model of the new generation and was at that time the best satellite-only solution. The radial error of a T/P orbit was estimated to have an uncertainty of 0.25 m with this model. GEM-T2 is the second in the series of Goddard Earth models of the terrestrial gravitational field [Marsh et al., 1990]. This model includes more than 600 coefficients above degree 36, the limit for GEM-T1, and like this one provides a dynamically determined model of the major tidal components (90 terms) [Christodoulidis et al., 1988; Cheng et al., 1992]. The accuracy assessment of its lower degree and order coefficients shows that their uncertainties have been reduced by 20 percent compared to the previous solution. The model shows an improvement by a factor of two over GEM-T1, bringing the estimated radial orbit error of T/P to about 0.12 m. In a parallel way, GRIM4 is the fourth completely new reprocessing within the series of previously European-derived gravity models. GRGS, cooperating with DGFI on geopotential modelling for many years, is responsible for the geodetic evaluation of the new DORIS Doppler tracking system (0.3 mm/s - accuracy) onboard the Spot-2 spacecraft and linked to a worldwide network of active ground stations (about 50 stations in 1992). At same time, DGFI is responsible for the precise orbit restitution of the ERS-1 spacecraft using laser tracking data and for the generation of global products from ERS-1 radar altimeter data. Two recent global Earth gravity models have been derived by these institutions: a satellite-only solution (-S1) and a combined one (-C1, with gravity anomalies and geoid undulations) [Schwintzer, Biancale et al., 1991]. The predicted radial orbit error of ERS-1 is still not considerably below the one meter level with the GRIM4-S1 gravity model. A further improvement might come from the inclusion of DORIS data (Spot-2), of Geosat Doppler tracking data and of course to a large extend from the ERS-1 tracking data during the mission. However, in an immediate future, the most productive high-accuracy tracking technique contributing to the improvement of global gravity coverage will be the new Doppler tracking systems (DORIS, PRARE, GPS). The major step will come from the fact that the data coverage is much higher than the current laser systems. Tapley et al. [1989] described the development of a subsequent University of Texas gravity model (TEG-2). An assessment of the formal errors identified this model as significantly better than previous Univ. of Texas solutions. Using the calibrated covariance matrix, the radial orbit error for Geosat was found to be 0.15-0.20 m. Two new models, OSU89 A and B, were recently described in Rapp and Pavlis [1990]. Complete to degree 360, these models incorporate an improved terrestrial gravity data base, Geosat altimeter data, and the GEM-T2 model with its full error covariance matrix.

Concerning the processing of altimeter data with these new gravity models, the initial objectives seem to have been reached. The error assessment of any new model is then particularly important to control the usefulness of the model for geophysical (or oceanographic) applications.

5. Accuracy Estimations

The processing of a large amount of observations and the combination of several types of gravity data for estimation of gravity field models make the use of rigorous least squares procedures complex. However, it is widely recognized that the least squares collocation is an optimal method for the solutions of many problems involving functionals of the gravity potential. But it is also widely known that in many cases the advantages of this method is balanced out by the problem of inverting large matrices. Several studies have been accomplished to find solutions which could improve the models and together with the improvement of the computer technology would be capable to handle the ever-increasing amount of observations and to produce accurate high degree solutions. In this context, one of the innovations was the use of a modified least squares procedure which minimized the sum of the observation residuals and the power of the coefficients [Lerch et al., 1979]. The procedure, still valid, consisted in including a set of "observations" of the harmonic coefficients themselves of zero with a noise value equal to a given power (Kaula's rule [1966]). Then, from the observations of each type of measurements, normal equations were formed and these were combined by a weighted least squares estimator. Because the satellites are insensitive to many of the coefficients in the geopotential expansion, it has been showed that without these constraints, which control the inversion of otherwise ill-conditioned matrices, the normal least squares adjustment diverges at high degrees and produces models with unrealistically large power at high degree.

In the 1980's, considerable discussion has taken place in the literature on the evaluation of the accuracy of the geopotential models. Despite the improvements, large discrepancies still existed between the various models, even between those derived from nearly the same observations [Wagner and Lerch, 1978; Lambeck and Coleman, 1983]. A general model which was equally adequate for orbit determinations and geodetic use was lacking [Szabo, 1986]. Today, this situation has been considerably improved by the development of recent global (general purpose) gravity models (OSU89, GRIM4-C1, TEG-2). An increasingly acute problem has been to find ways to validate the global results obtained by various group. In fact, the error assessment must transcend the formal error estimation of the solution which is too optimistic and reflect rather the precision of the estimated solve-for parameters than the accuracy with respect to the "real" values. So-called calibration methods were developed for this purpose. Wagner [1983] has considered the accuracy of the low-degree GEM-L2 potential coefficients in anticipation of their use in long-wavelength ocean circulation studies. Lerch et al. [1985] have discussed the accuracy of various Goddard models using data not included in the solutions. A calibration method using a dependent subset of the total data set is generally viewed as validating just the compatibility of the complementary subsets of the total. Absolute calibration of the total model relies on other tests with independent data (e.g., [Marsh et al., 1990]). Such data generally include terrestrial anomaly values, 24-Hour satellite accelerations, and resonant constraint equations. Furthermore, altimeter data are also used for comparison and internal test purposes; altimetric measurements of Δg are used for tests because they are not very contaminated by long wavelength ocean dynamic topography. The precision with which the satellite motion can be described is used as one of the criteria for evaluating gravity field models at long-wavelengths. The accuracy of the different solutions can also be illustrated by statistical comparisons of different results. However, these comparisons indicate only the order of magnitude of the errors in the various gravity field models. Klokocnik and Pospisilova [1981] have discussed one such test in which the effect of a linear combination of selected harmonics (lumped) on the satellite motion has been evaluated. However, various statistical tests are inappropriate in verifying of the low degree and order coefficient accuracies because there are few coefficients involved. In addition, all these tests are important because the most difficult aspect of combining several types of data is selecting relative weights, which are derived essentially by experiment.

The ratios of estimated commission to formal error in different solution since the 1970's range approximately from 1 to 6. For one example, an overall calibration value of $k = 4.5$ seems to be appropriate for the GRIM4-S1 gravity field model solution. After several tests (with independent data), this value has been adopted for scaling up a posteriori all formal standard deviations as estimated by the least squares adjustment, in order to approximate the accuracy of the parameters.

6. Long Period Studies

Until the development of Earth gravity field models which are based on a global tracking data, the determination of the long-wavelength geoid has been realized through "independent" low-degree solutions. Because only terms of order zero in the geopotential expansion give long-period perturbations (zonals) and secular effects (even zonals) in satellite orbits (in the absence of resonance effects), analyses have been conducted without solving for tesseral harmonic coefficients. One of the difficulties of using the zonal solutions, however, is that while they are usually well determined, there is generally a significant correlation between consecutive odd or even degree values. This is due to the dominance of just a few strong secular or long-period effects on a limited number of orbits in solutions for a large number of zonals. Until 1969, the best low-degree solution (up to J_{21}) was determined by Kozai [1969] from 11 satellites of different inclinations (28 to 110 deg.). With the launching in the 1970's of low-inclination satellites (Dial, Peole) the results have been substantially improved [Cazenave and Forestier,1972], simply due to the better geographical coverage (15 to 110 deg.). King-Hele et al.'s [1981] solution for odd zonal harmonic coefficients were also made analytically from a determination of the long periodic oscillation of perigee on 28 satellites. Their orbits have been carefully selected to provide a good coverage of inclinations. As a matter of fact, it is extremely difficult to determine a total and pure analytical theory of satellite motion. Therefore in the mid 1970's the determination of geophysical parameters from mean orbit analyses was based on numerical integration of orbit-averaged disturbances and differential correction procedure (e.g., [Goad and Douglas, 1977]). A seminumeric solution by Wagner [1973] utilized all measurable perturbations on the mean Kepler elements for 21 satellite orbits.

Since the end of the 1970's the long-wavelength geoid has been strengthened through incorporation of several years of 0.10 m laser tracking data on Lageos into global Earth models such as GEM-9 [Lerch et al., 1979] or GRIM3-B [Reigber et al., 1984] and then GEM-L2 [Lerch et al., 1982] and GRIM3-L1 [Reigber et al., 1985]. In the last two cases, Lageos tracking data already appeared to dominate the low-degree zonal information. The intensive use of satellite tracking data in the determination of recent earth gravity field models (GEM-T2, GRIM4-S1,...) reduced the commission error to a few 0.01m, at long wavelengths typical of ocean basins. For example, the differences in the geoid representation induced by the geopotential coefficient differences between GEM-T2 and GRIM4-S1 models are on the 0.05 m level for the terms up to degree and order 5. In this context, and by accounting for the requirements of oceanographic studies, the level of accuracy required of independent data to test the low-degree zonal estimates (errors) should be of the order of 0.01 m or less (geoid heights) or a few hundredths of 1 percent for these zonal coefficients. Although purely numerical integration of the satellite orbital equations is possible over a long period of time (e.g., [Cheng et al., 1990]) thanks to the improvement of the computer technology and the development of new softwares, the only independent data which seem capable of testing the recent models are derived from observations of long-period or secular satellite perturbations from zonal (or resonant) harmonic coefficients. For this purpose, we propose to use a semi-numerical method yet under development which is based on the concept of precise mean orbital motion [Exertier et al.,1991].

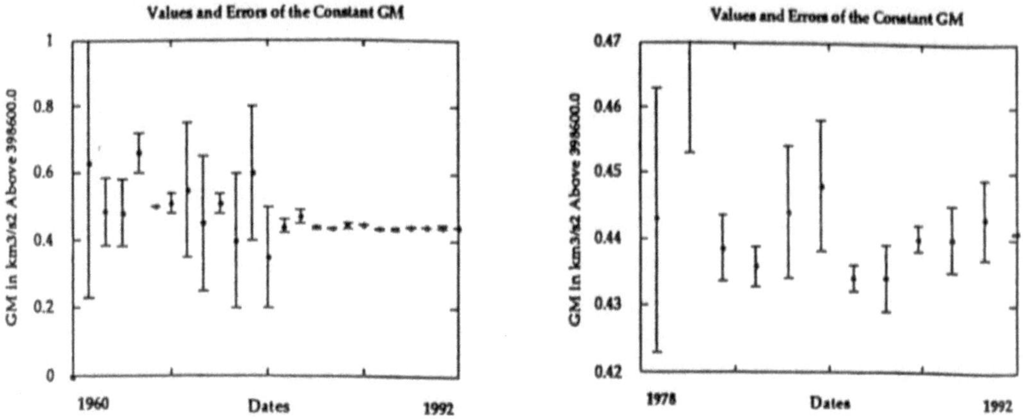

Fig. 2. Various values and error estimate of the gravitational coefficient of the Earth GM. Left: Values from space missions and LLR [Lerch, et al., 1978]. Recent values from near-earth satellite orbit determinations.

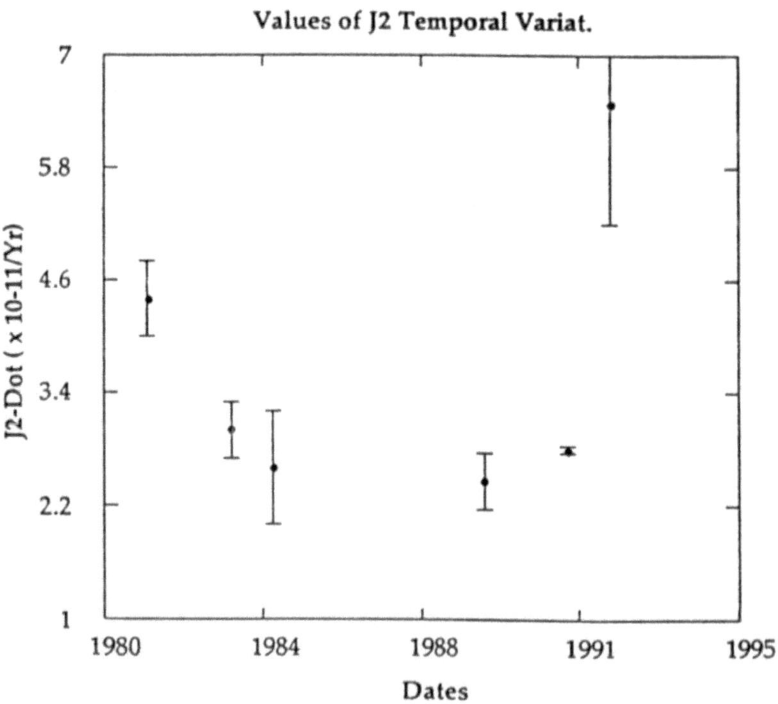

Fig. 3. Various values and error estimates of the temporal variation of the second degree zonal harmonic coefficient (J2), from recent Lageos acceleration node analyses.

A dynamical method based on a numerical integration of analytically (gravitational effects) and numerically (non-gravitational effects) orbit-averaged disturbances could also afford new possibilities to assess global force model parameters such as the product of the gravitational constant with the Earth's mass (GM). In this way, the estimate of (GM) over several years of satellite tracking data can yield important information about the variations in overall scale within a given solution [Ries et al., 1989, 1992]. The gravitational constant (GM) has been adjusted from laser ranging on near-earth satellites toward the end of the 1970's (Fig. 2, right) [Lerch et al., 1978, 1979]. The solution for (GM) using four near-earth satellites produced a value of 398600.61 $\pm 0.02 km^3/s^2$, while the adjustment of (GM), simultaneously determined with the geopotential GEM-9 model, produced a value of 398600.64 $\pm 0.02 km^3/s^2$ (with $c = 299792.5\, km/s$ in both cases). The first value was already in very good agreement with a later determination from eight years of laser ranging to Lageos [Tapley et al., 1985], although the accuracy of the initial determinations has been limited by both measurement and satellite dynamical model errors. The actual estimations of the value of (GM) (398600.4415 km^3/s^2) from satellite tracking data [Ries et al., 1992] appears to confirm the previous value determined from four satellites but further reduces the uncertainty by a factor 20. The determination of (GM) is now of sufficient accuracy to require careful consideration of all relativistic effects.

Improvements in global gravity modeling have "naturally" led to investigations of the secular and periodic changes in the field on regional and global scales. The gravity field changes in time because of tectonic deformations, post-glacial rebound and seasonal glacial and other water mass fluctuations. This subtle dynamical effect of temporal variations in the zonal harmonics has been observed over the 10 years using satellite laser ranging measurements (Fig. 3) to Lageos [Yoder et al., 1983; Rubincam, 1984; Gegout and Cazenave, 1991] and Starlette [Cheng et al., 1989]. The three independent analyses based on several years of laser data on Lageos have provided the "first" satellite-based estimate of $\dot{J}_2$. The solutions ((-3.0 $\pm$0.3, -2.6 $\pm$0.6, -2.8 $\pm$0.4) x 10^{-11}/yr) are in good agreement one with the other and with the Starlette independent value (-2.5 $\pm$0.3 x 10^{-11}/yr), excepted for the last accuracy estimate [Gegout and Cazenave, 1991] which could be attributed to the longer period of analysis (7 years) and the better observation material (1984 through 1990). Then, the observed acceleration in the node of the orbit of the Lageos satellite has been shown to provide a good constraint on the viscosity of the Earth's mantle beneath the 670-km seismic discontinuity [Peltier, 1983].

7. Conclusion

A fundamental requirement for understanding the nature, origin, dynamics, and evolution of the lithosphere is the accurate measurement of relative motions of selected points on the various tectonic plates as reference values for the dynamic models [Ricard and Vigny, 1989]. In fact, the dynamics of the Earth encompass a broad spectrum of motions including the orbital motion of the Earth and the Moon about the Sun, the rotation of the Earth about its center of mass, and the motions of the individual constituents of the Earth. This latter motion includes, in addition to the tectonic plate motion, free and forces oscillations of the solid Earth, the coupling of the Earth's crust with the motions of the ocean and atmospheric components, and the coupling to the inner core. The nature of the problem is such that it must be addressed in a global context. The development of the ability to range accurately from the Earth's surface to satellites carrying laser cube-corners reflectors has provided a unique capability for studying global solid Earth dynamics.

The orbital analyses over very long period of times (10 years and beyond) are of great interest for the determination of specific coefficients of the gravity field (e.g., zonal harmonics, tides) including their temporal variations or for global force modeling parameters (GM, general relativity). As a matter of fact, geophysical informations are generally contained in the secular and long-period variations of satellite orbits. Therefore it could be of great interest to directly measure these variations from precise "observations", such as mean orbital elements, performed on dedicated satellites like Lageos-1 and Lageos-2. In addition, this independent approach of the dynamical problem of defining the long-wavelength geoid could afford new conditional equations of great quality (accuracy estimate), to support many oceanographic and geophysical applications in the future.

However, without data from a dedicated satellite gravity mission such as Aristoteles/Gradio, results in sensing many of the short-wavelengths of the geoid from satellite orbit perturbations will remain very difficult. Several effects concerning the Physics of the Earth could be detected at high accuracy with such a space-borne gradiometer.

Acknowledgements

The author would like to acknowledge Eric Bois and Pascal Oberti, the scientific organizers of the *International Astronomical Symposium*, Pléneuf-Val-André, France, 21-28 June, 1992, for their competence and for providing helpful suggestions. I also would like to thank F. Barlier for helpful discussions and permanent encouragement. J. Kovalevsky and M. Lefevre are to be thanked for their suggestions about the objectives of this paper. S.M. Klosko is to be thanked for carefully reviewing this article.

References

Barlier, F., Bernard, J., Bouiri, O. and Exertier, P.: 1982, 'The Geoid of the Mediterranean Sea deduced from Seasat Data', *2nd Intern. Symp.on Geoid in Europe and the Mediterranean Area*, Rome, Italy

Biancale, R., Cazenave, A. and Dominh, K.: 1991, 'Tectonic plate motions derived from Lageos', *Earth and Planet. Sci. Lett.*, **103**, 379–394

Cazenave, A. and Forestier, F.: 1970, 'Determination of the equations of Condition for the Zonal Harmonics using the Dial Satellite', *Space Research*, **XI**, 521–524

Cazenave, A.: 1992, 'The geoid and oceanic lithosphere', in *Geophysical Interpretation of the Geoid*, P. Vanicek, editor, CRC Press Inc., in press

Cheng, M.K.: 1989, 'Temporal Variations in Low Degree Zonal Harmonics from Starlette Orbit Analysis', *Geophys. Res. Lett.*, **16**, 393–396

Cheng, M.K., Shum, C.K., Eanes, R.J., Schutz, B.E. and Tapley, B.D.: 1990, 'Long Period perturbations in Starlette orbit and Tide solution', *J. Geophys. Res.*, **95**, 8723–8736

Cheng, M.K., Eanes, R.J. and Tapley, B.D.: 1992, 'Tidal Deceleration of the Moon's mean motion', *Geophys. J. Int.*, **108**, 401–409

Christodoulidis, D.C., Smith, D.E., Kolenkiewicz, R., Klosko, S.M., Torrence, M.H. and Dunn, P.J.: 1985, 'Observing Tectonic Plate Motions and Deformations from Satellite laser ranging', *J. Geophys. Res.*, **90**, 9249–9263

Christodoulidis, D.C., Smith, D.E., Williamson, R.G. and Klosko, S.M.: 1988, 'Observed Tidal Bracking in the Earth/Moon/Sun System', *J. Geophys. Res.*, **93**, 6216–6236

Degnan, J.J.: 1985, *IEEE Trans on Geoscience and Remote Sensing*, **GE-23(4)**, 398–413

Denker, H. and Rapp, R.H.: 1990, 'Geodetic and Oceanographic Results from the Analysis of 1 Year of Geosat Data', *J. Geophys. Res.*, **95**, 13151–13168

Douglas, B.C., McAdoo, D.C. and Cheney, R.E.: 1987, 'Oceanographic and Geophysiocal Applications of Satellite Altimetry', *Rev. Geophys. Space Phys.*, **25**, 875–880

Exertier, P., Metris, G. and Barlier, F.: 1991, 'Long Term Evolution of Mean Orbital elements of Artificial Satellites', *XX General Assembly of the IUGG*, Vienna, Austria

Gaposchkin, E.M. and Lambeck, K.: 1971, 'The Earth's Gravity Field to 16th degree and Station Coordinates from Satellite and terrestrial Data', *J. Geophys. Res.*, **76**, 4855–4883

Gaposchkin, E.M.: 1980, 'Global Gravity Field to Degree and Order 30 from Geos-3 Satellite Altimetry and other Data', *J. Geophys. Res.*, **85**, 7221–7234

Gegout, P. and Cazenave, A.: 1991, 'Geodynamic parameters derived from 7 years of Laser data on Lageos', *Geophys. Res. Lett.*, **18**, 1739–1742

Kaula, W.M.: 1966, *Theory of satellite geodesy*, Blaisdell publ. Co., Waltham, Mass, pp124

King-Hele, D.G., Brookes C.J. and Cook G.E.: 1981, 'Odd zonal harmonics in the geopotential analysis of 28 satellite orbits', *Geophys. J. R. astr. Soc.*, **64**, 3–30

Klokocnik, J. and Pospisilova, L.: 1981, 'Intercomparisons of Earth Gravity Models by Means of Lumped Coefficients', *Planet. Space Sci.*, **29**, 653–671

Koblinsky, C.J., Gaspar, P. and Lagerloef, G.: 1992, 'The Future of Spaceborne Altimetry: Oceans and Climate Change', *Joint Oceanographic Institutions Inc.*, Washington D.C., pp75

Kovalevsky, J.: 1971, 'Champ de Pesanteur et Forme de la Terre', in *Geodesie Generale*, Tome 4, Geodesie Spatiale, Levallois, J.J. and Kovalevsky, J., Eyrolles

Kozai, Y.: 1969, 'Revised Values for coefficients of zonal spherical harmonics in the geopotential', *SAO Spec. Rep.*, **295**, 1–17

Lambeck, K. and Coleman, R.: 1983, 'The Earth's Shape and Gravity Field: A report of Progress from 1958 to 1982', *Geophys. J. R. astr. Soc.*, **74**, 25–54

Lerch, F.J., Laubscher, R.E., Smith, D.E., Klosko, Kolenkiewicz, R., Putney, B.H., Marsh, J.G. and Brownd, J.E.: 1978, 'Determination of the Geocentric Gravitational Constant from laser ranging on near-Erath satellites', *Geophys. Res. Lett.*, **5**, 1031–1034

Lerch, F.J., Klosko, S.M., Laubscher, R.E. and Wagner C.A.: 1979, 'Gravity model improvement using Geos-3 (GEM-9 and 10)', *J. Geophys. Res.*, **84**, 3897–3915

Lerch, F.J., Putney, B.H., Wagner, C.A. and Klosko, S.M.: 1981, 'Goddard Earth Models for oceanographic applications (GEM-10B and 10C)', *Marine Geod.*, **5**, 145–187

Lerch, F.J., Klosko, S.M. and Patel G.B.: 1982, 'A refined gravity model from Lageos (GEM-L2)', *Geophys. Res. Lett.*, **9**, 1263–1266

Lerch, F.J., Marsh, J.G., Klosko, S.M. and Williamson, R.G.: 1982(b), 'Gravity Model Improvement for Seasat', *J. Geophys. Res.*, **87**, 3281–3296

Lerch, F.J., Klosko, S.M., Wagner, C.A. and Patel, G.B.: 1985, 'On the Accuracy of recent Goddard Gravity Models', *J. Geophys. Res.*, **90**, 9312–9334

Marsh, J.G.: 1983, 'Satellite Altimetry', *Rev. Geophys. Space Phys.*, **21**, 574–580

Marsh, J.G., Lerch, F.J. and Williamson, R.G.: 1985, 'Precision Geodesy and Geodynamics using Starlette laser ranging', *J. Geophys. Res.*, **90**, 9335–9345

Marsh, J.G., Brenner, A.C., Beckley, B.D. and Martin, T.V.: 1986, 'Global mean sea surface based upon the Seasat Altimeter data', *J. Geophys. Res.*, **91**, 3501–3506

Marsh, J.G., Lerch, F.J., Putney, B.H., Christodoulidis, D.C., Sanchez, B.V., Felsentreger, T.L., Smith, D.E., Klosko, S.M., Martin, T.V., Pavlis, E.C., Robbins, J.W., Williamson, R.G., Colombo, O.L., Chandler, N.L., Rachlin, K.E., Patel, G.B., Bhati, S. and Chinn, D.S.: 1988, 'A New Gravitational Model for the Earth from Satellite tracking data: GEM-T1', *J. Geophys. Res.*, **93**, 6169–6215

Marsh, J.G., Lerch, F.J., Putney, B.H., Sanchez, B.V., Felsentreger, T.L., Klosko, S.M., Pavlis, E.C., Robbins, J.W., Williamson, R.G., Chandler, N.L., Rachlin, K.E., Patel, G.B., Chinn, D.S., Engelis, T.L., Eddy, W.F., Kapoor, S. and Braatz, L.E.: 1990, 'The GEM-T2 Gravitational Model', *J. Geophys. Res.*, **95**, 22043–22071

Marsh, J.G., Koblinsky, C.J., Lerch, F.J., Klosko, S.M., Robbins, J.W., Williamson, R.G. and Patel, G.B.: 1990(b), 'Dynamic Sea Surface Topography, Gravity, and Improved Orbit Accuracies from the Direct Evaluation of Seasat Altimeter Data', *J. Geophys. Res.*, **95**, 13129–13150

Nerem, R.S., Tapley, B.D. and Shum, C.K.: 1990,'Determination of the Ocean Circulation Using Geosat Altimetry', *J. Geophys. Res.*, **95**, 3163–3179

Peltier, W.R.: 1983, 'Constraint on Deep Mantle Viscosity from Lageos acceleration data', *Nature*, **304**, 434–436

Rapp, R.H.: 1969,'The Geopotential to (14,14) from a Combination of Satellite and Gravimetric Data', *Bull. Geod.*, **91**, 47–80

Rapp, R.H.: 1982, 'Degree Variances of the Earth's Potential, Topography and its Isostatic Compensation', *Bull. Géod.*, **56**, 84–94

Rapp, R.H.: 1983, 'The determination of geoid undulations and gravity anomalies from Seasat altimeter data', *J. Geophys. Res.*, **88**, 1552–1562

Rapp, R.H.: 1989, 'The Decay of the Spectrum of the Gravitational Potential and the Topography of the Earth', *Geophys. J. Int.*, **99**, 449–455

Rapp, R.H. and Pavlis, N.K.: 1990, 'The Development and Analysis of Geopotential Coefficient Models to Spherical Harmonic degree 360', *J. Geophys. Res.*, **95**, 21885–21911

Ray, J.R., Ma, C., Ryan, J.W., Clark, T.A., Eanes, R.J., Watkins, M.M., Schutz, B.E. and Tapley, B.D.: 1991, 'Comparison of VLBI and SLR Geocentric Site Coordinates', *Geophys. Res. Lett.*, **18(2)**, 231–234

Reigber, Ch., Balmino, G., Mueller, H., Bosch, W., Rizos, Ch. and Moynot, B.: 1984, 'An improved GRIM3 Earth Gravity Model (GRIM3-B)', in *Proceedings of the IAG*, Symposia, 18th General Assembly, IUGG, Hamburg, **1**, 388–415

Reigber, Ch., Balmino, G., Mueller, H., Bosch, W. and Moynot, B.: 1985, 'GRIM Gravity Model Improvement using Lageos (GRIM3-L1)', *J. Geophys. Res.*, **90**, 9285–9299

Ricard, Y. and Vigny, Ch.: 1989, 'Mantle Dynamics with Induced plate tectonics', *J. Geophys. Res.*, **94**, 17543–17559

Ries, J.C., Eanes, R.J., Huang, C., Schutz, B.E., Shum, C.K., Tapley, B.D., Watkins, M.M. and Yuan, D.N.: 1989, 'Determination of the Gravitational coefficient of the Earth from near-earth satellites', *Geophys. Res. Lett.*, **16**, 271–274

Ries, J.C., Eanes, R.J., Shum, C.K. and Watkins, M.M.: 1992,'Progress in the Determination of the Gravitational coefficient of the Earth', *Geophys. Res. Lett.*, **19**, 529–531

Rubincam, D.P.: 1984, 'Post-glacial rebond observed from Lageos and the effective viscosity of the lower mantle', *J. Geophys. Res.*, **89**, 1077–1087

Schwintzer, P., Reigber, Ch., Massmann, F.H., Barth, W., Raimondo, J.C., Gerstl, M., Li, H., Biancale, R., Balmino, G., Moynot, B., Lemoine, J.M., Marty, J.C., Boudon, Y. and Barlier, F.: 1991, 'A New Gravity Field Model in Support of ERS-1 and SPOT-2: GRIM4-S1/C1', Final Report to the German Space Agency (DARA) and the French Space Agency (CNES), DGFI Muenchen / GRGS Toulouse, pp177

Smith, D.E., Kolenkiewicz, R., Dunn, P.J., Robbins, J.W., Torrence, M.H., Klosko, S.M., Williamson, R.G., Pavlis, E.C., Douglas, N.B. and Fricke, S.K.: 1990, 'Tectonic Motion and Deformation from Satellite Laser Ranging to Lageos', *J. Geophys. Res.*, **95**, 22013–22041

Szabo, B.: 1986, 'The Estimation of the Earth's Gravity Field', Rep. **369**, Dep. of Geod. Sci. and Surv., The Ohio State Univ., Colombus, Ohio

Tapley, B.D., Schutz, B.E. and Eanes, R.J.: 1985, 'Station Coordinates, Baselines and Earth Rotation from Lageos Laser Ranging: 1976-1984', *J. Geophys. Res.*, **90**, 9235–9248

Tapley, B.D., Nerem, R.S., Shum, C.K., Ries, C.J. and Yuan, D.N.: 1988, 'Determination of the general ocean circulation from a joint gravity field solution', *Geophys. Res. Lett.*, **15**, 1109–1112

Tapley, B.D., Schutz, B.E., Eanes, R.J. and Watkins, M.M.: 1991, 'The University of Texas Earth Gravitational Field Model (TEG-2)', XX General Assembly of the IUGG, Vienna, Austria

Wagner, C.A.: 1973,' Zonal gravity harmonics from long satellite arcs by a seminumeric method', *J. Geophys. Res.*, **78**, 3271–3280

Wagner, C.A. and Lerch, F.J.: 1978, 'The accuracy of geopotential models', *Planet. Space Sci.*, **26**, 1081–1140

Wagner, C.A.: 1983, 'The accuracy of the low-degree potential: implications for ocean dynamics', *J. Geophys. Res.*, **88**, 5083–5090

Yoder, C.F., Williams J.G., Dickey J.D., Schutz, B.E., Eanes, R.J. and Tapley, B.D.: 1983, 'Secular variation of earth's gravitational harmonic J_2 coefficient from Lageos and nontidal acceleration of earth rotation', *Nature*, **303**, 757–762

LUNI-SOLAR EFFECTS OF GEOSYNCHRONOUS ORBITS AT THE CRITICAL INCLINATION

FABIENNE DELHAISE* and ALESSANDRO MORBIDELLI
Department of mathematics FUNDP
8, Rempart de la Vierge, B-5000 Namur, Belgium

Abstract. The luni-solar effects of a geosynchronous artificial satellite orbiting near the critical inclination is investigated. To tackle this four-degrees-of-freedom problem, a preliminary exploration separately analyzing each harmonic formed by a combination of the satellite longitude of the node and the Moon longitude of the node is opportune. This study demonstrates that the dynamics induced by these harmonics does not show resonance phenomena. In a second approach, the number of degrees of freedom is halved by averaging the total Hamiltonian over the two non-resonant angular variables. A semi-numerical method can now be applied as was done when considering solely the inhomogeneity of the geopotential (see Delhaise et Henrard, 1992). Approximate surfaces of section are constructed in the plane of the inclination and argument of perigee. The main effects of the Sun and Moon attractions compared to the terrestrial attraction alone are a strong increase in the amplitude of libration in inclination (from 0.6° to 3.2°) and a decrease of the corresponding libration period (from the order of 200 years to the order of 20 years).

Key words: artificial satellite theory– geosynchronous orbits– critical inclination–luni-solar effects– surface of section

1. Introduction

The published solutions of the main problem of artificial satellite theory (Brouwer 1959, Garfinkel 1959, Kozai 1959) contain an intrinsic singularity at the so-called critical inclination ($\cos^2 i = 1/5$). For these values of the inclination the secular J_2 effect in the argument of perigee is zero. The dynamics of this one-degree of freedom resonance problem has been extensively studied by many authors using various methods and variables. A detailed review of these investigations is given by Jupp (1988).

The perturbation effects on geosynchronous satellites orbiting near the critical inclination have however rarely been under investigation. In fact, this system requires a treatment totally different from the critical inclination problem. We deal with a double resonance problem: a mean motion commensurability between the satellite's mean motion and the Earth's rotation rate, combined with the critical inclination. In a preliminary study (Delhaise et Henrard, 1992), only the gravitational perturbations arising from an aspherical Earth were taken into consideration. The secular dynamics of this two-degrees-of-freedom system was explored in a global way. More specifically the 12-hr Molniya-type and the 24-hr Tundra-type orbits, both located at the critical inclination, were analyzed.

In this paper, the perturbation model is expanded through inclusion of the luni-solar gravitational effects. For the Molniya and Tundra type orbits which travel out to several Earth radii, the effects of the gravitational attractions of Sun and Moon are

* Research Assistant for the Belgian National Fund for Scientific Research

Celestial Mechanics and Dynamical Astronomy **57**: 155–173, 1993.

appreciable. The eccentricity can increase sufficiently to cause premature decay of the satellite by forcing the perigee down to the Earth surface. This will be illustrated for the case of the Molniya orbit.

First, the Hamiltonian of this system is developed. The lunar orbital elements are preferably referred with respect to an ecliptic reference system while the satellite's orbital elements are still related to the Earth equator. Obviously, taking into account the terrestrial together with the luni-solar perturbation effects compounds the difficulties of the problem. After averaging over the fast angular variables, the problem still has four degrees of freedom. Due to this complexity, the terrestrial and lunar effects are at first considered alone. A preliminary exploration separately analyzing each harmonic defined as a combination of the longitude of the ascending node of the satellite (x_4) and the longitude of the Moon (x_5), is appropriate for a subsequent simplification of the system.

In a second approach, the number of degrees of freedom is halved by averaging the total Hamiltonian over the two non-resonant angular variables x_4 and x_5. The resulting two-degrees-of-freedom Hamiltonian can then be studied by means of the semi-numerical method (Henrard, 1990). Approximate surfaces of section are constructed in the plane of the inclination and argument of perigee. Such figures yield a global view of the secular dynamics of such type of orbits. Finally, the solar effects are added to the Hamiltonian system.

2. Formulation of the Total Hamiltonian

The part of the Hamiltonian taking into account the oblateness of the Earth as well as the tesseral harmonics, is averaged over the mean anomaly of the satellite. Up to the second order in the second zonal harmonic J_2, it takes on the following form (see Brouwer, 1959 and Sochilina, 1982):

$$\begin{aligned}\mathcal{H}_{\text{Earth}} &= -\frac{\mu^2}{2L^2} - \omega_e\, H \\ &- \frac{\mu^4}{L^6}\frac{R_e^2 J_2}{4}\left(3\frac{H^2}{G^2}-1\right)\left(\frac{L}{G}\right)^3 \\ &- \frac{\mu^6}{L^{10}}\frac{R_e^4 J_2^2}{4}\,[A(L,G,H)\cos 2\omega + C(L,G,H)] \\ &- \sum_{l,m,p,q} B_{l,m,p,q}(a,e,i)\,\begin{matrix}\cos\\ \sin\end{matrix}\,\{\theta_{l,m,p,q}\}\ \begin{matrix}\text{for } l-m \text{ even}\\ \text{for } l-m \text{ odd,}\end{matrix}\end{aligned} \tag{1}$$

where μ is the product of the gravitational constant and the Earth's mass and

$$\theta_{l,m,p,q} = m(\frac{\alpha}{\beta}\ell + h - \lambda_{lm}) + (l-2p)\omega. \tag{2}$$

The angular variable $\frac{\alpha}{\beta}\ell\ +\ h$ is the critical argument associated to the resonance in mean motion where the satellite performs β revolutions while the Earth rotates

α times. The variable λ_{lm} is the longitude of major axis of symmetry of the (l, m) spherical harmonic. The set of variables $(\ell = M, g = \omega, h, L, G, H)$ is the usual set of the Delaunay variables with respect to a frame rotating with the Earth, which implies:

$$h = \Omega - \omega_e t, \tag{3}$$

where Ω is the right ascension of the ascending node in an inertial frame and t is the time. The coefficients A, B_{lmpq} and C are the same as in Delhaise et Henrard (1992). It is noteworthy that these coefficients are not expanded in powers of eccentricity or inclination permitting, for example, a very large value of the eccentricity. The term $\omega_e H$, where ω_e is the Earth's mean angular velocity, is added to the initial Hamiltonian in order to remove its explicit time dependence.

The indices of the summation l, m, p, q verify the following relation:

$$l - 2p + q = \frac{\alpha}{\beta} m, \tag{4}$$

so that the short periodic terms disappear of the Hamiltonian leaving only long period ones for commensurate or near-commensurate orbits. The main critical tesseral harmonics which are retained in this study are listed in Delhaise et Henrard (1992).

The units have been chosen such that the gravitational constant G, the mass of the Earth M_e and the mean equatorial radius of the Earth R_e are unity.

A third body gravitational force acting on the spacecraft can be described in a reference system defined at the Earth centre by the following potential:

$$V = \frac{-GM_k}{\rho_k} + GM_k \frac{\mathbf{r}_k \cdot \mathbf{r}}{r_k^3}. \tag{5}$$

The index k refers to the third body. The vectors $\mathbf{r}$ and $\mathbf{r}_k$ are the coordinates of the spacecraft and of the third body, respectively, $\rho_k = \mathbf{r} - \mathbf{r}_k$, GM_k is the Newtonian gravitational constant times the third body's mass. It is worth noting that the accelerating terms due to the Earth rotation around the Sun are neglected in this study. Kaula (1962) expanded this function in terms of the equatorial Keplerian elements of the satellite and of the third body:

$$\begin{aligned} V = & -\mu_k \sum_{l=2}^{\infty} \frac{a^l}{a_k^{l+1}} \sum_{m=0}^{l} k_m \frac{(l-m)!}{(l+m)!} \sum_{p=0}^{l} F_{lmp}(i) \sum_{s=0}^{l} F_{lms}(i_k) \\ & \sum_{q=-\infty}^{\infty} H_{lpq}(e) \sum_{j=-\infty}^{\infty} G_{lsj}(e_k) \cos\left[(l-2p)\omega + (l-2p+q)\,M - \right. \\ & \left. - (l-2s)\,\omega_k - (l-2s+j)\,M_k + m\,(\Omega - \Omega_k)\right], \end{aligned} \tag{6}$$

where $k_o = 1$ and $k_m = 2$ for $m \neq 0$. The eccentricity functions H_{lpq} and G_{lsj} can be related to the Hansen function $X_k^{n,m}$ (see Giacaglia, 1976). The inclination functions $F_{lmp}(i)$ are given in Kaula (1966). The form (6) is convenient for inclusion of the solar effects with the Earth gravitational effects.

Kozai (1973) noted that it is more desirable to develop a theory relying on Keplerian ephemerides for the Moon referred to the ecliptic plane. This renders the Moon's inclination roughly constant and the lunar longitude of the ascending node can be approximated by a single linear function of time. The orbital elements of the satellite should still be referred to the Earth equator. Lane (1989) assumes an ecliptic reference system for the lunar ephemerides while the satellite elements are related to the Earth equator. After several geometrical rotations, the following form of the lunar potential is obtained:

$$\begin{aligned} V = & -\sum_{l=2}^{\infty}\sum_{m=0}^{l}\sum_{s=0}^{l}\sum_{p=0}^{l}\sum_{q=0}^{l}\sum_{j=-\infty}^{\infty}\sum_{r=-\infty}^{\infty} (-1)^{k_1}\,\mu_k\,\frac{\varepsilon_m\,\varepsilon_s\,(l-s)!}{2\,a_k\,(l+m)!}\left(\frac{a}{a_k}\right)^l \\ & F_{lmp}(i)\,F_{lsq}(i_k)\,G_{lqr}(e_k)\,H_{lpj}(e) \\ & \Big\{(-1)^{k_2}\,U_l^{m,-s}(\varepsilon)\,\cos\left[\theta_{lmpj} + \theta'_{lsqr} - b_s\,\pi\right] \\ & (-1)^{k_3}\,U_l^{m,+s}(\varepsilon)\,\cos\left[\theta_{lmpj} - \theta'_{lsqr} - b_s\,\pi\right]\Big\}, \end{aligned} \tag{7}$$

where

$$\begin{aligned} \varepsilon_m &= \begin{cases} 1 & \text{if } m = 0, \\ 2 & \text{if } m \neq 0, \end{cases} \\ \varepsilon_s &= \begin{cases} 1 & \text{if } s = 0, \\ 2 & \text{if } s \neq 0, \end{cases} \\ k_1 &= \left\|\frac{m}{2}\right\|, \text{ the largest integer part of } \frac{m}{2}, \\ k_2 &= t\,(m+s-1)+1, \\ k_3 &= t\,(m+s), \\ t &= 0 \text{ if } l-1 \text{ is even and } t = 1 \text{ if } l-1 \text{ is odd}, \\ b_s &= 0 \text{ if } s \text{ is even and } b_s = 1/2 \text{ if } s \text{ is odd}, \\ \theta_{lmpj} &= (l-2p)\,\omega + (l-2p+j)\,M + m\,\Omega, \\ \theta'_{lsqr} &= (l-2q)\,\omega_k + (l-2q+r)\,M_k + s\,(\Omega_k - \pi/2), \end{aligned} \tag{8}$$

(9)

and

$$\begin{aligned} U_l^{m,s}(\varepsilon) = & (-1)^{m-s}\sum_{r=\max[0,-(m+s)]}^{\min[l-s,l-m]}(-1)^{l-m-r}\binom{l+m}{m+s+r}\binom{l-m}{r} \\ & \times \cos^{m+s+2r}\left(\frac{\varepsilon}{2}\right)\,\sin^{-m-s+2l-2r}\left(\frac{\varepsilon}{2}\right), \end{aligned} \tag{10}$$

with $\varepsilon = 23.4^\circ\,27'\,08.26''$, the ecliptic inclination at Epoch J.D. 2415020.0. The total Hamiltonian is the result of the summation of the geopotential (1), the solar potential (6) and the lunar potential as given by (7).

3. Inclusion of the Moon's Gravitational Effects

Due to the complexity of the system the terrestrial and lunar gravitational effects are at first taken into account alone. The solar perturbations are included into the model subsequently. This study is here limited to the analysis of the Molniya-type orbits. Identical methods could of course be applied for the Tundra-type orbits too.

3.1. SIMPLIFYING ASSUMPTIONS

The following simplifications and assumptions are introduced in the approximate study of the secular dynamics of the Molniya-type orbits subjected to the terrestrial and lunar attractions.

1. The summation over the index "l" of the lunar potential (7) is truncated to include just the second harmonic (where the first is absent). We thus have:
$$l = 2. \tag{11}$$
2. The Hamiltonian is averaged over the satellite mean anomaly up to the first order. The indices (l, p, j) verify:
$$l - 2p + j = 0. \tag{12}$$
3. The Hamiltonian is averaged over the Moon mean anomaly up to the first order. The indices (l, q, r) verify:
$$l - 2q + r = 0. \tag{13}$$
4. The Moon's orbital eccentricity is assumed as zero so
$$l - 2q = 0. \tag{14}$$
Note that assuming the Moon orbit as circular has nearly the same effect as averaging the Hamiltonian over the Moon argument of perigee at the first order since
$$G_{210}(e_M) = \left(1 - e_M^2\right)^{-3/2} = 1.00438 \simeq G_{210}(0). \tag{15}$$

The part of the Hamiltonian taking into account the lunar effects is then reduced to the following form:

$$\begin{aligned}\mathcal{H}_M = -\sum_{m=0}^{2}\sum_{s=0}^{2}\sum_{p=0}^{2} & D_M(a, e, i, a_M, i_M) \\ & \left\{(-1)^{k_2}\, U_2^{m,-s}(\varepsilon)\cos\left[(2-2p)\omega + m\Omega + s(\Omega_M - \frac{\pi}{2}) - b_s\pi\right]\right. \\ & \left.(-1)^{k_3} U_2^{m,+s}(\varepsilon)\cos\left[(2-2p)\omega + m\Omega - s(\Omega_M - \frac{\pi}{2}) - b_s\pi\right]\right\}\end{aligned} \tag{16}$$

where

$$D_M = \mu_M(-1)^{k_1}\,\frac{\varepsilon_m\,\varepsilon_s\,(2-s)!}{2\,a_M\,(2+m)!}\left(\frac{a}{a_M}\right)^2 F_{2,m,p}(i)\,F_{2,s,1}(i_M)\,H_{2,p,2p-2}(e) \tag{17}$$

where the index M refers to the Moon's parameter.

3.2. PRELIMINARY EXPLORATION

We are confronted with a three-degrees-of-freedom problem, the Hamiltonian of which is time-dependent through the two angular variables $g_e = \omega_e t$ and Ω_M. These can be approximated by linear functions of time:

$$\begin{aligned} g_e(t) &= g_e(0) - 5.8834 \cdot 10^{-2}\, t, \\ \Omega_M(t) &= \Omega_M(0) - 8.6298 \cdot 10^{-6}\, t. \end{aligned} \tag{18}$$

The time unit is given with the units defined in section 2 and is of the order of $13^{\mathrm{m}}44^{\mathrm{sec}}$.

The phase space is extended by two additional degrees of freedom in order to render the Hamiltonian autonomous. The set of the canonical variables is defined by:

$$\begin{array}{lll} \ell = M & L & = \sqrt{\mu a}, \\ g = \omega & G & = L\sqrt{(1-e^2)}, \\ \Omega & H & = G \cos i, \\ g_e = \omega_e t & \Lambda_e, & \\ \Omega_M & \Lambda_M. & \end{array} \tag{19}$$

Note that the phase space is in fact restricted to four degrees of freedom inasmuch as the Hamiltonian is averaged over the satellite's mean anomaly. The frequency of Ω (of order of $2 \cdot 10^{-5}$ due to the J_2 secular effects) and especially Ω_M (equals to $8.6 \cdot 10^{-6}$) being relatively small, it is interesting to analyze the effect produced by each harmonic of the type $m\Omega + s\Omega_M$, for $m \in \{0,2\}$ and for $s \in \{-2,2\}$, taking also into account the dynamics stemming from (x_3, y_3), simultaneously. Following Delaunay's idea (1867), this study is performed sequentially for each value of the indices m and s. The phase space corresponding to such a harmonic could present two types of dynamics: a resonant or a non-resonant one. A resonant topology including libration and circulation zones separated by a critical curve would preclude all averaging over the above mentioned harmonics. In this case, a more sophisticated method like the successive elimination of perturbation harmonics (Morbidelli, (1991)) could be applied. This would still meet with considerable difficulties due to the high number of degrees of freedom. In the other case if the topology induced by each of the harmonics presents only circulation with a reasonable amplitude of variations, a classical averaging technique can be applied to rid the Hamiltonian of these harmonics up to the desired order.

To proceed to the sequential analysis of the harmonics, a canonical transformation is performed to obtain a suitable set of variables containing the two angular variables x_3 ($x_3 = \frac{\alpha}{\beta}\ell + h$) and x_4 ($x_4 = m\Omega + s\Omega_M$). Three different transformations are performed depending on the values taken on by the indices m and s:

1. **for $m \neq 0$ and $s \neq 0$:**

$$
\begin{array}{lll}
x_1 = \ell & y_1 = L - \frac{\alpha}{\beta} H + \frac{\alpha}{\beta}\frac{m}{s}\Lambda_M, & \\
x_2 = g & y_2 = G, & \\
x_3 = \frac{\alpha}{\beta}\ell + \Omega - g_e & y_3 = H - \frac{m}{s}\Lambda_M, & (20) \\
x_4 = m\,\Omega + s\,\Omega_M & y_4 = \frac{1}{s}\Lambda_M, & \\
x_5 = g_e & y_5 = \Lambda_e + H - \frac{m}{s}\Lambda_M. &
\end{array}
$$

2. **for $m = 0$ and $s \neq 0$:**

$$
\begin{array}{lll}
x_1 = \ell & y_1 = L - \frac{\alpha}{\beta} H, & \\
x_2 = g & y_2 = G, & \\
x_3 = \frac{\alpha}{\beta}\ell + \Omega - g_e & y_3 = H, & (21) \\
x_4 = \Omega_M & y_4 = \Lambda_M & \\
x_5 = g_e & y_5 = \Lambda_e + H. &
\end{array}
$$

3. **for $m \neq 0$ and $s = 0$:**

$$
\begin{array}{lll}
x_1 = \ell & y_1 = L - \frac{\alpha}{\beta}\Lambda_e, & \\
x_2 = g & y_2 = G, & \\
x_3 = \frac{\alpha}{\beta}\ell + \Omega - g_e & y_3 = -\Lambda_e, & (22) \\
x_4 = \Omega & y_4 = H + \Lambda_e & \\
x_5 = \Omega_M & y_5 = \Lambda_M. &
\end{array}
$$

The total Hamiltonian taking into account the terrestrial (see (1)) and lunar attractions (given in (16)), is expanded up to the quadratic terms in $y_i - y_i^*$ around their values at the equilibrium y_i^*. The values y_i^* are computed at the critical inclination ($i^* = 63.4^o$) and at the value of the semi-major axis (a^*) which corresponds to the resonance in mean motion, i.e. such that $dx_3/dt = 0$ at the first order in J_2. The value of the eccentricity is arbitrarily chosen as $e^* = 0.7222$ (see Delhaise et Henrard, 1992).

For the indices m and s fixed to a particular value, this Hamiltonian takes on the form:

$$
\begin{aligned}
\mathcal{H} = & \left[s\,\dot{\Omega}_M + m\,\dot{\Omega}\Big|_{J_2} \right] y_4 + \frac{1}{2}\frac{\partial^2\mathcal{H}}{\partial y_2^2}\, y_2^2 + \frac{1}{2}\frac{\partial^2\mathcal{H}}{\partial y_3^2}\, y_3^2 \\
& + \frac{1}{2}\frac{\partial^2\mathcal{H}}{\partial y_4^2}\, y_4^2 + \frac{\partial^2\mathcal{H}}{\partial y_2 y_3}\, y_2\, y_3 + \frac{\partial^2\mathcal{H}}{\partial y_2 y_4}\, y_2\, y_4 + \frac{\partial^2\mathcal{H}}{\partial y_3 y_4}\, y_3\, y_4 \\
& - \sum_{l,r,p,q} B_{lrpq}(a^*, e^*, i^*) \begin{matrix}\cos\\ \sin\end{matrix} \{r(x_3 - \lambda_{lr}) + (l-2p)x_2\} \begin{matrix}\text{for } l-r \text{ even}\\ \text{for } l-r \text{ odd},\end{matrix} \\
& - \frac{\mu^6}{L^{*10}}\frac{R_e^4 J_2^2}{4} A(y_i^*)\cos 2x_2 \\
& - \sum_p D^*\,(-1)^{k_i}\, U_2^{m,s}(\varepsilon)\,\cos\left[(2-2p)x_2 + x_4 + s\frac{\pi}{2} - b_s\pi\right], \qquad (23)
\end{aligned}
$$

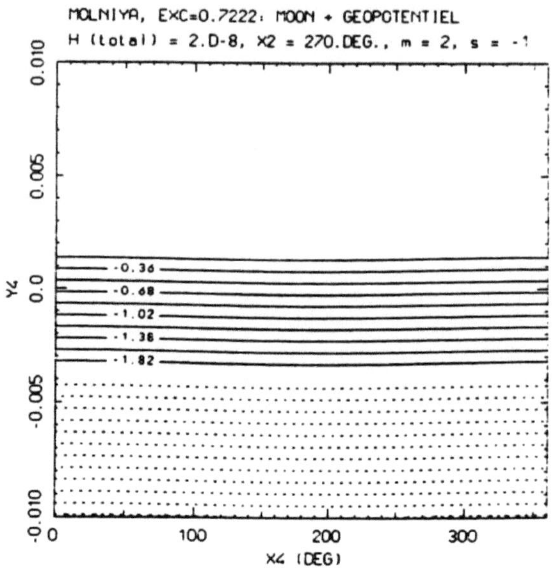

Fig. 1. Level curves of the pseudo-integral $\overline{J}$ illustrating the dynamics of the Molniya orbit in the plane (x_4, y_4) issued from the specific harmonics x_4 defined by $m = 2$ and $s = -1$. The full lines represent the cases when the point (x_3, y_3) is in libration mode and the dotted lines when it is in circulation mode. The results obtained for the harmonics: (m, s) = (1,1), (1,-2), (1,2), (2,-2), (2,1), (2,2) are similar to those represented here.

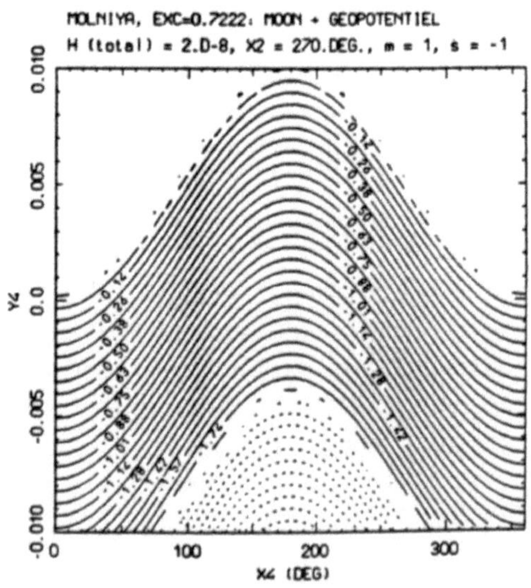

Fig. 2. Level curves of the pseudo-integral $\overline{J}$ illustrating the dynamics of the Molniya orbit in the plane (x_4, y_4) for the harmonics x_4 defined by $m = 1$ and $s = -1$. The results obtained for the harmonics: $m = 0$ and $s = 0$ are similar to those represented here.

where all partial derivatives are evaluated at $y_i = y_i^*$. Furthermore to shorten the notation, $y_i - y_i^*$ is simply written as y_i. The Hamiltonian can be decomposed into three parts as follows:

$$H_o(x_2, x_3, y_3, y_4) + H_1(x_2, x_4) + H_1'(y_2, y_3, y_4), \tag{24}$$

such that the semi-numerical method can be applied to $H_0 + H_1$ in the same way as performed in Delhaise et Henrard (1992). The terms of $H_1{}'$ containing y_2 are not considered in this preliminary approach to enable an examination of the dynamics of the two-degrees-of-freedom problem $H_0 + H_1$ in the field $< x_4, y_4 >$ with x_2 taken as parameter. The perturbation method applied here takes into account the full distortion of the invariant tori of the separable Hamiltonian H_0 which is resonant, instead of expanding it around the equilibrium point as is the case for a local analysis. This aim is achieved by introducing numerically a set of convenient action angle variables (J, ψ) in the three different topological regions of the phase space of the separable Hamiltonian H_0. The transformation from (x_3, y_3) to (J, ψ) has to be extended to a two-degrees of freedom canonical transformation in order to compute H_1 in the new set of variables. The total Hamiltonian is then averaged at the first order over the fast angular variable ψ. A supplementary quasi-integral results from this averaging. Indeed $\bar{J}$ now remains constant up to the first order. The action angle variables (J, ψ) and the pseudo-integral $\bar{J}$ are computed in a semi-numerical way (see Henrard, 1990): a grid of points is defined in the plane (x_4, y_4) while $x_3 = x_3^*$ defines the Poincaré surface of section and y_3 is computed such that $H_0 + H_1 = \text{const.}$. The level curves of the pseudo-integral $\bar{J}$ denote the trajectories on such Poincaré section. They are computed for the case of the Molniya-type orbits and are represented in the plane (x_4, y_4) in Figs (1) and (2) for two different harmonics x_4. The values of the constant energy level and of the parameter x_2 equal $2 \cdot 10^{-8}$ and 270°, respectively. From these figures, it can be deduced that the harmonics producing the most significant variations on y_4 are given by $(m, s) = (1, -1)$, $m = 0$ and $s = 0$. All other harmonics generate practically no variations over their conjugated momentum y_4.

The main conclusion emanating from this preliminary approach is that no libration zone is induced by any harmonic $m\Omega + s\Omega_M$. The amplitudes of variations of their corresponding conjugated momentum are non-negligible but not excessively large. This justifies further simplifications of the model by averaging over the two non-resonant angular variables Ω and Ω_M. The averaging process will be done up to the second order to take into account the effects of these harmonics as far as possible.

3.3. REDUCTION TO A TWO DEGREES OF FREEDOM SYSTEM

In this section, the secular dynamics of the Molniya-type orbits subjecting terrestrial and lunar gravitational perturbations is studied in a global way. The Hamiltonian will be averaged over the two non-resonant variables Ω and Ω_M. The dimension of the phase space is thus reduced from 8 to 4. Approximate surfaces of section are constructed in the plane of the inclination and argument of perigee. The qualitative and some quantitative aspects illustrating the global secular dynamics are deduced from these figures.

The following set of canonical variables is convenient for this study:

$$
\begin{aligned}
x_1 &= \ell & y_1 &= L + \tfrac{\alpha}{\beta}\Lambda_e, \\
x_2 &= g & y_2 &= G, \\
x_3 &= \tfrac{\alpha}{\beta}\ell + \Omega - g_e & y_3 &= -\Lambda_e, \\
x_4 &= \Omega & y_4 &= H + \Lambda_e, \\
x_5 &= \Omega_M & y_5 &= \Lambda_M.
\end{aligned}
\tag{25}
$$

The relations describing the dependence of the set (a, e, i) with respect to the variables (y_1, y_2, y_3) are the following:

$$
\begin{aligned}
a &= \frac{1}{\mu}\left(y_1 + \frac{\alpha}{\beta}y_3\right)^2, \\
e &= \sqrt{1 - \frac{y_2^2}{\left(y_1 + \frac{\alpha}{\beta}y_3\right)^2}}, \\
i &= \arccos\left(\frac{y_3 + y_4}{y_2}\right).
\end{aligned}
\tag{26}
$$

In this study the full Hamiltonian $\mathcal{H}(x_2, y_2, x_3, y_3, x_4, y_4, x_5, y_5)$, taking into account all harmonics is expanded into a Taylor series over the variables y_i around y_i^* up to the third order for the non-trigonometric terms and up to the first and second orders for the trigonometric terms derived from the inhomogeneity of the geopotential and from the Moon's gravitation, respectively. The higher degree of accuracy compared to the study considering solely the inhomogeneity of the geopotential (see Delhaise et Henrard, 1992) is necessary due to the larger considered interval in inclination and consequently in y_2. The selected extremal inclination values are:

$$
\begin{aligned}
i_{\text{min}} &= 59.835^o, \\
\text{and } i_{\text{max}} &= 67.035^o.
\end{aligned}
\tag{27}
$$

In this case, the maximum difference between y_2 and y_2^* equals 0.180. The corresponding interval in e_{mean} is:

$$
\begin{aligned}
e_{\text{min}} &\simeq 0.609, \\
e_{\text{max}} &\simeq 0.788.
\end{aligned}
\tag{28}
$$

Note that the value of e_{mean} has the same meaning as that given in Delhaise et Henrard (1992):

$$e_{\text{mean}} = \sqrt{1 - \frac{(y_3^*/\cos i)^2}{\left(y_1 + \frac{\alpha}{\beta} y_3^*\right)^2}}. \tag{29}$$

The form of the expanded full Hamiltonian is given in Delhaise (1992, p. 138). This Hamiltonian has four degrees of freedom. To halve the number of degrees of freedom of the phase space, the Hamiltonian is averaged over the two non-resonant variables x_4 and x_5. This course of action is supported by the results obtained in the preliminary analysis. As was demonstrated in section 5.3, the variations of the corresponding conjugated momentum y_4 produced by certain harmonics of the type $m\Omega + s\Omega_M$ are quite significant. For this reason the Hamiltonian is averaged up to the second order. The technique of Lie transforms is applied. The coefficients y_i are small quantities (of order of 10^{-1} for y_2 and 10^{-2} for the others). The coefficients of the trigonometric terms are of the order of y_i^2 or below. It is then natural to organize the inputs of the Lie triangle as follows:

- linear terms in y_i in $H_0^{(0)}$.
- quadratic terms in y_i and trigonometric terms independent of y_i in $H_1^{(0)}$.
- cubic terms in y_i and trigonometric terms dependent on y_i in $H_2^{(0)}$.

However, in our particular case the coefficient of y_3^2 is so large that the magnitude of this quadratic term is about that of the linear terms in y_4 and y_5. That is the reason why the libration period in x_3 turns out to be the same order as the circulation period in y_4 and y_5. Therefore, the quadratic term in y_3^2 must be included in $H_0^{(0)}$. In the summation of the terms of the kind $B_{i,j}\cos(ix_2 + jx_3)$, the term independent of x_2 is retained in $H_0^{(0)}$, which thus takes the form:

$$H_0^{(0)} = \omega_4 y_4 + \omega_5 y_5 + \frac{1}{2}\frac{\partial^2 \mathcal{H}}{\partial y_3^2} y_3^2 + B_{0,j}\cos(jx_3) \tag{30}$$

The Lie algorithm to average the Hamiltonian over x_4 and x_5 can not be directly applied with such a choice of $H_0^{(0)}$. To overcome this problem, suitable action angle variables (φ_3, I_3) are implicitly introduced by the canonical variable transformation:

$$\begin{aligned} x_3 &= X_3(\varphi_3, I_3), \\ y_3 &= Y_3(\varphi_3, I_3), \end{aligned} \tag{31}$$

So that the term $\frac{1}{2}\frac{\partial^2 \mathcal{H}}{\partial y_3^2} y_3^2 + B_{0,j}\cos(jx_3)$ can be rewritten as $K_3(I_3)$. Note that the functions X_3 and Y_3 are not known explicitly. Furthermore, the function K_3 is expanded locally around I_3^*, thus giving

$$K_3^* = \omega_3 \hat{I}_3 + \alpha \hat{I}_3^2 + \beta \hat{I}_3^3 + \ldots, \tag{32}$$

where $\hat{I}_3 = I_3 - I_3^*$ is considered small. Following this transformation, the inputs of the Lie triangle may now be written as:

$$\begin{aligned}
H_0^{(0)} &= \omega_4\, y_4 \,+\, \omega_5\, y_5 \,+\, \omega_3\, \hat{I}_3, \\
H_1^{(0)} &= \alpha\, \hat{I}_3^2 \,+\, \frac{1}{2}\frac{\partial^2 \mathcal{H}}{\partial y_2^2}\, y_2^2 + \frac{1}{2}\frac{\partial^2 \mathcal{H}}{\partial y_4^2}\, y_4^2 + \frac{\partial^2 \mathcal{H}}{\partial y_2 y_3}\, y_2\, Y_3(\varphi_3, I_3) + \frac{\partial^2 \mathcal{H}}{\partial y_2 y_4}\, y_2\, y_4 \\
&\quad + \frac{\partial^2 \mathcal{H}}{\partial y_3 y_4}\, Y_3(\varphi_3, I_3)\, y_4 - \sum_{lmpq} B_{lmpq} \cos\theta_{lmpq} + A \cos 2x_2 \qquad (33)\\
&\quad - \sum_{mps} D_M \cos\left((2-2p)x_2 \,+\, m\, x_4 \,+\, s\,(x_5 - \frac{\pi}{2}) - b_s \pi\right), \\
H_2^{(0)} &= \text{the remaining terms of the Hamiltonian with } y_3 \text{ replaced by} \\
&\quad Y_3(\varphi_3, I_3) \text{ and } x_3 \text{ by } X_3(\varphi_3, I_3).
\end{aligned}$$

With these settings, the Lie algorithm can be applied, in principle, up to an arbitrary order. The computations are performed here up to the second order to compute the mean values: $H_0^{(1)}$ and $H_0^{(2)}$ and the generating function W_1. This is feasible even if the functions X_3, Y_3 and K_3 are not given explicitly. The calculation of the Lie algorithm beyond the second order requires the explicit knowledge of x_3 and y_3 as function of the action angle variables (φ_3, I_3).

The averaged Hamiltonian turns out to be a two degrees of freedom one:

$$\overline{\mathcal{H}} \,=\, \overline{H}_o(\overline{x}_2, \overline{x}_3, \overline{y}_3, \overline{y}_4) \,+\, \overline{H}_1(\overline{x}_2, \overline{y}_2, \overline{x}_3, \overline{y}_3, \overline{y}_4). \qquad (34)$$

The semi-numerical method (Henrard, 1990) can thus be applied. The technique computing the level curves of the pseudo-integral $\overline{J}$ denoting the trajectories in the field of the inclination and argument of perigee for a constant energy level is the same as described in Delhaise et Henrard (1992). The surface of section is defined by $x_3 = x_3^*$ verifying:

$$\frac{dy_3}{dt} \,=\, \frac{-\partial \overline{H}_o}{\partial x_3} \,=\, 0, \qquad (35)$$

with x_2 fixed at its initial value. The results illustrating the global secular dynamics of the Molniya type orbits are displayed through Figures (3) and (4) for energy levels equal: 0 and $-3 \cdot 10^{-7}$, respectively. The continuous lines represent the cases where the angle x_3 librates, the dotted lines the cases where it circulates. This allows to locate the separatrix curve for the resonance of the first degree of freedom in the graphs. As will be verified by numerical integration, chaotic motion can be expected in the regions where the level curves cross this critical curve. The contributions of the Moon's attraction on the orbit evolution can be deduced from these figures. The lunar effects strongly increase the amplitude of libration

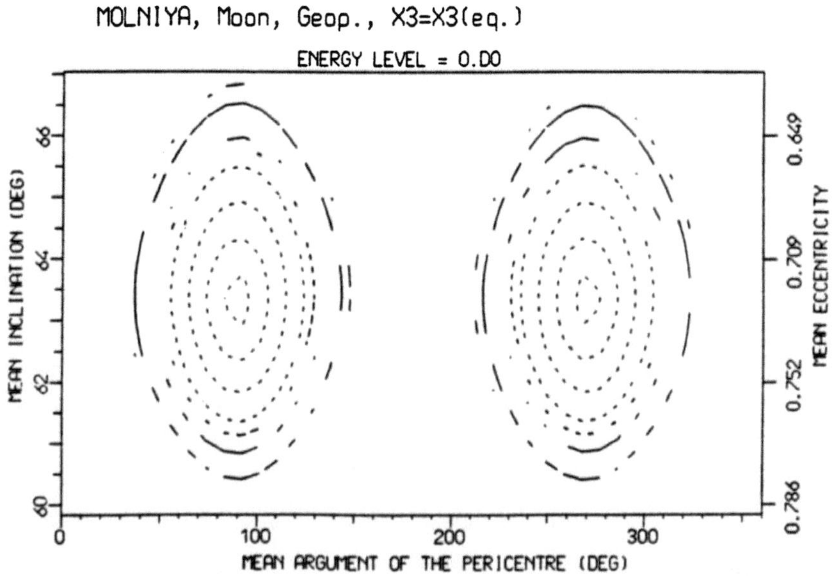

Fig. 3. Level curves of the pseudo-integral $\overline{J}$ for $\mathcal{H} = 0$ for the Molniya orbit. Terrestrial and lunar effects are taken into account. The continuous lines represent the cases where the angle x_3 librates, the dotted lines the cases where it circulates. Here, the continuous lines are "cut in pieces" since we are either close to the separatrix or close to the stable equilibrium point in the y_3, x_3 degree of freedom, which are both singularities for our action angle variables.

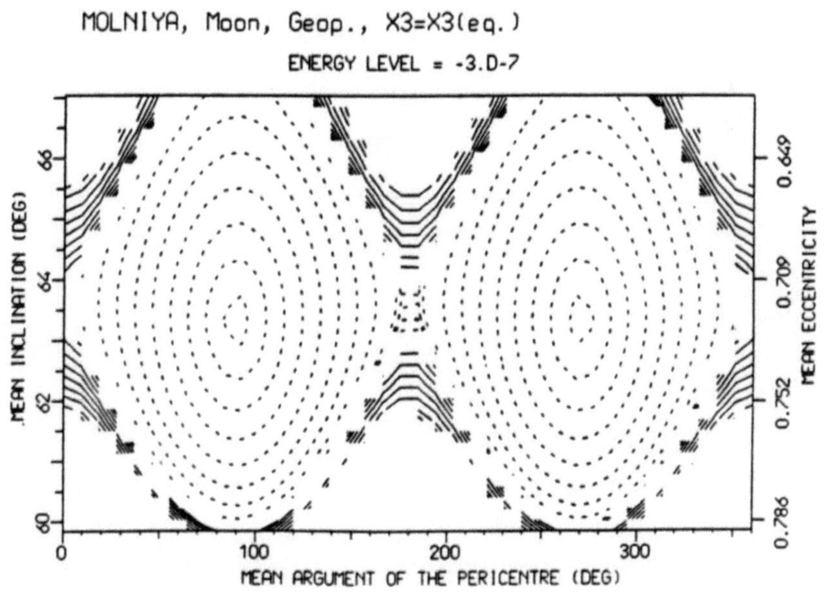

Fig. 4. Level curves of the pseudo-integral $\overline{J}$ for $\mathcal{H} = -3 \cdot 10^{-7}$ for the Molniya orbit. Terrestrial and lunar effects are taken into account.

in inclination, which reaches a value as high as 3.2^o while the geopotential effects alone cause a maximum libration amplitude of around 0.6^o. Another significant effect of the Moon's gravitation is the decrease of the libration period in (x_2, y_2) (numerically computed). This ranges between 25 and 50 years but attains more than 200 years when solely the inhomogeneity of the geopotential is considered. The value of the critical inclination is slightly different when the lunar gravitational perturbations are added to the model. Figures like (4) clearly show that this value is a function of x_2. The equilibrium value in inclination is maximal (63.8^o) for $x_2 = 0^o$ and 180°, the unstable equilibria, and minimal (63.2^o) for $x_2 = 90^o$ and 270°, the stable equilibria.

In conclusion, including the Moon's gravitational effects to the model boils down to adding a term of the form $\alpha \cos kx_2$ with a factor α approximately 10 times larger than when considering only the effects of the geopotential.

4. Inclusion of the Sun's Gravitational Effects

In this section the solar effects are included in the model. Therefore, all dominant gravitational effects from the Earth, Sun and Moon are taken into consideration.

4.1. SIMPLIFYING ASSUMPTIONS

The parameters of the apparent Sun orbit related to the Earth equator are given in Appendix 1. The form of the Hamiltonian taking into account the solar perturbations is given in equation (6). The following simplifications are introduced in this part of the Hamiltonian:

1. The summation over the index "l" of the solar potential (7) is truncated to include just the second harmonic.
$$l = 2. \tag{36}$$
2. The Hamiltonian is averaged over the satellite and Earth mean anomaly up to the first order. Therefore the indices (l, p, q) and (l, s, j) verify:
$$\begin{aligned} l - 2p + q &= 0, \\ l - 2s + j &= 0. \end{aligned} \tag{37}$$
3. The longitude of the ascending node Ω_s is assumed as invariant over the time interval of interest.
4. The Earth orbit is assumed being circular, so
$$l - 2s = 0. \tag{38}$$

4.2. RESULTS

Assuming Ω_s constant, no supplementary variable has to be added to the set of variables given in (25). The solar part of the Hamiltonian is also averaged over the satellite longitude of the ascending node x_4 up to the first order.

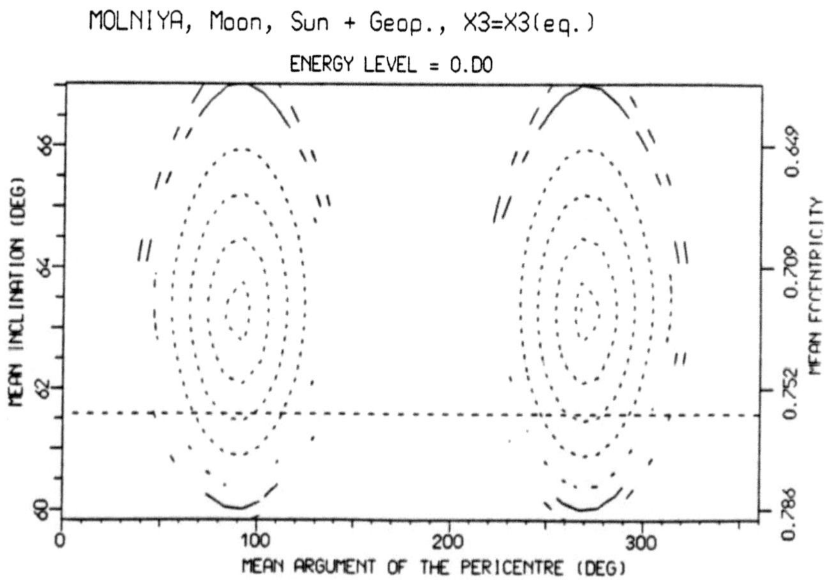

Fig. 5. Level curves of the pseudo-integral $\overline{J}$ for $\mathcal{H} = 0$ for the Molniya orbit. Terrestrial and luni-solar effects are taking into account. The continuous lines represent the cases where the angle x_3 librates, the dotted lines the cases where it circulates.

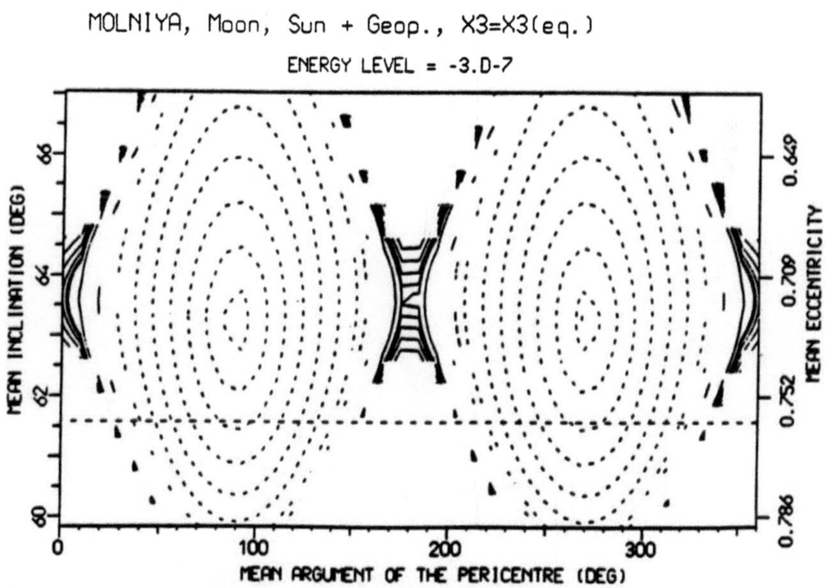

Fig. 6. Level curves of the pseudo-integral $\overline{J}$ for $\mathcal{H} = -3 \cdot 10^{-7}$ for the Molniya orbit. Terrestrial and luni-solar effects are taking into account.

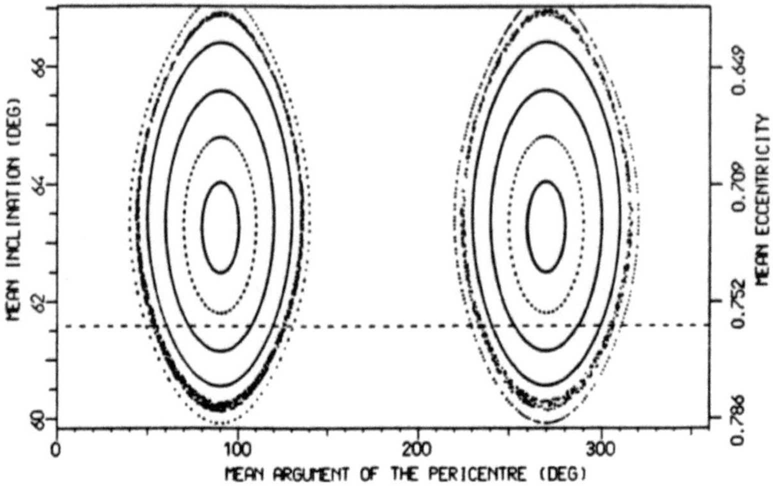

Fig. 7. Mapping resulting of the numerical integration of the Hamiltonian taking into account the terrestrial and luni-solar effects on a Molniya-type orbit. The energy level equals 0.

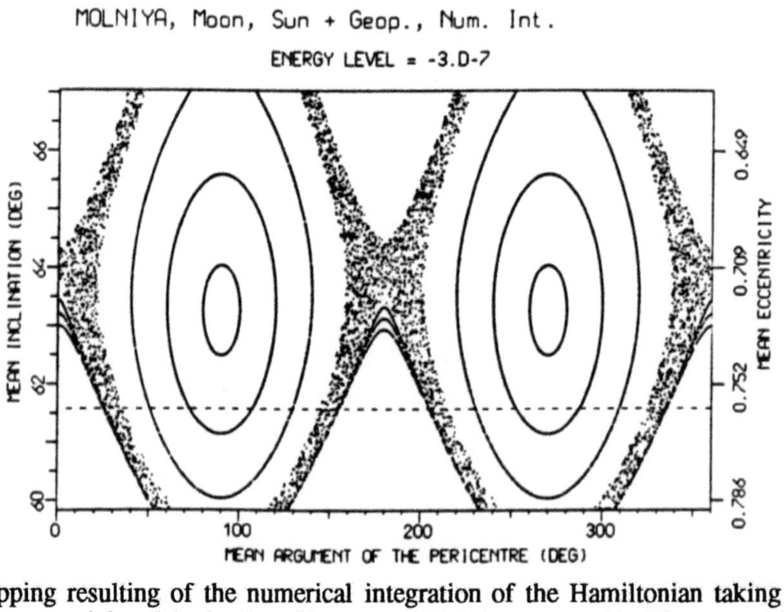

Fig. 8. Mapping resulting of the numerical integration of the Hamiltonian taking into account the terrestrial and luni-solar effects on a Molniya-type orbit. The energy level equals $-3 \cdot 10^{-7}$.

The contribution of the Sun to the total Hamiltonian is of the form:

$$\mathcal{H}_S = -\frac{\mu}{a_s} \sum_{p=o}^{2} \left(\frac{a}{a_s}\right)^2 F_{2,o,p}(i)\, F_{2,0,1}(i_s)\, H_{2,p,2p-l}(e)\, \cos[(2-2p)\,x_2]. \quad (39)$$

Each term of (39) is expanded around y_i^* up to the first order to obtain the following formulation:

$$[S_1 + S_2\, y_2 + S_3 y_3]\, \cos 2x_2 + S. \quad (40)$$

The semi-numerical method is again applied to yield the results of Figures (5) and (6) for the constant energy levels: 0 and $-3 \cdot 10^{-7}$, respectively. The Sun gravitation further increases the libration amplitude and shortens the libration period in (x_2, i). Close to the equilibrium point, the latter is reduced to about 20 years. The straight dashed line in Figures (5) and (6) represents the approximate limit where the eccentricity becomes so high that the perigee altitude descends below 100 km. The luni-solar effects are thus strong enough to reduce the satellite lifetime for particular initial conditions. Note that for a Tundra-type orbit, the initial perigee altitude being about 25000 km, there is no possibility of the luni-solar effects inducing a premature orbit decay.

5. Comparison with the Numerical Integrations

These results are confirmed by means of numerical integration of the differential equations derived from the averaged total Hamiltonian. The tests are performed for the constant energy levels 0 (see Fig. (7)) and $-3 \cdot 10^{-7}$ (see Fig. (8)). Comparing Figs (5) and (6) with Figs (7) and (8), we note that the results obtained by means of the semi-numerical method agree with those deriving of the numerical integration. As expected large chaotic regions appear along the separatrix curve of the resonance in (x_3, y_3).

6. Conclusion

The problem of a geosynchronous artificial satellite orbiting near the critical inclination is examined. In a preliminary study (Delhaise et Henrard, 1992), only the effects of the inhomogeneity of the geopotential were taken into consideration. This lead to a two-degrees-of-freedom problem. The perturbation model is now expanded through inclusion of the quite significant luni-solar gravitational effects. Obviously, this compounds the difficulties of the problem. Essentially, two additional degrees of freedom are introduced into the system. To tackle this four-degrees-of-freedom problem, a preliminary exploration separately analysing each harmonic stemming from the lunar potential is opportune. The considered harmonics are formed by a combination of the longitude of the ascending node of the satellite (x_4) and the longitude of the Moon (x_5). This study demonstrates that the topology induced by these harmonics does not show resonance phenomena (i.e. libration and circulation zones separated by a so-called critical curve). In all cases the dynamics remains that of a circulating angular variable. Three specific harmonics produce the most significant variations on their respective conjugated momenta; the others are of negligible influence. In a second approach, the number of degrees of freedom is halved by averaging the total Hamiltonian over the two non-resonant angular variables x_4 and x_5. As was demonstrated before, the variations produced by certain combinations of these two angles are significant. Therefore, the Hamiltonian is averaged up to the second order by the technique of Lie transforms. A semi-numerical method can now be applied as was done when considering solely the inhomogeneity of the geopotential. Approximate surfaces of section are constructed in the plane of the inclination and argument of perigee. The main effects of the Sun and Moon attractions compared to the terrestrial attraction alone are a strong increase in the amplitude of libration in inclination (from 0.6^o to 3.2^o) and a decrease of the corresponding libration period (from the order of 200 years to the order of 20 years).

Acknowledgements

The authors wish to thank Prof. Jacques Henrard for his helpful suggestions and advice.

Appendix

A. Moon and Sun's parameters

The Moon's orbital parameters at Epoch J.D. 2415020.0 and referred to the ecliptic plane are the following:

$$\begin{aligned} a_M &= 384399\,\text{km}, \\ e_M &= 0.0549, \\ i_M &= 5^o8'43.427'', \\ \dot{\Omega}_M &\simeq -0.05295^o/\text{days}, \\ &\quad \text{period} \simeq 18.6 \text{ years}, \\ \dot{\omega}_M &\simeq 0.1114^o/\text{days}, \\ &\quad \text{period} \simeq 8.85 \text{ years}, \\ \dot{M}_M &\simeq 12.8^o/\text{days}, \\ &\quad \text{period} \simeq 28 \text{ days}. \end{aligned} \tag{41}$$

The parameters of the apparent Sun orbit related to the Earth equator are given by:

$$\begin{aligned} a_s &= 1495979 \cdot 10^2 \text{ km}, \\ e_s &= 0.01673, \\ i_s &= 23.4^\circ 27' 08''.26 \\ \dot{\Omega}_s &\simeq 18''.8/\text{ year}, \\ \dot{M}_s &\simeq 1.01^\circ/\text{ days}, \\ &\quad \text{period } \simeq 1 \text{ year}. \end{aligned} \tag{42}$$

The motion of the argument of perigee with time is strongly nonlinear.

References

Brouwer, D.: 1959, "Solution of the Problem of Artificial Satellite Theory without Drag", *Astron. J* **64**, 378–397

Delhaise, F.: 1989, "The Effects of Luni-Solar Gravitation and Solar Radiation Pressure on the Molniya and Tundra Orbits", *ESA, MAS Working Paper*, **295**

Delhaise, F. and Henrard, J.: 1992, "The Problem of Critical Inclination Combined with a Resonance in Mean Motion in Artificial Satellite Theory", in press in *Celest. Mech.*

Delhaise, F.: 1992, "Analytical and Semi-Numerical Treatments of two Typical Problems in Artificial Satellite Theory: Air-Drag and Critical Inclination", *Ph. D. Thesis, FUNDP, Namur, Internal report.*

Exertier, P., 1989, "Extensions de la Théorie de Kaula aux Fonctions Perturbatrices Luni-Solaires et de Marées", *CNES, Internal Report.*

Giacaglia, G. E.: 1974, "Lunar Perturbations of Artificial Satellites of the Earth", *Celest. Mech.* **9**, 239–267

Giacaglia,G. E.: 1976, "A Note on Hansen's Coefficients in Satellite Theory", *Celest. Mech.* **14**, 515–523

Garfinkel, B.: 1959, "The orbit of a Satellite of an Oblate Planet", *Astron. J* **64**, 353

Henrard, J.: 1990, "A Semi-Numerical Perturbation Method for Separable Hamiltonian systems", *Celest. Mech.* **49**, 43–67

Hough, M.E.: 1981, "Orbits near Critical Inclination, Including Lunisolar Perturbations", *Celest. Mech.* **25**, 111–136

Jupp, A. H.: 1988, "The Critical Inclination Problem - 30 Years of Progress", *Celest. Mech.* **43**, 127–138

Kaula, W. M.: 1962, "Development of the Lunar and Solar Disturbing Functions for a Close Satellite", *Astron. J.* **67**, 300–303

Kaula, W. M.: 1966, "Theory of satellite geodesy", *Blaisdell Publishing Co.* ,

Kozai, Y.: 1959, "The Motion of a Close Earth Satellite", *Astron. J.* **64**, 367–377

Lane, M. T.: 1989, "On Analytic Modeling of Lunar Perturbations of Artificial Satellites of the Earth", *Celest. Mech.* **46**, 287–305

Lecohier, G., Guermonprez, V. and Delhaise, F.: 1989, "European Molniya and Tundra Orbit Control", *CNES, Mécanique Spatiale, Symposium International en Mécanique Spatiale*, Toulouse (France), 165–191

Morbidelli, A.: 1991, "On the Successive Eliminations of Perturbation Harmonics", in press in *Celest. Mech.*

LONG PERIOD VARIATIONS OF THE MOTION OF A SATELLITE DUE TO NON-RESONANT TESSERAL HARMONICS OF A GRAVITY POTENTIAL

G. METRIS, P. EXERTIER, Y. BOUDON, F. BARLIER
Observatoire de la Côte d'Azur, 06130, Grasse, France

Abstract. The main effects of tesseral harmonics of a gravity potential expansion on the motion of a satellite, are short period variations as well as long period variations due to resonances. However, other smaller long period and secular variations can arise from interactions between tesseral terms of the same order. The analytical integration of these effects is developed, using numerical evaluation of Kaula eccentricity and inclination functions. Examples for some Earth's geodetic satellites show that secular effects can reach a few decameters per year. The secular variations can even reach several hundred of meters per year for the Mars natural satellite Phobos.

Key words: Satellite theory – Tesseral harmonics interactions – Second order perturbations – Long period and secular effects

1. Introduction

Among all the forces acting on a satellite, the oblateness of the central body is often the most important one. When the gravity potential is developed in spherical harmonics, a zonal part and a tesseral part can be distinguished. The perturbations in the orbital elements induced by the zonal part include very large short period, long period and secular variations and have been extensively studied. Precise analytical solutions have been proposed for this problem (e.g. Brouwer, 1959; Kozai,1962; Deprit and Rom, 1970; Kinoshita, 1977; Coffey and Deprit, 1980).

The perturbations induced by the tesseral part have been less studied. Most of the works deal with the first order short period effects (e.g. Kaula, 1966; Wnuk, 1988; Wnuk and Breiter, 1990a; Elipe and Abad, 1991). However, short period variations due to interactions between zonal and tesseral harmonics were treated by Wnuk and Breiter (1990b). Long term effects arising from resonances have been also under investigation (Gedeon, 1969; Sochilina, 1982; Delhaise and Henrard, 1991). So far, interactions between tesseral harmonics have not been considered. However, if long period and secular variations are considered as essential, these phenomena can be of importance.

Thus, in this paper, we deal with secular and long period variations due to coupling effects between tesseral harmonics, for satellites of moderate eccentricity. In this aim, we proceed in two steps. First, we construct the averaged Hamiltonian of the problem, discarding resonant terms. In a second step, the averaged system is analytically integrated, assuming constant action variables and frequencies in the right hand member.

Celestial Mechanics and Dynamical Astronomy **57**: 175–188, 1993.

2. The Model

The development of the gravity field potential is limited to the tesseral harmonics although some effects of the zonal harmonics will be considered. In this context, direct perturbations due to tesseral harmonics are considered as first order perturbations and interactions between tesseral harmonics are considered as second order perturbations. The Hamiltonian taking into account the tesseral harmonics perturbations is (Delhaise and Henrard, 1991):

$$\begin{aligned} \mathcal{K} &= \mathcal{K}_0 + \mathcal{K}_1 \\ \mathcal{K}_0 &= -\frac{\mu^2}{2L^2} - \omega_e H \\ \mathcal{K}_1 &= -\sum_{n=2}^{+\infty}\sum_{m=1}^{n}\sum_{p=0}^{n}\sum_{q=-\infty}^{+\infty} \mathcal{B}_{nmpq}(a,e,I)\mathcal{S}_{nmpq}(l,g,h) \end{aligned} \tag{1}$$

With

$$\begin{aligned} \mathcal{B}_{nmpq} &= \frac{\mu}{a}\left(\frac{R_e}{a}\right)^n \mathcal{F}_{nmp}(I)\mathcal{G}_{npq}(e)J_{nm} \\ \mathcal{S}_{nmpq} &= \begin{bmatrix} \cos \\ \sin \end{bmatrix} (\psi_{nmpq}) \begin{matrix} \text{for } (n-m) \text{ even} \\ \text{for } (n-m) \text{ odd} \end{matrix} \\ \psi_{nmpq} &= (n-2p+q)l + (n-2p)g + m(h-\lambda_{nm}) \end{aligned} \tag{2}$$

$\mathcal{G}_{npq}(e)$ and $\mathcal{F}_{nmp}(I)$ are the standard Kaula functions (Kaula, 1966). μ is the geocentric gravitational constant and ω_e is the Earth's mean angular velocity. J_{nm} and λ_{nm} are respectively the tesseral coefficient and the longitude of the major axis of symmetry associated to the (n, m) spherical harmonic. The adopted set of canonical variables is the set of Delaunay variables (l, g, h, L, G, H) with respect to a frame rotating with the Earth; $h = \Omega - \omega_e t$ where Ω is the right ascension of the ascending node in the inertial frame.

We will use the integers:

$$\begin{aligned} P &= n - 2p \\ Q &= n - 2p + q \end{aligned}$$

3. Averaging

In this section, we construct a canonical transform whose aim is to remove short period terms of the Hamiltonian. These terms are related to the fast angles l and $h = \Omega - \omega_e t$. According to Deprit's algorithm (Deprit, 1969), we developed a Lie transform of generating function W which transforms the Hamiltonian $\mathcal{K}$ of the initial elements $(v_i \in \{l, g, h\}, V_i \in \{L, G, H\})$ to the Hamiltonian $\mathcal{K}'$ of the mean elements (v_i', V_i'). W is chosen so that $\mathcal{K}'$ is independent of l' and h'.

3.1. FIRST ORDER

The Hamiltonian $\mathcal{K}_1$ could contain resonant terms for indices n^*, m^*, p^*, q^* such as $\dot{\psi}_{n^*m^*p^*q^*}$ <1cycle/day. These terms are not considered in our problem and only short period terms are included in the summation (1). Consequently, the averaged Hamiltonian of the first order is readily:

$$\mathcal{K}_1' \equiv 0 \tag{3}$$

The first order of the generating function, W_1 is obtained from the equation (Deprit, 1969):

$$\{\mathcal{K}_0; W_1\} = \mathcal{K}_1' - \mathcal{K}_1 \tag{4}$$

where $\{\alpha; \beta\}$ stands for the Poisson brackets:

$$\{\alpha; \beta\} = \sum_{i=1}^{3} \frac{\partial \alpha}{\partial v_i} \frac{\partial \beta}{\partial V_i} - \frac{\partial \alpha}{\partial V_i} \frac{\partial \beta}{\partial v_i}$$

we get

$$\mathcal{N} \frac{\partial W_1}{\partial l} - \omega_e \frac{\partial W_1}{\partial h} = -\sum_{nmpq}{}' \mathcal{B}_{nmpq} \mathcal{S}_{nmpq} \tag{5}$$

where $\mathcal{N} = \mu^2/L^3$ and the prime on the summation sign indicates that only non-resonant, non-zonal $nmpq$ combinations are to be selected.

Considering the definition of $\mathcal{B}_{nmpq}$ and $\mathcal{S}_{nmpq}$, the solution is:

$$W_1 = -\sum_{nmpq}{}' \overline{\mathcal{B}}_{nmpq} \overline{\mathcal{S}}_{nmpq} + \mathcal{C} \tag{6}$$

where

$$\begin{aligned} \overline{\mathcal{B}}_{nmpq} &= \frac{\mathcal{B}_{nmpq}}{Q\mathcal{N} - m\omega_e} \\ \overline{\mathcal{S}}_{nmpq} &= \int \mathcal{S}_{nmpq} d\psi_{nmpq} = \begin{bmatrix} \sin \\ -\cos \end{bmatrix} (\psi_{nmpq}) \begin{matrix} (n-m) \text{ even} \\ (n-m) \text{ odd} \end{matrix} \end{aligned} \tag{7}$$

$\mathcal{C}$ is an arbitrary function independent of l and h. We choose $\mathcal{C} \equiv 0$ so that no long period term is involved at the first order of the transformation.

3.2. SECOND ORDER

The second order of the Lie transform has the expression (Deprit, 1969):

$$\begin{aligned} \mathcal{K}_2' &= \{\mathcal{K}_1 + \mathcal{K}_1'; W_1\} + \{\mathcal{K}_0; W_2\} \\ &= \{\mathcal{K}_1; W_1\} + \{\mathcal{K}_0; W_2\} \end{aligned} \tag{8}$$

$\mathcal{K}_2'$ and W_2 are left to our discretion; we choose $\mathcal{K}_2'$ independent of l' and h':

$$\mathcal{K}_2' = < \{\mathcal{K}_1; W_1\} >_{l',h'} \tag{9}$$

We have to remark that $\mathcal{K}_1$ and W_1 are functions of (v_i, V_i) whereas $\mathcal{K}_2'$ is function of (v_i', V_i'). But the Lie derivatives are evaluated for a null value of the perturbation i.e for $v_i = v_i'$ and $V_i = V_i'$.

The problem is now to calculate the Poisson bracket $\{\mathcal{K}_1; W_1\}$, expression which is given in appendix A. It is of the form:

$$\{\mathcal{K}_1; W_1\} = \sum_{1,2}{}' (D_{1,2}\mathcal{S}_1\mathcal{S}_2 + \overline{D}_{1,2}\overline{\mathcal{S}}_1\overline{\mathcal{S}}_2) \tag{10}$$

where the indices 1 and 2 symbolize two different sets of integers $(n_1 m_1 p_1 q_1)$ and $(n_2 m_2 p_2 q_2)$. $D_{1,2}$ and $\overline{D}_{1,2}$ are independent of the angles. It follows:

$$\mathcal{K}_2' = \sum_{1,2}{}' (D_{1,2} < \mathcal{S}_1\mathcal{S}_2 >_{l',h'} + \overline{D}_{1,2} < \overline{\mathcal{S}}_1\overline{\mathcal{S}}_2 >_{l',h'}) \tag{11}$$

Considering the expressions

$$\begin{aligned}
\mathcal{S}_1\mathcal{S}_2 &= \frac{1}{2}\begin{bmatrix} +\cos\psi^+ + \cos\psi^- \\ +\sin\psi^+ - \sin\psi^- \\ +\sin\psi^+ + \sin\psi^- \\ -\cos\psi^+ + \cos\psi^- \end{bmatrix} \begin{matrix} \text{for } (n_1 - m_1) \text{ even and } (n_2 - m_2) \text{ even} \\ \text{for } (n_1 - m_1) \text{ even and } (n_2 - m_2) \text{ odd} \\ \text{for } (n_1 - m_1) \text{ odd and } (n_2 - m_2) \text{ even} \\ \text{for } (n_1 - m_1) \text{ odd and } (n_2 - m_2) \text{ odd} \end{matrix} \\
\overline{\mathcal{S}}_1\overline{\mathcal{S}}_2 &= \frac{1}{2}\begin{bmatrix} -\cos\psi^+ + \cos\psi^- \\ -\sin\psi^+ - \sin\psi^- \\ -\sin\psi^+ + \sin\psi^- \\ +\cos\psi^+ + \cos\psi^- \end{bmatrix} \begin{matrix} \text{for } (n_1 - m_1) \text{ even and } (n_2 - m_2) \text{ even} \\ \text{for } (n_1 - m_1) \text{ even and } (n_2 - m_2) \text{ odd} \\ \text{for } (n_1 - m_1) \text{ odd and } (n_2 - m_2) \text{ even} \\ \text{for } (n_1 - m_1) \text{ odd and } (n_2 - m_2) \text{ odd} \end{matrix}
\end{aligned} \tag{12}$$

where we have noted

$$\begin{aligned}
\psi^+ &= (Q_1 + Q_2)l' + (P_1 + P_2)g' + (m_1 + m_2)h' - (m_1\lambda_{n_1 m_1} + m_2\lambda_{n_2 m_2}) \\
\psi^- &= (Q_1 - Q_2)l' + (P_1 - P_2)g' + (m_1 - m_2)h' - (m_1\lambda_{n_1 m_1} - m_2\lambda_{n_2 m_2})
\end{aligned} \tag{13}$$

the problem is now to select the indices $(n_1 m_1 p_1 q_1)$, $(n_2 m_2 p_2 q_2)$ for which ψ^+ or ψ^- are independent of l' and h'.

Since m_1 and m_2 are always strictly positive, the coefficient of h' in ψ^+ cannot be null.

ψ^- is independent of l' and h' if and only if:

$$\begin{cases} m_1 = m_2 \\ n_1 - 2p_1 + q_1 = n_2 - 2p_2 + q_2 \end{cases} \tag{14}$$

The solution of this Diophantine system is (Casotto, 1991):

$$\begin{cases} m_2 = m_1 \\ n_2 = n_1 + j \\ p_2 = p_1 + j + k \\ q_2 = q_1 + j + 2k \end{cases} \tag{15}$$

Given the bounds for n, m, p and q in formula (1), j and k can be theoretically infinite integers. In practice, the development is truncated and:

$$\begin{aligned} 2 \leq n &\leq n_{\max} \\ -q_{\max} \leq q &\leq q_{\max} \end{aligned} \tag{16}$$

$n_{\max}$ represents a truncations of the potential expansion whereas $q_{\max}$ is a truncation in eccentricity. With these truncation, the ranges $[j_{\min}, j_{\max}]$ and $[k_{\min}, k_{\max}]$ of j and k are obtained following Casotto (1991). We emphasize the fact that the set (n_1, m_1, p_1, q_1) refer to $\mathcal{K}_1$ and (n_2, m_2, p_2, q_2) refer to W_1. Therefore two solutions of (14) of the form $(n'_1, m'_1, p'_1, q'_1; n'_2, m'_2, p'_2, q'_2)$ and $(n'_2, m'_2, p'_2, q'_2; n'_1, m'_1, p'_1, q'_1)$ are not the same solutions unless $n'_1 = n'_2$, $m'_1 = m'_2$, $p'_1 = p'_2$ and $q'_1 = q'_2$. Imposing $2 \leq n_2 \leq n_{\max}$ yields:

$$\begin{aligned} j_{\min} &= 2 - n_1 \\ j_{\max} &= n_{\max} - n_1 \end{aligned} \tag{17}$$

The range of k is found by imposing $0 \leq p_2 \leq n_2$ and $-q_{\max} \leq q_2 \leq q_{\max}$ giving:

$$\begin{aligned} k_{\min} &= \max(-p_1 - j, E[(-q_{\max} - q_1 - j + 1)/2]) \\ k_{\max} &= \min(n_1 - p_1, E[(q_{\max} - q_1 - j)/2]) \end{aligned} \tag{18}$$

where $E[x]$ indicates the integral part function.

Then, setting

$$E_{nmpqjk} = D_{nmpq(n+j)m(p+j+k)(q+j+2k)} + \overline{D}_{nmpq(n+j)m(p+j+k)(q+j+2k)} \tag{19}$$

$$\mathcal{T}_{nmjk} = \frac{1}{2} \begin{bmatrix} +\cos \\ -\sin \\ +\sin \\ +\cos \end{bmatrix} \left(\theta_{nmjk}\right) \begin{matrix} \text{for } (n-m) \text{ even and } j \text{ even} \\ \text{for } (n-m) \text{ even and } j \text{ odd} \\ \text{for } (n-m) \text{ odd and } j \text{ even} \\ \text{for } (n-m) \text{ odd and } j \text{ odd} \end{matrix} \tag{20}$$

with

$$\theta_{nmjk} = (j + 2k) * g' - m(\lambda_{nm} - \lambda_{(n+j)m}) \tag{21}$$

we finally get:

$$\mathcal{K}'_2 = \sum_{nmpq}{}' \sum_{j=2-n}^{n_{\max}-n} \sum_{k=k_{\min}}^{k_{\max}} E_{nmpqjk} \mathcal{T}_{nmjk} \tag{22}$$

4. Secular and Long Period Effects

4.1. THE MEAN DIFFERENTIAL SYSTEM

The evolution of the new variables (v_i', V_i') is governed by the differential system:

$$\left\{\begin{array}{lllll} \dot{l}' & = \dfrac{\partial K'}{\partial L'} & = \dfrac{\mu^2}{L'^3} & + \dfrac{1}{2} \sum\limits_{nmpq}{}' \sum\limits_{j=2-n}^{n_{\max}-n} \sum\limits_{k=k_{\min}}^{k_{\max}} \dfrac{\partial E_{nmpqjk}}{\partial L'} \mathcal{T}_{nmjk} \\ \dot{g}' & = \dfrac{\partial K'}{\partial G'} & = & + \dfrac{1}{2} \sum\limits_{nmpq}{}' \sum\limits_{j=2-n}^{n_{\max}-n} \sum\limits_{k=k_{\min}}^{k_{\max}} \dfrac{\partial E_{nmpqjk}}{\partial G'} \mathcal{T}_{nmjk} \\ \dot{h}' & = \dfrac{\partial K'}{\partial H'} & = -\omega_e & + \dfrac{1}{2} \sum\limits_{nmpq}{}' \sum\limits_{j=2-n}^{n_{\max}-n} \sum\limits_{k=k_{\min}}^{k_{\max}} \dfrac{\partial E_{nmpqjk}}{\partial H'} \mathcal{T}_{nmjk} \\ \dot{L}' & = -\dfrac{\partial K'}{\partial l'} & = 0 & \\ \dot{G}' & = -\dfrac{\partial K'}{\partial g'} & = & - \dfrac{1}{2} \sum\limits_{nmpq}{}' \sum\limits_{j=2-n}^{n_{\max}-n} \sum\limits_{k=k_{\min}}^{k_{\max}} (j+2k) E_{nmpqjk} \mathcal{T}'_{nmjk} \\ \dot{H}' & = -\dfrac{\partial K'}{\partial h'} & = 0 & \end{array}\right. \tag{23}$$

where

$$\mathcal{T}'_{nmjk} = \frac{\partial \mathcal{T}_{nmjk}}{\partial \theta_{nmjk}} \tag{24}$$

The partial derivatives of E_{nmpqjk} are developed in appendix B.

It is worth noting that, despite the truncation (16), the summation can still involve thousands of terms. However, the smallness of the effects permits some simplifications. For a given satellite, nominal values (e.g. mean values) a_0, e_0 and I_0 can be used for evaluating each coefficient E_{nmpqjk} of the summation and its partial derivatives. Then, replacing λ_{nm} and $\lambda_{(n+j)m}$ in θ_{nmjk} by their values and adding the amplitudes for the same arguments leads to a very compact system of the form:

$$\left\{\begin{array}{lll} \dot{l}' & = -\dfrac{\mu^2}{L'^3} + & \sum\limits_{s\geq 0} \left(A^C_{1s} \cos sg' + A^S_{1s} \sin sg' \right) \\ \dot{g}' & = & \sum\limits_{s\geq 0} \left(A^C_{2s} \cos sg' + A^S_{2s} \sin sg' \right) \\ \dot{h}' & = -\omega_e + & \sum\limits_{s\geq 0} \left(A^C_{3s} \cos sg' + A^S_{3s} \sin sg' \right) \\ \dot{G}' & = & \sum\limits_{s\geq 1} \left(A^C_{4s} \cos sg' + A^S_{4s} \sin sg' \right) \end{array}\right. \tag{25}$$

where A_{is}^C and A_{is}^S are numerical coefficients. In the equation for $\dot{G}'$, the summation starts with $s = 1$ because the right hand side results from a derivation with respect to g' and then the coefficient of this argument can not be zero. The amplitudes A_{is}^C and A_{is}^S rapidly decrease when s increases and are generally negligible for $s > 2$ (it can be shown that A_{is}^C and A_{is}^S are proportional to e^{s-c} with $0 \leq c \leq 4$).

4.2. INTEGRATION OF THE MEAN DIFFERENTIAL SYSTEM

At this stage, the differential system (25) could be integrated by numerical methods, using a very large step of integration permitted by the absence of short periods. Such a method is especially interesting if this system is added to other mean systems describing long period variations induced by different perturbations: zonal harmonics of the geopotential, luni-solar effects, solid and oceanic tides (Borderies, 1976; Exertier, 1988; Metris, 1991). But, except in the case of the critical inclination, a full analytical integration can be easily derived. Indeed, in the second member of (25), it is sufficient to take into account the secular variations $\tilde{g}'$ of g' essentially deriving from the first zonal harmonic. For a given satellite we can use, for example:

$$\tilde{g}'(t)-\tilde{g}'(t_0) = \mathcal{N}_g(t-t_0) = \frac{3}{4}J_{20}\left(\frac{R_e}{a_0}\right)^2 \frac{\mathcal{N}_0}{(1-e_0^2)^2}(4-5\sin^2 I_0)(t-t_0) \quad (26)$$

Where $\mathcal{N}_0 = \mu^2/L_0^3 = \sqrt{\mu/a_0^3}$ and t and t_0 are times in seconds. Consequently:

$$\begin{cases} \Delta_{LT}l' = A_{10}^C(t-t_0) + \dfrac{1}{\mathcal{N}_g}\sum\limits_{s\geq 1}\left(A_{1s}^C \sin sg' - A_{1s}^S \cos sg'\right) \\ \Delta_{LT}g' = A_{20}^C(t-t_0) + \dfrac{1}{\mathcal{N}_g}\sum\limits_{s\geq 1}\left(A_{2s}^C \sin sg' - A_{2s}^S \cos sg'\right) \\ \Delta_{LT}h' = A_{30}^C(t-t_0) + \dfrac{1}{\mathcal{N}_g}\sum\limits_{s\geq 1}\left(A_{3s}^C \sin sg' - A_{3s}^S \cos sg'\right) \\ \Delta_{LT}G' = \dfrac{1}{\mathcal{N}_g}\sum\limits_{s\geq 1}\left(A_{4s}^C \sin sg' - A_{4s}^S \cos sg'\right) \end{cases} \quad (27)$$

$(\Delta_{LT}l', \Delta_{LT}g', \Delta_{LT}h', \Delta_{LT}G')$ denote the long period and secular variations of the mean elements (l', g', h', G') due to tesseral harmonics.

As $\mathcal{N}_g$ is of the order of J_{20}, the secular variations are proportional to $J_{n_1m}J_{n_2m}$ whereas the long period variations are proportional to $\dfrac{J_{n_1m}J_{n_2m}}{J_{20}}$.

4.3. LONG PERIOD VARIATIONS OF THE INITIAL ELEMENTS

The evolution of the initial elements (v_i, V_i) differs from the evolution of the mean elements (v_i', V_i') essentially by short period terms; more precisely by:

- short period terms proportional to J_{nm} $(m \neq 0)$ which could be computed using W_1,
- short period terms proportional to $J_{20}J_{nm}$ $(m \neq 0)$ discarded in this paper,
- short period and long period terms proportional to $J_{n_1m_1}J_{n_2m_2}$ (which could be computed using W_1 and W_2) or to $J_{20}^2 J_{nm}$ or $J_{n_10}J_{n_2m}$ $(m_1 \neq 0, m_2 \neq 0, m \neq 0)$. These terms are negligible.

Since the long period terms which appear in the transformation are smaller than the long period terms of the formula (27), we have with a very good approximation:

$$\begin{cases} \Delta_{LT}v_i &= \Delta_{LT}v_i' \\ \Delta_{LT}V_i &= \Delta_{LT}V_i' \end{cases} \tag{28}$$

And then (27) can be used directly.

5. Results for Geodetic Satellites

We applied the previous results to the Lageos and Starlette satellites which are tracked for a long time (more than 15 years) by laser telemetry.

For Lageos ($a_0 = 12270$ km, $e_0 = 0.0044$, $I_0 = 110°$, $\mathcal{N}_g = -4.335\ 10^{-8}$ rd/s), we used the truncation $n_{\max} = 10$ and $q_{\max} = 4$. However, the discrepancies between the results derived with these values and the results obtained with $n_{\max} = 6$ and $q_{\max} = 2$ do not exceed 5 mm/year for the secular terms and 1 cm for the amplitudes of the long period terms.

We give the variations of the Keplerian elements which are more usual than the Delaunay variables:

$$\begin{cases} a_0 * \Delta_{LT}e' &= & & 0.06\cos g' - 0.12\sin g' \\ a_0 * \Delta_{LT}I' &= 0 & & \\ a_0 * \Delta_{LT}h' &= -0.78(t-t_0)/y & & \\ a_0 * \Delta_{LT}(l'+g') &= -1.15(t-t_0)/y & -0.01\cos g' & \end{cases} \tag{29}$$

where the variations are in meters and $y = 315576$ seconds/year such as $(t-t_0)/y$ is in years.

The effects are not large, but we have to keep in mind that the RMS of the residuals between observed distances and the distances derived from a fitted orbit, is less than one meter for a time span of 12 years (Eanes and Watkins, 1991).

The effects are larger for Starlette ($a_0 = 7331$ km, $e_0 = 0.02$, $I_0 = 50°$, $\mathcal{N}_g = 6.69\ 10^{-7}$ rd/s) because of its lower altitude. For that reason, a higher degree of truncation in the harmonics must be used: $n_{\max} = 15$. We keep $q_{\max} = 4$.

$$\begin{cases} a_0 * \Delta_{LT}e' &= & 0.09\cos g' & +0.68\sin g' \\ a_0 * \Delta_{LT}I' &= & & -0.01\sin g' \\ a_0 * \Delta_{LT}h' &= 19.87(t-t_0)/y & -0.15\cos g' & -0.06\sin g' \\ a_0 * \Delta_{LT}(l'+g') &= 23.47(t-t_0)/y & +0.38\cos g' & +0.01\sin g' \end{cases} \tag{30}$$

6. Conclusion

The coupling effects between the tesseral harmonics induce non negligible long period and secular variations on the orbital elements of a satellite. Also our examples were limited to Earth's geodetic satellites, the same results could be applied to artificial and natural satellites of different planets. For instance, we computed the corresponding perturbations on satellites of Mars which has very strong tesseral harmonics coefficients (Balmino et al, 1982). We found secular effects of several hundreds of meters per year on Phobos' position and of several tens of kilometers per year on the positions of imaginary artificial satellites already used by Wnuk and Breiter (1990b).

Acknowledgements

The authors thank Dr. S. Casotto for suggesting important improvements. We also thank the French Centre National d'Etudes Spatiales for its support.

Appendix

A. Expression of the Poisson bracket $\{\mathcal{K}_1; W_1\}$

We give formulae to compute the Poisson bracket $\{\mathcal{K}_1; W_1\}$ involved in the expression of the averaged second order Hamiltonian $\mathcal{K}_2'$ (formula 9). From (1) and (6):

$$\begin{aligned}
\mathcal{K}_1 &= - \sum_{n_1 m_1 p_1 q_1}{}' \mathcal{B}_{n_1 m_1 p_1 q_1} \mathcal{S}_{n_1 m_1 p_1 q_1} \\
W_1 &= - \sum_{n_2 m_2 p_2 q_2}{}' \overline{\mathcal{B}}_{n_2 m_2 p_2 q_2} \overline{\mathcal{S}}_{n_2 m_2 p_2 q_2}
\end{aligned}$$

we get formula (10):

$$\{\mathcal{K}_1; W_1\} = \sum_{1,2}{}' (D_{1,2} \mathcal{S}_1 \mathcal{S}_2 + \overline{D}_{1,2} \overline{\mathcal{S}}_1 \overline{\mathcal{S}}_2)$$

where the indices 1 and 2 symbolize respectively the sets of indices $n_1 m_1 p_1 q_1$ and $n_2 m_2 p_2 q_2$.

The expressions of $\mathcal{S}_1 \mathcal{S}_2$ and $\overline{\mathcal{S}}_1 \overline{\mathcal{S}}_2$ are given by (12) whereas $D_{1,2}$ and $\overline{D}_{1,2}$ can be computed with the following sequential formulae in which $\mathcal{F}'(I)$ and $\mathcal{F}''(I)$ stand for the first and second derivatives of $\mathcal{F}(I)$ with respect to the inclination, $\mathcal{G}'(e)$ and $\mathcal{G}''(e)$ designate the first and second derivatives of $\mathcal{G}(e)$ with respect to the eccentricity:

$$\begin{aligned} P_1 &= n_1 - 2p_1 \\ P_2 &= n_2 - 2p_2 \\ Q_1 &= n_1 - 2p_1 + q_1 \\ Q_2 &= n_2 - 2p_2 + q_2 \\ \mathcal{P}_2 &= \frac{1}{Q_2\mathcal{N} - m_2\dot{\omega}_e} \\ R_2 &= Q_2\mathcal{N}\mathcal{P}_2 \\ \eta &= \frac{G}{L} \end{aligned} \tag{31}$$

$$\mathcal{A} = \left(\frac{\mu}{a}\right)^2 \left(\frac{R_e}{a}\right)^{n_1+n_2} \mathcal{P}_2 \tag{32}$$

$$\begin{aligned} A_1 &= \frac{\eta^2}{e}\mathcal{G}_1'(e) - 2(n_1+1)\mathcal{G}_1(e) \\ A_2 &= \frac{\eta^2}{e}\mathcal{G}_2'(e) - \left[2(n_2+1) - 3R_2\right]\mathcal{G}_2(e) \\ B_1 &= \frac{\cos I}{\sin I}\mathcal{F}_1'(I)\mathcal{G}_1(e) - \frac{\eta^2}{e}\mathcal{F}_1(I)\mathcal{G}_1'(e) \\ B_2 &= \frac{\cos I}{\sin I}\mathcal{F}_2'(I)\mathcal{G}_2(e) - \frac{\eta^2}{e}\mathcal{F}_2(I)\mathcal{G}_2'(e) \\ C_1 &= \frac{1}{\sin I}\mathcal{F}_1'(I)\mathcal{G}_1(e) \\ C_2 &= \frac{1}{\sin I}\mathcal{F}_2'(I)\mathcal{G}_2(e) \\ T_{1,2} &= \frac{1}{G}\left[Q_2\eta\mathcal{F}_1(I)A_1 + P_2B_1 - m_2C_1\right] \\ \overline{T}_{1,2} &= \frac{1}{G}\left[Q_1\eta\mathcal{F}_2(I)A_2 + P_1B_2 - m_1C_2\right] \end{aligned} \tag{33}$$

$$\begin{aligned} D_{1,2} &= -\mathcal{A}T_{1,2}\mathcal{F}_2(I)\mathcal{G}_2(e)J_{n_1m_1}J_{n_2m_2} \\ \overline{D}_{1,2} &= -\mathcal{A}\overline{T}_{1,2}\mathcal{F}_1(I)\mathcal{G}_1(e)J_{n_1m_1}J_{n_2m_2} \end{aligned} \tag{34}$$

B. Partial derivatives of E with respect to L, G, H.

The partial derivatives of

$$E_{nmpqjk} = D_{nmpq(n+j)m(p+j+k)(q+j+2k)} + \overline{D}_{nmpq(n+j)m(p+j+k)(q+j+2k)}$$

with respect to L, G and H are needed for the construction of the mean differential system (formula 23). In the following, we give the derivatives of $D_{1,2} = D_{n_1m_1p_1q_1n_2m_2p_2q_2}$ and $\overline{D}_{1,2} = \overline{D}_{n_1m_1p_1q_1n_2m_2p_2q_2}$ defined in appendix A, for general sets of indices.

Derivatives with respect to L

$$\frac{\partial A_1}{\partial L} = \frac{1}{L}\left[-\left(2(n_1+2)+\frac{\eta^2}{e^2}\right)\frac{\eta^2}{e}\mathcal{G}_1'(e)+\frac{\eta^4}{e^2}\mathcal{G}_1''(e)\right]$$

$$\frac{\partial A_2}{\partial L} = \frac{1}{L}\left[9\mathcal{R}_2(\mathcal{R}_2-1)\mathcal{G}_2(e)+\left(3\mathcal{R}_2-2n_2-4-\frac{\eta^2}{e^2}\right)\frac{\eta^2}{e}\mathcal{G}_2'(e)+\frac{\eta^4}{e^2}\mathcal{G}_2''(e)\right]$$

$$\frac{\partial B_1}{\partial L} = \frac{\eta^2}{eL}\left[\frac{\cos I}{\sin I}\mathcal{F}_1'(I)\mathcal{G}_1'(e)+\left(2+\frac{\eta^2}{e^2}\right)\mathcal{F}_1(I)\mathcal{G}_1'(e)-\frac{\eta^2}{e}\mathcal{F}_1(I)\mathcal{G}_1''(e)\right]$$

$$\frac{\partial B_2}{\partial L} = \frac{\eta^2}{eL}\left[\frac{\cos I}{\sin I}\mathcal{F}_2'(I)\mathcal{G}_2'(e)+\left(2+\frac{\eta^2}{e^2}\right)\mathcal{F}_2(I)\mathcal{G}_2'(e)-\frac{\eta^2}{e}\mathcal{F}_2(I)\mathcal{G}_2''(e)\right]$$

$$\frac{\partial C_1}{\partial L} = \frac{1}{G\sin I}\frac{\eta^3}{e}\mathcal{F}_1'(I)\mathcal{G}_1'(e)$$

$$\frac{\partial C_2}{\partial L} = \frac{1}{G\sin I}\frac{\eta^3}{e}\mathcal{F}_2'(I)\mathcal{G}_2'(e)$$

$$\frac{\partial T_{12}}{\partial L} = \frac{1}{G}\left[Q_2\eta\mathcal{F}_1(I)\left(\frac{\partial A_1}{\partial L}-\frac{A_1}{L}\right)+P_2\frac{\partial B_1}{\partial L}-m_2\frac{\partial C_1}{\partial L}\right]$$

$$\frac{\partial \overline{T}_{12}}{\partial L} = \frac{1}{G}\left[Q_1\eta\mathcal{F}_2(I)\left(\frac{\partial A_2}{\partial L}-\frac{A_2}{L}\right)+P_1\frac{\partial B_2}{\partial L}-m_1\frac{\partial C_2}{\partial L}\right]$$

$$\mathcal{R}_{12} = \left[3\mathcal{N}\mathcal{Q}_2\mathcal{P}_2-2(n_1+n_2+2)\right]$$

$$\frac{\partial D_{12}}{\partial L} = \frac{D_{12}}{L}\mathcal{R}_{12}-\mathcal{A}\mathcal{F}_2(I)\left[\frac{\eta^2}{eL}\mathcal{G}_2'(e)T_{12}+\mathcal{G}_2(e)\frac{\partial T_{12}}{\partial L}\right]$$

$$\frac{\partial \overline{D}_{12}}{\partial L} = \frac{\overline{D}_{12}}{L}\mathcal{R}_{12}-\mathcal{A}\mathcal{F}_1(I)\left[\frac{\eta^2}{eL}\mathcal{G}_1'(e)\overline{T}_{12}+\mathcal{G}_1(e)\frac{\partial \overline{T}_{12}}{\partial L}\right]$$

Derivatives with respect to G

$$\frac{\partial A_1}{\partial G} = \frac{\eta}{eL}\left[\left(2(n_1+2)+\frac{\eta^2}{e^2}\right)\mathcal{G}_1'(e) - \frac{\eta^2}{e}\mathcal{G}_1''(e)\right]$$

$$\frac{\partial A_2}{\partial G} = -\frac{\eta}{eL}\left[\left(3R_2-2(n_2+2)-\frac{\eta^2}{e^2}\right)\mathcal{G}_2'(e) + \frac{\eta^2}{e}\mathcal{G}_2''(e)\right]$$

$$\frac{\partial B_1}{\partial G} = \frac{\cos I}{G\sin I}\left[\frac{\cos I}{\sin I}\mathcal{F}_1''(I) - \frac{1}{\sin^2 I}\mathcal{F}_1'(I)\right]\mathcal{G}_1(e) - \frac{\eta}{eL}\left[2\frac{\cos I}{\sin I}\mathcal{F}_1'(I)+\left(2+\frac{\eta^2}{e^2}\right)\mathcal{F}_1(I)\right]\mathcal{G}_1'(e) + \frac{\eta^3}{e^2L}\mathcal{F}_1(I)\mathcal{G}_1''(e)$$

$$\frac{\partial B_2}{\partial G} = \frac{\cos I}{G\sin I}\left[\frac{\cos I}{\sin I}\mathcal{F}_2''(I) - \frac{1}{\sin^2 I}\mathcal{F}_2'(I)\right]\mathcal{G}_2(e) - \frac{\eta}{eL}\left[2\frac{\cos I}{\sin I}\mathcal{F}_2'(I)+\left(2+\frac{\eta^2}{e^2}\right)\mathcal{F}_2(I)\right]\mathcal{G}_2'(e) + \frac{\eta^3}{e^2L}\mathcal{F}_2(I)\mathcal{G}_2''(e)$$

$$\frac{\partial C_1}{\partial G} = \frac{1}{G\sin I}\left[\left(\mathcal{F}_1''(I)-\frac{\cos I}{\sin I}\mathcal{F}_1'(I)\right)\frac{\cos I}{\sin I}\mathcal{G}_1(e) - \frac{\eta^2}{e}\mathcal{F}_1'(I)\mathcal{G}_1'(e)\right]$$

$$\frac{\partial C_2}{\partial G} = \frac{1}{G\sin I}\left[\left(\mathcal{F}_2''(I)-\frac{\cos I}{\sin I}\mathcal{F}_2'(I)\right)\frac{\cos I}{\sin I}\mathcal{G}_2(e) - \frac{\eta^2}{e}\mathcal{F}_2'(I)\mathcal{G}_2'(e)\right]$$

$$\frac{\partial T_{12}}{\partial G} = \frac{1}{G}\left[Q_2\eta\left(\mathcal{F}_1(I)\frac{\partial A_1}{\partial G}+\frac{\cos I}{\sin I}\mathcal{F}_1'(I)\frac{A_1}{G}\right)+P_2\left(\frac{\partial B_1}{\partial G}-\frac{B_1}{G}\right) - m_2\left(\frac{\partial C_1}{\partial G}-\frac{C_1}{G}\right)\right]$$

$$\frac{\partial \overline{T}_{12}}{\partial G} = \frac{1}{G}\left[Q_1\eta\left(\mathcal{F}_2(I)\frac{\partial A_2}{\partial G}+\frac{\cos I}{\sin I}\mathcal{F}_2'(I)\frac{A_2}{G}\right)+P_1\left(\frac{\partial B_2}{\partial G}-\frac{B_2}{G}\right) - m_1\left(\frac{\partial C_2}{\partial G}-\frac{C_2}{G}\right)\right]$$

$$\frac{\partial D_{12}}{\partial G} = -\mathcal{A}\left[\left(\frac{\cos I}{\sin I}\mathcal{F}_2'(I)\mathcal{G}_2(e)-\frac{\eta^2}{e}\mathcal{F}_2(I)\mathcal{G}_2'(e)\right)\frac{T_{12}}{G}+\mathcal{F}_2(I)\mathcal{G}_2(e)\frac{\partial T_{12}}{\partial G}\right]$$

$$\frac{\partial \overline{D}_{12}}{\partial G} = -\mathcal{A}\left[\left(\frac{\cos I}{\sin I}\mathcal{F}_1'(I)\mathcal{G}_1(e)-\frac{\eta^2}{e}\mathcal{F}_1(I)\mathcal{G}_1'(e)\right)\frac{\overline{T}_{12}}{G}+\mathcal{F}_1(I)\mathcal{G}_1(e)\frac{\partial \overline{T}_{12}}{\partial G}\right]$$

Derivatives with respect to H

$$\frac{\partial B_1}{\partial H} = \frac{1}{G\sin I}\left[\left(\frac{\mathcal{F}_1'(I)}{\sin^2 I} - \frac{\cos I}{\sin I}\mathcal{F}_1''(I)\right)\mathcal{G}_1(e) + \frac{\eta^2}{e}\mathcal{F}_1'(I)\mathcal{G}_1'(e)\right]$$

$$\frac{\partial B_2}{\partial H} = \frac{1}{G\sin I}\left[\left(\frac{\mathcal{F}_2'(I)}{\sin^2 I} - \frac{\cos I}{\sin I}\mathcal{F}_2''(I)\right)\mathcal{G}_2(e) + \frac{\eta^2}{e}\mathcal{F}_2'(I)\mathcal{G}_2'(e)\right]$$

$$\frac{\partial C_1}{\partial H} = \frac{1}{G\sin^2 I}\left[\frac{\cos I}{\sin I}\mathcal{F}_1'(I) - \mathcal{F}_1''(I)\right]\mathcal{G}_1(e)$$

$$\frac{\partial C_2}{\partial H} = \frac{1}{G\sin^2 I}\left[+\frac{\cos I}{\sin I}\mathcal{F}_2'(I) - \mathcal{F}_2''(I)\right]\mathcal{G}_2(e)$$

$$\frac{\partial T_{12}}{\partial H} = \frac{1}{G}\left[-\frac{Q_2\eta}{G\sin I}\mathcal{F}_1'(I)A_1 + P_2\frac{\partial B_1}{\partial H} - m_2\frac{\partial C_1}{\partial H}\right]$$

$$\frac{\partial \overline{T}_{12}}{\partial H} = \frac{1}{G}\left[-\frac{Q_1\eta}{G\sin I}\mathcal{F}_2'(I)A_2 + P_1\frac{\partial B_2}{\partial H} - m_1\frac{\partial C_2}{\partial H}\right]$$

$$\frac{\partial D_{12}}{\partial H} = \mathcal{A}\left[\frac{1}{G\sin I}\mathcal{F}_2'(I)T_{12} - \mathcal{F}_2(I)\frac{\partial T_{12}}{\partial H}\right]\mathcal{G}_2(e)$$

$$\frac{\partial \overline{D}_{12}}{\partial H} = \mathcal{A}\left[\frac{1}{G\sin I}\mathcal{F}_1'(I)\overline{T}_{12} - \mathcal{F}_1(I)\frac{\partial \overline{T}_{12}}{\partial H}\right]\mathcal{G}_1(e)$$

References

Balmino, G., Moynot, B. and Valès, N.: 1982, 'Gravity Field Model of Mars in Spherical Harmonics up to Degree and Order Eighteen', *J. Geophys. Res.* **87(B12)**, 9735-9746.

Borderies, N.: 1976, 'La Méthode de Centrage en Géodésie Spatiale', Thesis, Université Paul-Sabatier, Toulouse.

Brouwer, D.: 1959, 'Solution of the Problem of Artificial Satellite Theory without Drag', *Astron. J.* **64**, 378-397.

Casotto, S.: 1991, 'Spectral Decomposition of Geopotential, Earth and Ocean Tidal Perturbations in Linear Satellite Theory', *Celest. Mech.* **50**, 125-141.

Coffey, S. and Deprit, A.: 1980, 'Short Period Elimination in Satellite Theory', AIAA/AAS Astrodynamics Conference, Denvers, Mass.

Delhaise, F. and Henrard, J.: 1993, 'The Problem of Critical Inclination Combined with a Resonance in Mean Motion in Artificial Satellite Theory', *in press in Celest. Mech.*,

Deprit, A.: 1969, 'Canonical Transformations Depending on a Small Parameter', *Celest. Mech.* **1**, 12-30.

Deprit, A. and Rom, A.: 1970, 'The Main Problem of Artificial Satellite Theory for Small and Moderate Eccentricities', *Celest. Mech.* **2**, 166-206.

Eanes R.J. and Watkins, M.M.: 1991, 'Temporal Variability of Earth's Gravitational Field from Satellite Laser Ranging Observations', presented at the XX General Assembly of the IUGG, Vienna, Austria, August 1991.

Elipe, A. and Abad, A.: 1991, 'Algebraic Manipulators and Lie Transforms in Artificial Satellite Theory', University of Saragoza, Internal Report.

Exertier, P.: 1988, 'Orbitographie des Satellites Artificiels sur de Grandes Périodes de Temps; Possibilités d'applications', Thesis, Observatoire de Paris.

Gedeon, G., S.: 1969, 'Tesseral Resonance Effects on Satellite Orbits', *Celest. Mech.* **1**, 167-189.

Kaula, W.M.: 1966, *Theory of Satellite Geodesy*, Blaisdell publ. Co., Walthman, Mass.

Kinoshita, H.: 1977, 'Third-Order Solution of an Artificial Satellite Theory', *S.A.O. Spec. Rep.* **379**,

Kozai, Y.: 1962, 'Second-Order Solution of Artificial Satellite Theory without Air Drag', *Astron. J.* **67-7**, 446-461.

Metris, G.: 1991, 'Théorie du Mouvement du Satellite Artificiel: Développement des Equations du Mouvement Moyen, Application à l'étude des Longues Périodes', Thesis, Observatoire de Paris.

Sochilina, A. S.: 1982, 'On the Motion of a Satellite in Resonance with its Rotating Planet', *Celest. Mech.* **26**, 337-352.

Wnuk, E.: 1988, 'Tesseral Harmonic Perturbations for High Order and Degree Harmonics', *Celest. Mech.* **44**, 179-191.

Wnuk, E. and Breiter, S.: 1990a, 'Tesseral Harmonic Perturbations in Radial Transverse and Binormal Components', *Celest. Mech.* **48**, 375-385.

Wnuk, E. and Breiter, S.: 1990b, 'The Motion of Natural and Artificial Satellites in Mars Gravity Field', *Adv. Space Res.* **Vol. 11, No 6**, 183-188.

ON THE ROTO–TRANSLATORY MOTION OF A SATELLITE OF AN OBLATE PRIMARY

J.M. FERRÁNDIZ, M.E. SANSATURIO AND R. CABALLERO

Dpto. de Matemática Aplicada a la Ingeniería, E.T.S. de Ingenieros Industriales, 47011 Valladolid, Spain

Abstract. We deal with the problem of the motion of a triaxial satellite of an oblate primary of larger mass. We show that the treatment is simplified by using a canonical set of variables previously introduced by the authors, that allows a drastic reduction in the expansions of the potential. A general method to avoid the appearance of virtual singularities when the angles between certain planes are small is designed. Our approach is applicable either to natural or artificial satellites.

Key words: Satellite Dynamics, orbit–rotation coupling, canonical transformations, symbolic manipulation.

1. Introduction

In this paper we address the general motion of rotation and translation of a triaxial satellite of an oblate primary of large mass, assuming that the rotation of the latter is not affected by the minor body. A complete treatment of the simpler problem in which the central body is spherical can be seen in Kinoshita (1972). A different version appears in Ferrándiz and Sansaturio (1988), this time using a new canonical set of variables, that will be employed in this paper, too. This system is similar to those of Hill and Andoyer, but in its definition the total angular moment – of rotation and translation – of the satellite plays a fundamental role, which permits a drastic reduction in the part of the perturbing potential caused by the triaxiality of the satellite. Similar systems, but non canonical, were used by Hitzl and Breakwell (1971).

In relation to the classical approach, in which all the variables are referred to an inertial frame, the expansions of the perturbing potential reduce from 1126 terms – using Hill-Andoyer variables – to 106, in the new set.

We also offer canonical systems of variables appropriate for avoiding virtual singularities. These arise from the appearance of divisors that become small with the eccentricity or with the relative inclinations between some planes on which the angular coordinates are measured.

The application of standard perturbation methods allows us to eliminate periodic terms. The resulting problem is considered in the absence of resonances and in the very frequent case of synchronous rotation. Attention is paid to the influences of the different perturbations on the orbital motion and the evolution of the angular momentum, indicating under which conditions the figure of the satellite can affect the mean motion in longitude. Likewise, it is shown that in most cases the orbital plane can be identified with the invariant plane used to refer the rotation of the rigid body to it, without altering the accuracy of the solution.

Celestial Mechanics and Dynamical Astronomy **57**: 189–202, 1993.

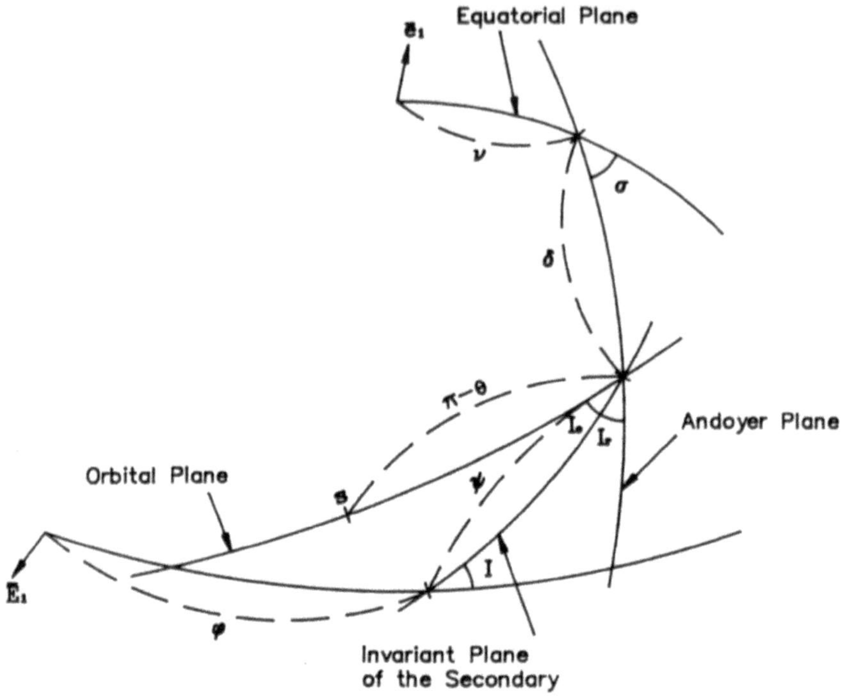

Fig. 1. Definition of the variables

2. Formulation of the Problem

Let us consider the problem of the general motion of a triaxial rigid body attracted by an oblate primary, assuming that, if m_s and m_p denote the respective masses of both bodies with $m_s << m_p$, the rotation of the latter is not affected by the minor body.

In order to obtain the Hamiltonian of such a problem we have chosen a set of canonical variables $(r, \varphi, \psi, \theta, \delta, \nu, R, \Phi, \Psi, \Theta, \Delta, N)$ of Hill-Andoyer type, which was introduced by the first two authors in 1987 (see Ferrándiz and Sansaturio, 1988, 1989), where r stands for the distance between the centers of mass of both bodies. The geometrical meaning of the rest of the coordinates is illustrated in Figure 1. However, let us draw attention to the fact that the variable θ is reckoned from the ascending node of the orbital plane with respect to the invariant one – that is, the opposite one to the node from which the angle δ is measured on the Andoyer plane. As for the momenta, we have

$R \equiv$ Radial velocity of the center of mass of the secondary.

$\Phi \equiv$ Third component of the total angular momentum vector of the secondary in space frame.

$\Psi \equiv$ Magnitude of the total angular momentum vector of the secondary.
$\Theta \equiv$ Magnitude of the angular momentum of the center of mass of the secondary.
$\Delta \equiv$ Magnitude of the rotational angular momentum vector of the secondary.
$N \equiv$ Third component of the angular momentum of the secondary in body frame (principal axes of inertia).

In this set of variables the Hamiltonian is written as

$$\mathcal{K} = \frac{1}{2m}\left(R^2 + \frac{\Theta^2}{r^2}\right) + \frac{1}{2}\left[(\Delta^2 - N^2)\left(\frac{\sin^2\nu}{A} + \frac{\cos^2\nu}{B}\right) + \frac{N^2}{C}\right] - \frac{\mu^* m}{r} + \mathcal{V}_s + \mathcal{V}_p, \tag{1}$$

where $\mu^* = \mathcal{G}(m_p + m_s)$, $\mathcal{G}$ is the gravitational constant, m is the reduced mass of the system, A, B, C stand for the principal moments of inertia of the secondary and $\mathcal{V}_s$ and $\mathcal{V}_p$ denote the non-Keplerian potentials due to the non–sphericity of the secondary and primary bodies respectively.

In order to avoid the appearance of the masses as much as possible we have found it convenient to perform a scaling of the problem by multiplying all the momenta by the constant

$$\frac{1}{m} = \frac{m_p + m_s}{m_p m_s},$$

which give us a new Hamiltonian $\mathcal{H} = \mathcal{K}/m$, whose expression is the following

$$\mathcal{H} = \frac{1}{2}\left(R^2 + \frac{\Theta^2}{r^2}\right) + \frac{1}{2}\frac{m_p}{m_s + m_p}\left[(\Delta^2 - N^2)\left(\frac{\sin^2\nu}{A} + \frac{\cos^2\nu}{B}\right) + \frac{N^2}{C}\right] - \frac{\mu^*}{r} + \frac{m_p + m_s}{m_p m_s}\left(\mathcal{V}_s + \mathcal{V}_p\right), \tag{2}$$

where, for the sake of simplicity, we have used the same notation for the new variables and A, B, C now stand for moments of inertia by unit of mass.

The developments of the perturbing potentials $\mathcal{V}_s$ and $\mathcal{V}_p$, up to a first order of perturbation, can be obtained by the standard procedures; for instance, by using relationships for the transformation by rotation of Legendre's functions, as in Kinoshita (1972), or by carrying out changes of coordinates starting from McCullagh's formulae (see Leimanis (1965)) as in Ferrándiz and Sansaturio (1989). Such formulae provide the following expressions for $\mathcal{V}_s$ and $\mathcal{V}_p$

$$\mathcal{V}_s = \frac{2C - A - B}{2}(1 - 3\cos^2\gamma) + (A - B)\left[1 - 3\cos^2\alpha + \frac{1}{2}(1 - 3\cos^2\gamma)\right],$$

$$\mathcal{V}_p = \frac{2C_p - A_p - B_p}{2}(1 - 3\cos^2\gamma),$$

where A_p, B_p, C_p are the moments of inertia of the primary and ($\cos\alpha$, $\cos\beta$, $\cos\gamma$) is the unit vector in the direction of the particle, which has to be referred to the principal axes of the secondary and expressed in the variables used. In the classical system of Hill–Andoyer this step implies 7 rotations for $\mathcal{V}_s$ as opposed to the 3 required to express $\mathcal{V}_p$, while in the canonical set that we have chosen it needs 5 rotations in both cases. In this way, the price to be paid to shorten the expansion of $\mathcal{V}_s$ is to enlarge that of $\mathcal{V}_p$ going to the orbital plane through the invariant plane, which involves two more rotations. Nevertheless, in any case, the total number of terms is much smaller than in the classical approach (1126 *vs.* 102).

Now, if we restrict the domain of the variables (r, θ, R, Θ) to that in which the Keplerian part of the previous Hamiltonian is negative and we carry out a complete canonical transformation from the variables (r, θ, R, Θ) to those of Delaunay (l, g, L, G) — see, for instance Deprit (1989) pp. 113-117 or Levi-Civita (1913) pp. 157-161 — the following equality holds

$$\frac{-\mu^*}{2L^2} = \frac{1}{2}\left(R^2 + \frac{\Theta^2}{r^2}\right) - \frac{\mu^*}{r}.$$

After some calculations which include elementary trigonometric relations to transform sines and cosines of the involved angles into sines and cosines of certain linear combinations of them and by using Hansen's formulae for elliptic motion providing Fourier series in the mean anomaly, we obtain the expressions for the perturbing potentials

$$\frac{1}{m}\mathcal{V}_s = k_1\mathcal{P} + k_2\mathcal{Q}\,, \qquad \frac{1}{m}\mathcal{V}_p = k_3\mathcal{R}\,,$$

where

$$\begin{aligned}
\mathcal{P} &= \sum_{k\in Z^+\cup\{0\}} C_k^*\cos kl \;+ \sum_{(i,j,k)\in J} D_{ijk}\cos(i\delta + j\,g + k\,l), \\
\mathcal{Q} &= \sum_{(i,j,k,q)\in J'} E_{ijkq}\cos(i\delta + j\,g + k\,l + q\,\nu), \\
\mathcal{R} &= \sum_{k\in Z^+\cup\{0\}} X_k^*\cos kl \;+ \sum_{(i,j,k)\in J} X_{ijk}\cos(i\psi + j\,g + k\,l), \\
J &= \{-2,-1,0,1,2\}\times\{0,2\}\times Z, \qquad J' = J\times\{-2,2\},
\end{aligned} \tag{3}$$

the coefficients C_k^*, D_{ijk}, E_{ijkq}, X_k^*, X_{ijk} being functions of the momenta and of Hansen's coefficients, and the small quantities k_i $(i = 1,2,3)$, which depend on the momentum L, are defined by the equations

$$k_1 = \frac{{\mu^*}^4}{L^6}J_2R_s^2, \quad k_2 = \frac{4{\mu^*}^4}{L^6}J_{22}R_s^2, \quad k_3 = \frac{{\mu^*}^4}{L^6}J_2^pR_p^2,$$

where, as usual, we have denoted

$$J_2 = \frac{2C - A - B}{2R_s^2}, \qquad J_{22} = \frac{A - B}{4R_s^2},$$

R_s being "mean-radius"of the secondary. Analogous expressions hold for the oblateness coefficient J_2^p of the primary, with mean radius R_p.

In order to obtain the order of magnitude of k_i, let us denote by a and n the semimajor axis and the mean motion in the unperturbed problem. Then, if we put $\mu^* \simeq n^2a^3$ and $L \simeq na^2$, it follows that

$$k_1 \simeq n^2R_s^2J_2, \qquad k_2 \simeq 4n^2R_s^2J_{22}, \qquad k_3 \simeq n^2R_p^2J_2^p\,. \tag{4}$$

In the Appendix we offer full expressions for $\mathcal{V}_s$ and $\mathcal{V}_p$, up to second order in the eccentricity for Hansen's coefficients, when $I_0 = \sigma = 0$, and which have been calculated by means of a symbolic manipulator developed by the third author to handle Poisson's series. The main features of such a manipulator can be summarized as follows:

1. Programmed in standard Fortran-77, it can run without changes on many different environments. The same code has been compiled on a Cyber 930, a HP-Series 700 workstation and a PC.
2. Dynamic storage, with the method of coded keys for the representation of the series as is described by Henrard (1989).
3. Implemented routines following the scheme built by Broucke (1989).
   - General purpose: multiplication, derivation, integration, averaging, norms, substitution, etc, of Poisson series.
   - Special purpose: Bessel, Legendre and Hansen functions, classical in all the expansions of Celestial Mechanics.
4. An optimized method for the multiplication of series, Caballero (1992).
5. Output in LaTeX form, as can be seen in the Appendix. The rational approximation of double precision numbers is carried out with the usual technique of continuous fractions.
6. Ability of communicating this processor with other general purpose Computer Algebra Systems, such as Mathematica.
7. User adaptable, the number of polynomial and trigonometric variables can be easily changed.

3. Elimination of Virtual Singularities

As pointed out in the previous section, the disturbing potential depends on the canonical momenta through the angles between the "fundamental"planes, more precisely, σ and $I_0 + I_r$ for the $\mathcal{V}_s$ term and, I and I_0 for $\mathcal{V}_p$. As the partial derivatives of the Hamiltonian with respect to the momenta contain divisors in the sines of the said angles, they become singular when any of them vanishes. Thus it is convenient to introduce sets of canonical variables, in which the equations of motion (or those of perturbation, where necessary) are free from those divisors, and therefore from the virtual singularities they create. The way to reach this is well known when using usual sets of variables, but in the present case a new approach is required, since the dependence of $I_0 + I_r$ and I_0 on the canonical moments is more involved. In fact, we have

$$\cos(I_0 + I_r) = \frac{\Psi^2 - \Delta^2 - \Theta^2}{2\Theta\Delta}, \qquad \cos I_r = \frac{\Psi^2 + \Delta^2 - \Theta^2}{2\Psi\Delta},$$
$$\cos I_0 = \frac{\Psi^2 + \Theta^2 - \Delta^2}{2\Psi\Delta}, \qquad \cos\sigma = \frac{N}{\Delta}, \qquad \cos I = \frac{\Phi}{\Psi}. \tag{5}$$

Next we will establish a general procedure to obtain systems of canonical variables free from singularities, under quite wide conditions. Let us suppose first the classical case in which $(q_1, q_2; p_1, p_2)$ belongs to a canonical set $(q_i; p_i)$, and k is a function depending on the former set through the components of the matrix $R_3(q_1)\,R_1(\alpha)\,R_3(q_2 - q_1)$, where $R_i(x)$ stands for a rotation of angle x around the (i)-axis.

Let us assume that α is an angle depending on the momenta through an equation of the form

$$\cos\alpha = 1 + \frac{p_1}{F(p_1, p_2, p_3)}, \tag{6}$$

F being an arbitrary function on the momenta, and $p_1 \leq 0$ if $F > 0$, so that $\alpha = 0$ corresponds to $p_1 = 0$.

Changing the pair (p_1, q_1) by their associate PoincarŽ variables (P_1, Q_1), defined as

$$P_1 = -\sqrt{-2p_1}\cos q_1, \quad Q_1 = \sqrt{-2p_1}\sin q_1\,, \tag{7}$$

a straightforward calculation leads to

$$d\alpha = \frac{1}{\sqrt{F}\cos\alpha/2}(-\cos q_1\, dP_1 + \sin q_1\, dQ_1) - \frac{\tan\alpha/2}{F}dF.$$
$$dq_1 = \frac{1}{\sqrt{-2p_1}}(\sin q_1\, dP_1 + \cos q_1\, dQ_1). \tag{8}$$

It is clear that the first equation does not present any divisor in $\sin\alpha$, provided that neither F nor dF/F contain such a divisor.

Then, it is obvious that the partial derivatives of k with respect to the new set do not contain such divisors multiplying $\partial k/\partial\alpha$. Nevertheless, transformation (7) gives rise to a term $\sin\alpha/2$ in the denominator affecting $\partial k/\partial q_1$, because of the identity

$$\sin\frac{\alpha}{2} = \frac{\sqrt{-2p_1}}{2\sqrt{F}}\,. \tag{9}$$

It is not difficult to prove that it cancels with a factor $\sin\alpha$ inside $\partial k/\partial q_1$, due to the functional dependence assumed for k. It is enough to realize that, among the terms in $R_3(q_1)\,R_1(\alpha)\,R_3(q_2-q_1)$, the only ones that do not contain a factor $\sin\alpha$ are precisely those not vanishing when making $\alpha = 0$. Actually, these terms do not depend on q_1, since $R_3(q_1)\,R_1(0)\,R_3(q_2-q_1) = R_3(q_2)$, and the result follows.

In the rest of this section we will apply the former procedure to eliminate the singularities arising in the equations of motion coming from the terms $\mathcal{V}_s$, when the angle $I_0 + I_r$ approaches 0. The first step is to substitute the variables $(\Psi,\Theta,\Delta;\psi,\theta,\delta)$ by

$$\begin{aligned} p_1 &= \Psi - \Theta - \Delta, & q_1 &= \psi, \\ p_2 &= \Theta + \Delta, & q_2 &= \psi + \theta + \pi, \\ p_3 &= \Delta, & q_3 &= \delta - \theta - \pi. \end{aligned} \tag{10}$$

Notice that $\Psi - \Theta - \Delta \le 0$, according to the meaning of each one of the implied momenta (total angular momentum, orbital and rotational ones, in that order) and that p_1 is close to zero as $I_0 + I_r$. Moreover, q_2 plays the role of an argument of longitude.

Next we perform the canonical change

$$P_1 = -\sqrt{2(\Theta+\Delta-\Psi)}\cos\psi\,, \qquad Q_1 = \sqrt{2(\Theta+\Delta-\Psi)}\sin\psi\,. \tag{11}$$

From the first identity in (5) we find

$$2\sin^2\frac{I_0+I_r}{2} = \frac{(\Delta+\Theta-\Psi)(\Psi+\Delta+\Theta)}{2\Delta\Theta} = -p_1\frac{p_1+2p_2}{2p_3(p_2-p_3)}\,, \tag{12}$$

and according to our aforementioned result the new system is free from virtual singularities in $I_0 + I_r$, arising from the term $\mathcal{V}_s$, which depends on functions of $R_3(\delta)\,R_1(I_0+I_r)\,R_3(\pi-\theta)$.

Similarly, we can derive the identity

$$2\sin^2\frac{I_0}{2} = \frac{(\Delta+\Theta-\Psi)(\Delta+\Psi-\Theta)}{2\Psi\Theta} = -p_1\frac{p_1+2p_3}{2(p_1+p_2)(p_2-p_3)}. \tag{13}$$

Then, the same variables are also suitable for preventing the appearance of singularities corresponding to the near annihilation of I_0, this parameter now arising from the term $\mathcal{V}_p$ depending on $R_3(\psi)\,R_1(-I_0)\,R_3(\pi-\theta)$.

The obtaining of canonical sets useful for small values of any special angle between the planes in which the angular variables are reckoned can be easily accomplished. For instance, to avoid divisors in all the angles, as well as in eccentricity, it suffices to construct first a canonical set

$$\begin{aligned}
P_1 &= \Phi - \Psi, & Q_1 &= \varphi, \\
P_2 &= \Psi - G - \Delta, & Q_2 &= \varphi + \psi, \\
P_3 &= G + \Delta, & Q_3 &= \varphi + \psi + g + l + \pi, \\
P_4 &= \Delta, & Q_4 &= \delta + \nu - g - l - \pi, \\
P_5 &= N - \Delta, & Q_5 &= \nu, \\
P_6 &= L - G, & Q_6 &= l
\end{aligned} \tag{14}$$

and then to introduce the Poincaré variables associated to the pairs (P_i, Q_i), for $i = 1, 2, 5$ and 6. The last one with suitable change of sign, since $P_6 \geq 0$. When the quoted angles and the eccentricity are small, Q_3 means the longitude of the secondary, while $Q_3 + Q_4$ corresponds to its sidereal rotation.

Let us recall that in general it is not advisable to express the Hamiltonian in the new set but to keep it in the old one. The required new partial derivatives have to be computed from those with respect to the old variables. For our present purposes it will be enough to write some of the pertinent operators of derivation. Those most frequently applied in the solar system are

$$\begin{aligned}
\frac{\partial}{\partial \Phi} + \frac{\partial}{\partial \Psi} + \frac{\partial}{\partial G} + \frac{\partial}{\partial L} = & \left(\frac{\partial}{\partial \Phi}\right) + \left(\frac{\partial}{\partial \Psi}\right) + \left(\frac{\partial}{\partial G}\right) + \left(\frac{\partial}{\partial L}\right) - \\
& - \frac{\eta e}{L(1+\eta)} \frac{\partial}{\partial e} - \frac{1}{\Psi} \tan \frac{I}{2} \frac{\partial}{\partial I} - \\
& - \frac{\Delta}{\Psi G} \left\{ \cos I_r \cot \frac{I_0}{2} + \sin I_r \right\} \frac{\partial}{\partial I_0} - \\
& - \frac{2\Psi}{\Delta G} \frac{\sin^2 I_0/2}{\sin(I_0 + I_r)} \frac{\partial}{\partial (I_0 + I_r)},
\end{aligned} \tag{15}$$

$$\begin{aligned}
\frac{\partial}{\partial \Phi} + \frac{\partial}{\partial \Psi} + \frac{\partial}{\partial \Delta} + \frac{\partial}{\partial N} = & \left(\frac{\partial}{\partial \Phi}\right) + \left(\frac{\partial}{\partial \Psi}\right) + \left(\frac{\partial}{\partial \Delta}\right) + \left(\frac{\partial}{\partial N}\right) - \\
& - \frac{1}{\Psi} \tan \frac{I}{2} \frac{\partial}{\partial I} + \frac{2\Delta}{\Psi G} \frac{\sin^2 I_r/2}{\sin I_0} \frac{\partial}{\partial I_0} - \\
& - \frac{1}{\Delta} \tan \frac{\sigma}{2} \frac{\partial}{\partial \sigma} - \frac{2\Psi}{\Delta G} \frac{\sin^2 I_r/2}{\sin(I_0 + I_r)} \frac{\partial}{\partial (I_0 + I_r)},
\end{aligned} \tag{16}$$

where e is the eccentricity of the orbit of the center of mass of the secondary, $\eta = \sqrt{1 - e^2}$, the symbol $(.)$ stands for the derivation with respect to the explicit appearance of the relevant variable and $\partial/\partial I_0$ means partial derivative with respect to I_0 when it does not appear in the combination (I_0+I_r). When these operators are applied to the secular Hamiltonian they provide the mean motion in longitude and the mean motion in longitude of the first principal axis of inertia of the secondary, respectively, when all the inclinations and the eccentricity are small, and they correspond to $\partial/\partial P_3$ and $\partial/\partial P_3 + \partial/\partial P_4$ computed in the old set.

In the case where I_r (the obliquity of the secondary with respect to its invariant plane) is not small, it is convenient to use

$$\begin{aligned}\frac{\partial}{\partial\Phi} + \frac{\partial}{\partial\Psi} = {} & \left(\frac{\partial}{\partial\Phi}\right) + \left(\frac{\partial}{\partial\Psi}\right) - \frac{1}{\Psi}\tan\frac{I}{2}\frac{\partial}{\partial I} - \\ & -\frac{\Delta}{\Psi G}\frac{\cos I_r}{\sin I_0}\frac{\partial}{\partial I_0} - \frac{\Psi}{\Delta G}\frac{1}{\sin(I_0+I_r)}\frac{\partial}{\partial(I_0+I_r)}\end{aligned} \tag{17}$$

and

$$\begin{aligned}\frac{\partial}{\partial\Delta} + \frac{\partial}{\partial N} = {} & \left(\frac{\partial}{\partial\Delta}\right) + \left(\frac{\partial}{\partial N}\right) - \frac{1}{\Delta}\tan\frac{\sigma}{2}\frac{\partial}{\partial\sigma} + \\ & +\frac{\Delta}{\Psi G}\frac{1}{\sin I_0}\frac{\partial}{\partial I_0} + \frac{\Psi}{\Delta G}\frac{\cos I_r}{\sin(I_0+I_r)}\frac{\partial}{\partial(I_0+I_r)}\end{aligned}$$

instead of (16). The operator (17) gives the mean motion in the longitude of the node of the Andoyer plane over the orbit and it is equal to $\partial/\partial P_2$ computed in the old set.

The operators required when the inclination of the orbital plane with respect to the horizontal one of the inertial reference is not small can be easily deduced taking into account the equality

$$\frac{\partial}{\partial\Phi} = \left(\frac{\partial}{\partial\Phi}\right) - \frac{1}{\Psi}\frac{1}{\sin I}\frac{\partial}{\partial I}.$$

4. Secular Motion

The Hamiltonian can be broken down in the form $\mathcal{H} = \mathcal{H}_0 + \mathcal{H}_1$, where the unperturbed term $\mathcal{H}_0$ includes both the Keplerian and rotational kinetic energy. Assuming quasi-sphericity for the secondary body, a standard Lie-series based perturbation technique – see for instance Hori (1966) or Deprit (1969) – can be applied to the elimination of the periodic terms up to the first order, in the absence of resonances. The transformed Hamiltonian will be $\mathcal{H}^* = \mathcal{H}_0^* + \mathcal{H}_1^*$ where $\mathcal{H}_0^*$ has a functional dependence identical to $\mathcal{H}_0$, and

$$
\begin{aligned}
\mathcal{H}_1^* &= k_1'\left[-\frac{1}{2}\left(1+\frac{3}{2}e^2\right)+\frac{3}{4}\sin^2(I_0+I_r)\left(1+\frac{3}{2}e^2\right)\right]+\\
&+k_3'\left[-\frac{1}{2}\left(1+\frac{3}{2}e^2\right)+\frac{3}{4}\sin^2 I\left(1+\frac{3}{2}e^2\right)+\right.\\
&\left.+\frac{3}{4}\sin^2 I_0\left(1+\frac{3}{2}e^2\right)-\frac{9}{8}\sin^2 I\sin^2 I_0\left(1+\frac{3}{2}e^2\right)\right],
\end{aligned}
\tag{18}
$$

where the expression has been approximated up to second order in e —close to zero— and the new coefficients k_i' that give the orders of magnitude are $k_i' = k_i L^3/G^3$, as a result of the average procedure. Moreover, we have made $\sigma = 0$ to shorten the expressions, limiting ourselves to a pure rotation around the third axis of inertia.

In the case in which the angles I and I_0 are small, the average orbital mean motion in longitude will be, in absence of resonances, the following:

$$
\begin{aligned}
\dot\psi+\dot g+\dot l+\dot\varphi &= \frac{3k_1'}{L}\left(1-\frac{3}{2}\sin^2(I_0+I_r)\right)+\frac{3k_3'}{L}\left(1-\frac{3}{2}\sin^2 I_0\right)-\\
&-3\frac{k_3'}{\Psi}\sin^2\frac{I}{2}\cos I\left(1-\frac{3}{2}\sin^2 I_0\right)-\\
&-\frac{3\Psi}{\Delta G}k_1'\sin^2\frac{I_0}{2}\cos(I_0+I_r)-\\
&-\frac{3k_3'}{2\Psi}\sin^2 I_0\cos I_0\left(1-\frac{3}{2}\sin^2 I\right)-\\
&-\frac{3\Delta}{G\Psi}k_3'\sin^2\frac{I_0}{2}\cos I_0\cos I_r\left(1-\frac{3}{2}\sin^2 I\right),
\end{aligned}
\tag{19}
$$

where the terms in e^2 have been ignored for the sake of brevity.

In order to find the magnitude of the different terms, let us note that $\Delta \simeq \omega k R_s^2$, where ω is the angular velocity of rotation and the coefficient $k = C/R_s^2$, which in the major planets oscillates around 0.3. In addition, $L = na^2$, and in general we can take $G \simeq na^2$, as the difference is of the order of e^2. On the other hand, the identities

$$
\Psi = G\cos I_0 + \Delta\cos I_r\,, \qquad \Delta\sin I_r = G\sin I_0
$$

provide the orders of magnitude

$$
\Psi = G[1+O(\varepsilon)]\,, \qquad \sin I_0 = \varepsilon\sin I_r\,,
$$

where we have introduced the parameter ε, whose order is

$$\varepsilon \sim \frac{\Delta}{G} \sim \frac{\Delta}{\Psi} \sim \frac{\Delta}{L} \sim k\left(\frac{\omega}{n}\right)\left(\frac{R_s}{a}\right)^2 .$$

In the case of artificial satellites, it is necessary to assume that $\varepsilon = 0$, while in the solar system, as the primary is considered to be either the Sun or a planet, this parameter is sufficiently small for it to be ignored in the majority of cases. The best-known exception is the Earth-Moon system, in which it reduces to $(R_s/a)^2 \simeq 2 \cdot 10^{-5}$. Something similar happens in the case of Neptune's Triton satellite, for which the forementioned amount is of the same order, although the practical effect is smaller given the lack of accuracy in the observations.

In fact, from the fourth term onwards in (19), which contains a factor ε, the rest all have a factor ε^2, so that the mean motion reduces to the first three terms, that is

$$\frac{3k_3'}{L} - \frac{3k_3'}{\Psi}\sin^2\frac{I}{2}\cos I + \frac{3k_1'}{L}\left(1 - \frac{3}{2}\sin^2 I_r\right),$$

which would result from making $I_0 = 0$ in (19).

It is worth emphasizing that this simplification $I_0 = 0$, which is equivalent to making the orbital plane of the satellite coincide with that orthogonal to its total angular moment, can be carried out before calculating the partial derivatives, so that the functional dependence of $\mathcal{V}_p$, which gives the oblateness perturbation, is the usual one in Hill or Delaunay variables. An inspection of the derivation operators (16), shows that all the terms which come from applying $\partial/\partial I_0$ (to $\mathcal{V}_p$) or $\partial/\partial(I_0 + I_r)$ (to $\mathcal{V}_s$), include the small parameter ε as a factor, so the said simplification is legitimate in relation to the calculation of the orbital motion. As for the rotational motion, the contribution of $\mathcal{V}_p$, through the partial with respect to I_0, contains a factor of the order of $\varepsilon k_3'$, which can, in general, again be ignored.

In relation to the mean orbital motion, it only remains to discuss the effect of the term k_1'/L, that is of the order of $n(R_s/a)^2 J_2$. As the contribution of the primary is of the order of $n(R_p/a)^2 J_2^p$, that of the secondary should not, a priori, be discarded when the values of $R_s^2 J_2$ and $R_p^2 J_2^p$ are not too different. Given that the J_2 of the principal planets usually vary between 10^{-2} and 10^{-5}, as a precaution it would seem to be interesting to consider the effect of the minor body when the ratio between the linear dimensions of both bodies does not exceed the three orders of magnitude. Thus, in the case of Triton and Neptune we have $(R_s/R_p)^2 \simeq 6 \cdot 10^{-3}$ and $(R_p/a)^2 J_2^p \simeq 4.7 \cdot 10^{-3} \times 3.3 \cdot 10^{-3} \simeq 1.5 \cdot 10^{-5}$ (Burns and Mathews, 1986), so it is not unreasonable to expect a contribution of the oblateness of the satellite to the average motion. Perhaps the inclusion of a J_2 of the satellite, to be determined, could contribute to lessening the uncertainty in the knowledge of Neptune's J_2, to the order of 4% in the simulations of Jacobson (1988).

Let us finally point out that in the case where the average orbital and rotational motion are in resonance 1:1, then we would have to add the following term to (18)

$$\begin{aligned}\mathcal{H}_{1r}^{*} &= k_2'\frac{3}{2}\left[-\frac{9}{8}e^2\sin^2(I_0+I_r)\cos(2\delta-2l+2\nu)+\right.\\ &\quad +\cos^2\frac{(I_0+I_r)}{2}\left(1-\frac{5}{2}e^2\right)\cos(2\delta-2g-2l+2\nu)-\\ &\quad \left.-\frac{1}{4}\sin^2(I_0+I_r)\left(1-\frac{5}{2}e^2\right)\cos(2\delta-2g-2l+2\nu)\right],\end{aligned}$$

and according to the previous arguments no appreciable contribution to the orbital motion can be expected from it. Moreover, the effect on the rotation does not seem to be observable, except for the case of the Moon.

5. Conclusions

In practice, the Andoyer-type canonical variables, defined starting from the invariant plane (rotation+translation) of the secondary, can be substituted by an analogous set defined simply with respect to the plane of the orbit, which naturally is not canonical, although it works as canonical, since the differences between the corresponding equations of motion are completely negligible. The main interest of our approach is that the expansion of the potential of the secondary is drastically shortened, without complicating that due to the oblateness of the primary, which keeps the same usual expression. Something similar could be said of other perturbations such as those of third body, which can be expressed as a development in polynomials of Legendre of the cosine of the angle between the direction of the vector radius and that of the perturbing body.

In this way the features of this problem do not essentially differ, at first order, from those corresponding to the rotation of a triaxial satellite around a mass point primary together with the motion of a point mass secondary orbiting around an oblate primary.

6. Appendix

In this Appendix we include full expressions for the perturbing potentials $\mathcal{V}_s$ and $\mathcal{V}_p$ in the particular case where $I_0 = 0$ and $\sigma = 0$. For the sake of brevity in the formulae we have used the notation $\cos a = C_a$ and $\sin a = S_a$.

$$\begin{aligned}\mathcal{V}_p &= -\frac{1}{2}k_3 - \frac{3}{4}k_3e^2 + \frac{3}{4}k_3S_I^2 + \frac{9}{8}k_3e^2S_I^2 - \frac{3}{2}k_3e\cos l + \frac{9}{4}k_3eS_I^2\cos l-\\ &-\frac{9}{4}k_3e^2\cos 2l + \frac{27}{8}k_3e^2S_I^2\cos 2l + \frac{3}{8}k_3eS_I^2\cos(2g+l+2\psi)-\\ &-\frac{3}{4}k_3S_I^2\cos(2g+2l+2\psi) + \frac{15}{8}k_3e^2S_I^2\cos(2g+2l+2\psi)-\end{aligned}$$

$$-\frac{21}{8}k_3 e S_I^2 \cos(2g+3l+2\psi) - \frac{51}{8}k_3 e^2 S_I^2 \cos(2g+4l+2\psi)$$

$$\mathcal{V}_s = -\frac{1}{2}k_1 - \frac{3}{4}k_1 e^2 + \frac{3}{4}k_1 S_{I_r}^2 + \frac{9}{8}k_1 e^2 S_{I_r}^2 -$$

$$-\frac{3}{2}k_1 e \cos l + \frac{9}{4}k_1 e S_{I_r}^2 \cos l - \frac{9}{4}k_1 e^2 \cos 2l + \frac{27}{8}k_1 e^2 S_{I_r}^2 \cos 2l +$$

$$+\frac{3}{4}k_2 S_{I_r}^2 \cos(2\delta+2\nu) + \frac{9}{8}k_2 e^2 S_{I_r}^2 \cos(2\delta+2\nu) +$$

$$+\frac{3}{8}k_1 e S_{I_r}^2 \cos(2g+l) - \frac{3}{4}k_1 S_{I_r}^2 \cos(2g+2l) +$$

$$+\frac{15}{8}k_1 e^2 S_{I_r}^2 \cos(2g+2l) - \frac{21}{8}k_1 e S_{I_r}^2 \cos(2g+3l) -$$

$$-\frac{51}{8}k_1 e^2 S_{I_r}^2 \cos(2g+4l) + \frac{9}{8}k_2 e S_{I_r}^2 \cos(2\delta+l+2\nu) +$$

$$+\frac{27}{16}k_2 e^2 S_{I_r}^2 \cos(2\delta+2l+2\nu) + \frac{9}{8}k_2 e S_{I_r}^2 \cos(2\delta-l+2\nu) +$$

$$+\frac{27}{16}k_2 e^2 S_{I_r}^2 \cos(2\delta-2l+2\nu) - \frac{3}{8}k_2 e \cos(2\delta+2g+l+2\nu) +$$

$$+\frac{3}{8}k_2 e C_{I_r} \cos(2\delta+2g+l+2\nu) + \frac{3}{16}k_2 e S_{I_r}^2 \cos(2\delta+2g+l+2\nu) +$$

$$+\frac{3}{4}k_2 \cos(2\delta+2g+2l+2\nu) - \frac{15}{8}k_2 e^2 \cos(2\delta+2g+2l+2\nu) -$$

$$-\frac{3}{4}k_2 C_{I_r} \cos(2\delta+2g+2l+2\nu) - \frac{3}{8}k_2 S_{I_r}^2 \cos(2\delta+2g+2l+2\nu) +$$

$$+\frac{15}{8}k_2 e^2 C_{I_r} \cos(2\delta+2g+2l+2\nu) + \frac{15}{16}k_2 e^2 S_{I_r}^2 \cos(2\delta+2g+2l+2\nu) +$$

$$+\frac{21}{8}k_2 e \cos(2\delta+2g+3l+2\nu) - \frac{21}{8}k_2 e C_{I_r} \cos(2\delta+2g+3l+2\nu) -$$

$$-\frac{21}{16}k_2 e S_{I_r}^2 \cos(2\delta+2g+3l+2\nu) + \frac{51}{8}k_2 e^2 \cos(2\delta+2g+4l+2\nu) -$$

$$-\frac{51}{8}k_2 e^2 C_{I_r} \cos(2\delta+2g+4l+2\nu) - \frac{51}{16}k_2 e^2 S_{I_r}^2 \cos(2\delta+2g+4l+2\nu) -$$

$$-\frac{3}{8}k_2 e \cos(2\delta-2g-l+2\nu) - \frac{3}{8}k_2 e C_{I_r} \cos(2\delta-2g-l+2\nu) +$$

$$+\frac{3}{16}k_2eS_{I_r}^2\cos(2\delta-2g-l+2\nu)+\frac{3}{4}k_2\cos(2\delta-2g-2l+2\nu)-$$

$$-\frac{15}{8}k_2e^2\cos(2\delta-2g-2l+2\nu)+\frac{3}{4}k_2C_{I_r}\cos(2\delta-2g-2l+2\nu)-$$

$$-\frac{3}{8}k_2S_{I_r}^2\cos(2\delta-2g-2l+2\nu)-\frac{15}{8}k_2e^2C_{I_r}\cos(2\delta-2g-2l+2\nu)+$$

$$+\frac{15}{16}k_2e^2S_{I_r}^2\cos(2\delta-2g-2l+2\nu)+\frac{21}{8}k_2e\cos(2\delta-2g-3l+2\nu)+$$

$$+\frac{21}{8}k_2eC_{I_r}\cos(2\delta-2g-3l+2\nu)-\frac{21}{16}k_2eS_{I_r}^2\cos(2\delta-2g-3l+2\nu)+$$

$$+\frac{51}{8}k_2e^2\cos(2\delta-2g-4l+2\nu)+\frac{51}{8}k_2e^2C_{I_r}\cos(2\delta-2g-4l+2\nu)-$$

$$-\frac{51}{16}k_2e^2S_{I_r}^2\cos(2\delta-2g-4l+2\nu)$$

7. Acknowledgements

The authors thank Prof. Broucke for providing them his Poisson series Processor which has constituted their first insight in this field. Financial support has been supplied by Spanish CICYT within the National Programme of Space Research under grant ESP88-541.

References

Broucke, R.A.: 1989, *Celest. Mech.*, **45**, 255-265.

Burns, J.A. and Mathews, M.S.: 1986, *Satellites*, The Univ. of Arizona Press.

Caballero, R.: 1992, *A note on the fast multiplication of Poisson Series.* (Submitted to Celestial Mechanics).

Deprit, A.: 1969, *Celest. Mech.*, **1**, 12-30.

Deprit, A.: 1981, *Celest. Mech.*, **24**, 111–153.

Ferrándiz, J.M. and Sansaturio, M.E.: 1988, General Motion of a Triaxial Rigid Body in a Newtonian Force Field. In *Long Term Dynamical Behaviour of Natural and Artificial N-Body Systems*, A. E. Roy ed., Kluwer Academic Publishers, NATO ASI Series C, Vol. **246**, pp. 181–187.

Ferrándiz, J.M. and Sansaturio, M.E.: 1989, *Celest. Mech.*, **46**, 307-320.

Henrard, J.: 1989, *Celest. Mech.*, **45**, 245-253.

Hitzl, D. and Breakwell, J.: 1971, *Celest. Mech.*, **3**, 346-383.

Hori, G.I.: 1966, *Publ. Astron. Soc. Japan*, **18**, 287-295.

Jacobson, R.A.: 1988, Satellite Ephemerides for the Voyager Neptune Encounter (AAS 87-464). In *Astrodynamics 1987*, J.K. Solder et al, eds, *Advances in the Astronautical Sciences*, Vol. **65**, pp. 657-680.

Leimanis, E.: 1965, *The General Problem of the Motion of Coupled Rigid Bodies about a Fixed Point*, Springer Verlag, Berlin.

Levi-Civita, T.: 1913, *Annali di Mat. Pura ed Applicata*, Serie III, **20**, 153-169.

Kinoshita, H.: 1972, *Pub. Astron. Soc. Japan*, **24**, 423-457.

CANONICAL ELEMENTS AND KEPLERIAN–LIKE SOLUTIONS FOR INTERMEDIARY ORBITS OF SATELLITES OF AN OBLATE PLANET

LUIS FLORíA
Departamento de Matemática Aplicada a la Ingeniería.
E. T. S. de Ingenieros Industriales. Universidad de Valladolid.
E – 47 011 Valladolid. SPAIN

Abstract. Within the framework of the Canonical Formalism in the extended phase space, *a general Hamiltonian* is investigated that covers a wide class of radial intermediaries accounting for the *major secular effects* due to a planet's oblateness perturbations.

An analytical, closed–form solution for this generic Hamiltonian is developed in terms of elementary functions via the corresponding Hamilton–Jacobi equation. The analytical solution so obtained can be contemplated according to a simple geometrical and dynamical interpretation in Keplerian language by means of the usual relations characterizing elliptic elements along a *hypothetic Keplerian motion.*

Appropriate choices for the terms appearing in the proposed Hamiltonian lead to recovering the analogues of some well–known, classical radial intermediaries (those introduced by Deprit and the one built by Alfriend and Coffey), but also certain new ones derived by Ferrándiz for the Main Problem in the Theory of Artificial Satellites of the Earth. In any case, the results are also applicable to problems dealing with orbital motion of other planetary satellites.

The generality of this pattern leads to a *systematic obtaining* of solutions to the considered intermediaries: special choices of the Hamiltonian yield the corresponding *analytical solution* to the respective intermediary problem.

Key words: Kepler Problem, Natural and Artificial Satellites, Oblate Spheroid, Radial Intermediaries, Hamilton–Jacobi Method.

1. Introduction

A *general,* homogeneous Hamiltonian is considered that, under special choices of its terms, allows one to *recover* certain well–known radial intermediaries for the problem of motion of a natural or artificial orbiter about an oblate central body when J_2–perturbations are taken into account.

In the extended, 8–dimensional phase space, a *simple analytical solution* for this generic Hamiltonian is obtained by means of the integration of the associated *Hamilton–Jacobi equation* by the technique of *separation* of variables. In the proposed development, carrying out this integration will not, *apparently,* require any resort to changes of independent variable, and so the obtained solution will seem to be parameterized by the physical time, t. However, the required quadratures over the radial variable r will be performed after the introduction of *auxiliary integration variables*; this process will be shown to amount to a sly introduction of an anomaly–type variable in terms of which the solution can be represented in *closed form.* With regard to this point, remember that equations of the kind of that of Kepler cannot be analytically solved to produce closed–form solutions in terms of t.

As a result of the treatment of the problem by Hamilton–Jacobi techniques, a *rectification* (see, e. g., Lanczos (1970), Chapter VIII, §2 and Scheck (1990),

Celestial Mechanics and Dynamical Astronomy **57**: 203–223, 1993.

§2.37.1) of the Hamiltonian flow of the autonomous system at issue will be obtained after the canonical transformation by means of which the solution is investigated.

The present research takes advantage of the approach followed in the paper by Franco and Palacios (1990), their results for the Alfriend and Coffey radial intermediary in the case of small eccentricities being generalized to the same and to other intermediaries, operating on the extended phase space and leaving aside the said restriction concerning the value of the eccentricity, and also considering terms proportional to other powers of the radial distance r.

The study of the problem with r^{-3}–perturbation (which, for instance, would lead to the integration of the equatorial Main Problem and that of intermediaries of the type introduced by Cid and his associates in Cid *et al.* (1986)) is not undertaken here: the analysis has already been carried out by Franco and Palacios in their aforesaid 1990 paper, and the developments here presented should be modified in accordance with their investigation of the roots of a cubic polynomial and the resulting analytical solution in terms of elliptic integrals, as also done in Jezewski (1983).

The transformation proposed in the next Section will be thoroughly investigated throughout the rest of this paper.

Starting from polar nodal variables, the analytical solution to the considered Hamiltonian will be arrived at by obtaining a complete solution to the associated Hamilton–Jacobi partial differential equation. In order to specify the transformation equations, certain *auxiliary quadratures* will have to be determined. As above stated, their determination will be based on the introduction of auxiliary integration variables for which a meaning will be obtained in terms of a *fictitious Keplerian orbit* whose main features concerning form and size will be described by *appropriately modified elliptic elements*.

Attention will be paid to the task of specifying an important group of radial intermediaries that fit into the framework here described; after some reckoning work, the corresponding solution can be derived from the general results developed in this paper. Some of these intermediaries (introduced by Deprit and his associates) are well known and have become *classical* in the field of the analytical satellite theories; two new first–order radial intermediaries derived by Ferrándiz are also alluded to; these new intermediaries were presented in Ferrándiz and Floría (1990), and their construction was a by–product of an extension of the results contained in the previous papers Ferrándiz and Floría (1989), (1991).

Concerning the subject of obtaining analytical solutions for radial intermediaries, while wording a part of the present paper on *systematically* solving these Hamiltonians, the author was informed (Ferrándiz, (1992)) that Ferrándiz had already constructed separate solutions for particular intermediaries, having then communicated his results at the Orbital Mechanics Seminar of the University of Texas at Austin in November 1990. Agreement between both approaches has been verified. At this point, the author expresses his gratitude to Prof. J. M. Ferrándiz for fruitful discussion and clarifying comments.

2. Construction of the Transformation

In the present Section the extended canonical set of *polar nodal variables*, also referred to as *Hill variables*, $(r, \theta, \nu, t; p_r, p_\theta, p_\nu, p_0)$, will be considered as the starting point for the subsequent developments. The meaning of these variables is the following: r denotes the radial distance from the primary's centre of mass to the moving mass; θ is the argument of latitude of the orbiter; ν represents the argument of longitude of the ascending node; t stands for the physical time; as for the canonical momenta, p_r is the radial velocity of the moving mass, p_θ denotes the modulus of the total angular momentum, p_ν designates the polar component of the angular momentum, and p_0 (the canonical momentum conjugate to the physical time t) is the negative of the total energy in the problem to which the variables are applied.

In the extended canonical set of *polar nodal variables*, the general expression for the Hamiltonian $\mathcal{H}$ analysed in the present paper will be assumed to be of the form

$$\mathcal{H} = \mathcal{H}_0(r; p_r, p_\theta) + V_{0,(n_0)}(p_\theta, p_\nu, p_0; \varepsilon)$$

$$+ \frac{1}{r} V_{1,(n_1)}(p_\theta, p_\nu, p_0; \varepsilon) + \frac{1}{r^2} V_{2,(n_2)}(p_\theta, p_\nu, p_0; \varepsilon) + p_0 ,$$

where

$$\mathcal{H}_0(r; p_r, p_\theta) = \frac{1}{2}\left[p_r{}^2 + \frac{p_\theta{}^2}{r^2}\right] - \frac{\mu}{r}$$

represents the Hamiltonian governing an unperturbed Kepler problem, and

$$V_{j,(n_j)} \equiv V_{j,(n_j)}(p_\theta, p_\nu, p_0; \varepsilon) = \sum_{l=1}^{n_j} \varepsilon^l \mathcal{V}_{j,l}(p_\theta, p_\nu, p_0)$$

will be an expansion (truncated at the higher order n_j corresponding to the intermediary) in ascending powers of a small parameter $\varepsilon = -J_2$, the coefficients

$$\mathcal{V}_{j,l}(p_\theta, p_\nu, p_0)$$

of such an expansion depending on the momenta p_ν, p_θ (through the function $I \equiv I(p_\theta, p_\nu)$ defined by the relation $\cos I = p_\nu/p_\theta$ and, in certain cases, negative powers of p_θ) and p_0. As a matter of fact, the said coefficients can be described as *polynomials in* $\sin^2 I$ affected by negative powers of p_θ^2 (if needed, such coefficients will also be read as polynomials in $\cos^2 I$ affected by the same negative powers of p_θ^2). However the functional form and explicit dependence of these coefficients will not be significant for the *validity and generality* of the following developments.

For the sake of conciseness, from now on the subscripts (n_j) will not be alluded to, unless such omissions and the subsequent simplification of the notations could cause misunderstanding. Consequently the symbol V_j is to be considered as a shortened form of $V_{j,(n_j)}$. In what follows, the construction will be carried forward in its utmost generality within the framework above established.

The functional form of the considered Hamiltonian

$$\mathcal{H} \equiv \mathcal{H}(r, -, -, -; p_r, p_\theta, p_\nu, p_0)$$

(where a dash is used in place of a variable to emphasize its explicit absence from $\mathcal{H}$) reveals that the canonical variables θ, ν and t are cyclic, and so the corresponding conjugate momenta are constants of the motion:

$$p_\theta = \Theta_0 , \quad p_\nu = N_0 , \quad p_0 = \Pi_0 .$$

Formulated in the enlarged phase space, the problem is a conservative one; therefore, the system possesses a first integral of the type of the energy integral

$$\mathcal{H} = K_0 ,$$

where K_0 denotes the energy constant. In this context, K_0 will be the initial value of $\mathcal{H}$; but the numerical value of $\mathcal{H}$ is identically zero along solutions. The situation is similar to that of Alfriend *et al.* (1977), pp. 443–444, concerning Vinti's intermediary.

The integration of this Hamiltonian system is equivalent to the determination of a canonical transformation leading to a canonical set for which all the variables (except one, that will be a linear function of the independent variable) turn out to be constants. This transformation will be obtained from an appropriate generating function S whose specification requires one to know a *complete solution* of the *Hamilton–Jacobi* partial differential *equation* associated to the Hamiltonian $\mathcal{H}$ at hand.

In the present case, the Hamilton–Jacobi equation reads

$$\frac{1}{2}\left[\left(\frac{\partial S}{\partial r}\right)^2 + \frac{1}{r^2}\left(\frac{\partial S}{\partial \theta}\right)^2\right] - \frac{\mu}{r} + V_0\left(\frac{\partial S}{\partial \theta}, \frac{\partial S}{\partial \nu}, \frac{\partial S}{\partial t}; \varepsilon\right) +$$

$$+\frac{1}{r}V_1\left(\frac{\partial S}{\partial \theta}, \frac{\partial S}{\partial \nu}, \frac{\partial S}{\partial t}; \varepsilon\right) + \frac{1}{r^2}V_2\left(\frac{\partial S}{\partial \theta}, \frac{\partial S}{\partial \nu}, \frac{\partial S}{\partial t}; \varepsilon\right) + \frac{\partial S}{\partial t} = K_0 ,$$

and the unknown function S generates a canonical transformation

$$(r, \theta, \nu, t; p_r, p_\theta, p_\nu, p_0) \xrightarrow{S} (Q_K, Q_\Theta, Q_N, Q_\Pi; K_0, \Theta_0, N_0, \Pi_0) ,$$

the constants K_0, Θ_0, N_0 and Π_0 having been chosen as the canonical momenta of the new system of variables. As for the notation, observe the mnemonic use of small letters for the old variables and capital letters for the new ones, and the way in which the role of the subscripts has been interchanged from the old momenta to the new coordinates.

Since $\mathcal{H}$ is a one–degree–of–freedom Hamiltonian in the radial variables (r, p_r), the corresponding Hamilton–Jacobi equation is *separable*, and its integration can be performed by trying out a solution

$$S \equiv S(r,\theta,\nu,t; K_0,\Theta_0,N_0,\Pi_0) = \theta\Theta_0 + \nu N_0 + t\Pi_0 + W(r);$$

introducing this function in the equation leads to the expression

$$\frac{1}{2}\left(\frac{dW}{dr}\right)^2 + \frac{{\Theta_0}^2}{2r^2} - \frac{\mu}{r} + V_0(\Theta_0, N_0, \Pi_0; \varepsilon)$$

$$+ \frac{1}{r} V_1(\Theta_0, N_0, \Pi_0; \varepsilon) + \frac{1}{r^2} V_2(\Theta_0, N_0, \Pi_0; \varepsilon) + \Pi_0 = K_0 ,$$

and consequently

$$\left(\frac{dW}{dr}\right)^2 = Q(r; K_0, \Theta_0, N_0, \Pi_0; \varepsilon) = a_0 + \frac{a_1}{r} + \frac{a_2}{r^2} ,$$

where the following notations have been used:

$$a_0 = 2K_0 - 2\Pi_0 - 2V_0(\Theta_0, N_0, \Pi_0; \varepsilon) ,$$

$$a_1 = 2\mu - 2V_1(\Theta_0, N_0, \Pi_0; \varepsilon) ,$$

$$a_2 = -{\Theta_0}^2 - 2V_2(\Theta_0, N_0, \Pi_0; \varepsilon) .$$

Since only *a* complete (but not the general) solution to the Hamilton–Jacobi equation is needed, after choosing the positive determination in the above square root, the expression

$$W(r) = \int_{r_0}^{r} \sqrt{Q}\, dr$$

$$= \int_{r_0}^{r} \sqrt{2(K_0 - \Pi_0 - V_0) + \frac{2}{r}(\mu - V_1) - \frac{1}{r^2}(\Theta_0^2 + 2V_2)}\; dr ,$$

is obtained for the function W, and the generating function is

$$S \equiv S(r, \theta, \nu, t; K_0, \Theta_0, N_0, \Pi_0)$$

$$= \theta\,\Theta_0 + \nu\, N_0 + t\,\Pi_0 + \int_{r_0}^{r} \sqrt{a_0 + \frac{a_1}{r} + \frac{a_2}{r^2}}\; dr ,$$

understanding that the lower limit r_0 in the integral is the lowest positive root of the equation $Q(r; K_0, \Theta_0, N_0, \Pi_0; \varepsilon) = 0$.

Bearing the above remarks in mind, the implicit equations of the transformation defined by S are now

$$p_r \equiv \frac{dr}{dt} = \frac{\partial S}{\partial r} = \sqrt{Q} = \frac{1}{r}\sqrt{a_0 r^2 + a_1 r + a_2} ,$$

$$p_\theta = \frac{\partial S}{\partial \theta} = \Theta_0 ,$$

$$p_\nu = \frac{\partial S}{\partial \nu} = N_0 ,$$

$$p_0 = \frac{\partial S}{\partial t} = \Pi_0 ,$$

$$Q_K = \frac{\partial S}{\partial K_0} = \hat{I}_0 ,$$

$$Q_\Theta = \frac{\partial S}{\partial \Theta_0} = \theta - \frac{\partial V_0}{\partial \Theta_0}\hat{I}_0 - \frac{\partial V_1}{\partial \Theta_0}\hat{I}_1 - \left[\Theta_0 + \frac{\partial V_2}{\partial \Theta_0}\right]\hat{I}_2 ,$$

$$Q_N = \frac{\partial S}{\partial N_0} = \nu - \frac{\partial V_0}{\partial N_0}\hat{I}_0 - \frac{\partial V_1}{\partial N_0}\hat{I}_1 - \frac{\partial V_2}{\partial N_0}\hat{I}_2 ,$$

$$Q_\Pi = \frac{\partial S}{\partial \Pi_0} = t - \left[1 + \frac{\partial V_0}{\partial \Pi_0}\right]\hat{I}_0 - \frac{\partial V_1}{\partial \Pi_0}\hat{I}_1 - \frac{\partial V_2}{\partial \Pi_0}\hat{I}_2 ,$$

where

$$\hat{I}_m = \int_{r_0}^{r} \frac{dr}{r^m \sqrt{Q}} , \quad m = 0, 1, 2 .$$

Notice that obtaining the expressions for the coordinates of the new canonical set depends on evaluating these quadratures, whose integrands contain the square root of a quadratic polynomial of the radial variable r in their denominators.

By virtue of the relation

$$\frac{dr}{dt} = \sqrt{Q} ,$$

one has

$$\frac{dr}{dt} = 0 \iff a_0 r^2 + a_1 r + a_2 = 0 ;$$

accordingly, the extrema of the function $r(t)$ coincide with the zeros of Q, and therefore with the zeros of the aforementioned polynomial in r.

For negative–energy problems the motion remains bounded and the above polynomial has *two positive roots* r_a and r_p in such a way that

$$0 < r_p \leq r \leq r_a ;$$

these roots are to be considered as the *perturbed pericentre and apocentre radii.* So the quadratic polynomial can be factorised in the form of a product of linear polynomials according to the expression

$$a_0 r^2 + a_1 r + a_2 = \gamma^2 (r_a - r)(r - r_p) = 0 ,$$

with

$$\gamma^2 = | a_0 | = | 2 K_0 - 2 \Pi_0 - 2 V_0 | ,$$

and so

$$\sqrt{Q} = \frac{1}{r} \sqrt{a_0 r^2 + a_1 r + a_2} = \frac{1}{r} \gamma \sqrt{(r_a - r)(r - r_p)} .$$

The preceding calculation allows one to express the required quadratures in a slightly different manner:

$$\hat{I}_0 = \frac{1}{\gamma} I_1 , \quad \hat{I}_1 = \frac{1}{\gamma} I_0 , \quad \hat{I}_2 = \frac{1}{\gamma} I_{-1} ,$$

where the quadratures

$$I_m = \int_{r_0}^{r} \frac{r^m \, dr}{\sqrt{(r_a - r)(r - r_p)}} , \quad m = -1, 0, 1 ,$$

have been used. In the following Section, the practical calculation of these expressions will be carried out by introducing a new variable whose meaning will be specified in Keplerian language in a later paragraph. (See Section 6).

By employing such notations, the coordinate transformation reads

$$Q_K = \frac{1}{\gamma} I_1 ,$$

$$Q_\Theta = \theta - \frac{1}{\gamma} \frac{\partial V_0}{\partial \Theta_0} I_1 - \frac{1}{\gamma} \frac{\partial V_1}{\partial \Theta_0} I_0 - \frac{1}{\gamma} \left[\Theta_0 + \frac{\partial V_2}{\partial \Theta_0} \right] I_{-1} ,$$

$$Q_N = \nu - \frac{1}{\gamma} \frac{\partial V_0}{\partial N_0} I_1 - \frac{1}{\gamma} \frac{\partial V_1}{\partial N_0} I_0 - \frac{1}{\gamma} \frac{\partial V_2}{\partial N_0} I_{-1} ,$$

$$Q_\Pi = t - \frac{1}{\gamma} \left[1 + \frac{\partial V_0}{\partial \Pi_0} \right] I_1 - \frac{1}{\gamma} \frac{\partial V_1}{\partial \Pi_0} I_0 - \frac{1}{\gamma} \frac{\partial V_2}{\partial \Pi_0} I_{-1} .$$

3. Evaluation of Auxiliary Quadratures

With the aim of performing the above integrals, at this stage the change of integration variable $r \longrightarrow x$ such that

$$\sqrt{(r_a - r)(r - r_p)} = (r_a - r) x , \quad x(r) = \sqrt{\frac{r - r_p}{r_a - r}} ,$$

$$dx = \frac{(r_a - r_p)}{2 x} \frac{dr}{(r_a - r)^2} ,$$

is considered. In view of the helpful relation

$$1 + x^2 = \frac{r_a - r_p}{r_a - r} ,$$

the following explicit expressions for r and dr in terms of x and dx are readily obtained:

$$r(x) = \frac{r_a x^2 + r_p}{1 + x^2} , \qquad d r = \frac{2 x (r_a - r_p)}{(1 + x^2)^2} \, d x .$$

The application of the preceding set of formulae leads immediately to

$$\frac{1}{\sqrt{(r_a - r)(r - r_p)}} = \frac{1}{x (r_a - r)} \frac{r_a - r_p}{r_a - r_p} = \frac{1 + x^2}{x (r_a - r_p)} .$$

Remembering the meaning of the zeros of Q one concludes that

$$r_0 = r_p \implies x(r_0) = 0 ,$$

and so the required, preliminary elements to determine I_1 are available:

$$I_1 = \int_0^x \frac{2 (r_a x^2 + r_p)}{(1 + x^2)^2} \, d x = 2 r_a \mathcal{Y}_2 + 2 r_p \mathcal{Y}_0 ,$$

where, to save writing, the abbreviations

$$\mathcal{Y}_j = \int_0^x \frac{x^j}{(1 + x^2)^2} \, d x , \quad j = 0, 2,$$

have been used. After looking up Gradshteyn and Ryzhik (1980), formulae (2.148.3) and (2.147.2), the latter with $n = m = 2$, the following expressions

$$\mathcal{Y}_0 = \int_0^x \frac{d x}{(1 + x^2)^2} = \frac{x}{2 (1 + x^2)} + \frac{1}{2} \arctan x ,$$

$$\mathcal{Y}_2 = \int_0^x \frac{x^2}{(1 + x^2)^2} \, d x = - \frac{x}{2 (1 + x^2)} + \frac{1}{2} \arctan x ,$$

will immediately furnish the desired auxiliary integrals I_1 and I_{-1}. Indeed, as a consequence of the previous calculations and returning to the original variable r one has

$$I_1 = (r_a + r_p) \arctan\left(\sqrt{\frac{r - r_p}{r_a - r}} \right) - \sqrt{(r - r_p)(r_a - r)} ,$$

$$I_{-1} = 2 \int_0^x \frac{d x}{r_a x^2 + r_p} = \frac{2}{\sqrt{r_a r_p}} \arctan\left(\sqrt{\frac{r_a}{r_p}} \, x \right) ,$$

and finally

$$I_{-1} = \frac{2}{\sqrt{r_a\, r_p}} \arctan\left(\sqrt{\frac{r_a\,(r - r_p)}{r_p\,(r_a - r)}} \right) .$$

As for the evaluation of I_0, one easily gets

$$I_0 = 2 \int_0^x \frac{d\,x}{1 + x^2} = 2 \arctan x = 2 \arctan\left(\sqrt{\frac{r - r_p}{r_a - r}} \right) .$$

4. Completing the Transformation Equations

Concerning the coordinates of the new system of variables, the equations of the considered canonical transformation, derived from a complete solution of the Hamilton–Jacobi equation corresponding to the Hamiltonian $\mathcal{H}$, are now specified with the help of the expressions obtained for the auxiliary quadratures:

Generalized Kepler Equation:

$$\gamma\, Q_K = (r_a + r_p) \arctan\left(\sqrt{\frac{r - r_p}{r_a - r}} \right) - \sqrt{(r - r_p)(r_a - r)} .$$

Argument–of–latitude Equation:

$$\theta = Q_\Theta + \frac{1}{\gamma} \frac{\partial V_1}{\partial \Theta_0}\, 2 \arctan\left(\sqrt{\frac{r - r_p}{r_a - r}} \right)$$

$$+ \frac{1}{\gamma} \left[\Theta_0 + \frac{\partial V_2}{\partial \Theta_0} \right] \frac{2}{\sqrt{r_a\, r_p}} \arctan\left(\sqrt{\frac{r_a\,(r - r_p)}{r_p\,(r_a - r)}} \right)$$

$$+ \frac{1}{\gamma} \frac{\partial V_0}{\partial \Theta_0} \left[(r_a + r_p) \arctan\left(\sqrt{\frac{r - r_p}{r_a - r}} \right) - \sqrt{(r - r_p)(r_a - r)} \right]$$

Node Equation:

$$\nu = Q_N + \frac{1}{\gamma} \frac{\partial V_1}{\partial N_0}\, 2 \arctan\left(\sqrt{\frac{r - r_p}{r_a - r}} \right)$$

$$+ \frac{1}{\gamma} \frac{\partial V_2}{\partial N_0} \frac{2}{\sqrt{r_a\, r_p}} \arctan\left(\sqrt{\frac{r_a\,(r - r_p)}{r_p\,(r_a - r)}} \right)$$

$$+ \frac{1}{\gamma} \frac{\partial V_0}{\partial N_0} \left[(r_a + r_p) \arctan\left(\sqrt{\frac{r - r_p}{r_a - r}} \right) - \sqrt{(r - r_p)(r_a - r)} \right]$$

Time–Element Equation:

$$t = Q_\Pi + \frac{1}{\gamma}\frac{\partial V_1}{\partial \Pi_0}\, 2 \arctan\left(\sqrt{\frac{r - r_p}{r_a - r}}\right)$$

$$+ \frac{1}{\gamma}\frac{\partial V_2}{\partial \Pi_0}\frac{2}{\sqrt{r_a\, r_p}} \arctan\left(\sqrt{\frac{r_a\,(r - r_p)}{r_p\,(r_a - r)}}\right)$$

$$+ \frac{1}{\gamma}\left[1 + \frac{\partial V_0}{\partial \Pi_0}\right]\left[(r_a + r_p)\arctan\left(\sqrt{\frac{r - r_p}{r_a - r}}\right) - \sqrt{(r - r_p)(r_a - r)}\right]$$

5. Interpretation of the Transformation in Keplerian–like Terms

In order to obtain a simple, geometrical and dynamically thought–provoking interpretation for the preceding developments in terms of Keplerian motion, a different approach is now taken with the aim of evaluating the quadratures

$$\hat{I}_m = \int_{r_0}^{r} \frac{d\,r}{r^m\,\sqrt{Q}}\ , \quad m = 0\,,1\,,2\ .$$

The way of proceeding is based on the idea of adapting a procedure (classically applied to a pure Kepler problem to derive Delaunay variables as done, e.g., in Deprit (1981), pp. 115–118, and for quasi–Keplerian systems in the same article Deprit (1981), pp. 124–126, to the considered, extended Hamiltonian $\mathcal{H}$ and taking advantage of the homogeneous canonical formalism.

Remember the canonical transformation

$$(r, \theta, \nu, t; p_r, p_\theta, p_\nu, p_0) \xrightarrow{S} (Q_K, Q_\Theta, Q_N, Q_\Pi; K_0, \Theta_0, N_0, \Pi_0)$$

defined by the generating function

$$S \equiv S(\,r\,,\theta\,,\nu\,,t\,;\ K_0\,,\Theta_0\,,N_0\,,\Pi_0\,)$$

$$= \theta\,\Theta_0 + \nu\,N_0 + t\,\Pi_0 + \int_{r_0}^{r} \sqrt{Q}\ d\,r\ ,$$

with

$$Q = a_0 + \frac{a_1}{r} + \frac{a_2}{r^2} =$$

$$= 2\,(\,K_0 - \Pi_0 - V_0\,) + \frac{2\,(\,\mu - V_1\,)}{r} - \frac{(\,{\Theta_0}^2 + 2\,V_2\,)}{r^2}\ .$$

The introduction of a set of appropriate quantities a, e, p, κ, μ^* by means of the relations

$$a = -\frac{\mu^*}{2(K_0 - \Pi_0 - V_0)}, \qquad \mu^* = \mu - V_1 ,$$

$$\Theta_0{}^2 + 2V_2 = \mu^* a(1-e^2) \equiv \kappa^2 , \quad p = a(1-e^2) = \frac{\kappa^2}{\mu^*} ,$$

$$e^2 = 1 - \frac{\kappa^2}{\mu^* a} = 1 + \frac{2(K_0 - \Pi_0 - V_0)\kappa^2}{(\mu^*)^2} ,$$

leads to an expression for Q in the form

$$Q = \frac{2\mu^*}{r} - \frac{\mu^*}{a} - \frac{\mu^* a(1-e^2)}{r^2}$$
$$= \frac{\mu^*}{a}\left[\frac{a(1+e)}{r} - 1\right]\left[1 - \frac{a(1-e)}{r}\right] .$$

These formulae are of the same type as those holding for a hypothetic, Keplerian motion characterized by the above *elliptic elements* (a, e, p) with κ as the *modified angular momentum magnitude*, provided that μ^* is taken as the gravitational parameter. Correspondingly, bearing in mind that p_θ, p_ν and p_0 are conserved quantities and translating the above expressions into the respective ones in terms of the original polar nodal variables, the moving mass can be regarded as *simulating a Kepler motion controlled by the Hamiltonian*

$$\mathcal{H}^* = \frac{1}{2}\left[p_r{}^2 + \frac{\kappa^2}{r^2}\right] - \frac{\mu^*}{r} .$$

The quantity e must not be identified with that denoted by e_k when considering the Alfriend and Coffey potential. (See Section 7).

This study presupposes that the quantity e is such that $0 \leq e < 1$, in which case the roots of the equation

$$Q(r;\, K_0, \Theta_0, N_0, \Pi_0;\, \varepsilon) = 0$$

are immediately calculated:

$$0 < r_0 \equiv r_p = a(1-e) \leq r_1 \equiv r_a = a(1+e) .$$

Consequently

$$Q = \frac{\mu^*}{a r^2}\{a(1+e) - r\}\{r - a(1-e)\} =$$

$$\frac{\mu^*}{a r^2}\{(ae)^2 - (a-r)^2\} = \frac{\mu^* a^2 e^2}{a r^2}\left[1 - \left(\frac{a-r}{ae}\right)^2\right] .$$

Defining an *auxiliary variable* E of the eccentric–anomaly–type through the relation

$$\frac{a - r}{a\,e} = \cos E \ , \quad r = a(1 - e\cos E) \ , \quad d\,r = a\,e\sin E\,dE \ ,$$

the expression of Q becomes

$$Q = \frac{\mu^* e^2}{a}\,\frac{\sin^2 E}{(1 - e\cos E)^2} \ ,$$

whence the quadrature $\hat{I}_0$ is found to be

$$\hat{I}_0 = \sqrt{\frac{a^3}{\mu^*}}\ (E - e\sin E) \ ,$$

which converts the equation for Q_K into an equation of the Kepler type

$$Q_K = \sqrt{\frac{a^3}{\mu^*}}\ (E - e\sin E) \ .$$

Notice that also

$$\hat{I}_0 = \sqrt{-\,\frac{(\mu^*)^2}{[2(K_0 - \Pi_0 - V_0)]^3}}\ (E - e\sin E) \ ,$$

which induces one to consider a quantity of the kind of a *mean motion* via the relation

$$\mu^* = n^2 a^3 \iff n = \frac{(-2K_0 + 2\Pi_0 + 2V_0)^{3/2}}{\mu^*} \ ,$$

and finally

$$Q_K = \frac{1}{n}\,(E - e\sin E) \ .$$

The introduction of another *auxiliary variable* f of the true–anomaly–type by means of

$$r = \frac{p}{1 + e\cos f} \ , \quad d\left(\frac{1}{r}\right) = -\,\frac{e\sin f}{p}\,d\,f$$

$$\Longrightarrow \quad Q = \frac{\mu^*}{a}\,\frac{e^2}{(1 - e^2)}\,\sin^2 f \doteq \frac{\mu^*}{p}\,e^2\sin^2 f \ ,$$

allows one to conclude that

$$\hat{I}_2 = \frac{f}{\sqrt{\mu^* p}} = \frac{f}{\kappa} \ .$$

In this way, these auxiliary variables can respectively be conceived as the *eccentric and true anomaly along the aforesaid fictitious Keplerian motion.*

The determination of the quadrature $\hat{I}_1$ can be accomplished with the help of the above variables E and f by a standard procedure: putting

$$\hat{I}_1 = \int_{r_0}^{r} \frac{dr}{r\sqrt{Q}} = \int_{r_0}^{r} \frac{r\,dr}{r^2\sqrt{Q}} = -\int_{r_0}^{r} \frac{r\,d(1/r)}{\sqrt{Q}}$$

$$\Longrightarrow \quad \hat{I}_1 = \sqrt{\frac{p}{\mu^*}} \int_0^f \frac{df}{1 + e\cos f} ,$$

and, after introducing the quantity

$$\eta = \sqrt{1 - e^2} ,$$

the last integral is calculated by means of the Keplerian–type relations

$$r\cos f = a(\cos E - e) , \quad r\sin f = a\eta\sin E , \quad r = a(1 - e\cos E) ,$$

which allows one to express $\cos f$ and df in terms of the variable E in such a way that

$$\int_0^f \frac{df}{1 + e\cos f} = \int_0^E \frac{dE}{\eta} = \frac{E}{\eta} ,$$

and finally

$$\hat{I}_1 = \sqrt{\frac{p}{\mu^*}}\, \frac{E}{\eta} = \frac{\kappa}{\mu^*\eta}\, E .$$

After these considerations, the equations of transformation are

$$p_r = \sqrt{Q} ,$$

$$p_\theta = \Theta_0 ,$$

$$p_\nu = N_0 ,$$

$$p_0 = \Pi_0 ,$$

$$Q_K = \frac{1}{n}\,(E - e\sin E) ,$$

$$Q_\Theta = \theta - \frac{\partial V_0}{\partial \Theta_0}\frac{1}{n}(E - e\sin E) - \frac{\partial V_1}{\partial \Theta_0}\sqrt{\frac{p}{\mu^*}}\frac{E}{\eta} - \left[\Theta_0 + \frac{\partial V_2}{\partial \Theta_0}\right]\frac{f}{\kappa} ,$$

$$Q_N = \nu - \frac{\partial V_0}{\partial N_0}\frac{1}{n}(E - e\sin E) - \frac{\partial V_1}{\partial N_0}\sqrt{\frac{p}{\mu^*}}\frac{E}{\eta} - \frac{\partial V_2}{\partial N_0}\frac{f}{\kappa} ,$$

$$Q_\Pi = t - \left[1 + \frac{\partial V_0}{\partial \Pi_0}\right]\frac{1}{n}(E - e\sin E) - \frac{\partial V_1}{\partial \Pi_0}\sqrt{\frac{p}{\mu^*}}\frac{E}{\eta} - \frac{\partial V_2}{\partial \Pi_0}\frac{f}{\kappa} .$$

The Hamiltonian $\mathcal{H}$ will now be subjected to the effect of this transformation, and the canonical equations of motion derived from the resulting, transformed Hamiltonian will be integrated in a straightforward manner:

$$\mathcal{H}(r,-,-,-;p_r,p_\theta,p_\nu,p_0) \longrightarrow \mathcal{K}(-,-,-,-;K_0,-,-,-) = K_0$$

$$\Longrightarrow \quad \frac{dQ_K}{dt} = \frac{\partial \mathcal{K}}{\partial K_0} = 1 \ ,$$

and so

$$Q_K = t + \text{const.} \ ,$$

while the remaining canonical variables are constants of the motion.

Remembering a remark from Section 1, as usual in Hamilton–Jacobi Theory (see, e. g., Lanczos (1970), Chapter VIII, §2, or Scheck (1990), §2.37.1), the following interpretation can be ascribed to the above results: in the neighbourhood of every point of phase space that is not an equilibrium position, the transformation *smooths the flow* of the autonomous Hamiltonian system generated by $\mathcal{H}$ to a *uniform, rectilinear flow*, producing a bundle of parallel straight lines inclined at an angle of 45 ^{o} to the time axis.

As expected, the structure of the above canonical solution shows that the new variables constitute a *set of canonical elements* (e.g., in the sense of Stiefel and Scheifele (1971), §18) for the dynamical problem governed by the Hamiltonian $\mathcal{H}$.

6. Concerning the Auxiliary Variables

The integration variable x (used in Section 3 to calculate the required quadratures that allow one to complete the set of transformation equations) will next receive a meaning in the light of the preceding results concerning the considered, fictitious Keplerian orbit.

Remember that $r_a = a(1+e)$, $r_p = a(1-e)$ and $r = a(1 - e\cos E)$ and the relation between r and f. Then, after some reckoning work:

$$x^2 = \frac{r - r_p}{r_a - r} = \tan^2\left(\frac{E}{2}\right) = \frac{1-e}{1+e}\tan^2\left(\frac{f}{2}\right) \ .$$

In the same way

$$\sqrt{\frac{r_a\,(r - r_p)}{r_p\,(r_a - r)}} = \sqrt{\frac{1+e}{1-e}}\tan\left(\frac{E}{2}\right) = \tan\left(\frac{f}{2}\right) \ .$$

As for the quadratures:

$$I_0 = E \ , \quad \hat{I}_1 = \frac{1}{\gamma}\,E \ ;$$

$$I_1 = a\,(E - e\sin E)\,, \quad \hat{I}_0 = \frac{a}{\gamma}\,(E - e\sin E)\,;$$

$$I_{-1} = \frac{f}{a\sqrt{1-e^2}}\,, \quad \hat{I}_2 = \frac{f}{\gamma\, a\sqrt{1-e^2}}\,.$$

and so the quantity γ can now be related to the "elements" of the auxiliary Keplerian orbit.

7. Application to Some Low–Order Radial Intermediaries

Starting from the preceding general development, in order to obtain the results and solution pertaining to particular specifications of the Hamiltonian $\mathcal{H}$ (for instance, when dealing with the radial intermediaries for the J_2–problem of motion of a satellite around an oblate planet), the respective set of partial derivatives should be calculated.

In this Section such a task will be illustrated by considering, without wishing to be exhaustive, the potentials corresponding to the intermediaries of the first and second order belonging to the chain of Deprit's radial intermediaries (Deprit (1981), pp. 138–139), the second–order radial intermediary introduced by Alfriend and Coffey (1984) (see also Cid *et al.* (1986), p. 202. Its expression for small eccentricity values, allowing one to neglect the $\varepsilon^2 e^2$–terms, is used in Franco and Palacios (1990)), and two new radial intermediaries proposed by Ferrándiz and presented in Ferrándiz and Floría (1990).

The potentials will first be expressed in extended Hill variables, and the following remarks concerning the notations are in order: the symbol $\varepsilon = -J_2$, which denotes the dimensionless oblateness parameter of the primary, will act as the (small) perturbation parameter; μ stands for the gravitational constant of the central body, and R_e represents its mean equatorial radius. Remember also that the functions of the inclination c and s are given by

$$c \equiv \cos I = \frac{p_\nu}{p_\theta}\,, \quad s \equiv \sin I\,, \quad s^2 = 1 - c^2\,.$$

For convenience, the expression for the potential will be given in terms of powers of c^2, instead of doing it in terms of powers of s^2.

Since the canonical momenta p_θ, p_ν, and p_0 remain invariant under the considered transformations, there is no difficulty in translating partial derivatives with respect to them into partial derivatives with respect to the corresponding new momenta, a mere change of notation being sufficient.

Deprit's Intermediary of the First Order:

The potential yielding this intermediary is obtained for

$$V_0(p_\theta, p_\nu, p_0; \varepsilon) \equiv 0 , \quad V_1(p_\theta, p_\nu, p_0; \varepsilon) \equiv 0 ,$$

$$V_2 \equiv V_{2,(1)}^{(D)}(p_\theta, p_\nu, -; \varepsilon) = \varepsilon \frac{\mu^2 R_e{}^2}{4 p_\theta{}^2} (3c^2 - 1) ,$$

and therefore

$$\frac{\partial V_2}{\partial \Theta_0} = \varepsilon \frac{\mu^2 R_e{}^2}{2 \Theta_0{}^3} \left[1 - \frac{6 N_0{}^2}{\Theta_0{}^2} \right] , \quad \frac{\partial V_2}{\partial N_0} = \varepsilon \frac{3 \mu^2 R_e{}^2 N_0}{2 \Theta_0{}^4} ,$$

while the remaining partial derivatives involved in the transformation vanish identically.

Deprit's Intermediary of the Second Order:

In this case, no contribution due to V_0 and V_1 is present in the Hamiltonian. As for V_2, two terms are to be considered:

$$V_{2,(2)}^{(D)} \equiv V_{2,(2)}^{(D)}(p_\theta, p_\nu, -; \varepsilon) = \varepsilon \mathcal{V}_{2,1} + \varepsilon^2 \mathcal{V}_{2,2}$$

$$= \varepsilon \frac{\mu^2 R_e{}^2}{4 p_\theta{}^2} (3c^2 - 1) + \varepsilon^2 \frac{\mu^4 R_e{}^4}{32 p_\theta{}^6} (1 - 21 c^4) ,$$

where $\varepsilon \mathcal{V}_{2,1}$ coincides with the above expression for the potential in the first order intermediary of Deprit type. Therefore, the only derivatives of interest are

$$\frac{\partial V_2}{\partial \Theta_0} = \varepsilon \frac{\mu^2 R_e{}^2}{2 \Theta_0{}^3} \left[1 - \frac{6 N_0{}^2}{\Theta_0{}^2} \right] + \varepsilon^2 \frac{\mu^4 R_e{}^4}{16 \Theta_0{}^7} \left[\frac{105 N_0{}^4}{\Theta_0{}^4} - 3 \right] ,$$

$$\frac{\partial V_2}{\partial N_0} = \varepsilon \frac{3 \mu^2 R_e{}^2 N_0}{2 \Theta_0{}^4} - \varepsilon^2 \frac{21 \mu^4 R_e{}^4 N_0{}^3}{8 \Theta_0{}^{10}} ,$$

and no contribution will stem from the other terms.

Alfriend and Coffey's Intermediary of the Second Order

(*Case of small eccentricities*):

For small values of the eccentricity this intermediary is characterized, after neglecting terms of the order of $\varepsilon^2 e^2$, by the potential

$$V_{2,(2)}^{(AC)} \equiv V_{2,(2)}^{(AC)}(p_\theta, p_\nu, -; \varepsilon) = \varepsilon \mathcal{V}_{2,1} + \varepsilon^2 \mathcal{V}_{2,2}$$

$$= \varepsilon \frac{\mu^2 R_e{}^2}{4 p_\theta{}^2} (3c^2 - 1) + \varepsilon^2 \frac{\mu^4 R_e{}^4}{64 p_\theta{}^6} (27 c^4 - 138 c^2 + 71) ,$$

and the partial derivatives are given by the expressions

$$\frac{\partial V_2}{\partial \Theta_0} = \varepsilon \frac{\mu^2 R_e{}^2}{2\,\Theta_0{}^3}\left[1 - \frac{6\,N_0{}^2}{\Theta_0{}^2}\right] - \varepsilon^2 \frac{3\,\mu^4 R_e{}^4}{32\,\Theta_0{}^7}\left[\frac{45\,N_0{}^4}{\Theta_0{}^4} - \frac{184\,N_0{}^2}{\Theta_0{}^2} + 71\right],$$

$$\frac{\partial V_2}{\partial N_0} = \varepsilon \frac{3\,\mu^2 R_e{}^2 N_0}{2\,\Theta_0{}^4} + \varepsilon^2 \frac{\mu^4 R_e{}^4}{16\,\Theta_0{}^7}\left[\frac{27\,N_0{}^3}{\Theta_0{}^3} - \frac{69\,N_0}{\Theta_0}\right].$$

Due to the way in which Alfriend and Coffey constructed this intermediary, its first–order part is that of any Deprit intermediary.

As already mentioned, this case of negligible $\varepsilon^2 e^2$ –terms was studied by Franco and Palacios (1990) in the 6–dimensional phase space coordinatized by the usual polar nodal variables.

Alfriend and Coffey's Intermediary of the Second Order

(Case of arbitrary eccentricities in bounded motion):

As in the above situation, only a potential of the type $V_{2\,,\,(2)}$ is to be taken into account. After modifying the notations, the expression given in Cid *et al.* (1986), p. 202, in terms of s^2 is translated into the corresponding one in terms of c^2 yielding

$$V^{(AC)}_{2\,,\,(2)} \equiv V^{(AC)}_{2\,,\,(2)}(p_\theta\,, p_\nu\,, p_0\,;\,\varepsilon) = \varepsilon\,\mathcal{V}_{2,1} + \varepsilon^2\,\mathcal{V}_{2,2}$$

$$= \varepsilon \frac{\mu^2 R_e{}^2}{4\,p_\theta{}^2}(3c^2 - 1) + \varepsilon^2 \frac{3\,\mu^4 R_e{}^4}{16\,p_\theta{}^6}\left[\frac{1}{12}(27c^4 - 138c^2 + 71) + \frac{e_k^2}{8}(5c^4 - 18c^2 + 5)\right].$$

The "Keplerian eccentricity" e_k must be written in canonical variables, which in terms of the extended Hill set is accomplished by the formula

$$e_k{}^2 \equiv 1 + \frac{2\,h_k\,p_\theta{}^2}{\mu^2},$$

where h_k stands for the Keplerian energy and so, bearing in mind that

$$h_k = -\,p_0 + \mathcal{O}(\varepsilon)\,,$$

the expression for the second order part of this Hamiltonian is, after disregarding terms of the third order in ε,

$$\varepsilon^2\,\mathcal{V}_{2,2} = \varepsilon^2 \frac{\mu^4 R_e{}^4}{128\,p_\theta{}^6}(69\,c^4 - 330\,c^2 + 157)$$

$$- \ \varepsilon^2 \ \frac{\mu^2 \, R_e{}^4}{64 \, p_\theta{}^4} \ p_0 \, (\, 15 \, c^4 \ - \ 54 \, c^2 \ + \ 15 \,) \ ,$$

and the required set of partial derivatives reduces to the following:

$$\begin{aligned}
\frac{\partial V_2}{\partial \Theta_0} \ = \ & \varepsilon \ \frac{\mu^2 \, R_e{}^2}{2 \, \Theta_0{}^3} \left[1 \ - \ \frac{6 \, N_0{}^2}{\Theta_0{}^2} \right] \\
& - \ \varepsilon^2 \ \frac{\mu^4 \, R_e{}^4}{64 \, \Theta_0{}^7} \left[\frac{345 \, N_0{}^4}{\Theta_0{}^4} \ - \ \frac{1320 \, N_0{}^2}{\Theta_0{}^2} \ + \ 471 \right] \\
& + \ \varepsilon^2 \ \frac{\mu^2 \, R_e{}^4}{16 \, \Theta_0{}^5} \left[\frac{30 \, N_0{}^4}{\Theta_0{}^4} \ - \ \frac{81 \, N_0{}^2}{\Theta_0{}^2} \ + \ 15 \right] \Pi_0 \ , \\
\frac{\partial V_2}{\partial N_0} \ = \ & \varepsilon \ \frac{3 \, \mu^2 \, R_e{}^2 \, N_0}{2 \, \Theta_0{}^4} \ + \ \varepsilon^2 \ \frac{\mu^4 \, R_e{}^4}{32 \, \Theta_0{}^7} \left[\frac{69 \, N_0{}^3}{\Theta_0{}^3} \ - \ \frac{165 \, N_0}{\Theta_0} \right] \\
& - \ \varepsilon^2 \ \frac{\mu^2 \, R_e{}^4}{16 \, \Theta_0{}^5} \left[\frac{15 \, N_0{}^3}{\Theta_0{}^3} \ - \ \frac{27 \, N_0}{\Theta_0} \right] \Pi_0 \ , \\
\frac{\partial V_2}{\partial \Pi_0} \ = \ & - \ \varepsilon^2 \ \frac{\mu^2 \, R_e{}^4}{64 \, \Theta_0{}^4} \left[\frac{15 \, N_0{}^4}{\Theta_0{}^4} \ - \ \frac{54 \, N_0{}^2}{\Theta_0{}^2} \ + \ 15 \right] \ .
\end{aligned}$$

A Brouwer–like Intermediary (Ferrándiz, 1990)

This intermediary, directly developed in the extended phase space and presented in Ferrándiz and Floría (1990), contains a first order contribution emanating from a potential V_0 of the form

$$\begin{aligned}
V_{0\,,\,(1)} \ \equiv \ & V_{0\,,\,(1)} \, (\, p_\theta \, , \, p_\nu \, , \, p_0 \ ; \ \varepsilon \,) \ = \ \varepsilon \ V_{0\,,1} \\
= \ & \varepsilon \ \frac{\mu \, R_e{}^2}{4 \, p_\theta{}^3} \ (\, 3 \, c^2 \ - \ 1 \,) \, (\, 2 \, p_0 \,)^{3/2} \ .
\end{aligned}$$

The corresponding Hamiltonian is close to the secular Hamiltonian of Brouwer's solution (1959), although not exactly the same, due to a different treatment of some quantities involved in the derivation. Further details can be found in the aforementioned paper by Ferrándiz and Floría (1990).

Because of a different choice of the perturbation parameter, this expression is slightly different from that given in the said paper: on that occasion, the choice was $\varepsilon \ = \ - \, \mu \, R_e{}^2 \, J_2$, while now $\varepsilon \ = \ - \, J_2$.

The list of partial derivatives involved in the transformation will now be

$$\frac{\partial V_0}{\partial \Theta_0} = \varepsilon \frac{3\,\mu\, R_e{}^2}{4\,\Theta_0{}^4} \left[1 - \frac{5\,N_0{}^2}{\Theta_0{}^2} \right] (2\,\Pi_0)^{3/2} ,$$

$$\frac{\partial V_0}{\partial N_0} = \varepsilon \frac{3\,\mu\, R_e{}^2\, N_0}{2\,\Theta_0{}^5} (2\,\Pi_0)^{3/2} ,$$

$$\frac{\partial V_0}{\partial \Pi_0} = \varepsilon \frac{3\,\mu\, R_e{}^2}{4\,\Theta_0{}^3} \left[\frac{3\,N_0{}^2}{\Theta_0{}^2} - 1 \right] (2\,\Pi_0)^{1/2} .$$

Ferrándiz' Intermediary (1990)

This is an r^{-1} – radial intermediary and was also presented in Ferrándiz and Floría (1990); remembering the above remark concerning the perturbation parameter, its functional form is

$$V_{1,(1)} \equiv V_{1,(1)}(p_\theta, p_\nu, p_0; \varepsilon) = \varepsilon\, \mathcal{V}_{1,1}$$

$$= \varepsilon \frac{\mu^2\, R_e{}^2}{4\, p_\theta{}^3} (3\,c^2 - 1) \sqrt{2\,p_0} ,$$

and the derivatives

$$\frac{\partial V_1}{\partial \Theta_0} = \varepsilon \frac{3\,\mu^2\, R_e{}^2}{4\,\Theta_0{}^4} \left[1 - \frac{5\,N_0{}^2}{\Theta_0{}^2} \right] \sqrt{2\,\Pi_0} ,$$

$$\frac{\partial V_1}{\partial N_0} = \varepsilon \frac{3\,\mu^2\, R_e{}^2\, N_0}{2\,\Theta_0{}^5} \sqrt{2\,\Pi_0} ,$$

$$\frac{\partial V_1}{\partial \Pi_0} = \varepsilon \frac{\mu^2\, R_e{}^2}{4\,\Theta_0{}^3} \left[\frac{3\,N_0{}^2}{\Theta_0{}^2} - 1 \right] (2\,\Pi_0)^{-1/2} .$$

8. Conclusion

According to the transformation formulae presented at the end of Section 5, by solving for the original polar nodal variables in terms of the auxiliary integration variables E and f and using the auxiliary Keplerian–kind quantities, a Keplerian–like solution to the dynamical problem posed by $\mathcal{H}$ can be set up by means of a parametric representation given in the form

$$r = a(1 - e\cos E) = \frac{p}{1 + e\cos f} ,$$

$$p_r = \sqrt{\frac{\mu^*}{a}} \frac{e\sin E}{(1 - e\cos E)} = \sqrt{\frac{\mu^*}{p}} e\sin f ,$$

$$\theta = Q_\Theta + \frac{\partial V_0}{\partial \Theta_0}\frac{1}{n}(E - e\sin E) + \frac{\partial V_1}{\partial \Theta_0}\sqrt{\frac{p}{\mu^*}}\frac{E}{\eta} + \left[\Theta_0 + \frac{\partial V_2}{\partial \Theta_0}\right]\frac{f}{\kappa} ,$$

$$p_\theta = \Theta_0 = \text{const.} ,$$

$$\nu = Q_N + \frac{\partial V_0}{\partial N_0}\frac{1}{n}(E - e\sin E) + \frac{\partial V_1}{\partial N_0}\sqrt{\frac{p}{\mu^*}}\frac{E}{\eta} + \frac{\partial V_2}{\partial N_0}\frac{f}{\kappa} ,$$

$$p_\nu = N_0 = \text{const.} ,$$

$$t = Q_\Pi + \left[1 + \frac{\partial V_0}{\partial \Pi_0}\right]\frac{1}{n}(E - e\sin E) + \frac{\partial V_1}{\partial \Pi_0}\sqrt{\frac{p}{\mu^*}}\frac{E}{\eta} + \frac{\partial V_2}{\partial \Pi_0}\frac{f}{\kappa} ,$$

$$p_0 = \Pi_0 = \text{const.} ,$$

together with the *generalized Kepler equation*

$$Q_K = \frac{1}{n}(E - e\sin E) = t - t_0 ,$$

where the integration constant involved in the equation for Q_K has been represented in terms of an epoch constant t_0 depending merely on the instant from which t is measured in the time scale. The choice of t_0 as the instant at which the moving mass performs *a pericentre passage* is a customary practice in Orbital Mechanics.

The generality of this pattern allows one to produce, as special cases under specific choices of the potential, the corresponding analytical solutions to the above radial intermediaries.

Acknowledgements

The author thanks Prof. J. M. Ferrándiz (University of Valladolid) for advice and helpful discussion, and Dr. A. Elipe (University of Zaragoza) for encouraging comments. Dr. M. Palacios (University of Zaragoza) is acknowledged for providing a copy of the paper by Franco and Palacios (1990), on the basis of which a part of the present research has been evolved. The author is also indebted to his Department colleagues Dr. P. Martín, Mr. R. Caballero and Mr. J. M. Farto for their kind attention to the author's LaTeX consultations.

Partial financial support was received from CICYT of Spain within the National Programme of Space Research under project ESP.88 – 0541.

References

Alfriend, K. T. and Coffey, S. L.: 1984, 'Elimination of the Perigee in the Satellite Problem', *Celest. Mech.* **32**, 163–172.

Alfriend, K. T. *et al.*: 1977, 'The Extended Phase Space Formulation of the Vinti Problem', *Celest. Mech.* **16**, 441–458.

Brouwer, D.: 1959, 'Solution of the Problem of Artificial Satellite Theory without Drag', *Astron. J.* **64**, 378–397.

Cid, R. *et al.*: 1986, 'On the Radial Intermediaries and the Time Transformation in Satellite Theory', *Celest. Mech.* **38**, 191–205.

Deprit, A.: 1981, 'The Elimination of the Parallax in Satellite Theory', *Celest. Mech.* **24**, 111–153.

Ferrándiz, J. M.: May 1992, private communication.

Ferrándiz, J. M. and Floría, L.: 1989, 'Generación sistemática de intermediarios en la Teoría del Satélite'. In: *Actas de las XIV Jornadas Hispano–Lusas de Matemáticas (Puerto de la Cruz, Spain, June 1989)*, vol. **III**, 1207–1211. (In Spanish language; abstract in English. Available from the authors.)

Ferrándiz, J. M. and Floría, L.: 1990, 'New Intermediaries for the Main Problem in Satellite Theory'. In: K. B. Bhatnagar (Editor), *Instability, Chaos and Predictability in Celestial Mechanics and Stellar Dynamics*, Nova Science Publishers, New York. (In Press). Manuscript available from the authors.

Ferrándiz, J. M. and Floría, L.: 1991, 'Towards a Systematic Definition of Intermediaries in the Theory of Artificial Satellites', *Bull. Astron. Inst. Czechosl.* **42**, 401–407.

Franco, J. M. and Palacios, M.: 1990, 'Analytical Solution for Some Second–Order Radial Intermediaries', *Bull. Astron. Inst. Czechosl.* **41**, 180–192.

Gradshteyn, I. S. and Ryzhik, I.M.: 1980, *Table of Integrals, Series, and Products*, Academic Press, 1980.

Jezewski, D. J.: 1983, 'An Analytic Solution for the J_2 Perturbed Equatorial Orbit', *Celest. Mech.* **30**, 363–371.

Lanczos, C.: 1970, *The Variational Principles of Mechanics* (Fourth Edition), Dover Publications, Inc., New York, 1986.

Scheck, F.: 1990, *Mechanics. From Newton's Laws to Deterministic Chaos*, Springer.

Stiefel, E. L. and Scheifele, G.: 1971, *Linear and Regular Celestial Mechanics*, Springer.

ALBEDO PERTURBATION MODELS : GENERAL FORMALISM AND APPLICATIONS TO LAGEOS

DAVID VOKROUHLICKÝ*
O.C.A., dept. C.E.R.G.A., Av N. Copernic, 06130 Grasse, France
and
PAOLO FARINELLA and DAVID LUCCHESI
Dip. di Matematica, Univ. di Pisa, Via Buonarroti 2, 56127 Pisa, Italy

Abstract. The force due to radiation pressure on a satellite of arbitrary shape is written in a general form within a formalism similar to that used in the theory of radiative transfer in atmospheres. Then the corresponding integrals are evaluated for the simple case of a spherically symmetric satellite, and applied to model the perturbation due to the Earth–reflected radiation flux on LAGEOS. For this purpose, the optical behaviour of the Earth's surface and atmosphere is described as a combination of Lambertian diffusion (continents), partial specular reflection consistent with Fresnel law (oceans) and anisotropic diffusion according to Chandrasekhar's radiative transfer theory (clouds). The in–plane Gauss components T and S vs. mean anomaly are computed for a simple orbital geometry and for different models of the Earth's optical properties. A sensitive dependence is found on the assumed cloud distribution, with significant perturbations possibly arising from oceanic specular reflection when the satellite is close to the Earth's shadow boundaries.

Key words: Artificial satellites – non-gravitational perturbations – radiation mechanisms.

1. Introduction

Non–gravitational perturbations on artificial satellites have been an active research subject for celestial mechanicians in the 50s and the early 60s, at the beginning of the space age; later on, the widespread use of purely numerical models for predicting and fitting the orbital evolution of satellites and probes decreased the interest in more general and/or abstract approaches. However, starting in the late 70s the situation has changed again. The main reason has been the availability of very accurate tracking techniques, specifically designed nearly spherical and "passive" satellites and drag–free technology — to be applied in particular to space geodesy and geodynamics, but also to other purposes, such as general relativity tests. In this context, new physical mechanisms have been analyzed and new formalisms and algorithms have been developed, both for predicting the instantaneous value of very tiny non–gravitational forces, and for modelling their long–term orbital effects (for a general review, see Milani *et al.*, 1987).

One of the most complex such forces is that due to radiation pressure resulting from light rediffused or reflected from the Earth's surface and atmosphere — commonly referred to as *albedo force* or *albedo perturbation*. Though the corresponding force is typically at least one order of magnitude smaller than that due to solar radiation pressure, the model uncertainties are normally higher, due to the

* On leave from Astronomical Institute, Charles University, Švédská 8, 15000 Prague 5, Czechoslovakia

Celestial Mechanics and Dynamical Astronomy **57**: 225–244, 1993.

difficulties involved in modelling the momentum flux associated with radiation for a source such as the Earth, geometrically extended and having a complex and variable optical behaviour. Often in the past simplifying approximations have been used — e.g., uniform albedo, purely diffusive behaviour, etc. — which neglect some critical features of the process resulting into detectable orbital perturbations. For this reason, in the last decade a significant number of papers have been devoted to discuss more general or refined treatments of the albedo force, with applications to spacecraft as diverse as LAGEOS, ERS–1, Magellan and MACEK (Anselmo *et al.*, 1983; Barlier *et al.*, 1986; Rubincam and Weiss, 1986; Rubincam *et al.*, 1987; Borderies, 1990; Borderies and Longaretti, 1990; Lucchesi and Farinella, 1992; Vokrouhlický and Sehnal, 1992a, 1992b). Still, however, these models are typically based on "paradigms" on the relevant physical mechanisms which limit their suitability in a more general framework.

The main purposes of this paper are the following : (i) to present very general formulae, inspired by the treatment of radiative transfer in planetary and stellar atmospheres, that give the force due to radiation pressure on a satellite of arbitrary shape moving within the radiative field of an arbitrary source, and that can then be specialized to deal with the simple case of a spherically symmetric satellite (Section 2); (ii) to discuss a physical model of the optical behaviour of the Earth's surface and atmosphere, which is relatively simple from a physical point of view but at the same time realistic enough to incorporate some subtle critical effects (Section 3); (iii) to apply these formulae and models to LAGEOS, a satellite for which the assumption about spherical symmetry provides a good approximation of its real structure and whose laser–tracked orbit has shown residual unmodelled accelerations probably caused for a significant part by the albedo force (Section 4).

2. Acceleration of a Satellite Plunged into a Radiative Field

In this section we will formulate in a general manner the mathematical tools needed for the derivation of the perturbative acceleration of a satellite, arising from the interaction with a radiative field. Of course, this is not the first attempt to deal with this problem (for another recent treatment see e.g. Borderies and Longaretti, 1990), but our aim is to present the theory in a different formalism. Our approach — which was already partially used by Vokrouhlický and Sehnal, 1992a,b) is inspired by the techniques widely used in the radiative transfer theory applied to planetary and stellar atmospheres (see e.g. Mihalas, 1978).

Our basic physical quantity is *the radiative intensity* I, which from the macroscopic point of view fully describes the radiative field interacting with the satellite. More precisely, this is true provided one neglects the effects of polarization, which are negligible as far as the radiative acceleration of satellites is concerned. Moreover, we use the frequency integrated radiative intensity, as we neglect any frequency dependence of all the relevant processes. This is actually a strong approximation, which *a priori* rules out the possibility of taking into account energy transfers due to frequency–dependent processes in the Earth's surface and atmosphere and thus, for instance, of describing by a unified theory the albedo effect (due to visible light) and the perturbations due to the Earth's thermal radiation. This can be justified *a posteriori* by considering that the dynamical effects of the infrared radiation field are very different, owing to its nearly–isotropic geometry with no significant day–night asymmetry (see e.g. Rubincam, 1987).

Many previous papers devoted to this subject use from the very beginning the *radiative flux*, which in our approach is a quantity derived from the radiative intensity (see Eq. (1) below). In our opinion, a clear formulation of the basic equations using as a fundamental quantity the radiative intensity is useful, in particular when deriving more complicated formulae (e.g. in the case of a complex satellite geometry or of a radiative field without any symmetry). The formulation of Borderies and Longaretti (1990) is close to our approach (note that they speak about *the radiance* rather than about radiative intensity; we adopt the latter terminology which is closer to that of the classical radiative transfer theory). They concentrate mainly on some subtle methods aiming at integrating their formulae analytically as far as possible (for another impressive analytical work see Rubincam and Weiss, 1986). Our goal here is different — to present a formulation (i) as compact and clear as possible; (ii) putting no restriction on the involved radiative field; and (iii) allowing to incorporate directly the results of the radiative transfer theory of the stellar/planetary atmospheres. Of course, when the equations have to be applied to specific concrete situations, we will need to adopt suitable numerical integration methods; on the other hand, the analytical formalisms such as those quoted above can be applied only after strongly restricting the generality of the model (e.g., by assuming purely diffusive light scattering by the Earth).

Let us assume that the satellite surface is a two–dimensional surface $\mathcal{S}$ parameterized by any two variables μ, φ (for instance, for a spherical satellite we can use $\mu = \cos\vartheta$, ϑ, φ being some spherical angles). Moreover, we assume that the satellite is plunged into a radiative field $\mathcal{R}$ parameterized by another pair of variables ν, ϕ (for instance, specifying a direction; in any specific situation both sets of parameters together with the corresponding reference frame will be explicitly given). Under this assumptions, the perturbative acceleration is given by the following formula

$$\mathbf{a} = \frac{1}{mc}\int_{\mathcal{S}(\mu,\varphi)} dA(\mu,\varphi)\left[(\mathbf{E}+\rho(\mu,\varphi)\mathbf{R}(\mathbf{N}(\mu,\varphi)))\cdot\mathbf{J}_1 + \frac{2}{3}\delta(\mu,\varphi)\mathbf{J}_2\right] , \quad (1)$$

where

$$\mathbf{J}_1 = -\int_{\mathcal{R}(\nu,\phi)} \Pi_{\mathbf{n}(\nu,\phi)}(\mathbf{N}(\mu,\varphi))\, I(\nu,\phi;\mu,\varphi)\,\tilde{\Theta}(\nu,\phi;\mu,\varphi)\, d\nu d\phi\,,$$

$$\mathbf{J}_2 = -\int_{\mathcal{R}(\nu,\phi)} \Pi_{\mathbf{N}(\mu,\varphi)}(\mathbf{n}(\nu,\phi))\, I(\nu,\phi;\mu,\varphi)\,\tilde{\Theta}(\nu,\phi;\mu,\varphi)\, d\nu d\phi\,,$$

$$\mathbf{R}(\mathbf{a}) = 2\Pi_{\mathbf{a}} - \mathbf{E} \quad , \quad \tilde{\Theta}(\nu, \phi; \mu, \varphi) = \Theta(-\mathbf{N}(\mu, \varphi) \cdot \mathbf{n}(\nu, \phi)) \, .$$

Here $dA(\mu, \varphi)$ is the satellite surface element, $\mathbf{E}$ is the unit matrix of dimension 3, Π is a projection operator defined by : $\Pi_{\mathbf{a}}(\mathbf{b}) = (\mathbf{a} \cdot \mathbf{b})\, \mathbf{a}$, $\Theta(x)$ is the Heaviside step function, c is the velocity of light and m is the satellite mass, $\mathbf{N}(\mu, \varphi)$ is the surface normal and $\mathbf{n}(\nu, \phi)$ the unit vector in the the direction of a chosen ray. The three terms in Eq. (1) describe respectively the linear momentum of radiation removed from the incident beam (both by absorption and scattering), specular reflection (phenomenologically described by the coefficient $\rho(\mu, \varphi)$) and diffusive scattering by the satellite's surface (phenomenologically described by the coefficient $\delta(\mu, \varphi)$). We stress that this decomposition of the interaction of radiation with the satellite surface is just another *paradigmatic* assumption of the radiation pressure model, whose validity may become questionable under specific circumstances. Let us also note, that we take into account only processes connected with direct momentum exchange between radiation and satellite, disregarding the emission of *thermal* radiation from it. The latter effect is due to the true absorption (i.e. contingent on energy exchange between the radiation and satellite). Although the two effects are physically related, their consequences are different : while the reflected radiation acts *immediately* and (roughly speaking) in a direction opposite to that of the source, the thermal re-emission of the absorbed radiation is normally subjected to some time delay (the so–called *thermal inertia*, which for a spinning satellite may show up as a phase delay) and also depends on other parameters characterizing the structure and orientation of the satellite itself. This effect has been studied in detail for LAGEOS (see Rubincam, 1987; Afonso *et al.*, 1989; Farinella *et al.*, 1990; Scharroo *et al.*, 1991) and we will not deal with it further in the present paper.

It is to be stressed that formula (1) does not take into account possible role of the photon multiple reflections on the satellite surface. As for the specular reflection, this rather complicated task was treated by tricky numerical models (e.g. Renard and Koeck, 1989). However, keeping in mind the LAGEOS application, the satellite shape is simple enough (spherical in a good approximation) so that formula (1) is sufficiently general. Moreover, since no shadowing of one satellite part by another can occur, we suppress the satellite surface element (μ, φ) dependence of the radiative intensity I (see formulas (2-5)).

We do not give here a full formal proof of Eq. (1), for which we refer to Vokrouhlický, 1991. The essential hints can also be found in Vokrouhlický and Sehnal, 1992a, who also give a proof of the full equivalence with the formulae of Borderies (1990). Instead, we will apply Eq. (1) to evaluate the total acceleration in a simple particular case, that of a spherical satellite with constant optical parameters (ρ, δ). Eq. (1) involves in fact a double integration (say, "over the satellite surface" and "over the radiation field"). We shall show that in the case of a spherical satellite the integral over the satellite surface can be evaluated analytically, leaving just one integral in the final formula. The introduction of a spherical angle parameterization of the satellite surface is advantageous in this case, though we shall fix later on the reference frame (in fact we will profit from this freedom to simplify the computations).

For practical reasons let us treat the three terms in Eq. (1) separately. The first one (**E** in the integrand brackets) is to be treated in the following way :[1]

$$
\begin{aligned}
\mathbf{a}_A &= -\frac{R_{\mathcal{L}}^2}{mc}\int_{\mathcal{S}(\mu,\varphi)} d\mu d\varphi \int_{\mathcal{R}(\nu,\phi)} d\nu d\phi \, (\mathbf{n}(\nu,\phi)\cdot\mathbf{N}(\mu,\varphi))\,\mathbf{n}(\nu,\phi) I(\nu,\phi) \\
&\qquad \tilde{\Theta}(\nu,\phi;\mu,\varphi) \\
&= -\frac{R_{\mathcal{L}}^2}{mc}\int_{\mathcal{R}(\nu,\phi)} d\nu d\phi \, I(\nu,\phi)\,\mathbf{n}(\nu,\phi)\int_{\mathcal{S}(\mu,\varphi)} d\mu d\varphi \, (\mathbf{n}(\nu,\phi)\cdot\mathbf{N}(\mu,\varphi)) \\
&\qquad \tilde{\Theta}(\nu,\phi;\mu,\varphi) \\
&= \frac{R_{\mathcal{L}}^2}{mc}\int_{\mathcal{R}(\nu,\phi)} d\nu d\phi \, I(\nu,\phi)\,\mathbf{n}(\nu,\phi)\int_{\mu=0}^{1}\int_{\varphi=0}^{2\pi} d\mu d\varphi \mu \\
&= \frac{A_\perp}{mc}\int_{\mathcal{R}(\nu,\phi)} d\nu d\phi \, I(\mathbf{n}(\nu,\phi))\,\mathbf{n}(\nu,\phi) \,. \qquad (2)
\end{aligned}
$$

where $R_{\mathcal{L}}$ and $A_\perp$ are the satellite radius and cross section, respectively. Note that we have fixed the local satellite frame (in which the spherical angles ϑ, φ are defined) in such a way to simplify the expression in the second line of Eq. (2), thus resulting in the third line. The quantities ν, ϕ parameterizing the radiative field (up to now unspecified, but normally also spherical angles in possibly another local frame) specify 'one ray' in it. The second line in Eq. (2) clearly shows that for every such ray we have to fulfill the integration over the satellite surface $\mathcal{S}$. In each of these cases we have rotated the satellite local frame in such a way that its z–axis coincides with the direction of the chosen ray. Thus, for each ray (ν, ϕ) we use a different satellite frame (and therefore different spherical angles ϑ, φ). This trick is allowed by the isotropy of satellite surface with respect to the parameters ρ, δ.

In the same manner we can treat the other two terms in Eq. (1). The second one, corresponding to the specularly reflected radiation, becomes

$$
\begin{aligned}
\mathbf{a}_R &= -\frac{R_{\mathcal{L}}^2}{mc}\int_{\mathcal{S}(\mu,\varphi)} d\mu d\varphi \int_{\mathcal{R}(\nu,\phi)} d\nu d\phi \, (\mathbf{n}(\nu,\phi)\cdot\mathbf{N}(\mu,\varphi)) \; I(\nu,\phi) \\
&\qquad (\mathbf{R}(\mathbf{n}(\mu,\varphi))\cdot\mathbf{n}(\nu,\phi)) \; \tilde{\Theta}(\nu,\phi;\mu,\varphi) \\
&= -\frac{R_{\mathcal{L}}^2}{mc}\int_{\mathcal{R}(\nu,\phi)} d\nu d\phi I(\nu,\phi)\int_{\mathcal{S}(\mu,\varphi)} d\mu d\varphi \, (\mathbf{n}(\nu,\phi)\cdot\mathbf{N}(\mu,\varphi)) \\
&\qquad \left(2\Pi_{\mathbf{N}(\mu,\varphi)}(\mathbf{n}(\nu,\phi)) - \mathbf{n}(\nu,\phi)\right)\tilde{\Theta}(\nu,\phi;\mu,\varphi) \\
&= \frac{R_{\mathcal{L}}^2}{mc}\int_{\mathcal{R}(\nu,\phi)} d\nu d\phi I(\nu,\phi)\mathbf{n}(\nu,\phi)\int_{\mu=0}^{1}\int_{\varphi=0}^{2\pi} d\mu d\varphi \mu(2\mu^2-1) = 0 \,. \qquad (3)
\end{aligned}
$$

[1] Note, that we implicitly neglect variations of the radiative field at different part of the satellite. Instead, we work with radiative field $I(\nu,\phi)$, which results from averaging over its surface. The same applies for formulas (3) and (4).

This result is a classical one, and in fact has its counterpart in the same result for the elastic specular reflection of the neutral atmosphere molecules in the theory of the atmospheric drag. Of course any, but very particular cases, deviation from the spherical shape or from the isotropy of the specular reflection coefficient ρ would lead to a nonzero result.

As for the third term (diffusively scattered radiation from the satellite surface), we have

$$\begin{aligned}
\mathbf{a}_D &= -\frac{2\delta R_{\mathcal{L}}^2}{3cm}\int_{\mathcal{S}(\mu,\varphi)} d\mu d\varphi \int_{\mathcal{R}(\nu,\phi)} d\nu d\phi\, (\mathbf{n}(\nu,\phi)\cdot\mathbf{N}(\mu,\varphi))\,\mathbf{N}(\mu,\varphi)\, I(\nu,\phi) \\
&\qquad \tilde{\Theta}(\nu,\phi;\mu,\varphi) \\
&= -\frac{2\delta R_{\mathcal{L}}^2}{3cm}\int_{\mathcal{R}(\nu,\phi)} d\nu d\phi\, I(\nu,\phi) \int_{\mathcal{S}(\mu,\varphi)} d\mu d\varphi\, (\mathbf{n}(\nu,\phi)\cdot\mathbf{N}(\mu,\varphi))\,\mathbf{N}(\mu,\varphi) \\
&\qquad \tilde{\Theta}(\nu,\phi;\mu,\varphi) \\
&= \frac{2\delta R_{\mathcal{L}}^2}{3cm}\int_{\mathcal{R}(\nu,\phi)} d\nu d\phi I(\nu,\phi)\mathbf{n}(\nu,\phi) \int_{\mu=0}^{1}\int_{\varphi=0}^{2\pi} d\mu d\varphi \mu^2 \\
&= \frac{4\delta A_\perp}{9cm}\int_{\mathcal{R}(\nu,\phi)} d\nu d\phi I(\mathbf{n}(\nu,\phi))\mathbf{n}(\nu,\phi)\,. \qquad (4)
\end{aligned}$$

Here note one subtle point – the arguments of the vectorial quantity have changed proceeding from the second line to the third. This is also luckily due to the spherical symmetry of the satellite surface (roughly speaking, this means that the integral of the normal vectors to the sphere over one hemisphere is parallel to its axis). In the case of nonspherical shapes, the result would thus be different.

The resulting perturbing acceleration is $\mathbf{a} = \mathbf{a}_A + \mathbf{a}_R + \mathbf{a}_D$, hence in total

$$\begin{aligned}
\mathbf{a} &= \frac{A_\perp}{mc}\left(1+\frac{4}{9}\delta\right)\int_{\mathcal{R}(\nu,\phi)} d\nu d\phi \mathbf{n}(\nu,\phi) I(\mathbf{n}(\nu,\phi)) = \frac{A_\perp}{mc}\left(1+\frac{4}{9}\delta\right)\mathbf{F} \\
&= \begin{pmatrix}\text{satellite}\\ \text{cross section}\end{pmatrix}\begin{pmatrix}\text{some coefficient characterizing the inte-}\\ \text{raction of light with the satellite surface}\end{pmatrix} \\
&\quad \begin{pmatrix}\text{radiative flux, i.e. energy transferred}\\ \text{through the perpendicular area unit}\end{pmatrix}\begin{pmatrix}\text{velocity}\\ \text{of light}\end{pmatrix}^{-1}\begin{pmatrix}\text{satellite}\\ \text{mass}\end{pmatrix}^{-1} \qquad (5)
\end{aligned}$$

The second and third lines give a clear "physical interpretation" of the relevant terms (notice that the product of the two terms in third line yields the momentum flux, namely the amount of linear momentum transferred through a perpendicular unit area). This result of course is not new; see e.g. Aksenov, 1977, Milani *et al.*, 1987 (pp. 74–75), Borderies and Longaretti, 1990 (where one has to substitute $\nu \rightarrow \frac{1}{3}\delta$); here it is just an example of the application of the more general Eq. (1).

The result given by Eq. (5) in general still holds for nearly–spherical satellites with *the quasi–isotropic* surface optical properties. This — admittedly, somewhat vague — notion is inspired by LAGEOS–type satellites, whose surface is covered by a large number of retroreflectors. We call the surface of a nearly–spherical satellite to be quasi–isotropic, if, roughly speaking, "it looks the same from all the directions to a high accuracy". For instance, let us imagine a parallel beam of radiation impinging on LAGEOS. Obviously, if the satellite is illuminated along the axis of any retroreflector the radiative force will be somewhat different with respect to the case of irradiation out of the retroreflector axis. But the difference will be very small, provided the retroreflector has a small size compared with the satellite. The quasi–isotropy assumption just neglects this small difference. More precisely, we have to assume that the results of the satellite surface integration along the different rays of the radiative field (as explained in the paragraph following Eq. (2)) in the formulae (3), (4) and (5) are the same, or at least that the differences are negligible.

Notice that the peculiar optical behaviour of the retroreflectors — perfect back-scattering of the impinging radiation — causes their contribution to the momentum transfer equation to be just twice that of the absorption term (2), so that for a nearly–spherical and quasi–isotropic satellite whose surface is covered by retroreflectors for a fraction f of the total area, the factor $\left(1+\frac{4}{9}\delta\right)$ in Eq. (5) needs just to be replaced by $\left[1+f+\frac{4}{9}\delta(1-f)\right]$. For LAGEOS, the effective value of f is much lower than the "geometrical" one (≈ 0.4), because the corner–cube retroreflectors backscatter light only within some restricted incidence cone about their axis. More in general, the classical factor $\left(1+\frac{4}{9}\delta\right)$ will be replaced by a coefficient $C_R(\rho,\delta)$, where the assumption of quasi–isotropy assures that $C_R(\rho,\delta)$ is a function of the surface coefficients ρ,δ only and not e.g. of the orientation of the satellite rotation axis. The acceleration is thus simply rescaled if, for instance, the optical parameters are not constant in time. For LAGEOS, C_R has been treated as a constant solve–for parameter in the orbit determination process, and its value has been found to be about 1.2.

Of course, if the quasi–isotropy assumption fails by a significant amount, the perturbative acceleration will be given by an expression more complicated than (5). Interestingly, the possibility of a sizeable difference between the optical properties of two different hemispheres of LAGEOS has been recently proposed by Scharroo *et al.* (1991) as a way of explaining some unmodelled peaks in LAGEOS' residual along–track acceleration. In the present context, however, we shall stick to the quasi–isotropy assumption.

Finally, note that the character of the radiative field is not limited in any way in the previous discussion. It may be either the nearly parallel direct solar radiation or the geometrically more complex radiative field coming from the Earth's surface and atmosphere. In what follows, we shall apply Eq. (5) to the latter case. In the next section we shall describe a model of this radiative field as well as its (ν,ϕ) parameterization in Eq. (5).

3. Models of the Earth's Radiative Field

We shall now apply the preceding formulae to the particular case when the radiative field is originated by the reflection/diffusion of the sunlight from the Earth — the so–called *albedo effect.*

The Earth is assumed to be spherical. No general *a priori* assumption is made on the optical characteristics of the Earth surface, but they are just subject to a set of different model assumptions, with the aim of studying their influence on the resulting orbital perturbations of LAGEOS. In particular, we are going to consider the specific role of the scattering of light from the cloud coverage. Actually, one should always speak about the reflection/diffusion from the Earth–atmosphere system rather than from the Earth surface only.

Also, we assume that the Earth–scattered radiation does not interact with any exoatmospheric near–Earth medium, so that the intensity of the radiation is constant along the photon rays out of the atmosphere (Mihalas, 1978). This assumption is essential to obtain a geometrical link between the Earth's optical models (treated basically in the Earth surface/upper atmosphere local frames) and the radiative field which appears in the satellite acceleration formulae (1) or (5). Recall that in the previous Section we considered processes occurring in a satellite–fixed local frame, while throughout this section we will formulate the Earth's optical models in the Earth surface/atmosphere local frames. The previous assumption bridges the gap between them.

Similarly to Vokrouhlický and Sehnal (1992b), we will consider three different *modes* of reflection/diffusion from the Earth's surface/atmosphere. They are :

- diffusion of light from the continents, which will be assumed to occur in a simple isotropic way (*Lambert diffusion*) with a multiplying albedo value developed in spherical harmonics of the geographical latitude and longitude (as given by Sehnal, 1979);
- reflection/diffusion from the oceans : here together with a diffusive behaviour part of the light is assumed to be reflected in a specular way;
- scattering of light from compact atmosphere formations (which we will simply name *clouds*), treated according to Chandrasekhar's solutions of the radiative transfer equations for a planar atmosphere. The detailed assumptions will be specified in what follows.

We introduce the spherical angles ϑ, φ (and also $\mu = \cos\vartheta$) in the local Earth surface/atmosphere frame (these angles should not be confused with those used in the previous section). The z–axis of this local frame coincides with the outer normal of the surface element. The orientation of the other two axes does not appear in the formulae and consequently it is irrelevant. We reserve $\mu_\odot$, $\varphi_\odot$ for the Solar direction, while μ, φ specify the satellite direction in this system (see Fig. 1).

Using these notations, one can write the intensity of the isotropically diffused

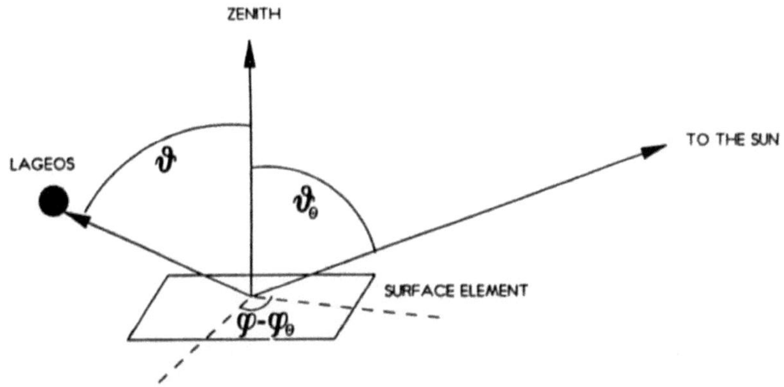

Fig. 1. The spherical coordinate system centered at an Earth surface element. The angle ϑ is measures the distance from the zenith direction of the chosen element. The origin of the φ coordinate is irrelevant in our models.

radiation according to Lambert's law as :

$$I_{\text{dif}}^{(\text{cont})} = \frac{1}{\pi} \Phi_0 \mathcal{A} \mu_\odot \ , \tag{6}$$

where Φ_0 is the flux (W/m^2) of the incoming solar radiation and $\mathcal{A}$ is *the albedo* coefficient. The $\mu_\odot$ factor is just due to geometrical reasons, namely is proportional to the cross–section with respect to the solar rays. Some criticisms to this purely geometrical approach can be found as early as in Levin, 1962, whose arguments were explicitly quoted by Vokrouhlický and Sehnal (1992b). Later on, the deviations from linearity with respect to $\mu_\odot$ were referred to as dependence of the albedo $\mathcal{A}$ on $\mu_\odot$ (see e.g. Lála et al., 1978; Rubincam et al., 1987; Knocke, 1989; and from an observational point of view Stephens et al., 1981). We feel that in spite of these studies, which usually empirically fit some sets of observed data, the theoretical understanding of this phenomenon is not satisfactory. Both light scattering in the atmosphere and details of the diffusion mechanisms on the Earth's solid surface (e.g., small–scale mutual shadowing) are plausible causes for a $\mu_\odot$ dependence of the albedo, and more in general for a dependence on all the involved angular parameters $(\mu_\odot, \varphi_\odot \ ; \mu, \varphi)$. But little is known about the separate importance of each of these effects and their detailed physics. In view of this, we have decided just to use the normal albedo model as given by Sehnal (1979). The albedo $\mathcal{A}$ is here a general function of the geographical latitude and longitude (expanded in spherical harmonics), but is constant with respect to the solar zenith angle $\mu_\odot$. In fact, it may be understood as an *averaged* value over $\mu_\odot$.

As for the oceans, we assume that besides Lambertian diffusion their surface can cause specular reflection of light. Following Barlier *et al.* (1986), we then decompose the intensity of the scattered radiation in two parts :

$$\begin{aligned} I_{\rm dif}^{(\rm oc)} &= \frac{1}{\pi}\Phi_0\mathcal{A}\mu_\odot(1-\mathcal{R}(\mu_\odot))\,, \\ I_{\rm spec}^{(\rm oc)} &= \Phi_0\mathcal{A}\mathcal{R}(\mu_\odot)\psi_{\rm sc}(\mu_\odot)\delta(\mu-\mu_\odot)\delta(\phi-\phi_\odot+\pi)\,, \end{aligned} \tag{7}$$

where

$$\begin{aligned} \psi_{sc}(\vartheta_\odot,\alpha) &= \frac{\alpha^2\sin\vartheta_\odot\cos\vartheta_\odot}{\sin\beta\,(2\cos(2\vartheta_\odot-\beta)-\alpha\cos\vartheta_\odot)}\,, \\ \alpha &= R_\oplus/r\,, \end{aligned} \tag{8}$$

r being the current satellite geocentric distance and the angle β fulfills the equation

$$\alpha\,\sin\vartheta_\odot = \sin(2\vartheta_\odot-\beta)\,.$$

Note that, in agreement with the definition of $\mathcal{A}$ as a fraction of incident sunlight which is not absorbed, but either diffused or reflected, the factor $\mathcal{A}$ appears also in the intensity of the specularly reflected component. Barlier *et al.* (1986; see their Eq. (10))on the contrary included $\mathcal{A}$ only in the intensity of the diffused component, thus implicitly adopting a different definition of albedo and amplifying the effects of radiation pressure from specularly reflected sunlight when compared with the present work.

For the derivation of Eq. (8) see Barlier *et al.*, 1986, where this factor is denoted as $J(\theta)$ (it actually simulates the divergence of the reflected radiation field due to the fact that the assumed mirror surface is spherical). The δ function in Eq. (7) is an obvious consequence of the specular reflection assumption. Of course, this is a somewhat idealized model : small scale wave phenomena on the ocean surfaces will actually cause some more divergence of the reflected light (as with the "sword of the sun" observed from a beach at sunset). A phenomenological model of this beam divergence can be found in Rubincam et al., 1987, who instead of the δ function adopt an exponential behaviour, and in Lucchesi and Farinella, 1992, who assume that the light intensity is uniformly distributed inside a narrow cone centered at $\mu=\mu_\odot$. The latter paper did show that the resulting perturbations depend weakly on the aperture of the reflection cone (or the corresponding exponential decay coefficient) provided it is less than several degrees. Therefore, here we shall stick to the simpler specular reflection assumption.

The "partitioning function" $\mathcal{R}$ in Eq. (7) is derived from the Fresnel reflection law (see Barlier *et al.*, 1986). The implicit assumptions here include a sufficiently short wavelength of the radiation and a negligible polarization (for a detailed treatment consult e.g. Jackson, 1975, and Ditchburn, 1976). Note that the Fresnel function $\mathcal{R}$ fulfills the following relation

$$(1-\mathcal{R}(\mu_\odot))_{\mu_\odot\simeq 0} \simeq \mu_\odot + \mathcal{O}(\mu_\odot^2)\,,$$

which results into a rapid increasing importance of the specularly reflected part of radiation with respect to the diffusively reflected part when the satellite is close to the Earth shadow's boundary, namely it "sees" a narrow Earth crescent. This is due to the fact that when $\mu_\odot \simeq 0$, $I^{(oc)}_{\rm dif} \propto \mu^2_\odot$ but $I^{(oc)}_{\rm spec} \propto \mu_\odot$ only. The corresponding sudden "kick" due to the specularly reflected radiation is in a way the result of extreme assumption, as the decomposition of the scattered radiation in the two parts is not unique (see Vokrouhlický and Sehnal, 1992b) and not all such decompositions give this feature. It should be noted that in accepting this model we just follow another common argument made in recent literature on LAGEOS perturbations : that of testing the effects of extreme assumptions whenever a reliable physical understanding is missing. In Sec. 4 we shall comment on some consequences of this assumption when describing the results of our model.

As for the scattering of light from the clouds, we shall use the model developed by Chandrasekhar (see e.g. Chandrasekhar, 1950; van de Hulst, 1980). The intensity of the scattered radiation is given by

$$
\begin{aligned}
I^{(cl)}_{\rm scat} &= \frac{1}{\pi}\frac{\omega_0\mu_\odot\Phi_0}{\mu+\mu_\odot}\big(H(\mu;\omega_0,x)H(\mu_\odot;\omega_0,x)\Psi(\mu,\mu_\odot;\omega_0,x) - \\
&\quad x\sqrt{(1-\mu^2)(1-{\mu_\odot}^2)}H^{(1)}(\mu;\omega_0,x)H^{(1)}(\mu_\odot;\omega_0,x)\cos(\varphi-\varphi_\odot)\big)\,,
\end{aligned}
\tag{9}
$$

where

$$
\Psi(\mu,\mu_\odot;\omega_0,x) = 1 - c(\omega_0)(\mu+\mu_\odot) - x(1-\omega_0)\mu\mu_\odot\ .
$$

These formulae solve the radiative transfer problem in a planar atmosphere with infinite optical thickness and scattering centers characterized by the phase function

$$
p(\Theta) = \omega_0(1 + x\cos\Theta),\quad 0 \le \omega_0 \le 1,\quad -1 \le x \le 1\,, \tag{10}
$$

where $H(\mu)$ and $H^{(1)}(\mu)$ are the Chandrasekhar H-functions and the associated μ–moments. They are defined as solutions of integral equations, but several numerical techniques for their evaluation are known (van de Hulst, 1980). Θ is the angle between the incident and the scattered ray at one scattering center in the atmosphere, and the phase function $p(\Theta)$ weighs the scattering probability at this angle.

The solution (9) is two–parametric, with the parameters ω_0 and x controlling physically different effects :

- ω_0 can just be interpreted as the albedo of the scattering centers in the atmosphere (not to be confused with the global atmospheric albedo A_{eff}; their relation is discussed in Vokrouhlický and Sehnal, 1992b, e.g. in the case $x = 0$ one gets $A_{eff} = \frac{4\omega_0}{(1+\omega_0)^2}$, while in the case $x \neq 0$ no such simple relation exists),
- the parameter x is related to the asymmetry of the light scattering from separate centers in the atmosphere : $x = 0$ implies symmetry, while $x = 1$ and $x = -1$ correspond to the extreme forward– and back–scattering cases (see Eq. (10)).

Since we are interested in testing models with local asymmetry of the cloud scattering behaviour (as it will be explained later), we will often put $x = 1$ (note however, that due to simplicity of the phase function (10), we cannot reach "the extreme asymmetry" as described by the asymmetry parameter g defined e.g in van de Hulst, 1980; for a review of more realistic phase functions see the brilliant book van de Hulst, 1980 and references in).

As with many theoretical models, that given by Eq. (9) suffers from several over-simplifications when compared to reality, e.g. :

- infinite optical thickness of the atmosphere (i.e., clouds);
- simplified phase function;
- both vertical and horizontal homogeneity of the atmosphere.

These simplifying assumptions are shortly commented in Vokrouhlický and Sehnal, 1992b, while more detailed discussions can be found in the specialized literature (Chandrasekhar, 1950; Irvine, 1975; van de Hulst, 1980). Here we are not going to include possible generalizations according any of the items listed above. But we stress that, as far as the first two items are concerned, these generalizations can be introduced if the comparison with the data shows that they are necessary to obtain a better fit.

We also note that all the modes of the light scattering/diffusion/reflection lead to formulae for the resulting radiative intensity depending only on the angle $\varphi - \varphi_\odot$ and not on φ, $\varphi_\odot$ independently (this property justifies the choice not to define the orientation of the local x–, y–axes, as noted earlier). This is well understandable, since the (x, y) local plane was chosen as tangent to the Earth's surface/atmosphere. We do not have any reason to distinguish symmetric directions with respect to the incident solar radiation. As far as we know, also all the models inferred from empirical data (e.g. Taylor and Stowe, 1984; Rubincam *et al.*, 1987) have the same property.

We now turn to description of the geographical features of the Earth's surface/atmosphere that we have used in the computation of the integrals in Eq.(5). In all our models, we used a realistic distribution of the continents, modelled by a $2^\circ \times 2^\circ$ grid in latitude and longitude. Then the final Earth *mask* giving the morphology of the different optical modes is created by covering the continental grid with an atmospheric grid, which gives the assumed cloud distribution. Of course, the latter is rapidly variable, and a realistic description could only be derived *a posteriori* from the available knowledge of global meteorological processes. This is well beyond the scopes of this work. Here we shall just distinguish among several possible idealized cases :

- the atmospheric grid is empty (we shall refer to this model as the *no cloud model*);
- the atmospheric grid is full (*full cloud model*);
- the atmospheric grid is empty on the Southern hemisphere and full on the Northern one (*hemisphere model*);

- the atmospheric grid is full in a belt surrounding the equator up to $\pm 10°$ in geographical latitude and empty elsewhere (*cloud belt model*);
- 60% of the atmospheric grid has been filled at random (*stochastic cloud model*).

4. Albedo Perturbations on LAGEOS

In this section we will discuss the perturbative accelerations resulting from the different models introduced earlier, and how they vary along one satellite revolution around the Earth. For the sake of simplicity we shall use a simplified orbit geometry, as already done by Rubincam *et al.* (1987) and Lucchesi and Farinella (1992) : LAGEOS' orbit is assumed to be polar ($I = 90°$; the real orbit has an inclination of about 110°) and the Sun lies in the LAGEOS orbit ascending node. We keep the true LAGEOS eccentricity ($e = 0.00444$) and longitude of perigee at epoch May 15, 1976 ($\omega = 235.35°$, see Barlier *et al.*, 1986), which determines the origin of anomalies. This configuration will allow us to estimate the maximum amplitude of the components of the albedo effect perturbative accelerations in the radial (S) and in–plane transverse (T) directions. These two components are important because they can be used to derive both the short– and the long–term perturbations on semimajor axis and mean anomaly; notice, however, that the assumed geometry minimizes the out–of–plane component W, which is important if one aims at estimating the inclination and node perturbations.

The most time–consuming part of the numerical treatment of the problem is the integration over the radiative field contributing to the radiative flux at the satellite. In order not to lose the detailed information on the Earth's surface/atmosphere optical behaviour, for each satellite position along the orbit we divided the visible Earth cap into a grid of 3025 surface elements, chosen in such a way as to be of equal area. This latter requirement implies that the grid is somewhat denser near the edge of the visible cap, and this may be important to properly model the acceleration when the satellite is close to the Earth's shadow boundary and just "sees" a thin illuminated Earth crescent. A similar method was used by Lucchesi and Farinella (1992), who showed that at least ≈ 100 surface elements are required for an accurate calculation; in our case the presence of specular reflection makes this requirement even more stringent. Similar accuracy problems have been taken into account in choosing the step size along the orbit; we have used 360 steps per revolution, i.e. 1° mean anomaly steps.

The results of several models are presented in Fig. 2, where the transverse component T is given as function of the mean anomaly M (whose origin is taken in such a way that at the subsolar point $M = 100°$). The curve(s) labeled 1 correspond to the no cloud model. More precisely, the solid curve 1a corresponds to the no cloud model as described in the Sec. 3, which takes into account the real distribution of continents, while the dashed curves 1b and 1c "ignore" the distribution of the continents : curve 1b corresponds to the ideal case when the

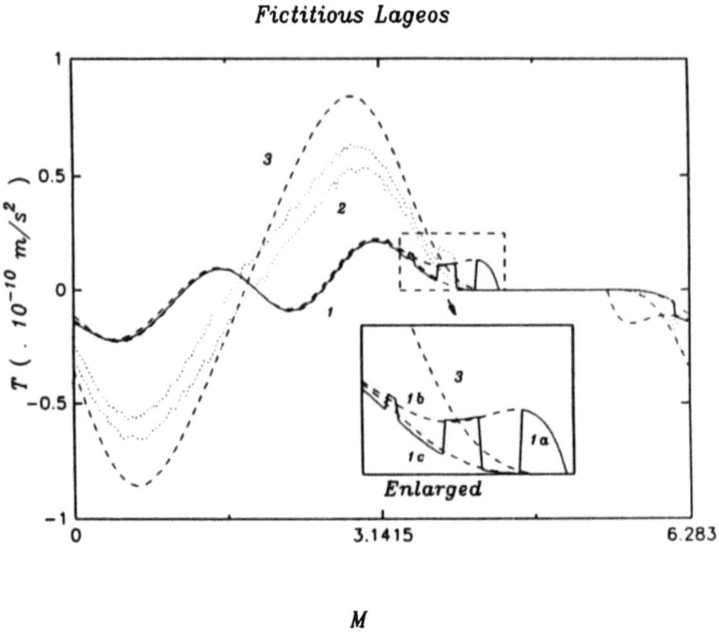

Fig. 2. The transverse component T of the albedo acceleration as a function of the mean anomaly M over one LAGEOS revolution, for the orbit geometry described in the text. Different *no cloud* models have been used to derive curves 1 : the (1a) model includes the real distribution of the continents on the Earth surface, the (1b) model considers an idealized Earth totally covered by the oceans, and the (1c) model assumes only diffusive (continental) reflection from the whole Earth surface. Note the essential role of the specular reflection effects when the satellite is close to the boundaries of the Earth's shadow. The dotted curves 2 bound the results of 50 different runs of the *stochastic* cloud model. Curves 3 correspond to the full cloud model. the adopted cloud parameters in both cases are : $\omega_0 = 1,\ x = 1$.

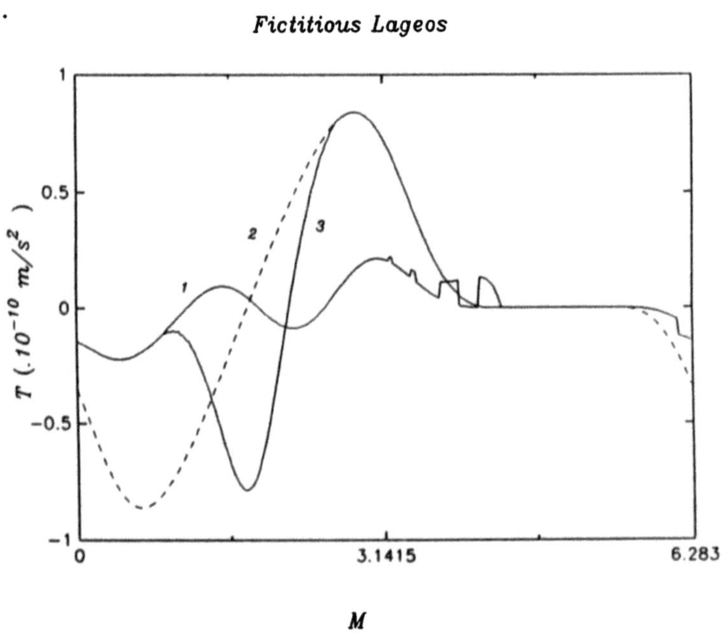

Fig. 3. The same as in Fig. 2, but comparing results of the *no cloud* model (curve 1), the *hemisphere* model (curve 2), and the *full cloud* model (curve 3). Cloud model parameters are kept as follows : $\omega_0 = 1,\ x = 1$.

whole Earth is covered by an ocean with the optical behaviour described in Sec. 3, and curve 1a corresponds to the ideal case when the whole Earth surface is covered by a continent with diffusive reflection. As expected, the largest differences occur near the satellite entry to/exit from the Earth shadow. Curve 1c shows there a smooth decrease, while curve 1b exhibits clearly the specular reflection "flash", as discussed earlier. The no cloud model 1a, combining 1b and 1c according to the actual distribution of the continents, shows sudden jumps due to interchanging of the two "pure" modes of reflection (due to the chosen orbit geometry, in the enlarged window in Fig. 2 the zone of the Earth possibly relevant for reflection is the Arctic). Such structures have a significant influence on the averaged value $< T >$ over an entire orbit, which is important to yield long–term perturbations. Such a potentially important role of specular reflection from relatively small zones of the Earth's surface was already pointed out by Barlier *et al.* (1986). It may be noted that our assumption that the Arctic polar cap behaves as an ocean is somewhat artificial; however, according to the data reported by Taylor and Stowe (1984), snow–covered surfaces do generate some specular reflection features, albeit less marked than liquid water, and thus may be seen as "intermediate" between the oceans and the continents.

Curve 3 corresponds to the full cloud model with parameters ($\omega_0 = 1,\ x = 1$). As no specular reflection occurs in this case, the curve is smooth. Actually, as noted already by Vokrouhlický and Sehnal (1992b), this case closely resembles the simple one with just a diffusion with a constant effective albedo $\mathcal{A}_{eff} = 1$ (see Lucchesi and Farinella, 1992, Fig. 5a, where the value $\mathcal{A}_{eff} = 0.2$ was used); only the anisotropy of the local scattering behaviour slightly disturbs the diffusion pattern. The amplitude of curve 1 is thus consistent with an average value of the albedo for the Earth's surface of approximately 0.3.

The region bounded by the dotted curves 2 corresponds to the stochastic cloud model; in this case, there is no unique curve, because the filling of the atmosphere grid is made at random. Therefore, we performed 50 runs of the program computing 50 different cloud distributions (always with a filling fraction of 60%), and then plotted the maximum and minimum values of the along-track accelerations. The typical difference of $10^{-11}\ m/s$ gives an idea of the consequences of the variability in the real cloud distribution (note in particular that the occurrence of the "Arctic reflection flash" is subject to the local cloud coverage). A similar method was used the case of the lower–altitude satellite ERS–1 by Vokrouhlický and Sehnal (1992b). An interesting observation can be done by comparing the ERS–1 results with those in Fig. 2 — the three curves 1–3 in Fig. 2 are well separated, as the region between the dotted curves 2 does not span the whole "gap" between curves 1 and 3, while the opposite was true for ERS–1. The reason has to do both with the simpler satellite geometry and with the higher orbit of LAGEOS.

In Fig. 3 we have replotted the mean anomaly dependence of the T acceleration component in the case of the no cloud model (curve 1), and the full cloud model (dashed curve 2), but we have added the curve (3) corresponding to the hemisphere model. As expected, the hemisphere model signal coincides with either one of the two previous cases when the satellite is over the poles (as its "visible cap" of the Earth behaves according to only one mode of reflection), while differs from them in the transient region, when the satellite is over the equator.

Curves 3 in Fig. 4 compare with the no cloud (curve 1) and full cloud (curve 2) models the results of the cloud belt model. The dashed and dotted curves allow an estimate of the effects of changing the x isotropy parameter : dashed curves correspond to the $x = 1$ choice, dotted curves to the $x = 0$ one. No qualitative new feature appears, though there is a significant quantitative difference. However, this is not a general result, due to the assumed symmetrical orientation of the orbital plane with respect to the Sun. Curves 3a–c also show the dependence of the results on the effective cloud albedo : 3a corresponds to $\mathcal{A}_{eff} = 1$, 3b to $\mathcal{A}_{eff} = 0.8$, and 3c to $\mathcal{A}_{eff} = 0.6$. A decreasing effective albedo of the clouds clearly diminishes their role.

Other remarks follow from the inspection of the radial component of the perturbative acceleration. Its dependence on the mean anomaly is shown in Fig. 5 in the case for the some models discussed in the previous paragraphs. Curve 1 corresponds to the no cloud model, curves 2 to the full cloud model (with different values of x) and curves 3 to the cloud belt model (with different values of both x and $\mathcal{A}_{eff}$). Similar graphs can be plotted for the other models. The observed differences are clearly of the order of several times 10^{-10} m/s^2. As the real cloud distribution is highly variable and indeed unknown when analyzing the real tracking data (unless a special effort is made to recover global meteorological data), one can expect an intrinsic uncertainty on the magnitude of the instantaneous albedo acceleration for LAGEOS of the order of 10^{-10} m/s^2. The corresponding error in LAGEOS' position predictions (or determinations) over a fraction of one orbital period (≈ 4 hr) is of the order of some mm. This is not much less than the current satellite laser–tracking capabilities, thus providing some motivation for modelling the albedo effect even when short orbital arcs are analyzed (see e.g. Milani *et al.*, 1992).

We do not discuss in detail here the $< T >$ values arising from averaging the resulting values of T over one orbital revolution, contrary to Rubincam *et al.*, 1987 and Farinella and Lucchesi, 1992. Although these $< T >$ values may provide some rough estimates of the long–term effects associated with the different models (see some remarks on this in Sec. 5), in most cases they sensitively depend on the assumed orientation of the satellite orbit, the Sun and Greenwich meridian. In view of this, we believe that using just *one* particular choice of the above parameters may mislead the interpretation. Consider for instance the dominant effect of specular reflection from the oceans when the satellite is close to the Earth's shadow boundaries. As it was pointed out earlier, any asymmetry caused in these zones by the actual occurrence of continents and/or clouds (not reflecting specularly) versus oceans may become a strong contributor to a nonzero $< T >$ value. Moreover, the albedo distribution model shows a significant longitude dependence with the

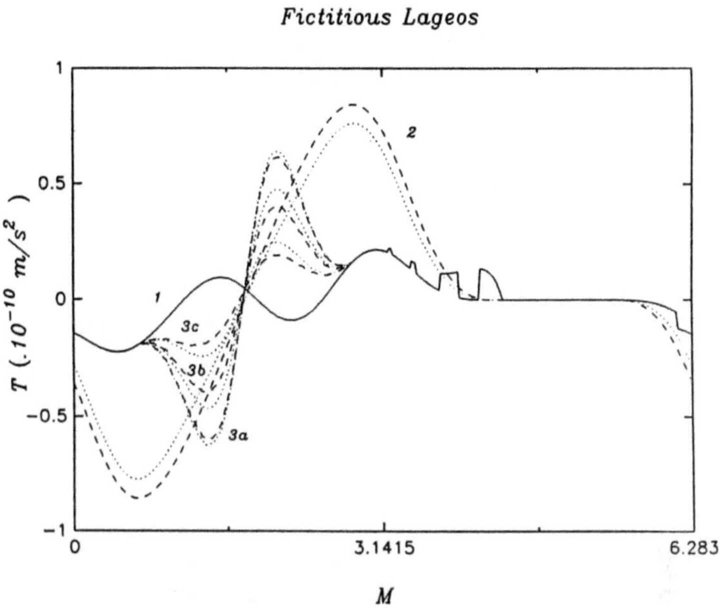

Fig. 4. Here the results of the *no cloud* model (solid curve 1) and of the *full cloud* models (curves 2) are compared with those of the *cloud belt* model (curves 3). Dashed curves correspond to strongly anisotropic light scattering from the atmosphere centers ($x = 1$), dotted curves to symmetric scattering ($x = 0$). The dependence on the effective cloud albedo is tested as well : curves (3a) correspond to $\mathcal{A}_{eff} = 1$, curves (3b) to $\mathcal{A}_{eff} = 0.8$, and curves (3c) to $\mathcal{A}_{eff} = 0.6$.

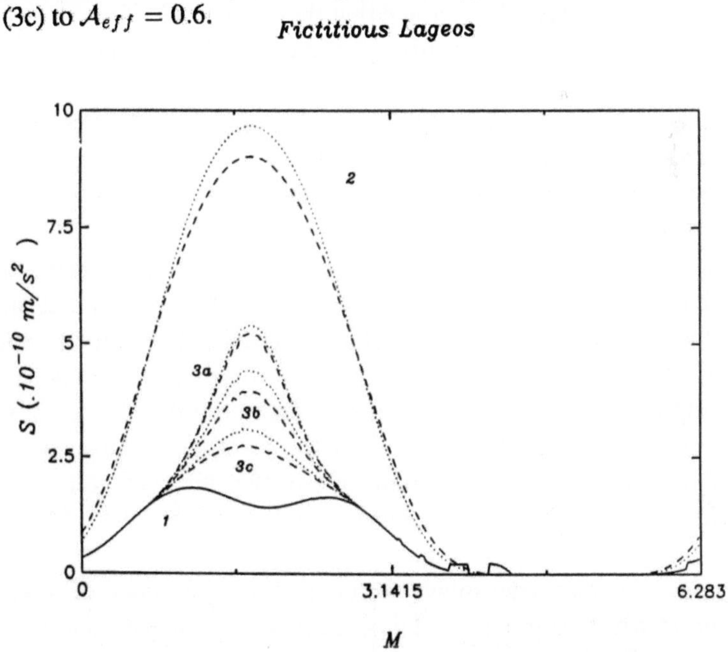

Fig. 5. The same as in Fig. 4, but for the radial component S of the albedo acceleration.

presence of localized structures (see Sehnal, 1979, Figs. 1 and 2), whose position could also somewhat affect the shape of the T curves. Thus the real value of $< T >$ may change in a substantial way even in successive revolutions. As a consequence, a reliable estimate of the long–term albedo effects requires a long time span integration (many thousands of revolutions) of the perturbation equations, followed by a Fourier transform analysis to point out periodicities.

It is also important to note that the previous discussion applies only to the intervals when LAGEOS' orbit crosses the Earth's shadow. Out of these periods the significant "kicks" caused by specularly reflected radiation are suppressed, since the Fresnel function $\mathcal{R}(\mu)$ undergoes a rapid decrease to very small values for non–tangential illumination of the reflecting surface elements. Actually, the LAGEOS orbital residuals show sharp peaks and dips correlated with the Earth's shadow crossing periods. While other effects may be responsible for this correlation (Afonso *et al.*, 1989; Scharroo *et al.*, 1991), we plan to carry out a detailed study of the long–term albedo effects in a forthcoming paper.

5. Conclusions

The main results obtained in this paper can be summarized as follows.

1. We have provided general integral formulae (1), similar to the ones appearing in the context of radiative transfer theory, that express the perturbing acceleration due to radiation pressure for a satellite of arbitrary shape moving in an arbitrary radiation field. We have then shown that they can applied in a straightforward way to the simple case of a nearly spherical, quasi–isotropic satellite such as LAGEOS.
2. In order to study the albedo orbital perturbations on LAGEOS, we have discussed in some detail the available models for the optical behaviour of the Earth's surface, including reflection from the oceans and possibly anisotropic scattering from clouds.
3. The Gauss components T and S vs. mean anomaly have been numerically computed for a simple orbital geometry (90° inclination, Sun at an equinox and in the orbital plane) and for different models of the cloud geographical distribution and optical properties. A sensitive dependence has been found for both components, with significant along–track perturbations possibly arising from oceanic specular reflection when the satellite is close to the Earth's shadow boundaries. This may contribute to the observed eclipse–correlated spikes in the unmodelled T acceleration residuals found from LAGEOS orbital determination.
4. The radial S component shows differences of the order of several times 10^{-10} m/s^2 when different cloud models are used, implying short–term errors (when arcs of a fraction of an orbital period are considered) in LAGEOS' position predictions/determinations of the order of some mm.

5. In spite of the *caveats* mentioned in Sec. 4, it can be noted that the difference in the orbit–averaged $< T >$ values between the no cloud model and the hemisphere model is of the order $10^{-12}\ m/s^2$. This is consistent with the findings of Lucchesi and Farinella (1992), who used somewhat different optical models and another numerical algorithm. As discussed by these authors, such a small along–track acceleration may cause on the long term comparatively large errors in LAGEOS' orbital predictions (or determinations). Moreover, this value is of just the same order of magnitude as those corresponding to other non–gravitational perturbations, including both drag–like forces and radiative effects (see the papers referred in Sec. 1). All these effects are intrinsically complex and uncertain, and the resulting long–term perturbations are probably mixed up in the observed acceleration residuals. As a consequence, we feel that it is dangerous to try to fit these residuals by semi–empirical models including a number of different mechanisms at the same time, as the modeling errors can cause large biases and/or aliasing effects. Our plan is thus to further study the albedo effect, investigating the long–term perturbations associated with the different models introduced in this paper.

Acknowledgements

We are grateful to F. Mignard for his careful review and many helpful comments. This work was partially supported by the Italian Space Agency (ASI) and the EEC research contract no. SC1–0011–C(GDF). One of us (D. V.) wishes to thank CERGA (Grasse, France) and the University of Pisa (Italy) for kind hospitality.

References

Aksenov, E.P.: 1977, *Theory of motion of artificial satellites*, Nauka (Moscow), in Russian

Barlier, F., Carpino, M., Farinella, P., Mignard, F., Milani, A., Nobili, A.M.: 1986, *Ann. Geophys.* **4**, 193

Borderies, N.: 1990, *Celest. Mech.* **49**, 99

Borderies, N. and Longaretti, P.-Y.: 1990, *Celest. Mech.* **49**, 69

Chandrasekhar, S.: 1950, *Radiative Transfer*, Oxford Univ. Press, Oxford

Ditchburn, R.W.: 1976, *Light*, Academic Press, London

Farinella, P., Nobili, A.M., Barlier, F., Mignard, F.: 1990, *A&A* **234**, 546

van de Hulst, H.C.: 1980, *Multiple Light Scattering : Tables, Formulas, and Applications*, Academic Press, New York

Irvine, W.M.: 1975, *Icarus* **25**, 175

Jackson, J.D.: 1975, *Classical Electrodynamics*, J. Wiley, New York

Knocke, P.Ch.: 1989, *Earth Radiation Pressure Effects on Satellites*, Dissertation, The University of Texas at Austin

Lála, P., Barlier, F. and Oyharcabal, G.: 1978, *Bull. Astron. Inst. Czechosl.* **29**, 238

Levin, E.: 1962, *ARS Journal*, 1328

Lucchesi, D. and Farinella, P.: 1992, *JGR* **97**, 7121

Mihalas, D.: 1978, *Stellar Atmospheres*, Freeman & co., New York

Milani, A., Nobili, A.M. and Farinella, P.: 1987, *Non–Gravitational Perturbations and Satellite Geodesy*, A. Hilger, Bristol

Milani, A., Carpino, M., Rossi, A. and Catastini, G.: 1992, *JGR*, submitted

Renard, P. and Koeck, C.: 1989, *Skin forces analysis report*, ESA Contract Report S374/NT/24/89
Rubincam, D.P.: 1987, *JGR* **92**, 1287
Rubincam, D.P. and Weiss N.R.: 1986, *Celest. Mech.* **38**, 233
Rubincam, D.P., Knocke, P.Ch., Taylor, V.R., Blackwell, S.: 1987, *JGR* **92**, 11 662
Sehnal, L.: 1979, *Bull. Astron. Inst. Czechosl.* **30**, 199
Scharroo, R., Wakker, K.F., Ambrosius, B.A.C. and Noomen, R.: 1991, *JGR* **96**, 729
Stephens, G.L., Campbell, G.G. and Vodner Haar, T.H.: 1981, *JGR* **86**, 9737
Taylor, V.R. and Stowe, L.L.: 1984, *JGR* **89**, 4987
Vokrouhlický, D.: 1991, *Notes on the Albedo Effect*, Dissertation, Charles University, Prague, in Czech
Vokrouhlický, D. and Sehnal L.: 1992a, *Celest. Mech.*, in press
Vokrouhlický, D. and Sehnal L.: 1992b, *Celest. Mech.*, submitted

STUDY OF A MOON'S ARTIFICIAL SATELLITE DYNAMICS VALID FOR ALL ECCENTRICITIES AND INCLINATIONS

DANIEL STEICHEN

Service des Calculs et de Mécanique Céleste du Bureau des Longitudes URA 707, 77 Avenue Denfert Rochereau, 75014 PARIS

This work deals with the study of trajectories of artificial satellites around the Moon to the end of missions. It means to study all the trajectories of lunar satellites below 10000 km, only considering for now actions that derive from a gravitational potential.

The perturbative function depends on the attraction of the Earth, the solar attraction, the lunar potential described by spherical harmonics and the libration of the equatorial plane of the Moon on the ecliptic. The fact that the perturbative functions due to the exterior body and to the harmonics of the interior potential are at the same order increases the difficulty of the study and forces us to use a detailed expansion. The perturbations of the Earth are given by the true longitude, the latitude and radius vector using a semi-analytic theory of the motion of the Moon (Chapront,J. and Chapront-Touzé,M. : 1982), the action of the Moon potential is described by the first hamonics (C20, C22, C30, C31, C32, C33, S31, S32 and S33), the motion of the Sun is simplified; metric elements are supposed to be constant and the angular linear of time and we introduce a semi analytic theory of the libration of the lunar equatorial plane in rotation to the ecliptic (Eckhart :1981).

Since the chosen orbits are not a priori classified, we introduce equations that remain as general as possible so that the solution can be found whatever the initial condition is. A direct numerical integration might be considered the most appropriate method, but it turns out to have two drabacks : the short periods of the satellite (that is, its periods of revolution around the Moon) require a step of integration in the region of ten minutes; which avoids an important error propagation for long duration and a prohibitive calculation time and the complexity of the equations prevents any classification of the orbits according to the initial conditions.

Hence, we chose averaging methods, in our case Lie's. It provides the elimination of rapid variations angles (longitudes of the satellite and the Moon) and finally gives long periods and secular evolutions of the orbits (analysing of the missions). Variables transformations will nevertheless give the longitude of the satellite.

The originality of this work is to consider the satellite orbits for any eccentricity or inclination at any given semi-major axis. We do not develop the expressions in terms of the eccentricity and the inclination for these two reasons: the first; it is not more expensive to put a satellite around the Moon on a eccentric or inclined orbit than on a circular or planar one, the second; the variation of the eccentricity is large because of the Earth perturbation and the harmonics ones. Furthemore, it was necessary to use Poincaré's variables which are not singular for an orbit whether it is a circular or located in the reference plan.

Celestial Mechanics and Dynamical Astronomy **57**: 245–246, 1993.

This purpose made it necessary to calculate the partial derivatives of the classical variables (trues longitude, "excentric longitude", radius vector) in relation to that. The secular parts which provide us a second order theory will be discussed too.

The next part will show the elimination of the satellite mean longitude and the introduction of a generative function by Lie's method. Then, by eliminating mean period angles, we obtain a twice averaged set whose periods are over 28 days.

The first solutions are described next. They concern a limited choice of initial conditions and are compared with a direct numerical integration. We necessite only a step of integration in the region of ten hours and find more rapidly the solution. We validate also this study .

An extention of this work will lead to research the resonances and a classification of the orbits in a theoritical way by the analysis of the twice averaged equations.

Acknowledgements

This research was carried out with a financial contribution by the "Centre National de Recherche Spatial" in Toulouse.

References

Chapront,J and Chapront-Touzé,M.: 1982, 'Tables simplifiées du mouvement de la Lune issues de ELP-2000 ', *Notes scientifiques et techniques du bureau des longitudes.*,

Eckhart,Donald H.: 1981, 'Theory of the libration of the Moon', *The Moon and the planets* **21**,

ON THE FLUID AND VISCOELASTIC DEFORMATIONS OF THE PLANETS

HILAIRE LEGROS and MARIANNE LEFFTZ
Institut de Physique du Globe de Strasbourg,
5 rue René Descartes, 67084 Strasbourg Cedex - FRANCE

Abstract. The shape of the Earth and of planets depends on the exciting forces and on the rheology of the medium. From the equilibrium equation, we present the main modelisations of the viscous and inviscid fluids and we essentially describe the characteristics of linear viscoelastic deformations, for the Maxwell viscoelastic model of rheology. We use the elastic, viscoelastic and fluid Love numbers in order to investigate the associated relaxation modes. For these various kinds of rheology of the planets interior, we compute the geoid and the topography induced by an internal mass distribution. Finally, we show the importance of this viscoelastic deformation calculations in the study of the celestial body rotations.

Key words: Fluid dynamics – deformation of planets – shape of planets – rotation of planets

1. Introduction

The Earth and the other telluric planets are astronomical bodies which have a radius of about some thousands of kilometers. Although the observations of theirs surfaces show high and low reliefs, theirs shapes are spherical or nearly spherical. The aim of this article is to show how the dynamical internal processes within a planet are responsible for its quasi spherical shape and for the temporal variations which may perturb the measurements related to its surface. The dynamical study of the figure of the Earth, presented hereafter by geodynamicists, has been investigated after Newton in particular, by Clairaut in 1743 working in celestial mechanics and known for his theory on the hydrostatic figure of the Earth which corresponds to an equilibrium figure of fluid gravitating and rotating bodies. In addition, the influence of an internal fluid part within a planet on the precession has been investigated by Hopkins (1839). It was the first time that the effects of internal processes as the ones coming from internal thermal mechanisms (geothermal gradient) were taken into account. Later, the elastic and viscous deformations have been calculated for a homogeneous Earth model respectively by Kelvin (1863) and by Darwin (1878), in order to model the deformation induced by the tidal potential. Finally, the composition of the Earth interior containing silicates and iron as deduced by studies of meteorites composition, has been taken into account by Wiechert (1897) for his study of elasticity following the Hopkins model. The compatibility between the core-mantle discontinuity, as founded by seismologists in 1915, and the fluidity of the core has been demonstrated by Jeffreys (1926). On the one hand, in addition to the deformational phenomena related to tides or rotation, the influence of geological phenomena on the internal stresses and on the rotation has been studied by Darwin (1881) and Schapiarelli (1889). In 1889, Dutton has considered the isostasy to be related to dynamical process. At the end of the 19th century and in the beginning of the 20th one, Poincaré has developed his theory

Celestial Mechanics and Dynamical Astronomy **57**: 247–278, 1993.

of equilibrium figures (1885, 1902) and the theory of the core rotation (1910), and Love (1909) has proposed his theory on elastic deformations. On the other hand, the geodetical and gravimetrical measurements have allowed the improvement of the knowledge of the Earth's flattening and of the geoid. Nevertheless, it is only in 1935 that Pekeris has developed a model of convection and Haskell a model of post-glacial rebound.

This brief historical review shows the large numbers of existing theories. It stresses also some problems concerning the study of the Earth figure and its temporal variations. One notices the large variety of the equations incise and of the assumed rheologies. We will present, from the fundamental equations and using a Love numbers formalism, a synthesis of the previous modeling and from the study of the relaxation modes, we will explain their ability to describe a lot of geodynamical processes.

2. Equations and Love Numbers

Before the introduction of the equations describing the Earth figure and its temporal variations and of the solutions using a Love numbers formalism, it is useful to specify some elements of the physics of the planets and to define some fundamental notions which appear in each model of planetary dynamics.

2.1. Fundamental Notions

The data coming from space geodesy, the geochemical study of the Earth and of meteorites, the physics at high pressures, the results obtained from seismology on the Earth and on the Moon, the rock thermodynamic properties, and the knowledge of the surface heat flux have allowed to propose a model of the interiors of the Earth and of the planets. It is a static description, i.e. a cut at a given constant time, of the dynamic of the planet, so that there is immediately the question how this present view is integrated within the historical past of the planet and how this present state is significant of a mean static state of the planet. However, this static description is necessary and we want to present it before to investigate the dynamical problem.

The telluric planets are essentially composed with silicates and iron separated between the mantle and the core. The relative quantity of these two components and the dimension of the radius determine, in a first approximation, the gravity and the internal pressure (static state). The internal temperature depends essentially on the heat resulting from accretion and on the heat due to radioactive disintegration. Because of the surface heat flux and of its resulting cooling, the present temperature depends also significantly on the size of the radius. These thermal and pressure conditions determine the mineralogical structures of silicates and iron and their rheological behaviour. The internal structure of the planet is described by the radial distribution of the physical parameters as the density, the pressure, the temperature

TABLE I
Terrestrial planets parameters. Indicative values after Zarkov (1983) and Toksoz et al. (1977).

	Parameters	Surface	Core mantle Boundary	Centre
	r in 10^6 m	2.4	1.7	0
	ρ in 10^3 kg	3.3	3.5 / 8.3	9.8
Mercury	P in 10^{11} Pa	0	0.1	0.5
	T in °K	–	–	2400 ?
	r	6.0	3.0	0
	ρ	3.3	5.4 / 9.6	11.7
Venus	P	0	1.0	2.9
	T	–	–	6000 ?
	r	1.7	0.3 ?	0
	ρ	0.3	0.35 / 7 ?	7 ?
Moon	P	0		0.05
	T	–	–	1200 ?
	r	6.4	3.5	0
	ρ	3.2	5.5 / 10	13
Earth	P	0	1.4	3.6
	T	1300	2500-3200/ 4000-5000	5500-6500
	r	3.4	1.7	0
	ρ	3.2	4.6 / 8.3	8.5
Mars	P	0	0.2	0.37
	T	–	–	2400 ?

and the elastic, viscous, thermodynamic coefficients (Fig. 1). The values of these parameters are given in Table I for the Earth and for some planets.

The primitive history of each planet may also have allowed a differentiation or a diffusion (equations which will not be used hereafter) so that the dense materials are in the central part of the body and the light materials in the external part. Moreover a differentiation is responsible for the residual superficial crusts. Thus, in a model of planet interior, we usually separate the stratification of composition (crust, mantle, core), the thermal stratification (lithosphere, viscous mantle, fluid and solid core) and the mineralogical stratification (olivine, spinelle, perovskite ...).

This static structure of the planet given by seismology may be dangerous especially for the rheology. As a matter of fact, the laws used for the equilibrium equations show similitudes but the behaviours are depending on the time-scale. For example, for the Earth, on the one hand, at the time-scale of the vibration modes (periods smaller than one hour) the Earth is considered to be elastic; on the other hand, at a long time-scale of about some millions years, it is more or less viscous, and at the geological time-scale, it is viscous and a quasi inviscid fluid.

This brief physical description of the interior of the Earth and planets allows us

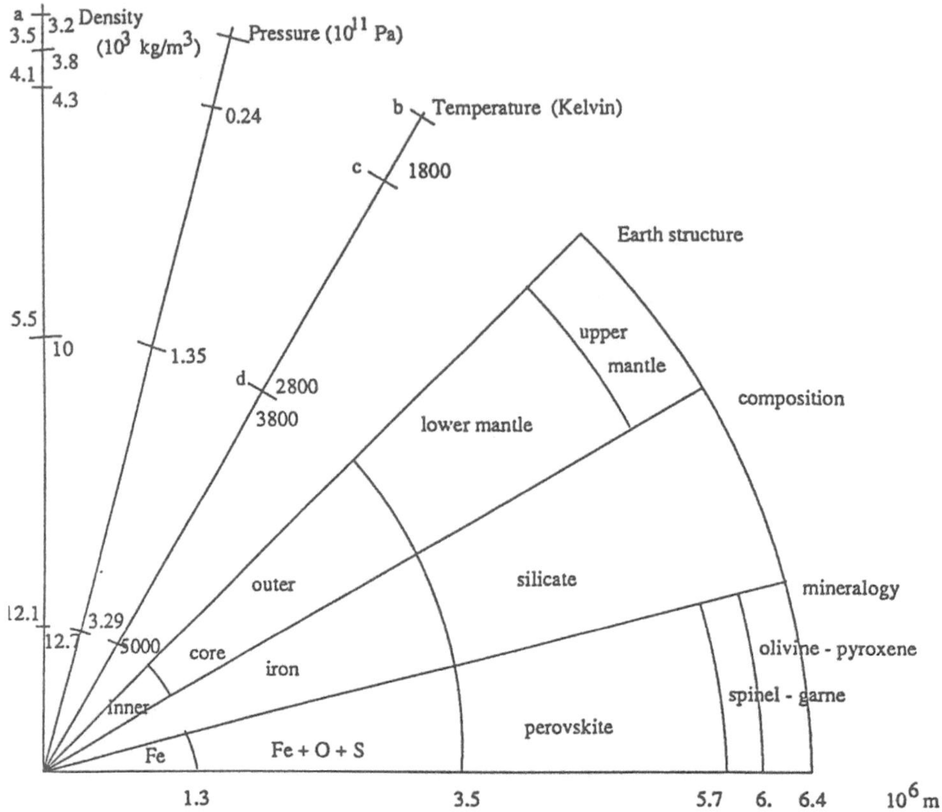

a: crustal boundary layer $\simeq$ 40 km
b: thermal boundary layer $\simeq$ 1300°K at 120 km
c: thermal boundary layer
d: thermal boundary layer
Central values
$P = 3.63\ 10^{11}$ Pa
$\rho = 1.3\ 10^4$ kg/m^3
$T \simeq 6000$°K

Fig. 1. Stratified Earth's model: mineralogy, chemical composition, Earth structure, temperature, pressure and density distribution. The radius within the Earth is represented on the horizontal axis.

to introduce the principal notions relative to the study of their figure.

The shape of the external surface boundary of a planet may be defined as a geometrical quantity (mathematical geodesy and topography) or as a mechanical form of equipotential. In the problem of the hydrostatic equilibrium figures, these two shapes are identical. For the hydrostatic forms, the equipotential is defined as the sum of the rotational and gravitational potential but some other exciting sources of deformations, with internal origin, could also be significant in such a problem. If there is an internal elastic or viscous stress field, this field induces a deformation of the boundary surface which perturbs the equipotential.

As a consequence, we will introduce:

- a topographic surface; in a reference frame, this surface could be defined as a value of the radius as a function of the colatitude θ and the longitude λ. From a spherical harmonics expansion, it may be written:

$$r_t(\theta,\lambda) = a\left[1 + \sum_{n=1}^{\infty}\sum_{m=0}^{n} a_{mn}R_{nm}(\theta,\lambda) + b_{mn}S_{nm}(\theta,\lambda)\right] \tag{2.1}$$

where a is the mean radius (the radius of the equi-volumic sphere) and with $R_{nm} = P_{nm}(\cos\theta)\cos(m\lambda)$ and $S_{nm} = P_{nm}(\cos\theta)\sin(m\lambda)$. $P_{nm}(\cos\theta)$ are the Legendre associated functions, P_{no} the Legendre polynomial.

The hypsometric map of the planet (with or without the oceanic part for the Earth) may be associated to this function $r_t(\theta,\lambda)$. This decomposition may also produce a "figure of geometrical reference", keeping only the terms in degrees 2 and 0, but the physical interpretation would be difficult.

- a geoid surface; it is the shape of the equipotential at a given point of the planet. For a nearly spherical body, the external potential may be written in a planetocentric system:

$$V(\theta,\lambda) = \frac{GM}{a}\left[1 - \sum_{n=2}^{\infty} J_n P_{no}(\cos\theta) \right. \tag{2.2}$$

$$\left. + \sum_{n=2}^{\infty}\sum_{m=1}^{n} c_{mn}R_{nm}(\theta,\lambda) + s_{mn}S_{nm}(\theta,\lambda)\right]$$

where G is the gravitational constant and J_n, c_{nm} and s_{nm} are the geopotential coefficients; they are time-dependent. With respect to a given reference figure, the height of the equipotential at a given point of this surface determines the geoid anomalies.

- a figure of reference surface; this surface is a rotating axisymmetric ellipsoid (combination of a homogeneous ellipsoid with a superficial layer) in which the two first terms of the potential are identical with the two first terms of the geopotential (GM and J_2). Thus a temporal variation of J_2 perturbs the figure of reference.

Since the perturbation of the stresses changes the shape or induces a topography, it modifies the topographic surface, that is to say the previously defined a_{nm} and b_{nm} coefficients change.

After the description of the physics and the definition of the shape, we have to specify the phenomena which are able to modify the shape and which are responsible for dynamical processes. With respect to the static physics, these phenomena are generally weak and appear as a perturbation; however, the planets are not rigid bodies and consequently theirs influences are extremely important for precise measurements. There exist a large diversity of excitation sources which vary according to the planets; to specify them, we will refer to the Earth.

The Earth has fluid superficial layers, the atmosphere and the ocean, a mantle, a fluid core and an inner core. The perturbations may be presented in some classes:

- volume forces such as the rotational perturbations of the luni-solar tidal forces.
- boundary conditions involved by hydrodynamical motions in the fluid parts, due to the solar energy in the atmosphere or to the magnetic field in the core.
- gravitational instabilities due to temperature and to compressibility which are responsible for the convective motions.

All these conditions appearing in the solutions of the equations play a significant role in the dynamics of the planet and in the perturbations of its topographic and its geoid.

To end the presentation of these fundamental notions, it seems to be useful to do some remarks concerning the superficial solid part of the planets, that is to say the crust. This is, on one hand, a boundary layer of compositional nature, and, on the other hand, it plays a complex role in the equations of the problem and it is often necessary to smooth its effects.

In a way, this layer is the seat of a particular dynamic due to tectonic processes and to geomorphologic phenomena of erosion and sedimentation induced by hydrologic and atmospheric actions. This paper will be concerned by these dynamical processes at large wavelengths relative to internal dynamical phenomena.

2.2. Rheologies and Equations

In this section, we want to model the physical and dynamical processes introduced in section 2-1, and describe the deformations of the topography and the temporal evolution of the geoid. In order to compare the various proposed theories on this topic and to find a common approach, we will start from the conservation equations of impulsion, mass and energy and we will investigate the various hypothesis which allow some simplifications within these equations. We also want to analyze the state equations and the relations between the different rheologies.

The equation of motion governing the equilibrium, whatever the point within the planet is, may be written:

$$\operatorname{div}\Sigma + \rho\nabla(\Phi + \Psi) + \rho\nabla V = \rho\gamma \tag{2.3}$$

Σ is the stress tensor, ρ the density, Φ the gravific potential, Ψ the rotational potential, V a volume potential (e.g. due to tidal forces) and γ the acceleration minus the inertial rotational acceleration

At the time-scale of this work, the term $\partial \mathbf{v}/\partial t$ and the non-linear term $\mathbf{v}.\nabla\mathbf{v}$, where $\mathbf{v}$ denotes the velocity, are negligible. The Coriolis term may also be neglected; in these conditions, γ is equal to zero. We are in the quasi-static approximation.

We distinguish the static state from the perturbed state which describes the dynamic. We use the eulerian perturbation with respect to the static state denoted by o. We have:

$\Sigma = -P_o I + \sigma^e$ where P_o is the initial hydrostatic pressure

$\rho = \rho_o + \rho_1^e$

$\Phi = \Phi_o + \Phi_1^e$

$\Psi = \Psi_o + \Psi_1^e$

$V = V_1^e$ a perturbation.

To specify the structure of the motion equation, we have to chose a rheology, i.e. a behaviour law relating the stress and the strain and theirs time derivatives. An example is Burger's rheology in which, for simplicity, we will only take the deviatoric part of the tensors (spherical part of strain supposed to be equal to zero). The dot denoting the time-derivative, we have:

$$\ddot{\sigma}_{kl} + \left(\frac{\mu_1 + \mu_2}{\eta_2} + \frac{\mu_1}{\eta_1}\right)\dot{\sigma}_{kl} + \frac{\mu_1\mu_2}{\eta_1\eta_2}\sigma_{kl} = 2\mu_1\ddot{\epsilon}_{kl} + \frac{2\mu_1\mu_2}{\eta_2}\dot{\epsilon}_{kl} \tag{2.4}$$

where μ_1, μ_2 are shear modulii, and η_1, η_2 viscosities. The analogy in terms of elasticity and piston is represented in the scheme 1.

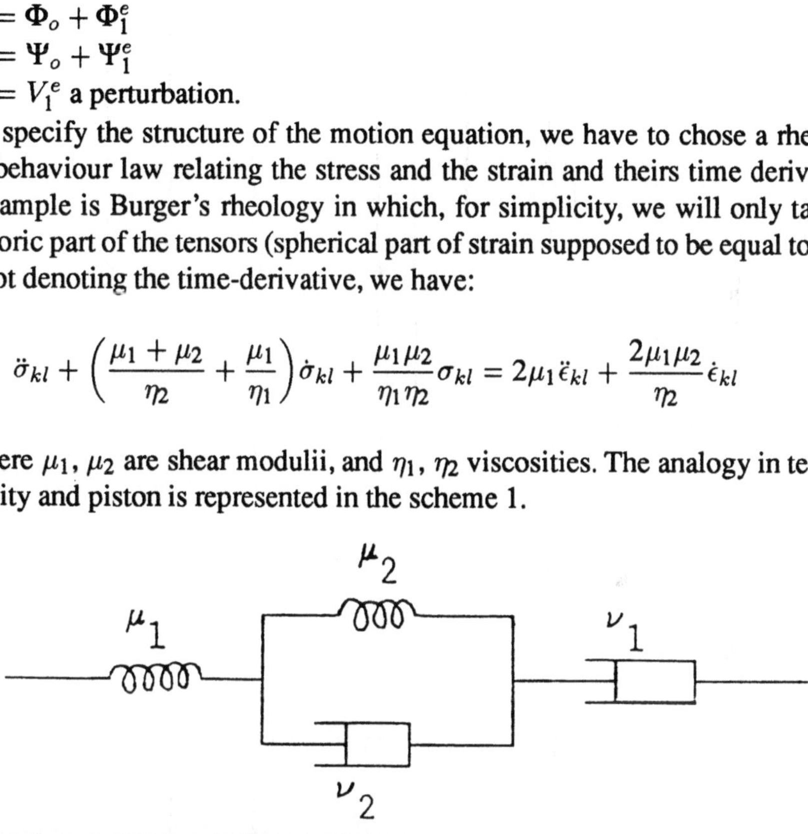

Some interesting particular cases are:

• if $\mu_1 \to \infty$ and $\mu_2 \to \infty$, then $\sigma_{kl} = 2\eta_1\dot{\epsilon}_{kl}$ (2.5)

In the Fourier domain, $\sigma_{kl} = 2\eta_1 i\lambda\epsilon_{kl}$, where λ is the frequency. It is the Newtonian fluid.

• if $\eta_1 \to \infty$ and $\eta_2 \to \infty$, then $\sigma_{kl} = 2\mu_1\epsilon_{kl}$ (2.6)

It is the Hookean elastic body.

• if $\eta_2 \to \infty$, then $\ddot{\sigma}_{kl} + \frac{\mu_1}{\eta_1}\dot{\sigma}_{kl} = 2\mu_1\ddot{\epsilon}_{kl}$ (2.7)

In the frequential domain, eq. (2.7) is $\sigma_{kl} = 2\mu_1 i\lambda/(i\lambda + \mu_1/\eta_1)$; it is the Maxwell viscoelastic body.

• if $\mu_1 \to \infty$ and $\eta_1 \to \infty$, then we obtain the Kelvin body or delayed elasticity.

These particular cases are the most used; it was necessary to introduce them because, according to the chosen rheology, the motion equations have usually different forms.

For the Newtonian fluid model of rheology, the variable in the motion equation is the velocity and the perturbation used is eulerian. If the spherical part of the strain tensor is equal to zero, we will write, from (2.4) and (2.5):

$$\operatorname{div}\Sigma = -\nabla(P_o + P_1^e) + \eta\Delta\mathbf{v} \tag{2.8}$$

and taking into account the previous comments, equation (2.3) will be written, neglecting the second order terms:

$$\begin{aligned} -\nabla P_o &+ \rho_o\nabla(\Phi_o + \Psi_o) - \nabla P_1^e + \eta\Delta\mathbf{v} \\ &+ \rho_o\nabla(\Phi_1^e + \Psi_1^e) + \rho_1^e\nabla(\Phi_o + \Psi_o) + \rho_o\nabla V = \mathbf{0} \end{aligned} \tag{2.9}$$

In equation (2.9), the static part is separated from the perturbed one.

In another way, for the elastic or viscoelastic rheology, the variable of motion is the displacement $\mathbf{u}$ and the considered perturbation is a lagrangian perturbation (except for the potential expressed in eulerian variable); the static state is considered as spherical. The lagrangian perturbations are noted with a "l" superscript:

$\rho_1^e = \rho_1^l - \mathbf{u}.\nabla\rho_o$

$\sigma^e = \sigma^l + \mathbf{u}.\nabla P_o$

If the spherical part of the strain tensor is equal to zero, from (2.4), we will have:

$$\operatorname{div}\Sigma = -\nabla(P_o + P_1^e) + \nabla(\mathbf{u}.\nabla P_o) + \mu\Delta\mathbf{u} \tag{2.10}$$

The equilibrium equation (2.3) becomes then:

$$\begin{aligned} -\nabla P_o + \rho_o\nabla(\Phi_o + \Psi_o) + \nabla(\mathbf{u}.\nabla P_o) + \mu(\lambda)\Delta\mathbf{u} + \rho_1^e\nabla(\Phi_o + \Psi_o) \\ + \rho_o\nabla(\Phi_1^e + \Psi_1^e) + \rho_o\nabla V - \nabla P_1^e = \mathbf{O} \end{aligned} \tag{2.11}$$

$\mu(\lambda)$ is the proportionality coefficient given by (2.6) and (2.7).

Now, we will present the mass conservation equation. Like in equation (2.9) and (2.11), we always distinguish the static part from the perturbed one.

The mass conservation equations are classically written:

$$\frac{\partial\rho}{\partial t} + \operatorname{div}(\rho\mathbf{v}) = 0$$

Following the chosen rheology, taking into account the static initial state and the perturbed state, neglecting at long time-scale the temporal variation of ρ_1^e, we have, for a newtonian fluid:

$$\operatorname{div}[(\rho_o + \rho_1^e)\mathbf{v}] = 0 \tag{2.12}$$

Practically, various applications of this formula have been done, and in particular, the incompressibility assumption

$$\mathrm{div}\,\mathbf{v} = 0. \tag{2.13}$$

Even in this case, a perturbation ρ_1^e defined by a state equation (Boussinesq approximation) may be conserved, either as a a-priori fixed perturbation (for example given by the seismological tomography) or as a temperature function (pseudo-convection).

$$\rho_1^e = -\rho_o \alpha \theta \tag{2.14}$$

α is the thermal dilatation coefficient, θ the difference between the temperature T and the adiabatic temperature T_s.

Finally, the equation (2.12) has been used in the frame of a non-linear theory (the terms in $\rho_1^e \mathbf{v}$ are kept). We obtain the theory of convection or the famous Eddington currents theory which has followed the equilibrium figures theory after the demonstration of Von Zeipel about the non-existence of the equilibrium figure when the internal temperature is taken into account (Mc Kenzie, 1968).

For an elastic model of rheology, the mass conservation equation is written:

$$\rho_1^e = -\,\mathrm{div}(\rho_o \mathbf{u}) \tag{2.15}$$

The previous equations (2.9) and (2.11) are the equations mostly used in dynamical studies of the Earth. However, we are going to present the heat equation which is necessary to the convection.

The heat equation describes the temperature evolution increasing when there are internal sources and cooling by conduction as given by the Fourier law:

$$\frac{\partial T}{\partial t} + \mathbf{v}.\nabla T - \frac{\alpha T}{C_p \rho_o}\left(\frac{\partial P}{\partial t} + \mathbf{v}.\nabla P\right) = \frac{1}{\rho_o C_p}\frac{\partial}{\partial x_i}\left(K\frac{\partial T}{\partial x_i}\right) + S + H \tag{2.16}$$

T is the temperature and P the pressure; C_p is the specific heat. The third term in the left hand side of this equation describes the variation of the adiabatic temperature given by:

$$\nabla T_s = \frac{\alpha T}{\rho C_p}\nabla P$$

where α is the thermal dilatation coefficient, K is the thermal conductivity, S the energy created by the viscous stresses and H the energy due to the disintegration of the radioactive elements.

This equation appears only in the problems of convection for a newtonian model of rheology.

To end the study of the equations governing the dynamic of the Earth and planets, we have to present the state equations. These last ones establish the relations between the different parameters of the system and are the equations describing the behaviour of the parameters them-self. One has, on the one hand, the Poisson equation :

$$\Delta\Phi_o = -4\pi G\rho_o \quad or \quad \Delta(\Phi_o + \Psi_o) = -4\pi G\rho_o + 2\omega^2$$

$$\Delta\Phi_1^e = -4\pi G\rho_1^e \tag{2.17}$$

and on the other hand, the other state equations appear essentially in the definition of ρ_1^e.

Thus we have presented the main equations governing the dynamic of the planet. Although the more interesting system is the one describing the internal convective motions (e.g. Machetel and Weber, 1991; Bercovici, Schubert and Glatzmeier, 1989), it is difficult to solve this system and it is not able to relate it to the dynamic of the realistic Earth. The most used equations are then the equations (2-9), (2.13) and (2.17) to investigate the mantle circulation (e.g. Richards and Hager, 1984; Ricard et al., 1984) or the equations (2.11), (2.15) and (2.17) for the elastic (Alterman, Jarosch and Pekeris, 1959) or viscoelastic (Peltier, 1974; Wu and Peltier, 1982) mantle.

These two models are formally very close and are presented, in literature, in the following way (with some different notations for some authors).

The vectors field **u** and **v** are expanded in spherical spheroidal vectors (the toroidal vectors are not interesting in these problems). We note :

$$\mathbf{u} \quad or \quad \mathbf{v} = \sum_{n=2}^{\infty}\sum_{m=0}^{n} y_{1n}(r)Y_n^m(\theta,\lambda)\frac{\mathbf{r}}{r} + ry_{3n}(r)\nabla Y_n^m(\theta,\lambda) \tag{2.18}$$

The $Y_n^m(\theta,\lambda)$ are spherical harmonics (combination between $R_{nm}(\theta,\lambda)$ and $S_{nm}(\theta,\lambda)$). We note also the tractions:

$$\mathbf{T} = \sum_{n=2}^{\infty}\sum_{m=0}^{n} y_{2n}(r)Y_n^m(\theta,\lambda)\frac{\mathbf{r}}{r} + ry_{4n}(r)\nabla Y_n^m(\theta,\lambda) \tag{2.19}$$

Finally, for the potential:

$$U = \sum_{n=2}^{\infty}\sum_{m=0}^{n} y_{5n}(r)Y_n^m(\theta,\lambda) \tag{2.20}$$

For the radial derivative of the potential, we introduce for each y_{5n} a function y_{6n} defines by:

-for the viscoelastic system

$$y_{6n} = \frac{dy_{5n}}{dr} - 4\pi G\rho_o y_{1n}(r)$$

-for the newtonian system

$$y_{6n} = \frac{dy_{5n}}{dr} \tag{2.21}$$

In these conditions the systems (2.9), (2.13) and (2.17) or (2.11), (2.15) and (2.17) are transformed in a system with 6 ordinary differential equations in $\frac{\partial}{\partial r}$ with 6 variables: y_{in} $(i = 1, ...6)$. In the above section, the subscript n will be suppressed for simplicity of notation. In its complete form, this system is the system of the gravito-elasticity (incompressible medium), or if we suppress the underlined terms, we obtain the newtonian incompressible fluid system. In the first case, y_{1n} and y_{3n} are the radial and tangential displacements and in the second y_{1n} and y_{3n} the radial and tangential velocities.

$$\begin{aligned}
\dot{y}_1 &= -\frac{2y_1}{r} + \frac{n(n+1)}{r} y_3 \\
\dot{y}_2 &= \left[\underline{\underline{-4\rho_o g_o}} + \frac{12\mu}{r}\right]\frac{y_1}{r} + \left[\underline{\underline{n(n+1)\rho_o g_o}} - \frac{6\mu n(n+1)}{r}\right]\frac{y_3}{r} + \frac{n(n+1)y_4}{r} - \rho_o y_6 \\
\dot{y}_3 &= -\frac{y_1}{r} + \frac{y_3}{r} + \frac{y_4}{\mu} \\
\dot{y}_4 &= \left[\underline{\underline{\rho_o g_o}} - \frac{6\mu}{r}\right]\frac{y_1}{r} - \frac{y_2}{r} - \frac{2\mu(2n^2+2n-1)}{r}\frac{y_3}{r} - \frac{3y_4}{r} - \rho_o \frac{y_5}{r} \\
\dot{y}_5 &= \underline{\underline{4\pi G \rho_o y_1}} + y_6 \\
\dot{y}_6 &= \underline{\underline{-4\pi G \rho_o n(n+1)}}\frac{y_3}{r} + \frac{n(n+1)}{r}\frac{y_5}{r} - \frac{2y_6}{r}
\end{aligned} \tag{2.22}$$

where we have introduced g_0 such as $\nabla\Phi_0 = -g_0\mathbf{r}/r$. To find a unique solution of this system (2.22), it is necessary to introduce boundary conditions. This is the aim of the following section, which also presents the generalized Love numbers.

2.3. Boundary Conditions and Love Numbers

The Love numbers have been introduced to express the solution of the elasto-gravitational system when the Earth was deformed by the tidal potential. There are 3 kinds of Love numbers : h_n related to the vertical displacement, l_n related to the horizontal displacement and k_n to the mass redistribution potential induced by defomation. Such Love numbers may also be defined for various excitation sources and also for other kind of rheology. Consequently, these numbers are very useful for the study of the defomation of the Earth and planets. But, before their introduction with the propagator of the system, the various boundary conditions at each interface of the model will be specified.

First, we present the boundary conditions at the external surface. For a newtonian fluid model, the classical conditions are the following (e.g. Richards and Hager, 1984):

- the radial velocity is equal to zero: $y_1(a) = 0$
- the tangential traction is equal to zero: $y_4(a) = 0$

This last condition, called sometimes free board condition, could be substituted by the condition of no-slip; in this case, $y_3(a) = 0$.

- there is a relation between the potential and its radial derivative:

$$y_6(a) + \frac{n+1}{a} y_5(a) = 0$$

These conditions are related to a problem without excitation. If an excitation source exists, it can appear, for example, as a pressure, noted P_n, which acts on the surface.

We have then $y_2(a) = -P_n$ and the condition with the potential and its derivative becomes:

$$y_6(a) + \frac{n+1}{a} y_5(a) = \frac{3P_n}{\rho a} \tag{2.23}$$

For the viscoelastic model of rheology, the boundary conditions are :
- the radial and tangential stresses are equal to zero: $y_2(a) = 0$, $y_4(a) = 0$
- there is a relation between the potential and its radial derivative:

$$y_6(a) + \frac{n+1}{a} y_5(a) = 0$$

The excitations appear with: (2.24)
$y_2(a) = -P_n$ and the other unchanged conditions
or $y_4(a) = S_n$ and the other unchanged conditions
or $y_6(a) + \frac{n+1}{a} y_5(a) = \frac{2n+1}{a} V_n$
P_n is a pressure excitation, S_n a tangential traction, V_n a volume potential. If there is a superficial mass density σ_n, there will be a pressure $P_n = \sigma_n g$ and a potential $V_n' = 3g\sigma_n/(2n+1)\rho$, where ρ is the mean density of the Earth and g the gravity at the surface. We just have to add these two effects to obtain the solutions due to a surface load.

Then, we want to investigate the boundary conditions at an interface within the mantle. This is an interface where some parameters such as the density, the shear modulus or the viscosity are discontinuous, or an interface where a mass superficial density, source of added radial stress and gravitational potential is located . We chose this interface within the mantle. We will consider the inviscid fluid core-mantle boundary later.

For a newtonian fluid mantle, if we note a_i the distance between the interface and the center of the Earth, we have the following conditions:
- continuity in radial and tangential velocities:

$$y_1(a_i^-) = y_1(a_i^+)$$

$$y_3(a_i^-) = y_3(a_i^+)$$

- continuity in tangential traction:

$$y_4(a_i^-) = y_4(a_i^+)$$

Noting ϵ a real number equal to 1 when there is a superficial mass density $\sigma_n(a_i)$ and equal to zero if there is no such a density.
- condition about the radial traction:

$$y_2(a_i^-) = y_2(a_i^+) - \epsilon\sigma_n(a_i)g(a_i) \tag{2.25}$$

where $g(a_i)$ is the gravity at the radius a_i.

- condition about the derivative of the potential:

$$y_6(a_i^-) = y_6(a_i^+) + 4\pi G\epsilon\sigma_n(a_i)$$

For an elastic model of rheology, the boundary solutions are very similar. Remember that in this case, y_1 and y_3 are displacements and y_6 has another definition.

We want now to study the last kind of boundary conditions which may exist within the planet, and especially for the Earth: the interface between the mantle, with its given rheology, and an inviscid fluid part as the core or the oceans. We note b the radius of this interface. This kind of limit introduces some particular problems. In a first approach, one part is an inviscid fluid and the perturbations are eulerian perturbations; if there is no shear modulus and viscosity, the equilibrium equation will be simplified. In a second approach, we add the assumption that if there is, for energetic reason, a dynamic within the fluid part, motions induced by density or thermal phenomena within the oceans, or magnetic process within the fluid core, we consider that the current lines of these motions are weakly perturbed by the viscous or viscoelastic deformation of the interface. In these circumstances, with the possible load at each interface, we may have some other conditions such as a pressure condition, noted P_n^c or a condition in potential, noted V_n^c, induced by these motions. These kinds of conditions are classical for the elastic model (Legros, 1987). This modelisation permits the interpretation of observations such as the free core nutation.

For a viscoelastic model, we have the following conditions:

$$\begin{aligned}
&y_2(b^+) = \rho^c(b^-)g(b^-)y_1(b^-) - P_n^c \\
&y_4(b^+) = 0 \\
&y_5(b^+) = V_n^c(b^-) \\
&y_6(b^+) = -4\pi G\rho^c(b^-)y_1(b^-) + \tfrac{dV_n^c}{dr}(b^-)
\end{aligned} \tag{2.26}$$

The functions $y_1(b)$ and $y_3(b)$ are the unknown of the system. There are obviously the same kind of conditions at the interface inviscid fluid core-solid inner core, for example.

For a newtonian rheology, the boundary conditions at the interface with an inviscid fluid are the same than at the surface for the velocities and the stresses and the same than at an internal interface within the mantle for the potential and its derivative.

All these conditions being precised, we may introduce the generalized Love numbers.

For a linear rheology, the stress and the strain are proportional in the frequency domain. The proportionality coefficient is a combination between the frequency, the shear modulus and the viscosity. Thus, in the frequency domain, the rheological laws have formally the same structure; this is the correspondence principle (Lee, Radok and Woodward, 1959). The solutions of the system (2.22) may be expressed, in general, in the frequential domain.

We introduce then, in each layer bounded by two interfaces, the propagator associated with the y_i system. Let a_i and a_{i+1} be the radius of each interfaces and $r_{ik}(a_i, a_{i+1})$ the propagator. We have:

$$y_k(a_i) = r_{kl}(a_i, a_{i+1}) y_l(a_{i+1}) \qquad for \quad k, l = 1, 6 \tag{2.27}$$

From the previous boundary conditions, there is, at each interface, a relation of continuity between $y_k(a_i^-)$ and $y_k(a_i^+)$ (e.g. relation (2.23), (2.24) or (2.25)).

This relation may have some source terms such as an internal load or a pressure or a traction or a potential.

Propagating from the surface down to the Earth's centre, we may progressively build up a propagator $r_{ij}(0, a)$ so that:

$$y_i(0) = r_{ij}(0, a) y_j(a) \tag{2.28}$$

To this end, we need to write the unknown components of the solutions vector of the problem (e.g. the displacements or the velocities, the potential ...) at each interface, particularly at the planet surface (this case will be detailed hereafter) as being proportional to the various excitation sources (because of the linearity of the problem, there is a superposition of the effects). These coefficients of proportionality are called generalized Love numbers. Thus, from the equation (2.28), taking $y_1(0) = y_3(0) = 0$ and $y_6(0) = y_5(0) = 0$ at the Earth's centre, we may calculate by inversion of the formula (2.28) the 3 coefficients of proportionality defined at the surface. For the study of the figure of the Earth and the geoid, we introduce, for a viscoelastic model of rheology, a Love number h (respectively l) associated with $y_1(a)$ (respectively $y_3(a)$) which gives the radial (respectively tangential) displacement for various sources. A Love number k associated with $y_5(a)$ (in fact to the mass redistribution part of $y_5(a)$) will give the perturbation of the gravific potential induced by excitation sources. As previously indicated, these Love numbers are computed from the propagators of the system; the following formula gives, for example, some h Love numbers for a 3-layered Earth model (with a mantle, an inviscid fluid core and an inner core). We may also introduce the perturbation of these numbers induced by a quasi-static fluid superficial layer (atmosphere or oceans) overlying the mantle. Notice that these Love numbers are defined in the frequency domain because the propagator $r_{ij}(0, a)$ depends on the frequency.

$$y_{1n}(a) = u_r(a) = \sum_{n=1}^{\infty} u_{rn}(a) Y_n \tag{2.29}$$

with

$$u_{rn}(a, \lambda) = h_n(\lambda) \frac{V_n(\lambda)}{g} + h_n^{'}(\lambda) \frac{V_n^{'}(\lambda)}{g}$$
$$+ \bar{h}_n(\lambda) \frac{P_n(\lambda)}{\rho g} + h_n^{*}(\lambda) \frac{S_n(\lambda)}{\rho g}$$

$$+h_n^1(\lambda)\frac{V_n^c(\lambda)}{g} + h_n'^1(\lambda)\frac{V_n'^c(\lambda)}{g}$$
$$+\bar{h}_n^1(\lambda)\frac{P_n^c(\lambda)}{\rho g} + h_n^{*1}(\lambda)\frac{S_n^c(\lambda)}{\rho g}$$
$$+h_n^2(\lambda)\frac{V_n^g(\lambda)}{g} + h_n'^2(\lambda)\frac{V_n'^g(\lambda)}{g}$$
$$+\bar{h}_n^2(\lambda)\frac{P_n^g(\lambda)}{\rho g} + h_n^{*2}(\lambda)\frac{S_n^g(\lambda)}{\rho g}$$
$$+h_n'^i(\lambda)\frac{V_n'^i(\lambda)}{g}$$

where the superscript c (respectively g) denotes sources at the CMB (respectively at the fluid core-inner core interface) and the superscript i is relative to internal loads. Overline Love numbers denote pressure Love numbers, whereas the $'$ denotes loading Love numbers.

We may obviously apply the same method for a newtonian fluid computing the propagator relating the interesting unknowns at the surface with the conditions at the centre. Notice that, until now, the viscous Love numbers were rarely expressed, excepted the Love number k for the perturbation of the gravific potential by Richards and Hager (1984). Notice also that for the newtonian fluid system, authors, for example Ricard et al. (1984), have introduced a stress tensor, directly proportional to the strain rate; as a consequence, the propagator is not depending on the frequency and the associated Love numbers will be "statics". These authors, instead of introducing a Love number associated with $y_2(a)$ and the excitation source, introduce a number H which is the height of the topography defined by:

$$H = \frac{y_2(a)}{\rho g} \tag{2.30}$$

And similarly a Love number N is defined from $y_5(a)$ associating the height of the geoid to the perturbation of the total potential.

$$N = \frac{y_5(a)}{g} \tag{2.31}$$

The topography, defined with H, is not, like in viscoelasticity, a surface displacement, but expresses the surface deformation which may be maintained by the radial stress.

We have shown how various Love numbers may be defined as the solutions of the viscoelastic or newtonian system. The advantage of writing the solution in function of various excitation sources is firstly, the simplicity of this formalism and secondly, the possibility of using different excitation amplitudes.

TABLE II
Geometrical and physical parameters of the four layered Earth model, with an elastic lithosphere, a viscoelastic upper and lower mantle and an inviscid fluid core.

Radius in km	Density in $kg.m^{-3}$	Rigidity in Pa	Viscosity in Pa.s
$a_1 = 6221 < r < a$	$\rho_1 = 3232$	$\mu_1 = 6.114 \times 10^{10}$	$\nu_1 = inf$
$a_2 = 5701 < r < a_1$	$\rho_2 = 3666$	$\mu_2 = 9.169 \times 10^{10}$	$\nu_2 = 10^{21}$
$a_3 = 3480 < r < a_2$	$\rho_3 = 4904$	$\mu_3 = 2.225 \times 10^{11}$	$\nu_3 = 5 \times 10^{21}$
$0 < r < a_3$	$\rho_4 = 10987$	$\mu_4 = 0.$	$\nu_4 = 0.$

These numbers are thus very useful to study the dynamical effects induced by the various excitation sources which deform the Earth and the planets.

In the next section, we want to investigate some examples for a viscoelastic body, taking into account the temporal perturbations induced by the rheology.

3. Relaxation Modes and Earth Deformations

We have investigated in section 2 the general equations useful for the theoretical study of the Earth and planets. We want now to present some more precise applications. As seen in the presentation of the physical processes perturbing the dynamic, if it is necessary to know the elastic or viscous instantaneous responses induced by variations in the boundary conditions, it is also important to know the deformations resulting from the relaxation of events in the past.

We will then firstly investigate the relaxation modes for a given Earth model, secondly we shall show the fluid response, and thirdly we will present some examples with, sometimes, a planetological generalization.

3.1. Relaxation Modes and Viscoelastic Love Numbers

The propagator of the y_i system (2.22), noted $r_{ij}(0, a)$ is a function of the frequency λ for a Maxwell viscoelastic model of rheology. The determinant associated with this propagator is a frequency polynomial which cancels for some roots which are the eigenmodes of the system. These modes are purely imaginary and consequently are relaxation modes (Peltier, 1974).

We have computed the relaxation modes associated with the viscoelastic Love numbers for a 4-layered Earth model. These relaxation modes are generated at each interface by the discontinuity of geometric and physical parameters. We shall describe them for a four layered Earth model with an elastic lithosphere, a viscoelastic upper and lower mantle, and an inviscid fluid core. The parameters of this model are given in the Table II. We have various relaxation modes:

— due to a density jump:

M_o, due to the density contrast at r=a

TABLE III
Relaxation modes associated with the viscoelastic Love numbers for the Earth's model defined in Table II.

	Relaxation modes in s^{-1}	Relaxation times in 10^3 years
T_1	0.8997×10^{-10}	0.35
T_2	0.8429×10^{-10}	0.38
M_o	0.1839×10^{-10}	1.7
C	0.4849×10^{-11}	6.5
L	0.1682×10^{-11}	18.8
M_1	0.2521×10^{-13}	1257

M_1, due to the density jump between the upper and the lower mantle

C, because of the density discontinuity between the mantle and the fluid core;

— due to a discontinuity in rigidity or in viscosity:

L, due to the viscosity contrast between the elastic lithosphere and the viscoelastic mantle

T_1, and T_2, due to a Maxwellian time jump between the upper and the lower mantle, and called transition modes (Peltier, 1974), because they relax rapidly and are weakly excited.

The values of these relaxation modes are given in Table III.

We show that the relaxation times are about some thousands years, excepted for the M_1 mode which is about one million years and which may create viscoelastic deformations resulting from events in the past and not completely relaxed at the present time.

The relaxation modes are given by the roots of the characteristic determinant. To compute the viscoelastic Love numbers, we have to inverse the system. For example, we present the results for a surface load or internal loads located at the core-mantle boundary and at 670 km depth between the upper and lower mantle.

Peltier (1974) or Wu and Peltier (1982), have shown that these Love numbers can be expressed as following (for example, $k(r,\lambda)$ and $h(r,\lambda)$):

$$k(r,\lambda) = k^E(r) + \sum_{i=1}^{M} \frac{r_i}{i\lambda + 1/\tau_i} \tag{3.1}$$

$$h(r,\lambda) = h^E(r) + \sum_{i=1}^{M} \frac{s_i}{i\lambda + 1/\tau_i} \tag{3.2}$$

where $1/\tau_i$ are the M relaxation modes, relative to the model of the planet, and $k^E(r)$ and $h^E(r)$ the elastic Love numbers. We emphasize that the $k^E(r)$ and $h^E(r)$ are independent on the frequency λ, and are consequently a Dirac function in the temporal domain: the elastic response in deformation is instantaneous.

The coefficients r_i and s_i are dependent on the excitation sources.

We compute now these Love numbers when the Earth is submitted to internal loads, located at the core-mantle interface, as well as at the 670 km depth discontinuity between the upper and the lower mantle, and at the Earth's surface.

We suppose a surficial mass density σ at the radius r_o. It creates two effects:

— a pressure acting at $r = r_o$:

$$P(t) = -\sigma g(r_o)$$

— an attraction potential acting in the whole Earth:

$$V(t) = \frac{4\pi G}{5}\sigma r_o \begin{pmatrix} (\frac{r_o}{r})^3 \ if \ r \geq r_o \\ (\frac{r}{r_o})^2 \ if \ r \leq r_o \end{pmatrix} \tag{3.3}$$

The mass redistribution potential $\chi(a)$ and the radial displacement $u_r(a)$ at the Earth's surface can be written as a function of these combined sources, using the loading Love numbers :

$$\chi(a,t) = k(a,t) * V(t) + \bar{k}(a,t) * \frac{P(t)}{\rho} = k'(a,t) * V(t)$$

$$u_r(a,t) = h(a,t) * \frac{V(t)}{g} + \bar{h}(a,t) * \frac{P(t)}{\rho g} = h'(a,t) * \frac{V(t)}{g} \tag{3.4}$$

where $*$ denotes the temporal convolution, and ρ is the mean density of the Earth and g the gravity at the surface. We have noted $k'(a,\lambda)$ and $h'(a,\lambda)$ the loading Love numbers.

For a constant internal load, which provides $\sigma(t) = \sigma_o H(t)$ or $V(t) = V_o H(t)$, we have, in the temporal domain:

$$\chi(a,t) = \left(k'^f - r_i \tau_i e^{-\frac{t}{\tau_i}}\right) \quad V_o H(t)$$

$$u_r(a,t) = \left(h'^f - s_i \tau_i e^{-\frac{t}{\tau_i}}\right) \quad \frac{V_o H(t)}{g} \tag{3.5}$$

where

$$k'^f = k'^E + \sum_{i=1}^{M} r_i \tau_i \quad \text{and} \quad h'^f = h'^E + \sum_{i=1}^{M} s_i \tau_i \tag{3.6}$$

We have represented in the figures 2-a, 2-b and 2-c respectively, the percentage of excitation of each relaxation modes (i.e. the ratio $\frac{r_i \tau_i}{\sum_{j=1}^{M} r_j \tau_j}$) when the internal load is located at the Earth's surface ($r_o = a$), at 670 km depth ($r_o = 5701$km) and at the core-mantle boundary ($r_o = 3480$km).

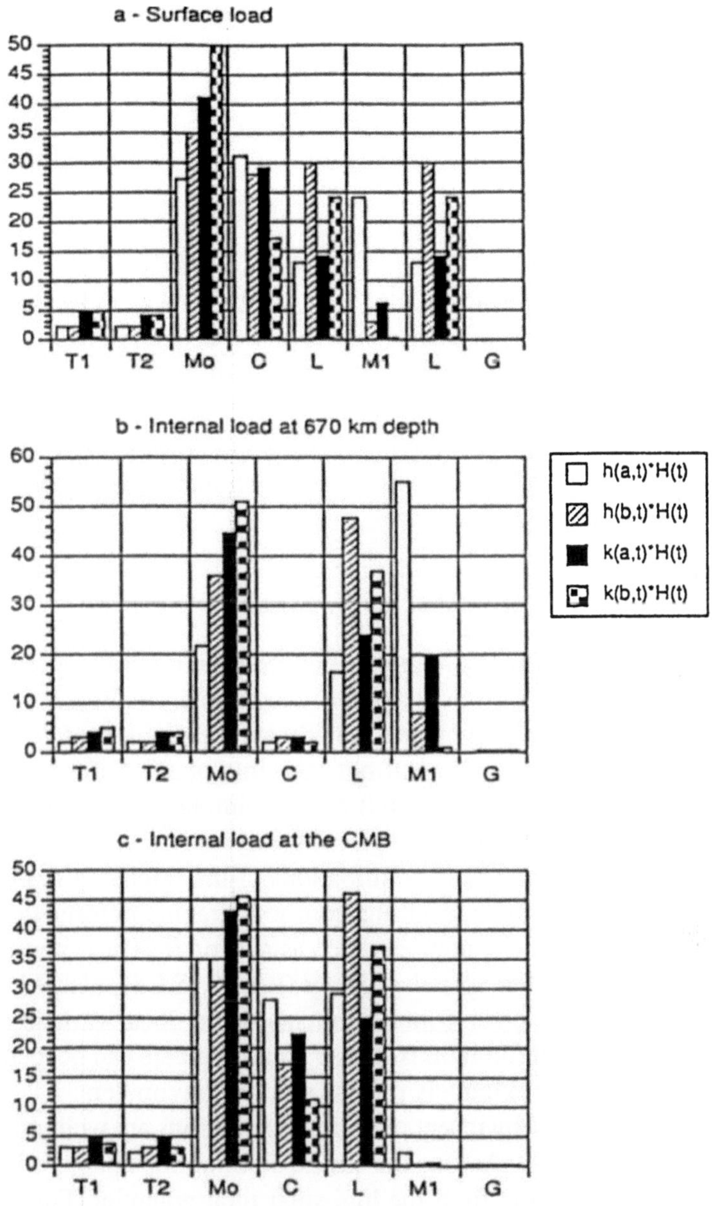

Fig. 2. The percentages of excitation of the relaxation modes $r_i \tau_i / \sum_{j=1}^{M} r_j \tau_j$ are presented in these charts for various excitation sources: (a) a surface load, (b) an internal load at 670 km depth, (c) an internal load at the CMB. The vertical axis indicates the respective percentages of each relaxation mode written in the horizontal axis, for various Love numbers: h(a,t) and h(b,t) proportional to the radial displacement at the surface and at the CMB (r=b), and k(a,t) and k(b,t) proportional to the mass redistribution potential at r=a and r=b.

From the elasto-gravitational set of equations, for an internal load located at an interface with a density jump, the following relations hold, in the fluid limit :

$$k'^f(a) = -(\frac{r_o}{a})^3 \quad \delta(t) \tag{3.7}$$

$$h'^f(a) = \begin{pmatrix} 0, & if \quad r_o \neq a \\ -\frac{5}{3}\frac{\rho_1}{\rho} \quad \delta(t), & if \quad r_o = a \end{pmatrix} \tag{3.8}$$

If the load is located within the Earth, at r_o depth (where there is a density discontinuity), there is a radial displacement (that is a deflection of the interface where the load is located) which exactly compensates the mass density (isostasy), so that the displacement at the Earth surface is zero.

This last result concerning the fluid limits of the viscoelastic Love numbers is very interesting because it explains why after sufficient time has elapsed ($t >> \tau_i$), one can reach the isostatic equilibrium.

3.2. Deformations and Geodynamical Problem

We have investigated the Earth deformations induced by surface and internal loads and their temporal evolution as a function of the relaxation times. Before presenting an application to some known phenomena, we will put the problem of deformation within a geodynamical context. We will not take into account the tectonic and morphologic problems which do not result from a simple modelisation, and we will just study some specific problems of global geodynamics.

The first problem is related to the perturbations of the reference frame by the variation of the stations positions resulting from topographic deformation of the Earth. We have to know the Love numbers such as h, l, or H. These perturbations are essentially observed by the spatial geodesy (L.L.R., S.L.R, V.L.B.I. ...) and may also be correlated with precise measurements of the local variations of the gravity associated with the deformations. These measurements can be recorded on the global Earth's surface by extremely precise relative gravimeters.

The deformations induced by recent or ancient excitations are well modeled by the viscoelastic Love numbers. Among the elastic deformations, there is evidently the Earth deformation resulting from the luni-solar tidal potential (Fig. 3) which produces a height variation of about 30 cm per day. Another elastic deformation is the one due to the atmospheric loading effect (Fig. 4). We have computed the local deformation induced by a global pressure expanded in spherical harmonics (monthly means of ECMWF surface pressure data have been kindly provided to us by the GSFC). The amplitude of the displacement is of about some millimeters. This is a preliminary study in order to give the order of magnitude of this effect. The post glacial rebound of the Pleistocene deglaciation is a well-known result of an event in the past which perturbs presently the Earth, but some internal loads, at

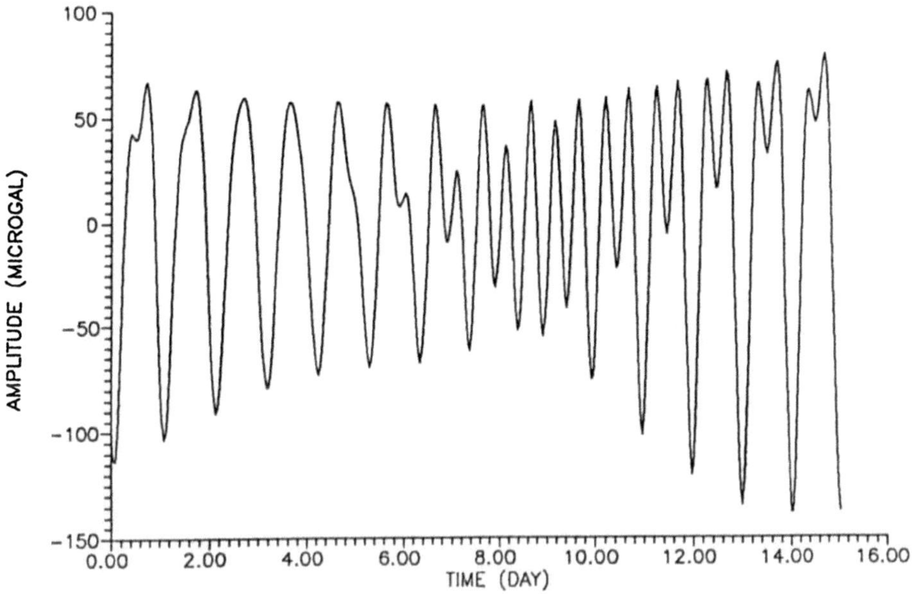

Fig. 3. Gravity variation induced by terrestrial tides during 15 days. $100\mu\text{gal} = 10^{-6}\text{m/s}^2 \longrightarrow 17\text{cm}$

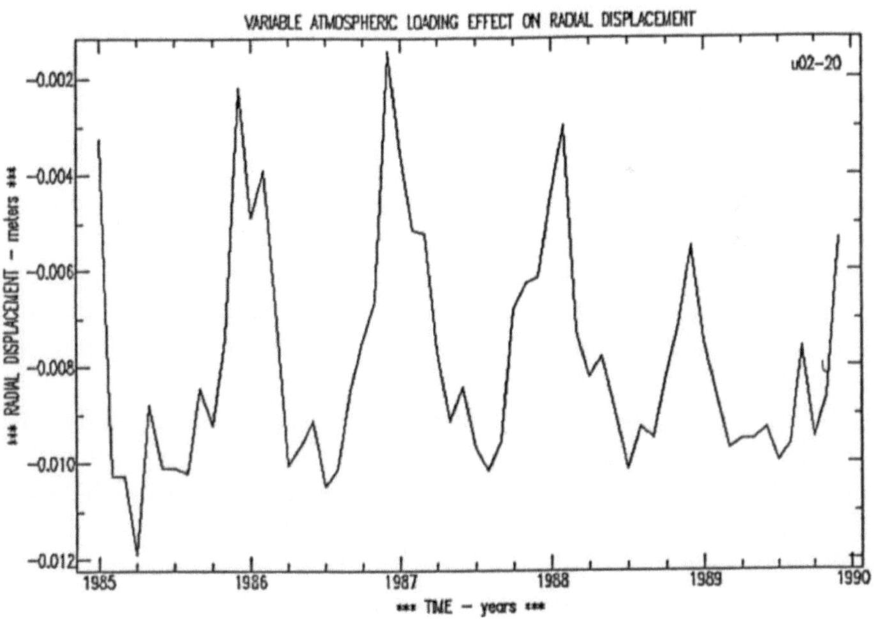

Fig. 4. Radial displacement at Strasbourg (France) induced by the global atmospheric pressure expanded in spherical harmonics until the degree 20 (Legros, Lefftz, Hinderer, Gegout and Florsch, 1992).

the core-mantle boundary or within the mantle, could also produce some surface variations (cf section 3-1, Lefftz and Legros, 1992-a, 1992-b).

Another interesting geodynamical problem is the one of the interpretation of the geoid anomalies and of the secular temporal evolution of the geopotential coefficients. To really understand this problem and to relate the theoretical results to the observations, it would be necessary to take into account the various relaxation processes involved at this time-scale: the long period 18,6 years tide, the evolution of the atmospheric and oceanic pressure at long time-scale and the post-glacial rebound. The Figure 5 shows the variation of the time derivative of the J_2 coefficient, the radial displacement at the surface and at the core-mantle boundary due to the post glacial rebound.

A third geodynamical application of the previous equations is directly related to the plate tectonic. It is the study of Ricard and Vigny (1989). For a newtonian model of rheology, from the internal mass density distribution given by the seismological tomography, they compute, on one hand, the geoid, and on the other hand, the plate motion, following the present geometry. This motion is induced by the tangential stresses under the plates resulting from the internal mass distribution. Figures 6 and 7 show the correlation between the observed and computed geoid and between the observed and computed plate motion.

Another application of the relaxation modes and their excitation by internal loads may be applied to the study of the figure of the Moon (Lefftz and Legros, 1993). For a realistic model of the interior of the Moon, we can show that the relaxation modes excited by the lunar rotational potential and by the tidal potential due to the action of the Earth on the Moon when the distance between these two planets was about 25 terrestrial radii, have to be presently relaxed and consequently can not explain the observed non-hydrostatic form of the Moon. However, a surface load, in combination with an internal load, located at 500 km depth (between the upper and the lower part of the lunar lithosphere) and resulting from the differentiation and the cooling of the magma ocean, produces some viscoelastic deformations in the selenoid and in the topography which are not completely relaxed at the present time and which could explain simultaneously the observed degree 2 terms in the selenoid and in the topography.

Finally, a last application of the temporal evolution of the deformation induced by a surface load may be done to the isostatic equilibrium which is observed on the telluric planets as Mars and Earth. If we consider that the crust of the planet is formed either with anorthositic primary crust or with light continental tertiary crust and with dense oceanic crust, after the stress relaxation described by the relaxation modes, it will be logical, independently on tectonic process, that the high reliefs are formed with light crust and that the low reliefs are formed with a more dense basaltic crust. It is a simple application of the fluid Love numbers given by the formula (3.8).

These various applications show the diversity of the geodynamical phenomena which may be explained with the proposed models.

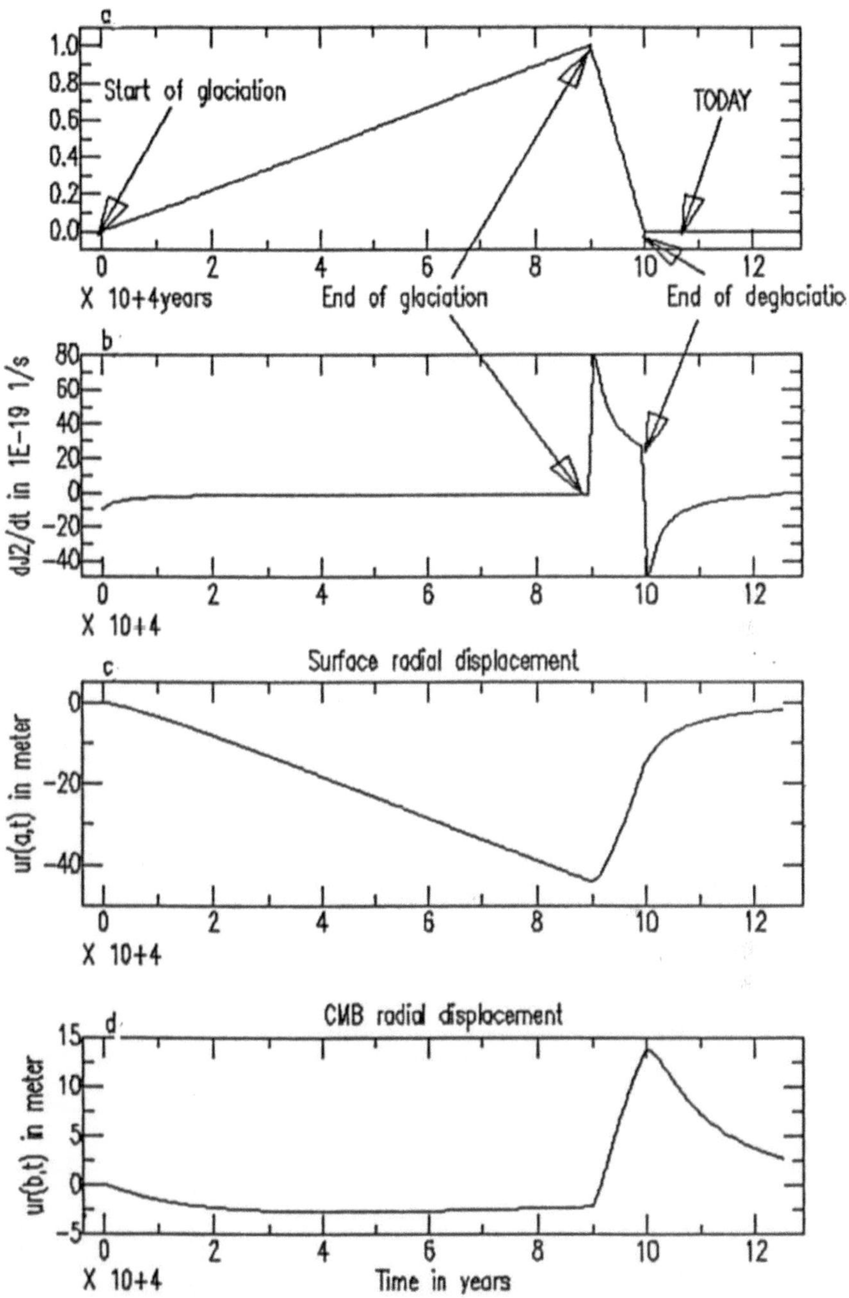

Fig. 5. a) Temporal evolution of the ice sheet. The glaciation is characterised by a slow build up over a period of about 90000 years, and the deglaciation by a fast collaps lasting 10000 years. The present time is almost 7000 years after the end of the deglaciation. b) Temporal variation of the $\dot{J}_2$ coefficient induced by the glaciation - deglaciation model described in 5-a, for the 4-layered Earth model given in Table II. The amplitude of the zonal loading potential is $V'_{20} = 235\text{m}^2/\text{s}^2$. c) Radial displacement at the Earth surface. d) Radial displacement at the core-mantle boundary.

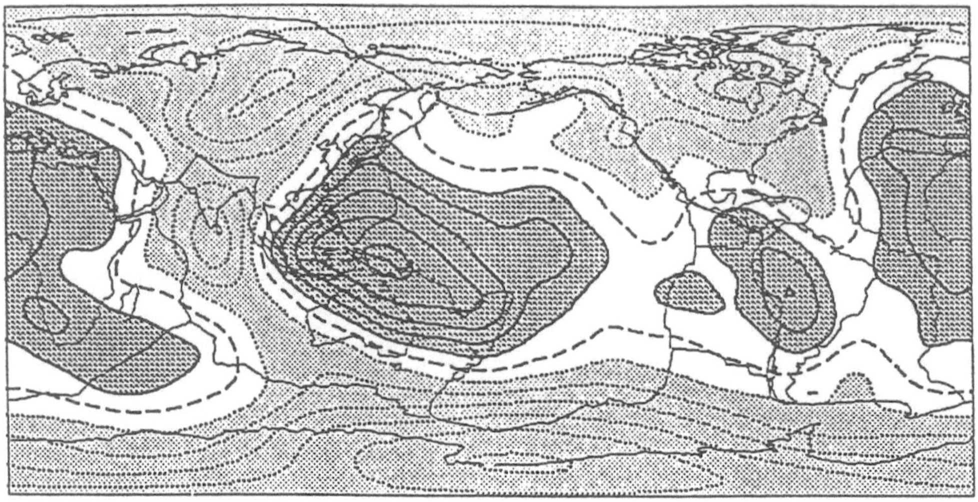

a

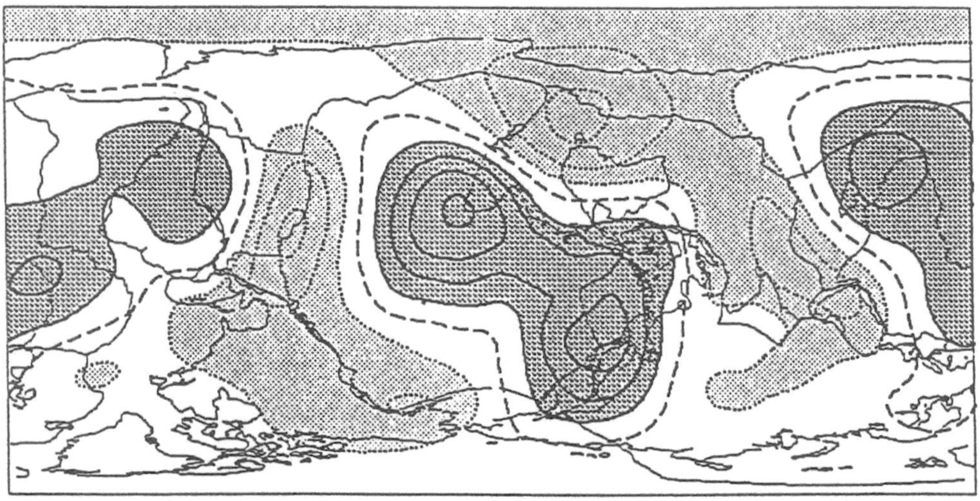

b

Fig. 6. a) Observed geoid. b) Geoid computed from an internal mass density distribution given by seismology, after Ricard and Vigny (1989).

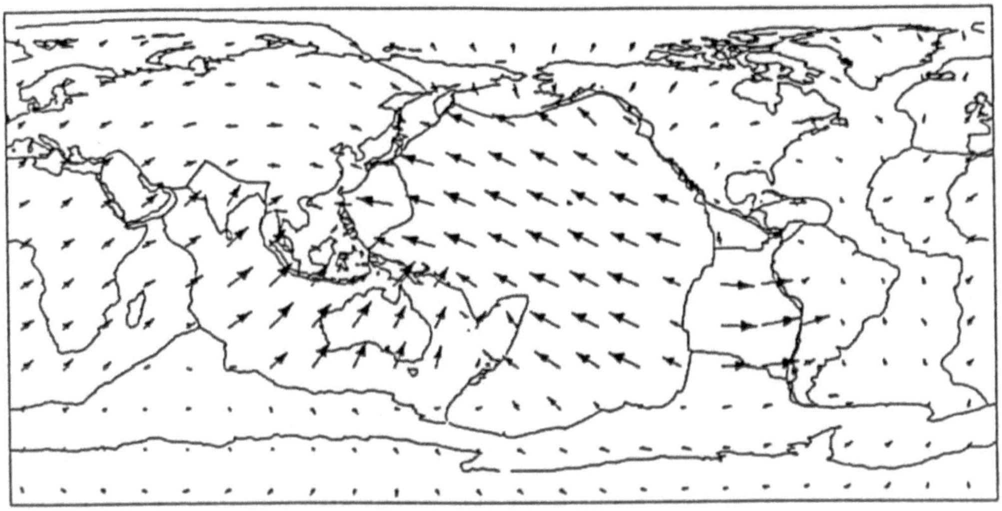

a

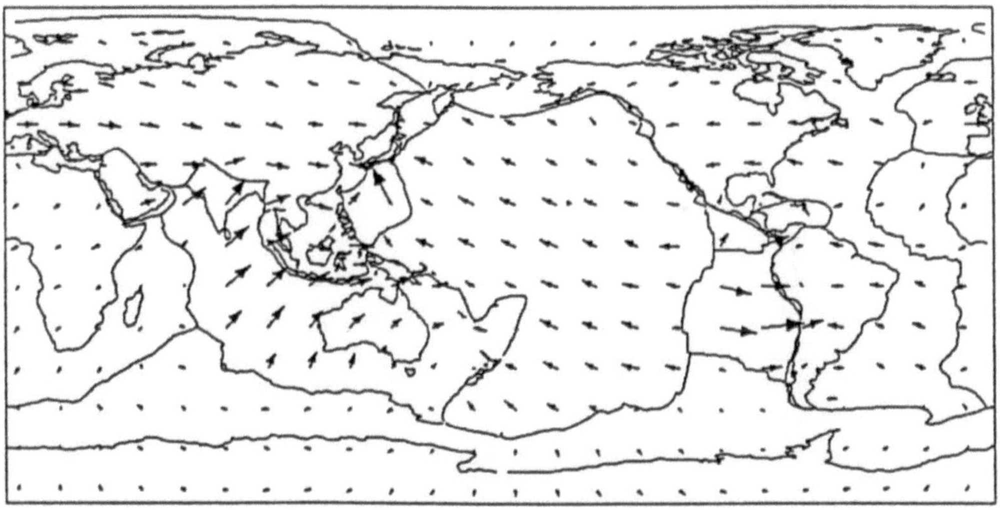

b

Fig. 7. a) Observed plate motion. b) Plate motion computed from the surface stresses induced by the internal density given by seismology, after Ricard and Vigny (1989).

3.3. Deformations and Rotation of Planets

Finally, we describe, in a more detailed way, another application of the deformation theory: its influence on the rotation of planets. We will not do a general study (e.g. Sabadini and Peltier, 1981; Lefftz, Legros and Hinderer, 1991; Spada et al., 1992) but we propose to investigate some interesting problems.

We are hereafter interested with the rotation of a quasi-spherical planet in a frame of axes related to the mantle. The angular momentum theorem is written:

$$\frac{d}{dt}(C_{ij}\omega_j) + \epsilon_{ijk} \quad \omega_j C_{kl}\omega_l = L_i \tag{3.9}$$

C_{ij} is the inertia tensor, ω_j the instantaneous rotation vector, L_i the torque of external forces which will be supposed equal to zero for the following applications.

For a reference frame initially along the axes of the principal moments of inertia, the C_{ij} tensor can be written as the sum of a diagonal matrix containing the principal moments of inertia on its diagonal plus a perturbation matrix.

From the Mc Cullagh formula, the perturbation of the inertia tensor from a static state may be computed from the deformations using a k Love number (identification between the potential of a quasi-spherical body with the potential of the deformed body given by $y_5(a)$).

In the frequency domain, the perturbations C_{13} and C_{33} , for example, may be written:

$$C_{13}(\lambda) = -\frac{3\alpha A}{\Omega^2 a^2 k_f}\left(k(\lambda)V_{21}(\lambda) + k'(\lambda)V'_{21}(\lambda) + ...\right) \tag{3.10}$$

$$C_{33}(\lambda) = -\frac{2\alpha A}{\Omega^2 a^2 k_f}\left(k(\lambda)V_{20}(\lambda) + k'(\lambda)V'_{20}(\lambda) + ...\right) \tag{3.11}$$

V_{21} and V_{20} are the coefficients of the spherical functions R_{21} and R_{20} in the expansion of the potential; α is the flattening of the Earth, A the moment of inertia, and Ω the mean velocity.

Using the volume Love number $k(\lambda)$, the C_{ij} of the equation (3.10) and (3.11) are function of the frequency.

For a planet model with a fluid core bounded by an ellipsoidal surface, it is necessary to introduce a vector of differential rotation of this core with respect to the mantle and to write the conservation of the angular momentum of the core. This equation is related to the first one (3.9) by torques of interaction at the core-mantle boundary. We will give an example of this problem.

The equation (3.9) is essentially interesting for the viscoelastic or viscous bodies. As a matter of fact, if the body is solid, we obtain the very classical Euler equation and if the body is an inviscid fluid the shape of the body is an equipotential (the problem is then a celestial mechanics one for equilibrium figure of rotating body).

The equation (3.9) is not linear because the inertia perturbations C_{ij} depend also on the rotation ω_j. However, we consider two possible simplifications of these equations: on one hand, one can consider that the planet is quasi-fluid, and on the other hand one can consider a linear approximation where the inertia tensor perturbations are small with respect to the principal moments of inertia [A,A and C] i.e., $C_{ij} << A$ and when the perturbations in rotation are small with respect to the mean axial rotation i.e., $\omega_j << \Omega$. First, we analyze this last case.

In the case of a linear approximation, the equation (3.9) may be separated into two independent systems: the axial equation in ω_3 and C_{33} and the equatorial system in ω_1, ω_2, C_{13} and C_{23}. The perturbations being functions of λ and of the relaxation modes, the linearized systems are also a function of λ; thus there is a coupling between the relaxation modes (purely imaginary modes) and the rotational eigenfrequencies (real frequencies for an elastic Earth). One can also conclude the same remarks for the equation associated to the fluid core. We obtain then a first interesting result: the rotational eigenfrequencies are complex for a viscoelastic Earth. For example, we write the axial system, which is more simple for the Earth and the core, the core being coupled with the mantle by a visco-magnetic torque Γ_3^c (e.g. Rochester et al., 1975) proportional to the differential rotation of the core (noted ω_3^c) with respect to the mantle ($\Gamma_3^c = -K''C^c\Omega\omega_3^c$). This system may be written, without external or internal torque: (3.12)

for the Earth $\quad C\dot{\omega}_3 + \Omega\dot{C}_{33} + C^c\dot{\omega}_3^c = 0$

for the core $\quad C^c\left(\dot{\omega}_3 + \dot{\omega}_3^c\right) + \Omega\dot{C}_{33}^c + K''C^c\Omega\omega_3^c = 0$

C^c is the principal moment of inertia along the 3-axis of the fluid core and C_{33}^c is the perturbation of the inertia tensor of the fluid core.

If we chose a simple Earth model with a viscoelastic homogeneous mantle and an inviscid fluid core, the system has 3 rotational eigenfrequencies: the first one is related to the frictional coefficient K'' and the two other one are, in the first order of approximation, the relaxation modes associated with the viscoelastic Love numbers. Thus, if this model is excited by a surface load as the Pleistocene deglaciation (Fig. 8), the responses of the Earth and its core have a temporal variation which is a combination of these eigenfrequencies. The evolutions of these rotations are shown in the Figure 8. Notice that, at the present time, that is almost 7000 years after the end of the deglaciation, there is a secular acceleration of the rotation velocity and that, for some realistic value of the frictional coefficient K" (of about 10^{-7}), the core has a negative rotation with respect to the mantle. There is consequently a westward drift of the core with respect to the mantle which may be correlated with the observed westward drift of the geomagnetic field of the Earth (the core containing the magnetic sources) (Lefftz and Legros, 1992-b).

Another application of modelling with Love numbers computed for a mantle viscosity associated to a viscoelastic model of rheology is the computation of a possible secular shift of the polar motion with respect to the figure axis (e.g. Sabadini and Peltier, 1981).

To end these remarks about the influence of the viscoelastic defomations on the rotation, we present some results for the case of a quasi-fluid approximation

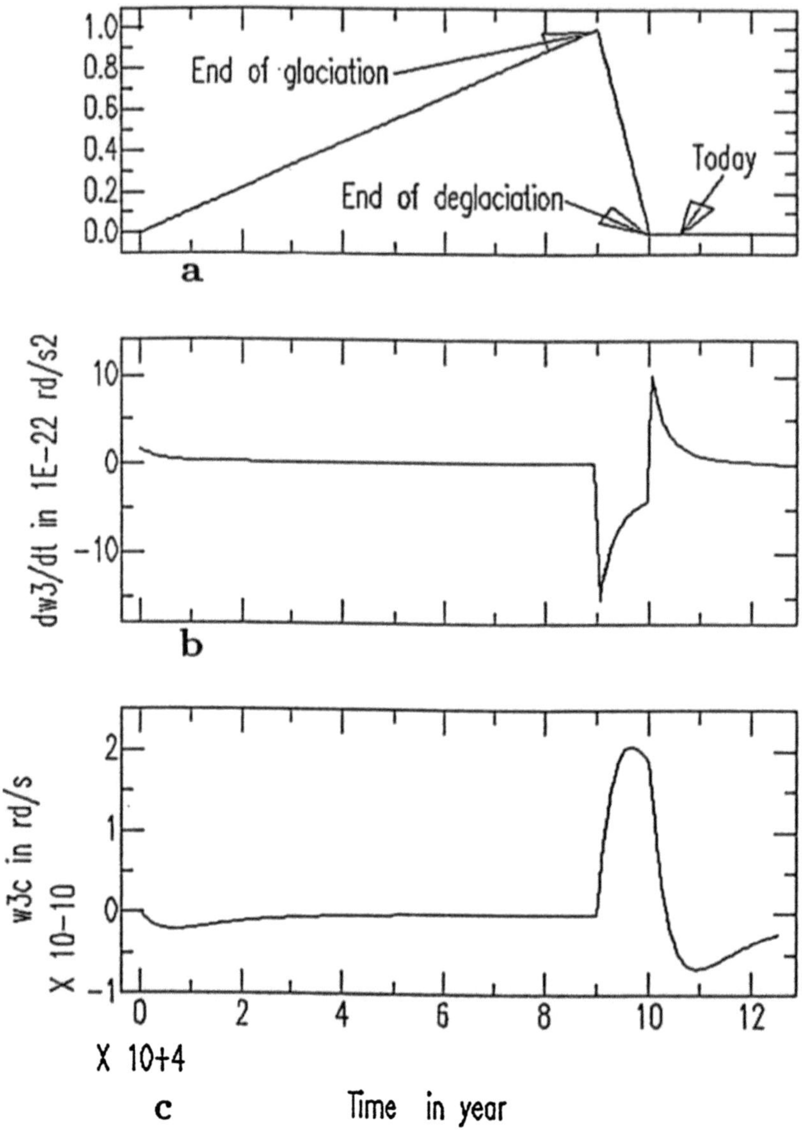

Fig. 8. a) The same as in Figure 5-a. b) Time derivative of the axial rotation of the Earth due to the glaciation-deglaciation model described in 5-a, for the 4-layered model of the Earth given in Table II, and for a frictional coefficient $K" = 10^{-7}$. The amplitude of the zonal loading potential is $V'_{20} = 235\mathrm{m}^2/\mathrm{s}^2$. c) Axial differential rotation of the core with respect to the mantle due to the last glaciation-deglaciation.

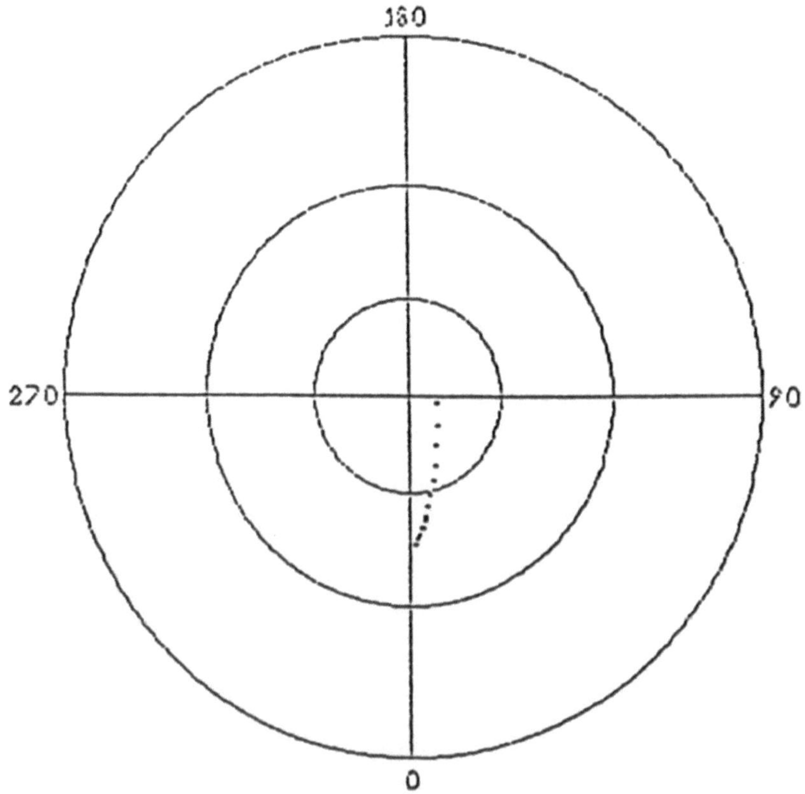

Fig. 9. Shift of the polar motion when the homogeneous planet is perturbed by a constant inertia product $C_{13} > 0$. The rotation axis is initially in $\theta_o = 10\,°$and $\lambda_o = 80\,°$. It tends to the limit $\theta = 45\,°$and $\lambda = 0\,°$, direction of the principal moment inertia axis. The temporal evolution of the rotation axis is represented on a stereographic projection: the time elapsed between two crosses is 20τ where τ is the Maxwell time of the planet. The numerical application is done for an homogeneous Moon model of planet ($C_{13}/A = 3 \times 10^{-6}$; $\bar{\mu} = 48$; $\alpha = 9.7 \times 10^{-6}$ for the Moon).

in equation (3.9). We will take a simple example where the excitation source is a constant at long time-scale, inducing tesseral inertia perturbation C_{13} and where the planet is homogeneous. The time derivative of the angular momentum is equal to zero in a celestial reference frame and thus the angular momentum is constant. As ω varies in time with respect to the reference frame, the planet moves with respect to the celestial frame (Fig. 9).

If we keep only the important terms in a quasi-fluid approximation, the equation (3.9) becomes:

$$\epsilon_{ijk}(-\beta\dot{\omega}_k\omega_j + C_{kl}\omega_l\omega_j) = 0 \tag{3.13}$$

where β is a coefficient related to the planet model; in the homogeneous case

$\beta = \alpha A\bar{\mu}\tau$ where $\bar{\mu}$ is related to elasticity and τ is the ratio of the viscosity over the rigidity.

From an initial position, noted θ_o and λ_o the evolution of the rotation axis is given in terms of latitude θ and longitude λ by :

$$\tan\lambda = \frac{\beta \exp(\frac{C_{13}}{\beta}t)}{2A\exp(2\frac{C_{13}}{\beta}t) - 1} \tag{3.14}$$

$$\tan\theta = \left(\frac{2A\exp(2\frac{C_{13}}{\beta}t) - 1}{2A\exp(2\frac{C_{13}}{\beta}t) + 1}\right)\left(1 + \frac{\beta\exp(\frac{C_{13}}{\beta}t)}{2A\exp(2\frac{C_{13}}{\beta}t) - 1}\right)^{\frac{1}{2}}$$

$$\text{with} \quad 2A = \frac{\tan\theta_o\cos\theta_o + 1}{1 - \tan\theta_o\cos\theta_o} \quad ; \quad B = \frac{2\tan\theta_o\sin\theta_o}{1 - \tan\theta_o\cos\theta_o}$$

The evolution of the position of the axis of rotation is shown in Figure 9.

4. Conclusions and Discussion

In this paper we have developed a formalism which is able to model most of the phenomena in geodynamics and planetology (excepted for the convection) and which allows to interpret the temporal variations of the topography and the geoid at different time-scales. We have shown the interest and the effects of these models on some examples of deformations and on the rotation of the planets.

We have noted that these modelings are in fact formally very similar and we have kept the same variables y_i to represent the displacements or the velocities. We have demonstrated that, in these circumstances, the generalized Love numbers are very useful to study the dynamics of the planets. These numbers are generalized for various excitation sources, at the surface or within the planet and also generalized for various models of rheology. They have been computed from the propagator of the y_i system, and they depend on the relaxation modes related to the rheology; we have investigated the excitation of these last ones for various sources.

For these cases, the main problems seem to be the following ones:

Which criteria may be used to chose a model of rheology ? If simplicity is an a-priori reasonable choice, it is perhaps not sufficient. It seems to be useful to have a rheology which is able to describe phenomena at various time-scales, for example, the elastic, viscous fluid and inviscid fluid for the Maxwell model of rheology.

How to constrain the parameters such as the shear modulii and the viscosities which appear in the rheologic law ? From the laboratory studies in adequate thermodynamic conditions and/or by inversion of the solutions from surface observations ? The more complex is the rheology the more large is the number of parameters and thus the less they are constrained.

The present modelings are quasi-static. Although some authors define current lines for the velocity field within the mantle, it is necessary to say that in opposite to the stationary convective motion, each particle does not vary very much (with a maximal order of magnitude of some kilometers) : it goes from its initial state to the final one which is reached at the end of the relaxation.

In order to have a complete description of the motion, it is obviously necessary to use the convection equations. But, nevertheless, the viscoelastic or newtonian modeling allows us to explain some dynamical phenomena, and it is sufficiently general to study simultaneously the motions of the stations at the Earth's surface and the orbitographic perturbations of the satellites.

Acknowledgements

We thank Veronique Dehant for important suggestions and review of this paper, and Pascal Gegout for computation of the radial displacement in Strasbourg induced by the global atmospheric pressure. This research has been supported by CNRS-INSU-DBT grant (contribution number 574).

References

Alterman, Z., Jarosch, H. & Pekeris, C.H.: 1959, 'Oscillation of the Earth', *Proc. R. Soc. London*, **A 252**, 80–95.

Bertovici, D., Schubert, G. & Glatzmaier, G.A.: 1989, 'Three dimensional spherical models of convection in the Earth's mantle', *Science*, **244**, 893–1016.

Clairaut, A.C.: 1743, *La theorie de la figure de la terre, tirée des principes de l'hydrostatique*, 2nde édition, Courcier, Imprimeur-Libraire, Paris.

Darwin, G.H.: 1881, 'On the Stresses caused in the Interior of the Earth by the Weight of Continents', *Phil. Trans. R.A.S.*, **173**,187–230.

Darwin, G.H.: 1878, 'On the Bodily Tides of Viscous and Semi-elastic Spheroids, and on Ocean Tides upon a Yielding Nucleus', *Phil. Trans. R.A.S.*, **170**, 1–35.

Darwin, G.H.: 1878, 'On the Precession of a Viscous Spheroid, and on the remote History of the Earth', *Phil. Trans. R.A.S.*, **170**, 447–538.

Darwin, G.H.: 1878, 'Problems connected with the Tides of a Viscous Spheroid', *Phil. Trans. R.A.S.*, **170**, 539-593.

Dutton, C.E.: 1889, 'On some of the greater problems of physical geology', *Bulletin of the Philosophical Society of Washington*, **11**, 51–64.

Haskell, N.A.: 1935, 'The motion of a viscous fluid under a surface load',*Physics*, **6**, 265.

Hopkins, W.: 1839, 'Researches in Physical Geology', *Phil. Trans. R.A.S.*, **129**, 381-425.

Jeffreys, H.: 1926, 'The rigidity of the Earth's Central Core', *M.N.R.A.S. Geophys. Suppl.*, 371–383.

Lee, E.H., Radok, J.R.M. & Woodward, W.B.: 1959, 'Stress analysis for linear viscoelasticity materials', *Transactions of the Society of Rheology III*, 41–59.

Lefftz, M., Legros, H. & Linderer, J.: 1991, 'Non-linear equations for the rotation of a viscoelastic planet taking into account the influence of a liquid core', *Celest. Mech.*, **52**, 14-43.

Lefftz, M. & Legros, H.: 1992a, 'Some remarks about the rotations of a viscous planet and its homogeneous liquid core: linear theory', *Geophys. J. Int.*, **108**, 705-724.

Lefftz, M. & Legros, H.: 1992b, 'Influence of viscoelastic coupling on the axial rotation of the Earth and its fluid core', *Geophys. J. Int.*, **108**, 725-739.

Lefftz, M. & Legros, H.: 1993, 'Some remarks about the non-hydrostatic form of the Moon', *P.E.P.I.*, in press.

Legros, H.: 1987, 'Sur quelques problèmes de dynamique planétaire', *Thesis, Université Strasbourg I*, France, 229 pp.

Legros, H., Lefftz, M., Hinderer, J., Gegout, P. & Florsh, N.: 1992, 'Quelques exemples de la déformation de la Terre à partir des nombres de Love',. *Actes des Journées 1992, Systèmes de référence spatio-temporels*. 97–101.

Love, A.E.H.: 1909, 'The Yielding of the Earth to the Disturbing Forces', *Proc. Roy. Soc. Lond.*, Ser. A, LXXXII, 73–88.

Machetel, P. & Weber, P.: 1991, 'Intermittent layered convection in a model mantle with an endothermic phase change at 60 km', *Nature*, **350**, 55–57.

Mackenzie, D.P.: 1968, 'The influence of the boundary conditions and rotation on convection in the Earth's mantle', *Geophys. J. R. astr. Soc.*, **15**, 457–500.

Pekeris, C.L.: 1935, 'Thermal convection in the interior of the Earth', *M.N.R.A.S. Geophys. Suppl.*, 343–367.

Peltier, W.R.: 1974, 'The impulse response of a Maxwell Earth', *Rev. Geophys.*, **12**, 649–669.

Poincaré, H.: 1885, 'Sur l'équilibre d'une masse fluide animée d'un mouvement de rotation', *Acta Mathematica*, 259–380.

Poincaré, H.: 1902, 'Figure d'équilibre d'une masse fluide', Paris, *C. Naud editeur*.

Poincaré, H.: 1910, 'Sur la précession des corps déformables', *Bull. Astr.*, **27**, 321–356.

Ricard, Y., Fleitout, L. & Froidevaux, C.: 1984, 'Geoid heights and lithospheric stresses for a dynamic Earth', *Ann. Geophys.*, **2**, 267–286.

Ricard, Y. & Vigny, C.: 1989,' Mantle dynamics with induced plate tectonics', *JGR*, **94**, B12, 17543–17559.

Richards, M.A. & Hager, B.: 1984, 'Geoid anomaly in a dynamic Earth', *J.G.R.*, **89**, 5987–6002.

Rochester, M.G., Jacobs, J.A., Smylie, D.E. & Chong, K.F.: 1975, 'Can precession power the geomagnetic dynamo ?', *Geophys. J. R. astr. Soc.* **43**, 661–678.

Sabadini, R. & Peltier, W.R.: 1981, 'Pleistocene deglaciation and the Earth's rotation: implications for mantle viscosity', *Geophys. J. R. astr. Soc.*, **66**, 553–578.

Schiaparelli, I.V.: 1889, *De la rotation de la terre sous l'influence des actions géologiques*, Saint-Petersbourg, Imprimerie de l'Académie impériale des Sciences, Vass. Ostr., 9^{ieme} ligne, 12.

Spada, G., Sabadini, R., Yuen, D.A. & Ricard, Y.: 1992, 'Effects on post-glacial rebound from the hard rheology in the transition zone', *Geophys. J. Int.*, **109**, 683–700.

Thomson, W. (Lord Kelvin): 1863, 'On the Rigidity of the Earth', *Phil.Trans. R.A.S.*, **153**, 573–582.

Thomson, W. (Lord Kelvin): 1863, 'Dynamical Problems regarding Elastic Spheroidal Schells and Spheroids of Incompressible Liquid', *Phil. Trans. R.A.S.*, **153**, 583–616.

Wiechert, E.: 1897, 'Uber die Massenverteilung im Innern der Erde', *Nachr. Ges. Wiss.Gottingen Math. physik. Kl.* 221–243.

Wu, P. & Peltier, W.R.: 1982, 'Glacial isostatic adjustment and the free air gravity anomaly as a constraint of deep mantle viscosity', *Geophys. J. R. astr. Soc.*, **74**, 377–449.

ON THE TIDAL VARIATION OF THE GEOPOTENTIAL

JOSÉ M. FERRÁNDIZ

Departamento de Matematica Aplicada a la Ingeniería.
E.T.S. de Ingenieros Industriales. 47011 Valladolid. Spain

and

JUAN GETINO

Departamento de Matemática Aplicada Fundamental.
Facultad de Ciencias. 47005 Valladolid. Spain

Abstract. In this paper analytical expressions are derived for the temporal variations of J_2 and J_{22} due to the tides of the solid Earth, taking into account only the deformation of the mantle, and employing a procedure already used by the authors in their Hamiltonian theory of the Earth's rotation, which obtain the necessary parameters in a direct way by integration of those provided by a selected model of Earth interior.

Numerical tables giving the periodic variation of coefficients are given, as well as a new prediction for $\Delta UT1$. For δJ_2 and δJ_{22} the amplitudes reach such a magnitude that both two variations should not be ignored in studies involving the analysis of highly precise satellite tracking data. Moreover, the possibility of improving our knowledge of the value of those harmonic coefficients in only a more exact digit appears as to be strongly dependent on the limitations in the theoretical modeling of the variations of the inertia tensor due to solid tides.

Key words: Solid tides – Earth potential – artificial satellites

1. Introduction

Over the last decade there has been a new interest in the analysis of the variation in time of the geopotential, due mainly to the availability of satellite tracking data with a high intrinsic accuracy, and to the need to improve the dynamic models used in their analysis. Of course, since the Earth is not a rigid solid, there is no reason to presuppose that the coefficients of the geopotential can be treated as constants when they reach certain levels of precision. In 1983, Bursa predicted secular variations of the J_2 due to the tides and whose values are consistent with the relative uncertainty which is usually connected to the said parameter. In the same year, Yoder et al. were successful in providing the first experimental evidence of temporal variations of the J_2 , using the analysis of LAGEOS tracking data, and particularly the motion of the node. The secular change detected of $-3 \times 10^{-11} yr^{-1}$ was explained through a phenomenon of viscous rebound following the last deglaciation, while the origin of the seasonal changes (mainly annual and semiannual) remained more uncertain. In 1986, Carpino et al. investigated the viability of detecting changes in the J_{22} coefficient, starting from the analysis of the orbit of LAGEOS, as well as that of achieving a relative accuracy of 10^{-4} or better in the value of this parameter. The difficulties of this problem are considerable, and more so considering that it is closely connected to other questions such as the motion of the pole or the variations in the angular velocity of the Earth. In any case, the improvement in the values of the different coefficients of the geopotential, including possible periodic variations,

Celestial Mechanics and Dynamical Astronomy **57**: 279–292, 1993.

is necessary if we wish to model the motion of satellites with truly centimetric or subcentimetric accuracy.

In this paper we concentrate on the problem of modeling the changes in the coefficients of the second order harmonics due to the tides of the solid Earth, taking into consideration only the deformation of the mantle. The concrete hypotheses are the same as those made by the authors in their recent Hamiltonian theory of the Earth's rotation $(1990, 1991a, 1991b, 1991c)$. The procedure followed is also similar: the variations of the coefficients are calculated starting from the integral which define them, supposing that the points of the mantle are deformed as described in the solution of Herglotz-Takeuchi. The analytical variations are obtained as a function of the variables which fix the position of the body that causes the tide (Moon or Sun) and of the numerical coefficients which depend on the model chosen for the internal constitution of the Earth. In this paper those coefficients have been computed using the rheological parameters from Earth Model 2 of Takeuchi (1950). In the next section, the tidal variations of inertia tensor, are included. These variations have been obtained using the same method and Earth model, since the effect of that change in the Earth angular velocity (or the universal time) cannot be forgotten.

In section 3 numerical expressions are given for the most important periodic variations of J_2 and J_{22} , whose order of magnitude corresponds exactly to the order of the absolute error usually admitted for the corresponding coefficient.

2. Tidal Variations of the Harmonic Coefficients of Second Order

Unlike the rigid solid, for which the positional vectors of its particles, $\mathbf{r}_0$, are constant for a reference connected to it, in the deformable body these vectors, $\mathbf{r}(t)$, will be functions of time. For our study it will be sufficient to assume that the particles of the deformable solid deviate slightly from the position they would occupy in the absence of deformation. The vectors $\mathbf{r}_0, \mathbf{r}(t)$, will be related to each other by the so called *displacement vector*, $\mathbf{u}(\mathbf{r}, t)$, in the following way

$$\mathbf{r}(t) = \mathbf{r}_0 + \mathbf{u}(\mathbf{r}_0, t) \rightarrow (x, y, z)(t) = (x_0, y_0, z_0) + (u, v, w)(x_0, y_0, z_0; t) \quad (1)$$

where (x, y, z) are the positional coordinates of the particle in the deformable body, and (x_0, y_0, z_0) those that the same particle would have in the absence of deformation, (u, v, w) being the components of the displacement vector. In this way it is possible to express the coordinates of the deformed body as a function of those corresponding to the undeformed one, that is to say, we can relate the deformed domain of dependence $\mathcal{V}$ as a function of the undeformed domain $\mathcal{V}_0$. As for the actual expression of the displacement vector, this will depend on each particular case of deformation being studied.

>From now on, we shall study the deformation produced in the Earth's mantle due to the gravitational attraction of the Moon and Sun, which is known as *tidal*

deformation. The potential cause for this deformation is (Kaula, 1964):

$$W = \frac{Gm^*}{4r^*} \sum_{n=2}^{\infty} \left(\frac{r}{r^*}\right)^n P_n(\cos S) , \tag{2}$$

where G is the gravitational constant, r the distance between the origin and the point within the Earth where the potential is evaluated, m^*, r^* are the mass and the distance to the perturbing body (Moon, Sun), and S is the angle between the vectors r and r^*. Since the factor r/r^* is of the order of $1/60$ for the Moon and $1/23000$ for the Sun, in our study we shall limit ourselves to considering the case $n = 2$.

Due to this potential, the elastic mantle of the Earth is deformed, and the corresponding displacement vector admits the expression (Takeuchi, 1950):

$$(u, v, w) = F_2(r_0)\frac{\partial W_2}{\partial(x, y, z)} + G_2(r_0)(x, y, z)W_2 . \tag{3}$$

On the other hand, in the process of deformation, the density will also suffer a modification, which, according to the mentioned work of Takeuchi, can be expressed as

$$\rho = \rho_0 - \mathcal{U}\frac{d\rho_0}{dr} - \rho_0\Delta , \tag{4}$$

where ρ_0 is the density corresponding to the undeformed state, $\mathcal{U} = (xu + yv + zw)/r$ being the radial displacement and $\Delta = \mathbf{div} \cdot \mathbf{u}$ the volume divergence.

2.1. Expression of J_2 for an Elastic Earth

Let us start from the well known expression of the term in the development of the geopotential:

$$J_2 = -\frac{1}{MR^2} \iiint \rho r^2 P_2(\cos\beta) d\sigma , \tag{5}$$

where M and R are the mass and the equatorial radius of the Earth and β the colatitude, the integral being extended to the whole volume. Introducing in the equivalent expression in Cartesian coordinates the relations (1) we have

$$J_2 = -\frac{1}{2MR^2} \iiint (\rho_0 + \delta\rho_0) \left[(2z_0^2 - x_0^2 - y_0^2) + 2(2z_0 w - x_0 u - y_0 v)\right] d\sigma , \tag{6}$$

where the products and squares of the components of the displacement vector have been neglected due to the hypothesis of linear elasticity with which we are working.

Now, this integral is extended to the whole volume of the deformed body, so it is convenient to transform the domain of integration to the undeformed body. Taking the relations (1) into account, the Jacobian of this transformation, neglecting the quadratic terms, will be in the form (Getino and Ferrándiz, 1990) $J \simeq 1 + \Delta$, with which we can break down the J_2 in the form

$$J_2 = J_{2)r} + \delta J_2 \,, \tag{7}$$

where

$$J_{2)r} = -\frac{1}{2MR^2} \iiint \rho_0 \left(2z_0^2 - x_0^2 - y_0^2\right) d\sigma_0 \,, \tag{8}$$

$$\delta J_2 = -\frac{1}{2MR^2} \iiint [2\rho_0 (2z_0 w - x_0 u - y_0 v) + + (\rho_0 \Delta + \delta\rho_0) \left(2z_0^2 - x_0^2 - y_0^2\right)\Big] d\sigma_0 \,, \tag{9}$$

where the integrals are extended to the volume of the Earth in the absence of deformation. It is clearly verifiable that the term $J_{2)r}$ is that corresponding to a rigid Earth, while δJ_2 is the increment due to the deformation.

In the system of principal axes, the unperturbed term is reduced to the well known expression for a rigid Earth

$$J_{2)r} = \frac{2C - A - B}{2MR^2} \,. \tag{10}$$

As for the perturbing term, in the first place we have, according to (3) and (4):

$$\rho_0 \Delta + \delta\rho_0 = -\mathcal{U}\frac{d\rho_0}{dr} = -\frac{d\rho_0}{dr}\left(\frac{2}{r}F_2(r) + rG_2(r)\right) W_2 \,, \tag{11}$$

with which, we can perform the integration. To do so we shall assume that the Earth be geometrically spherical in the undeformed state, that is, we shall neglect the Earth's ellipticity, to the order of *1/300* in the perturbing term. After some algebra we obtain:

$$\delta J_2 = -\frac{1}{2MR^2}\frac{Gm^*}{r^{*3}}\frac{4\pi}{5} 2P_2(\cos\beta^*)\, I \,, \tag{12}$$

where

$$I = \int_r \left[2\rho_0 r^4 \left(5F_2(r) + r^2 G_2(r)\right) - \frac{d\rho_0}{dr} r^5 \left(2F_2(r) + r^2 G_2(r)\right)\right] dr \,. \tag{13}$$

This integral is extended to the elastic mantle and depends on its internal structure, that is, on the chosen Earth model. In Getino and Ferrándiz $(1991a)$ its value is obtained for Model 2 of Takeuchi (1950), resulting $I = 1.917290 \times 10^{50}$ c.g.s.

It is convenient to express (12) as a function of the coefficient

$$D_t = \frac{Gm^*}{a^{*3}}\frac{2\pi}{15} I \,, \tag{14}$$

introduced by the authors (1991*b*), which expresses the order of magnitude of the tidal perturbation, where a^* is the major semi-axis of the perturbing body's orbit. Thus:

$$\delta J_2 = -\frac{12}{2MR^2} D_t \left(\frac{a^*}{r^*}\right)^3 P_2(\cos\beta^*) = -\frac{12}{2MR^2} D_t \left(\frac{a^*}{r^*}\right)^3 P_2(\sin\delta^*) \ , \quad (15)$$

where δ^* is the latitude of the perturbing body. With (10) and (15) we finally obtain the total expression of the term J_2 :

$$J_2 = \frac{1}{2MR^2} \left[(2C - A - B) - 12 D_t \left(\frac{a^*}{r^*}\right)^3 P_2(\sin\delta^*) \right] . \quad (16)$$

2.2. Expression of C_2^2 and S_2^2 for an Elastic Earth

The expression of the first coefficient is

$$C_2^2 = -\frac{1}{12MR^2} \iiint \rho r^2 P_2^2(\cos\beta) \cos 2\alpha d\sigma \ , \quad (17)$$

Following a process analogous to that described in the previous section, it can be broken down into a rigid part and perturbation for deformation as follows:

$$C_2^2 = C_{2)r}^2 + \delta C_2^2 \ , \quad (18)$$

with

$$C_{2)r}^2 = -\frac{1}{4MR^2} \iiint \rho_0 (x_0^2 - y_0^2) d\sigma_0 \ , \quad (19)$$

$$\delta C_2^2 = -\frac{1}{4MR^2} \iiint \left[2\rho_0 (x_0 u - y_0 v) + (\rho_0 \Delta + \delta\rho_0)(x_0^2 - y_0^2) \right] d\sigma_0 \ , \quad (20)$$

obtaining finally, in the system of principal axes:

$$C_2^2 = \frac{1}{4MR^2} \left[(A - B) - 2 D_t \left(\frac{a^*}{r^*}\right)^3 P_2^2(\sin\delta^*) \cos 2\alpha^* \right] . \quad (21)$$

As for the second coefficient, we have

$$S_2^2 = -\frac{1}{12MR^2} \iiint \rho r^2 P_2^2(\cos\beta) \sin 2\alpha d\sigma \ , \quad (22)$$

whose corresponding rigid and perturbed parts are

$$S_{2)r}^2 = -\frac{1}{2MR^2} \iiint \rho_0 x_0 y_0 d\sigma_0 \ , \quad (23)$$

$$\delta S_2^2 = -\frac{1}{2MR^2} \iiint \left[\rho_0 (x_0 v + y_0 u) + (\rho_0 \Delta + \delta\rho_0) x_0 y_0 \right] d\sigma_0 \ , \quad (24)$$

obtaining, with respect to the said axes:

$$S_2^2 = \frac{1}{2MR^2}\left[-D_t\left(\frac{a^*}{r^*}\right)^3 P_2^2(\sin\delta^*)\sin 2\alpha^*\right] . \tag{25}$$

For the rigid Earth it is customary to use the parameter $J_{22} = \sqrt{C_{22}^2 + S_{22}^2}$ that represents the amplitude of the perturbation parameter corresponding to the $(2,2)$ term in the potential expansion. In that sense, the periodic variations of C_{22} and S_{22} may be referred to as periodic variations of J_{22}, using a rough brief expression.

2.3. Expression of C_2^1 and S_2^1

Following a similar method, we obtain the following expressions for the perturbations of C_2^1 and S_2^1 :

$$C_2^1 = \frac{1}{MR^2}\left[-2D_t\left(\frac{a^*}{r^*}\right)^3 P_2^1(\sin\delta^*)\cos\alpha^*\right] , \tag{26}$$

$$S_2^1 = \frac{1}{MR^2}\left[-2D_t\left(\frac{a^*}{r^*}\right)^3 P_2^1(\sin\delta^*)\sin\alpha^*\right] . \tag{27}$$

3. Tidal Variation of the Tensor of Inertia

In the previous sections the coefficients of the geopotential have been calculated, starting from its own definition and introducing the displacement vector which marks the differences between a rigid and a deformable solid, to finish by performing the corresponding integrations.

The procedure is clearly valid for the harmonics of any order. On the other hand, in the case with which we are dealing, there exists a simpler form of obtaining these coefficients, using results already obtained by Getino and Ferrándiz (1990, 1991a, 1991b) in their theory of the rotation of an elastic Earth. As described in these papers, the effect of the tidal deformation is translated into a modification of the inertia tensor as follows:

$$\Pi = \Pi_0 + D_t\left(\frac{a^*}{r^*}\right)^3 \delta\Pi_0 , \tag{28}$$

where Π_0 is the tensor corresponding to the Earth in the absence of deformation, which, referred to its principal axes is the well known

$$\Pi_0 = \begin{pmatrix} A & 0 & 0 \\ 0 & B & 0 \\ 0 & 0 & C \end{pmatrix} , \tag{29}$$

while the perturbation by tidal effect is $D_t\,(a^*/r^*)^3$ times the matrix

$$\delta\Pi_0 = \begin{pmatrix} t_{11} & t_{12} & t_{13} \\ t_{12} & t_{22} & t_{23} \\ t_{13} & t_{23} & t_{33} \end{pmatrix} , \tag{30}$$

with

$$\begin{aligned} t_{11} &= 2P_2(\sin\delta^*) - P_2^2(\sin\delta^*)\cos 2\alpha^* , & t_{12} &= -P_2^2(\sin\delta^*)\sin 2\alpha^* , \\ t_{22} &= 2P_2(\sin\delta^*) + P_2^2(\sin\delta^*)\cos 2\alpha^* , & t_{13} &= -2P_2^1(\sin\delta^*)\cos\alpha^* , \\ t_{33} &= -4P_2(\sin\delta^*) , & t_{23} &= -2P_2^1(\sin\delta^*)\sin\alpha^* , \end{aligned}$$

Through these expressions it is easy to calculate the increments in the coefficients of the geopotential. Thus, for the J_2 and the C_2^2 we shall have, taking into account only the matrix (30) which gives us the increments of the inertia tensor:

$$\begin{aligned} \delta J_2 &= \frac{1}{2MR^2}(2\delta C - \delta A - \delta B) = \frac{1}{2MR^2}\left[-12D_t\left(\frac{a^*}{r^*}\right)^3 P_2(\sin\delta^*)\right] , \\ \delta C_2^2 &= \frac{1}{4MR^2}(\delta A - \delta B) = \frac{1}{2MR^2}\left[-2D_t\left(\frac{a^*}{r^*}\right)^3 P_2^2(\sin\delta^*)\cos 2\alpha^*\right] , \end{aligned}$$

that coincide with (16) and (21) above. In a similar way, via the matrix (30), the remaining coefficients are obtained.

As for the variation of the major moment of inertia, it is given by:

$$\delta C = -4D_t\left(\frac{a^*}{r^*}\right)^3 P_2(\sin\delta^*) = -\frac{Gm^*}{a^{*3}}\frac{4\pi}{15}\,I\left(\frac{a^*}{r^*}\right)^3(3\sin^2\delta^* - 1) \tag{31}$$

I being the integral introduced in (13). This expression can be written in a form probably more familiar to the reader, that is

$$\delta I_{33} = \frac{-k_{eq}m^*R^5}{r^{*3}}(\sin^2\delta - 1/3) \tag{32}$$

simply by defining a suitable coefficient, k_{eq}, that may be understood as a equivalent Love number corresponding to the tidal deformation of the mantle and computable from the geophysical data.

4. Numerical Values of the Variations of Harmonic Coefficients and $\Delta UT1$

In order to obtain the numerical values of the coefficient of the geopotential, from expressions (16), (21), (25), (26) and (27) we can see that it is necessary to develop the spherical functions which give us the position of the perturbing body.

TABLE I
Main Tidal Periodic Perturbations of Geopotential Coefficients
(Unit: 10^{-9})

Period	l	l'	F	D	Ω	K2	K22a	K22b	K21a	K21b
-31.81	1	0	0	-2	0	0.2033	-0.0053	-0.0053	-0.0486	-0.0486
27.55	1	0	0	0	0	1.0631	-0.0276	-0.0276	-0.2543	-0.2543
14.77	0	0	0	2	0	0.1764	-0.0046	-0.0046	-0.0422	-0.0422
9.12	1	0	2	0	1	0.1597	0.0278	-0.0012	0.1167	-0.0171
13.63	0	0	2	0	1	0.8342	0.1453	-0.0063	0.6099	-0.0891
-6798.36	0	0	0	0	1	-0.8433	-0.1469	0.0063	-0.6167	0.0901
9.56	-1	0	2	2	2	0.0732	-0.1417	-0.0003	0.1176	-0.0051
27.09	-1	0	2	0	2	-0.0569	0.1102	0.0002	-0.0914	0.0039
9.13	1	0	2	0	2	0.3854	-0.7462	-0.0014	0.6192	-0.0266
7.10	0	0	2	2	2	0.0615	-0.1192	-0.0002	0.0989	-0.0043
13.66	0	0	2	0	2	2.0124	-3.8976	-0.0072	3.2339	-0.1391
365.26	0	1	0	0	0	0.1487	-0.0039	-0.0039	-0.0356	-0.0356
121.75	0	1	2	-2	2	0.0548	-0.1060	-0.0002	0.0880	0.0009
182.62	0	0	2	-2	2	0.9342	-1.8090	-0.0034	1.5011	0.0160

As functions of Andoyer's canonical variables, these developments are (Kinoshita, 1977, Getino and Ferrándiz, 1991b):

$$\left(\frac{a^*}{r^*}\right)^3 P_2(\sin\delta^*) \simeq 3\sum_i B_i \cos\Theta_i \ ,$$
$$\left(\frac{a^*}{r^*}\right)^3 P_2^2(\sin\delta^*) \left\{ \begin{array}{c} \cos 2\alpha^* \\ \sin 2\alpha^* \end{array} \right\} \simeq 3\sum_i \sum_\tau D_i(\tau) \left\{ \begin{array}{c} -\cos \\ +\sin \end{array} \right\} (2\mu + 2\nu - \tau\Theta_i) \ ,$$
$$\left(\frac{a^*}{r^*}\right)^3 P_1^2(\sin\delta^*) \left\{ \begin{array}{c} \cos \alpha^* \\ \sin \alpha^* \end{array} \right\} \simeq 3\sum_i \sum_\tau C_i(\tau) \left\{ \begin{array}{c} \sin \\ \cos \end{array} \right\} (\mu + \nu - \tau\Theta_i) \ ,$$

where

$$\begin{aligned} B_i &= -\frac{1}{6}(3\cos^2\epsilon - 1)A_i^{0)} - \frac{1}{2}\sin 2\epsilon A_i^{1)} - \frac{1}{4}\sin^2\epsilon A_i^{2)} \ , \\ C_i(\tau) &= -\frac{1}{4}\sin 2\epsilon A_i^{0)} + \frac{1}{2}(1 + \tau\cos\epsilon)(-1 + 2\tau\cos\epsilon)A_i^{1)} \\ &\quad + \frac{\tau}{4}\sin\epsilon(1 + \tau\cos\epsilon)A_i^{2)} \ , \\ D_i(\tau) &= -\frac{1}{2}\sin^2\epsilon A_i^{0)} + \tau\sin\epsilon(1 + \tau\cos\epsilon)A_i^{1)} - \frac{1}{4}(1 + \tau\cos\epsilon)^2 A_i^{2)} \ , \end{aligned}$$

with $\tau = \pm 1$. The numerical values of the coefficients $A_i^{j)}$ are given in Kinoshita

(1977). As for the argument Θ_i , we have:

$$\Theta_i = m_1 l_{(} + m_2 l_{\odot} + m_3 F + m_4 D + m_5 \Omega \ , \quad \text{with } i = (m_1, m_2, m_3, m_4, m_5) \ ,$$

$$F = l_{(} + g_{(} \ ,$$

$$D = l_{(} + g_{(} + h_{(} - l_{\odot} - g_{\odot} - h_{\odot} \ ,$$

$l_{(}$, $g_{(}$, $h_{(}$ are the Delaunay variables for the Moon ,
$l_{\odot}$, $g_{\odot}$, $h_{\odot}$ are the Delaunay variables for the Sun .

In these expressions the simplifications $\sin\sigma \simeq 0$, $\cos\sigma \simeq 1$, have been assumed, since the angle $\sigma \simeq 10^{-6}$.

Using the following data in the c.g.s. system (Getino and Ferrándiz, 1991b):

$$M = 5.9742 \times 10^{27} \ , \ R = 6.378 \times 10^{8} \ , \ D_t = \begin{cases} 6.953379 \times 10^{36} (\text{Moon}) \\ 3.185508 \times 10^{36} (\text{Sun}) \end{cases} ,$$

we finally obtain

$$\delta J_2 = \sum_i K_2(i) \cos\Theta_i \ ,$$

$$\delta C_2^2 = \sum_i K_{22a}(i) \cos(2\mu + 2\nu - \Theta_i) + \sum_i K_{22b}(i) \cos(2\mu + 2\nu + \Theta_i) \ ,$$

$$\delta S_2^2 = \sum_i -K_{22a}(i) \sin(2\mu + 2\nu - \Theta_i) + \sum_i -K_{22b}(i) \sin(2\mu + 2\nu + \Theta_i) \ ,$$

$$\delta C_2^1 = \sum_i K_{21a}(i) \sin(\mu + \nu - \Theta_i) + \sum_i K_{21b}(i) \sin(\mu + \nu + \Theta_i) \ ,$$

$$\delta S_2^1 = \sum_i K_{21a}(i) \cos(\mu + \nu - \Theta_i) + \sum_i K_{21b}(i) \cos(\mu + \nu + \Theta_i) \ ,$$

In Table I the values of the main coefficients $K_2, K_{21a}, K_{21b}, K_{22a}$ and K_{22b} are listed as well as their respective arguments Θ_i.

With respect to the coefficient J_2 , the dominant periodic term corresponds to the argument $(2F + 2\Omega)$, and with an amplitude to the order of 2×10^{-9}. The terms with arguments (l), monthly, and $(2F + \Omega)$, semimonthly, have amplitudes very close to this, as do the ones corresponding to (Ω), whose period corresponds to that of the lunar node, and to $(2F - 2D + 2\Omega)$, semiannual, whereas the amplitude of the annual term, in (l'), is a little smaller, to the order of 10^{-10}. It is interesting to note the close relation among these orders of magnitude and the uncertainty in the experimental determination of the J_2 . According to Moritz (1987), the oblateness parameter is $(108263 \pm 0.2) \times 10^{-8}$, with an absolute error of precisely 2×10^{-9}. On the other hand, Milani et al. give a relative accuracy of 6×10^{-7}(corresponding to an absolute error of 6×10^{-10}, as a result of the differences between the GEM$-L2$(Lerch et al., 1983) and the GEM-9 (Lerch et al., 1979) geopotential models. In any case, it seems natural to consider that the

TABLE II
Nutation Series for $\Delta UT1$

No.	Period (days)	Argument l	l'	F	D	Ω	Amplitude $\times 10^{-7}$ sec
1	9.18	3	0	0	0	0	-18
2	131.67	2	1	0	-2	0	-12
3	-1095.17	2	0	-2	0	0	-134
4	205.89	2	0	0	-2	0	-539
5	-15.91	2	0	0	-4	0	15
6	7.13	2	0	0	2	0	-7
7	13.78	2	0	0	0	0	-334
8	-3232.85	1	-1	0	-1	0	40
9	-29.26	1	-1	0	-2	0	-11
10	29.80	1	-1	0	0	0	-54
11	-34.85	1	1	0	-2	0	84
12	25.62	1	1	0	0	0	39
13	-9.53	1	0	-2	-2	0	7
14	32.76	1	0	-2	2	0	7
15	-26.88	1	0	-2	0	0	-46
16	23.77	1	0	2	-2	0	9
17	9.11	1	0	2	0	0	-38
18	411.78	1	0	0	-1	0	34
19	-31.81	1	0	0	-2	0	1796
20	-10.08	1	0	0	-4	0	15
21	9.61	1	0	0	2	0	-75
22	27.55	1	0	0	0	0	-8136
23	-329.79	0	1	-2	2	0	9
24	117.54	0	1	2	-2	0	8
25	-15.39	0	1	0	-2	0	50
26	14.19	0	1	0	2	0	8
27	27.32	0	1	0	1	0	-6
28	173.31	0	0	2	-2	0	247
29	13.61	0	0	2	0	0	-295
30	14.77	0	0	0	2	0	-724
31	29.53	0	0	0	1	0	46
32	35.03	-1	-1	0	2	1	6
33	-32.61	-1	0	2	-2	1	-18
34	9.54	-1	0	2	2	1	-80
35	26.98	-1	0	2	0	1	173
36	-9.60	-1	0	0	-2	1	-5
37	31.96	-1	0	0	2	1	130
38	-388.27	-1	0	0	1	1	6
39	-27.44	-1	0	0	0	1	-526
40	1305.47	-2	0	2	0	1	415
41	-199.84	-2	0	0	2	1	-48
42	-13.75	-2	0	0	0	1	-22
43	-943.22	2	0	-2	0	1	6
44	12.79	2	0	2	-2	1	8
45	6.85	2	0	2	0	1	-40
46	212.32	2	0	0	-2	1	36

TABLE II (Suite)
Nutation Series for $\Delta UT1$

No.	Period (days)	l	l'	F	D	Ω	Amplitude $\times 10^{-7}$ sec
47	13.81	2	0	0	0	1	18
48	-34.67	1	1	0	-2	1	-5
49	23.86	1	0	2	-2	1	49
50	5.64	1	0	2	2	1	-10
51	9.12	1	0	2	0	1	-405
52	-31.66	1	0	0	-2	1	-116
53	9.63	1	0	0	2	1	-8
54	27.67	1	0	0	0	1	536
55	346.60	0	-1	2	-2	1	-45
56	14.16	0	-1	2	0	1	-5
57	-346.64	0	-1	0	0	1	-91
58	119.61	0	1	2	-2	1	32
59	13.14	0	1	2	0	1	7
60	386.00	0	1	0	0	1	-136
61	-169.00	0	0	-2	2	1	7
62	-13.58	0	0	-2	0	1	-6
63	177.84	0	0	2	-2	1	1153
64	7.09	0	0	2	2	1	-50
65	13.63	0	0	2	0	1	-3158
66	-14.73	0	0	0	-2	1	-46
67	14.80	0	0	0	2	1	-52
68	-6798.36	0	0	0	0	1	-1592617
69	9.81	-1	-1	2	2	2	-9
70	9.06	-1	0	4	0	2	4
71	5.80	-1	0	2	4	2	-5
72	9.56	-1	0	2	2	2	-194
73	27.09	-1	0	2	0	2	428
74	-27.33	-1	0	0	0	2	5
75	7.35	-2	0	2	4	2	-4
76	14.63	-2	0	2	2	2	5

main part of this error comes from the imperfect modelling of the solid Earth tides, specially when there is no general agreement about the first two figures of the Love number k - or the the contribution of the mantle to it, k_{eq} in our notation.

As for the values of the amplitudes of periodic variations in J_{22} , the greater ones appear in the semiannual and semimonthly terms, with orders of magnitude of $2 \sim 4 \times 10^{-9}$, followed closely by others draconian and monthly, next to new periods of 7.10, 9.56 and 121.75 days. We can see again the agreement among the values of these amplitudes and the experimental uncertainty in the determination of J_{22} .

On the other hand, it is well known (Yoder et al., 1981, Kubo, 1991) that the tidal

TABLE II (Suite)
Nutation Series for $\Delta UT1$

No.	Period (days)	l	l'	F	D	Ω	Amplitude $\times 10^{-7}$ sec
77	1615.74	-2	0	2	0	2	-12
78	8.75	3	0	2	-2	2	3
79	5.49	3	0	2	0	2	-9
80	12.81	2	0	2	-2	2	22
81	4.68	2	0	2	2	2	-3
82	6.86	2	0	2	0	2	-97
83	9.37	1	-1	2	0	2	-9
84	22.47	1	1	2	-2	2	4
85	8.91	1	1	2	0	2	8
86	23.94	1	0	2	-2	2	99
87	5.64	1	0	2	2	2	-23
88	9.13	1	0	2	0	2	-977
89	27.78	1	0	0	0	2	-7
90	7.24	0	-1	2	2	2	-8
91	14.19	0	-1	2	0	2	-24
92	13.17	0	1	2	0	2	25
93	409.23	0	1	0	0	2	3
94	12.66	0	0	4	-2	2	3
95	25.42	0	0	2	-1	2	-3
96	4.79	0	0	2	4	2	-2
97	7.10	0	0	2	2	2	-121
98	9.34	0	0	2	1	2	5
99	13.66	0	0	2	0	2	-7636
100	-3399.18	0	0	0	0	2	7779
101	182.63	0	2	0	0	0	-190
102	365.26	0	1	0	0	0	-15089
103	365.22	0	-1	2	-2	2	795
104	91.31	0	2	2	-2	2	-56
105	121.75	0	1	2	-2	2	-1852
106	182.62	0	0	2	-2	2	-47394

variation in $UT1$ can be obtained through the integration of δC in the following way

$$\Delta UT1 = -\int \frac{\delta C}{C} dt \, .$$

In Table II the resulting periodic perturbations are listed as series of sinus. To compare with the values given by Yoder et al. it must be taken into account that those must be multiplied by an experimental coefficient $k/C = 0.99$ that it is used to replace the theoretical $k/C_{mantle} = 0.94$. This last value was recommended because determinations from observations (mostly monthly and fortnightly tidal

TABLE III
Comparison of Amplitudes of $-\Delta UT1$ (in msec.)

Period	l	l'	F	D	Ω	Yoder	Kubo	Authors	IERS
-6798.36	0	0	0	0	1	172.0498	150.6	159.2617	161.727
182.62	0	0	2	-2	2	5.1327	4.51	4.7394	4.825
13.66	0	0	2	0	2	0.8252	0.73	0.7636	0.776
13.63	0	0	2	0	1	0.3413	0.30	0.3158	0.321

components) still show rather large variations ranging from 0.79 to 1.04 (Moritz, 1987). In any way, the different values are proportional, since they correspond to different values of parameters which play the role of numbers of Love. The coefficient k_{eq} used in our approach (computed from the Takeuchi Model 2) take the value 0.3047, in good agreement with the one given by Farrell (1972). This corresponds to a value of 0.923 for the coefficient k/C used to scale the series by Yoder.

In order to make the comparison easier, we write in Table III our dominant terms next to those from Yoder, Kubo and the IERS standard (Mc.Carthy, 1989), with the amplitude in msec. It can be seen that our solution is closer to that from IERS. The differences reach 2.47 ms. for the term with period of 18.6 years, and only 0.013 ms. for the period of 13.66 days, $(2\Omega + 2F)$, which correspond to differences in distance of about 113. cm. and 0.6 cm. respectively.

5. Conclusions

This paper shows a general procedure to predict the variation in time of the geopotential coefficients due to the mantle tides, that relies on the geophysical data about its internal structure provided by the Earth Models at hand. The approach is supported, a priori, by the good result that the basic technique has produced in the most complex general problem of the rotation of a deformable Earth. The method is applied to predict the periodic variation of J_2 and J_{22} , as well as that of the inertia tensor because of its significance on the determination of the universal time.

The numerical values of the amplitudes are obtained by using the Earth Model 2 of Takeuchi. For the variation of $\delta UT1$, the results are similar to those of other authors and are closer to those of IERS, from which they differ in less than 2% . On the other hand, the consideration of the series corresponding to the variations of the J_2 and J_{22} , and the relation of the dominant amplitudes to the experimental uncertainty in the determination of these parameters, would be seen to indicate that it is necessary to obtain two exact figures in the modelling of the effect of the solid tides as a necessary step to achieving relative errors less than 10^{-7} and 10^{-4} for these parameters (that is to say, one digit more). This coincides with the lack of agreement among the numerous authors on the values of the factors k or k/C, corresponding either to the whole Earth or solely the mantle.

In any case, the possibilities for improving knowledge about the most difficult effects to model - core or oceanic tides, etc, - depend in great part on our ability to improve the theoretical predictions of the effects of the mantle, whose internal constitution is better known starting from the geophysical data.

Acknowledgements

The first author acknowledges financial support received from Spanish CYCIT, under grant ESP88-541.

References

Bursa, M.: 1987, *Secular deceleration of the Moon and of the Earth's rotation and variation in the zonal geopotential harmonic*, Bull. Astron. Inst. Czechosl. **38**, 309-313.

Carpino, M. et al.: 1986, *Sensitivity of LAGEOS to changes in Earth's gravity coefficients*, Celes. Mech. **39**, 1-13.

Farrell, W.E.: 1972, *Deformation of the Earth by surface loads*, Rev. Geophys. Space Phys., **10**, 761-797.

Getino, J. and Ferràndiz, J.M.: 1990, *A Hamiltonian theory for an elastic Earth: canonical variables and kinetic energy*, Celes. Mech. **49**, 303-326.

Getino, J. and Ferràndiz, J.M.: 1991a, *A Hamiltonian theory for an elastic Earth: Elastic energy of deformation*, Celes. Mech. **51**, 17-34.

Getino, J. and Ferràndiz, J.M.: 1991b, *A Hamiltonian theory for an elastic Earth: First Order Analytical Integration*, Celes. Mech. **51**, 35-65.

Getino, J. and Ferràndiz, J.M.: 1991c, *A Hamiltonian theory for an elastic Earth: secular rotational acceleration*, Celes. Mech. **52**, 381-396.

Kinoshita, H.: 1977, *Theory of the rotation of the rigid Earth*, Celes. Mech. **15**, 277-326.

Kubo, Y.: 1991, *Solution to the rotation of the elastic Earth by method of rigid dynamics*, Celes. Mech. **50**, 165-187.

Lerch et al.: 1979, *Gravity model improvement using GEOS 3 (GEM 9 and 10)*, J. Geophys. Res. **84**, 3897-3916.

Lerch et al.: 1983, *A refined gravity model from LAGEOS (GEM$-L2$)*, NASA, Technical Memorandum 84986.

Milani et al.: 1987, *Non-gravitational perturbations and satellite geodesy*, ed. Adam Hilger.

Moritz, H. and Mueller, I.: 1987, *Earth rotation*, ed. Ungar, New York.

Takeuchi, H.: 1950, *On the Earth tide of the compressible Earth of variable density and elasticity*, Trans. Am. Geophys. Union, **31** (5), 651-689.

Yoder, C.F. et al.: 1981, *Tidal variations of Earth rotation*, J. Geophys. Res., **86** (B2), 881-891.

Yoder, C.F. et al.: 1983, *Secular variation of Earth's gravitational harmonic J_2 coefficient from LAGEOS and nontidal acceleration of Earth rotation*, Nature **303**, 757-762.

ORBITAL, ROTATIONAL, AND CLIMATE INTERACTIONS

JACQUES LASKAR and FREDERIC JOUTEL
Astronomie et Systèmes Dynamiques, Bureau des Longitudes, 77 Avenue Denfert-Rochereau, 75014 Paris, France

Key words: Precession - Paleoclimates - The Earth - Resonances - Stability

The climate of the Earth results from very complex interactions, including atmosphere, oceans, ice masses, and biosphere, which are far from being completely understood. Another important aspect, is the change of insolation at the surface of the Earth, resulting from changes in the orbit and orientation of the Earth, which act like a forcing term on the climate system of the Earth. Indeed, according to Milankovitch theory, the ice ages of the Quaternary, are essentially due to the changes of insolation at high latitude resulting from orbital and orientation changes of the Earth under planetary perturbations.

Recently, using huge algebraic manipulation, J. Laskar was able to integrate the averaged equations numerically from -200 Myr to +200 Myr. This revealed that the motion of the inner solar system is chaotic, with a Lyapounov exponent of about 1/5 Myr^{-1} (Laskar, 1989). This chaotic behavior was then identified as resulting from the presence of secular resonances between the Earth and Mars, and also between Mercury, Venus, and Jupiter (Laskar, 1990, 1992). These results imply that it is impossible to obtain a precise solution for the orbit of the Earth over about 100 Myr. On the other side, it is still possible to have ephemerides for the Earth for times ranging up to 10 Myr to 20 Myr, when the effect of the chaotic behaviour is not yet too sensible. Together with the orbital elements, the precession and obliquity of the Earth was integrated. The orbital solution (La90) was presented in (Laskar, 1990) and the precession, obliquity, and insolation solution in (Laskar *et al.*, 1992a).

A precise direct numerical integration of the whole solar system (QTD) has also been conducted over 3 Myr (Quinn *et al.*, 1991) and was compared to La90, revealing a very good agreement between these two solutions obtained in very different manners, and confirming the existence of the secular resonance between Mars and the Earth (Laskar *et al.*, 1992b). The excellent agreement of these different solutions insure that now we have at our disposal a reliable solution for the orbital evolution of the Earth over about 10 Myr.The two solutions La90 and QTD for the precession and obliquity are also in very good agreement when adding to the La90 solution the same tidal terms which were used in QTD (Laskar *et al.*, 1992a).

In (Laskar *et al.*, 1992a) the possible changes of obliquity of the Earth reached during an ice age were also studied. Indeed, in the frequency decomposition of the orbital planetary forcing terms, there exists a periodic term of frequency $s_6 - g_6 + g_5 = -50.30207''/yr$, which could enter into resonance with the precession frequency ($50.47118''/yr$). The amplitude of this term is small, ($0.104\ 10^{-8}$), but this term is very close to resonance.

Celestial Mechanics and Dynamical Astronomy **57**: 293–294, 1993.

The effect of an exact resonance with this small periodic term was investigated by changing slightly the value of the dynamical ellipticity of the Earth, keeping fixed the angular momentum. This is what can happen for example during an ice age, because of the redistribution of the ice on the Earth. The passage through the exact resonance increases the maximum obliquity of the Earth of about 0.5 degrees. This small term could thus be of great importance in the computation of the past insolation of the Earth, and appears to be a major difficulty for the determination of a precise insolation solution for the Earth over the last few millions years.

The problem of the global stability of the orientation of the Earth for all possible values of the initial obliquity was also studied (Laskar *et al.*, 1992c). A large chaotic zone was found in the phase space which extends from 60° to 90° and results from overlapping of secular resonances. In its present state, the Earth avoids this large chaotic zone and its obliquity is essentially stable. But if the Moon were not present, the torque exerted on the Earth would have been smaller, and for a large set of admissible initial spin rates of the Earth, the chaotic zone would have extended from nearly 0°, up to about 85°. This would have changed drastically the climate on the Earth. It can thus be claimed that the Moon acts as a climate regulator for the Earth, by maintaining its path outside from this large chaotic zone. Moreover, this study gives some new light on the origine of the obliquity of the Earth and on the possible origine of the Moon.

References

Laskar, J.: 1989, A numerical experiment on the chaotic behaviour of the Solar System *Nature*, **338**, 237–238

Laskar, J.: 1990, The chaotic motion of the solar system. A numerical estimate of the size of the chaotic zones, *Icarus*, **88**,266–291

Laskar, J.: 1992, ' A few points on the stability of the solar system', in *Chaos, Resonance and collective dynamical phenomena in the Solar System, S. Ferraz-Mello ed.*, 1–16

Laskar, J., Joutel, F., Boudin, F.: 1992a, Orbital, Precessional, and insolation quantities for the Earth from -20Myr to +10Myr *Astron. Astrophys., in press*

Laskar, J., Quinn, T., Tremaine, S.: 1992b, Confirmation of Resonant Structure in the Solar System, *Icarus*, **95**, 148-152

Laskar, J., Joutel, F., Robutel, P.: 1992c, 'The Moon: a climate regulator for the Earth', *submitted*

Quinn, T.R., Tremaine, S., Duncan, M.: 1991, 'A three million year integration of the Earth's orbit'*Astron. J.* **101**, 2287–2305

LUNAR AND TERRESTRIAL TIDAL EFFECTS ON THE MOON'S ROTATIONAL MOTION

E. BOIS and A. JOURNET
Observatoire de la Côte d'Azur, Dept. C.E.R.G.A.
Avenue Copernic, 06130 Grasse, France

Abstract. An accurate model of the rotation of the Moon, constructed by numerical integration, has been presented in a previous paper. All direct perturbations capable of producing at least 10^{-4} seconds of arc on the Moon's rotational motion have been included, and the physical librations resulting from planetary effects and Earth-Moon figure-figure interactions have been presented. The present study deals with the Moon's physical librations resulting from the non-rigidities of the Moon and the Earth. The effects of the Moon's elasticity and of a lunar phase lag are analyzed. Physical librations due to lunar tides and those due to terrestrial tides are presented and described.

Key words: Moon – rotation – librations – elasticity – anelasticity – tides

1. Introduction

Following the study presented in Bois *et al.* (1992) of the direct perturbations acting on the Moon's rotational motion, such as the Earth-Moon figure-figure interactions, the present part concerns the rotational behavior of a non-rigid Moon disturbed by a non-rigid Earth. The method permits the separation of various families of librations and the identification of relationships between causes and effects including the interactions between physics and dynamics. The theory may then give the qualitative behavior of each libration effect included in the Moon's rotational motion. Fitting the initial conditions, angles and angular velocities never change by more than 0.5 per cent of the total effect, so the method also gives a good order of magnitude estimate of each effect. After this necessary stage, the entire model will be accurately adjusted to the observations to provide a new determination of the parameters of the Earth-Moon system. The model uses the dynamical parameters of the JPL DE303 ephemeris. However, different numerical tests and experiments have shown that the global behavior of the present curves is quasi-insensitive to a refinement of the values of the dynamical parameters, a result that will follow in a forthcoming paper. The current uncertainty in the determination of the k lunar Love number is ± 0.003, while the value of the kT lunar rotational dissipation coefficient is 0.0048 ± 0.002 days (Veillet and Bois, 1990). We may then use a realistic range of values for these parameters to describe the non-rigidity effects on the Moon's rotational motion.

The lunar laser technique permits the measurement of the distance between a terrestrial station and a lunar reflector with a range uncertainty of 3 cm for the three last years of observations, and 1 cm for one night. The smallest amplitudes of lunar librations that can be detected are then, respectively, 0.01 and 0.001 seconds of arc (Veillet and Bois, 1990). These results are being improved, and all perturbations whose libration effects are at least of the order of 10^{-4} arc seconds in amplitude

Celestial Mechanics and Dynamical Astronomy **57**: 295–305, 1993.

must be included in the model (Bois *et al.*, 1992).
A non-rigid Moon model has been built with two alternatives:
- purely elastic deformations alone,
- anelastic deformations,
including deformations due the external attraction of a disturbing point mass and those due the rotational motion. The tidal effect of an extended external mass on the Moon's physical librations has been also studied. Expansions of direct perturbations due to purely elastic lunar deformations, due to a lunar phase shift, due to anelastic lunar tides, and due to anelastic terrestrial tides are presented in this paper as well as the curves of their effects on the Moon's rotational motion. All curves are presented here in the classical sequence 3-1-3 of Euler angles according to the following decomposition and notation: the precession ψ (around a fixed axis OZ, from a reference axis OX), the nutation θ (around an intermediate axis pointing toward the ascending node and representing the inclination of the Oz body-fixed axis with respect to the OZ-axis), and the rotation ϕ (around Oz). More details and conventions can be found in the previous paper Bois *et al.* (1992).

2. Elasticity and Anelasticity Effects

The rotation of a rigid or non-rigid homogeneous solid body satisfies the following equations of Liouville (Lambeck, 1980):

$$\frac{\partial}{\partial t}\left[(I)\omega + \mathbf{l}\right] + \omega \times \left[(I)\omega + \mathbf{l}\right] = \sum_i \mathbf{N}_i \tag{1}$$

where ω is the instantaneous rotational vector and N_i represents the torques acting on the rotation. (I) is the tensor of inertia and in the non-rigid body case, (I) is no longer constant. Consequently, because of the difficulty of keeping the definition of the body-fixed axes, there exits a relative angular momentum $\mathbf{l}$. However, the axes are defined so that $\mathbf{l} = 0$; they coincide with the average of the deformations, and in the assumption of complete rigidity they are the principal axes of inertia.

Start from

$$(I) = (I_0) + (\delta I), \tag{2}$$

where (I_0) represents the best tensor obtained in the assumption of rigidity and expressed in the principal axes of inertia, and let (δI) be the variable part of the tensor. Two kinds of internal deformations may occur. Some of them are specifically due to the rotational motion (rotational deformations); these forces stem from a centrifugal potential. The other deformations are due to the tides (tidal deformations) and the corresponding forces stem from a tidal potential. The non-rigidity of the body introduces then an additive potential, a function of the sum of these two potentials, which is proportional, in the case of an elastic body, to the Love number k for the potential perturbation. Consequently, the inertia tensor contains two parts as follows:

$$(\delta I) = (\delta I)^R + (\delta I)^T \tag{3}$$

where R and T designate, respectively, the rotational part and the tidal one.
To construct a non-rigid lunar model, the following assumption has been followed: the surface integral of the radial deformations of the Moon is always zero so that *trace*$(\delta I) = 0$. Consequently, the rotational part is (Eckhardt, 1981)

$$(\delta I)^R = K\left[\omega\omega^t - (1/3)(\omega^t\omega)(I_d)\right] \tag{4}$$

and the tidal part is (Eckhardt, 1981)

$$(\delta I)^T = -K\left[\left(\frac{3Gm}{a^3}\right)\left(\frac{a}{r}\right)^3\left(\mathbf{u}\mathbf{u}^t - (1/3)(I_d)\right)\right] \tag{5}$$

with

$$K = \frac{k(MR^2)(R/a)^3}{(M/m)(3Gm/a^3)} \tag{6}$$

where $\mathbf{u}$ is the selenographic unit vector toward the Earth, M the mass of the Moon and R its mean radius, m the mass of the Earth and a the mean distance between the two bodies while r is the instantaneous one, G being the gravitational constant. (I_d) designates the identity matrix and t the transpose operator. These two kinds of deformations produce two kinds of librations in the rotational behavior of non-rigid bodies, *tidal librations* and *centrifugal librations* (Bois, 1991). The two qualitative behaviors are wholly different and, moreover, the centrifugal librations are negligible here with respect to the tidal ones (Bois *et al.*, 1990).

In case of an anelastic body, k becomes a complex operator, *i.e.* the Love number with a phase shift. Knowing that for purely elastical deformations, its imaginary part $Im(k) = 0$, one gets from the equations (1) the following relation:

$$\omega \times (\delta I)\omega = \delta \mathbf{N} \tag{7}$$

Consequently, the additive elastic torque is written as follows:

$$\mathbf{N}_{\text{el.}} = -\frac{\partial}{\partial t}\left[(\delta I)\omega\right]. \tag{8}$$

For anelastic deformations, $Im(k) \neq 0$, the torque is then written as follows:

$$\mathbf{N}_{\text{an.}} = -\frac{\partial}{\partial t}\left[(\delta I)\omega\right] + \delta\mathbf{N} - \omega \times (\delta I)\omega \tag{9}$$

where expanding (9) through the second degree in spherical harmonics is sufficient, and one may reduce $\delta\mathbf{N}$ to $\delta\mathbf{N}_2$ as a function of (δI), according to the fundamental expression of calculation of a torque exerted on a body of potential V by a point mass m at position $r\mathbf{u}$ with respect to the body (Schutz 1981, Eckhardt 1981, Bois *et al.* 1992). $\delta\mathrm{N}_2$ is then written as follows:

$$\delta\mathrm{N}_2 = \left(\frac{3Gm}{a^3}\right)\left(\frac{a}{r}\right)^3 \mathbf{u}\times(\delta I)\mathbf{u} \tag{10}$$

with k reduced to k_2 .

Using the expression of the elastic torque given in (8), figure 1 represents the effects of purely elastic deformations on the three angles of librations. Amplitudes of the resulting physical librations plotted over 24 years are of the order of few milliseconds of arc, say 4 mas in ψ and ϕ, and therefore are sensitive to the observations.

In a second step, the expression for the torque (8) is replaced by the one in (9). After computations of the anelastic model effects, differences have been calculated between these effects and those of the previous elastic model. Figure 2 then shows the effects of the phase χ of k, *i.e.* the effects of a lunar dissipation factor Q on the Moon's physical librations. χ, which is considered close to the reciprocal of Q, is represented in our computation by a time delay, a phase shift on the position of the Earth. The values used are $k_2 = 0.027633$ and $\chi = 0.0497$ rd. ($Q \simeq 20$). Amplitudes of the resulting physical librations plotted over 24 years are of the order of 5.10^{-4} seconds of arc and therefore greater than our basic criterion. Notice in particular that the resulting amplitude in θ is of the same order of magnitude as the θ of figure 1. On doubling the lunar dissipation factor, *i.e.* $Q \simeq 40$, amplitudes of the long period effects become twice as small with the same periodic behavior.

3. Lunar Tidal Effects (k_2 and χ of the Moon)

An estimation of the magnitude of the librations induced by a lunar tidal bulge may also be required in order to achieve the accuracy level we are seeking. The Earth has a roughly fixed position as seen from the Moon. However we have made the assumption that the harmonics coefficients C_{2i} and S_{2j} used in the calculation of the principal part of the physical librations are free of tidal effects, and correspond to a rigid Moon. The slight variations of these second degree harmonics coefficients induced by the tides due to the Earth can easily be calculated and expressed as follows (Lambeck, 1980):

$$\begin{aligned}
\Delta C_{20} &= k_2\frac{m}{M}\frac{R}{r}P_{20}(\sin\varphi)\\
\Delta C_{21} &= k_2\frac{m}{M}\frac{R}{r}P_{21}(\sin\varphi)\cos(\lambda)\\
\Delta S_{21} &= k_2\frac{m}{M}\frac{R}{r}P_{21}(\sin\varphi)\sin(\lambda)\\
\Delta C_{22} &= k_2\frac{m}{M}\frac{R}{r}P_{22}(\sin\varphi)\cos(2\lambda)
\end{aligned} \tag{11}$$

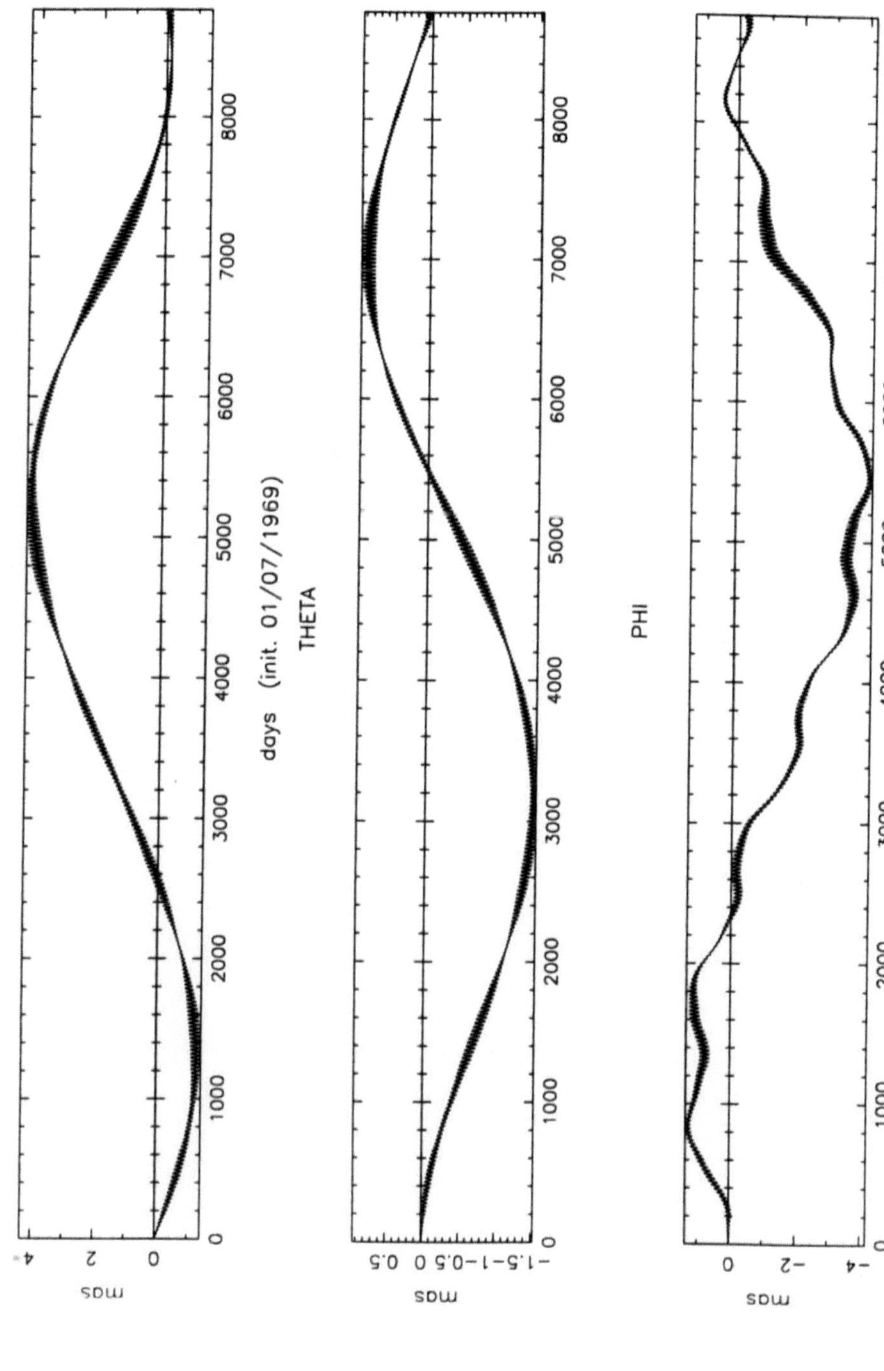

Fig. 1. Effect of elasticity

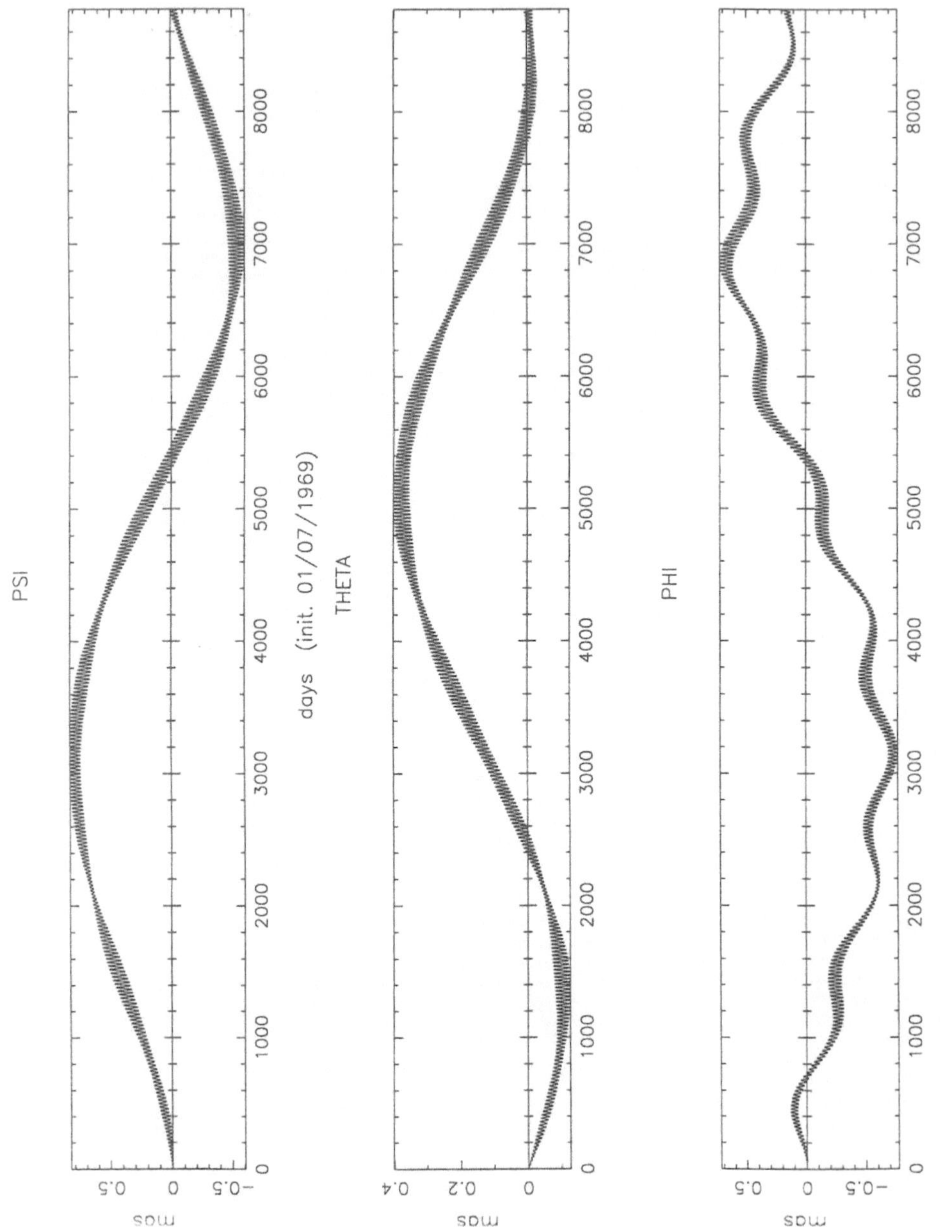

Fig. 2. Effect of a lunar dissipation factor

$$\Delta S_{22} = k_2 \frac{m}{M} \frac{R}{r} P_{22}(\sin\varphi)\sin(2\lambda)$$

where k_2 is the lunar Love number, m and M the masses of the Moon and of the Earth, R the mean radius of the Moon, and r, φ, λ the Earth center spherical coordinates in the lunar principal frame.

The supplementary torque caused by the Earth on the tidal bulge is calculated as the "point mass Earth–Moon 2" torque and added to it. The expression for the torque can be obtained in (9) in the case of the anelastic model but with (δI) reduced to $(\delta I)^T$. The magnitude of the corresponding librations can be seen on figure 3.

Questions of the internal structure and the "free" librations are obviously connected through the dissipation factor Q. Let us consider the librations due to the rotational deformations only. That means, we consider Moon alone in space, from the gravitational point of view. The effects of the rotational deformations contained in the librations of figure 1 are so faint that figure 1 represents in fact the librations due to the tidal deformations, with purely elastic deformations. Moreover, the effects of the lunar phase lag being ten times smaller in ψ and ϕ than the effects of elasticity for the same angles, the tidal librations obtained with an anelastic model (figure 3) are very close to those that would be induced by the elastic model. The most visible difference is on the θ angle. This confirms the studies and expectations of Migus (1977). Amplitudes of these tidal librations are of the order of a few thousandth seconds of arc. The ϕ angle of figure 3, with its three main modulations of libration (the proper rotation of 27.3 days, the resonant frequency of 2.9 years, and the precession of the nodes of 18.6 years) shows the effect of an equatorial bulge in the mean direction of the Earth. But because of the inclination of the equator of the Moon with respect to its orbit, the lunar equatorial bulge shifts slowly around the lunar equator. It appears a great libration with an amplitude of 4 mas in ϕ, 1 mas in θ (*i.e.* 3 cm in longitude and 1 cm in latitude). These amplitudes are relative to the input value of the Love number in the model. At this level, in order to precisely determine the radial amplitude of the lunar tides, it is absolutely necessary to adjust the theoretical model, including anelastic deformations, to the observed tides. But, in spite of the very good accuracy of the lunar laser technology, reduction of the observations can not give the value of the tidal amplitude. The reason for this is simply ascribable to the places of the Moon where the five reflectors have been set. The reflector highest in latitude does not work; consequently the greatest difference in latitude between two reflectors is about 25 degrees which is not sufficient, at present, to detect the amplitude of the tides.

4. Terrestrial Tidal Effects (k_2 and ϵ of the Earth)

The Moon is responsible of an appreciable bulge on the Earth, and this bulge has a lag angle with reference to the lunar direction, this lag being due to viscosity effects. The expansion of the induced tidal variations of the harmonics coefficients is exactly similar to the lunar one:

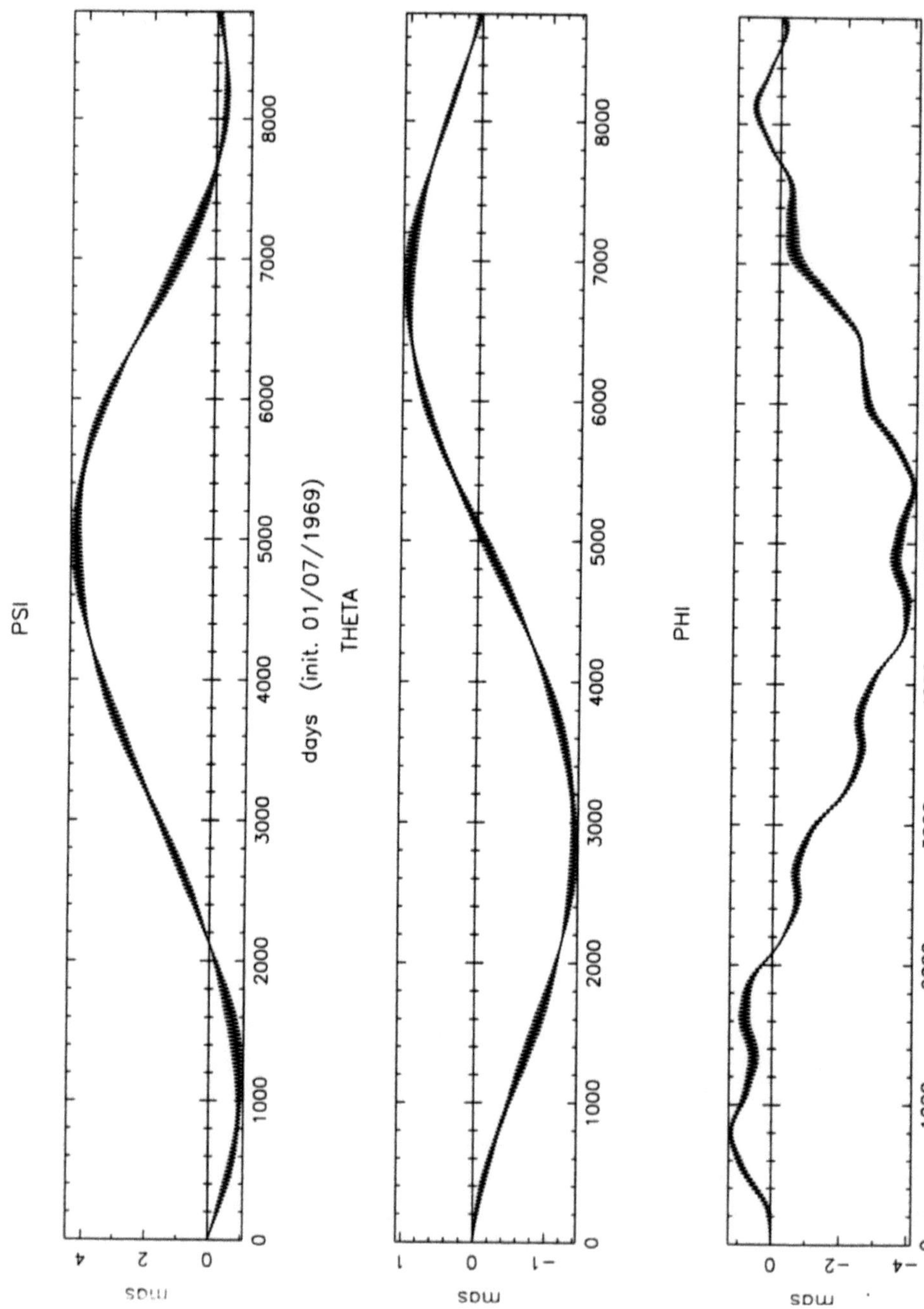

Fig. 3. Lunar Tidal Librations

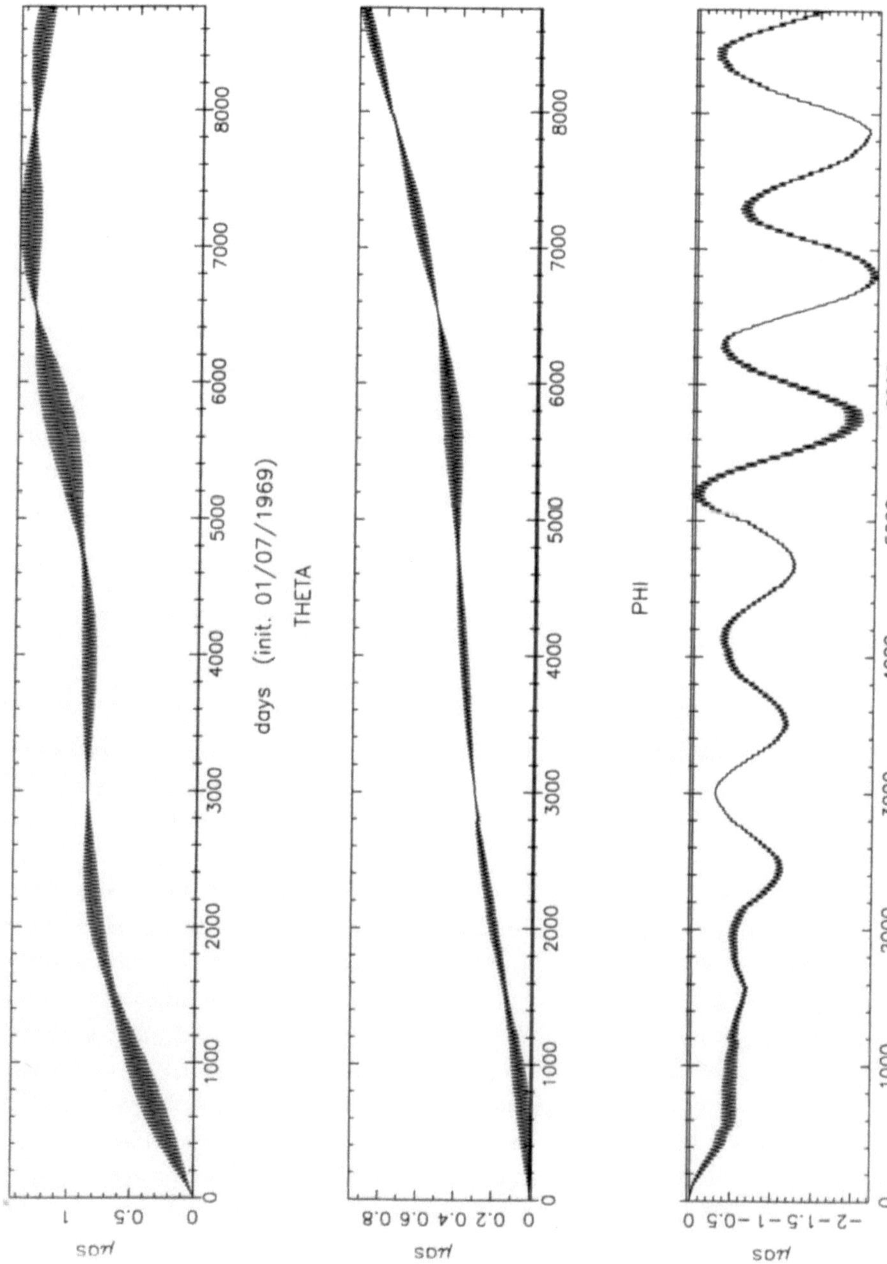

Fig. 4. Effect of Terrestrial Tides

$$\begin{aligned}
\Delta C_{20} &= k_2 \frac{M}{m}\frac{R_\oplus}{r} P_{20}(\sin\delta) \\
\Delta C_{21} &= k_2 \frac{M}{m}\frac{R_\oplus}{r} P_{21}(\sin\delta)\cos(\Theta-\alpha-\epsilon) \\
\Delta S_{21} &= k_2 \frac{M}{m}\frac{R_\oplus}{r} P_{21}(\sin\delta)\sin(\Theta-\alpha-\epsilon) \\
\Delta C_{22} &= k_2 \frac{M}{m}\frac{R_\oplus}{r} P_{22}(\sin\delta)\cos(2(\Theta-\alpha-\epsilon)) \\
\Delta S_{22} &= k_2 \frac{M}{m}\frac{R_\oplus}{r} P_{22}(\sin\delta)\sin(2(\Theta-\alpha-\epsilon))
\end{aligned} \tag{12}$$

where k_2 is the terrestrial potential Love number, r, α, δ the equatorial coordinates of the Moon, Θ the Greenwich sideral time, and ϵ the phase lag caused by the Earth viscosity. We have assumed that coefficient k_2 and lag ϵ are the same for all second degree harmonics. Theses coefficients are calculated in a terrestrial coordinate system. They should be rotated in a frame parallel to the principal lunar frame. The rotation formulae are well known. Their general forms are written as follows (Borderies, 1978):

$$P_{lm}(\sin\varphi) \quad e^{im\lambda} = \frac{1}{(l-m)!}\sum_{m'=-l}^{l}(l-m')!E_{lm}^{m'}(\psi,\theta,\phi)\times P_{lm'}(\sin\varphi')e^{im'\lambda'} \tag{13}$$

ψ, θ, ϕ being the Euler angles of the rotation, and $E_{lm}^{m'}$ the Euler functions defined in the cited paper. Afterward the coefficients are introduced into the expressions T_{Ox}, T_{Oy}, T_{Oz} of the paper of Schutz (1981) giving the figure-figure effects of rigid bodies.

No assumptions about the physical constitution of the Earth were made at this level of calculation. A conventional potential Love number $k_2 = 0.3$, and a terrestrial lag angle of 3 degrees (12 minutes of time) have been used. Other values more accurately fitted to the observed results could easily be introduced. The resulting torque has been added to the figure-figure torque already calculated. The behavior of the librations caused by this bulge can be seen in figure 4. Over this integration time interval, the amplitudes are negligible but the effects of long period terms are not yet reached. In particular a possible secular or pseudo-secular term, with respect to our nominal duration of 24 years, seems to appear in θ. Then, there would be a nutation secular variation corresponding to the Moon's secular acceleration due to the same cause, the equatorial terrestrial bulge.

5. Conclusion

An accurate lunar rotation model has been built up by numerical integration. The nature, cause and behavior of the Moon's physical librations have been isolated and described. The effects on the Moon's rotational motion, of purely elastic deformations, of a lunar phase lag, of anelastic lunar tides, and of terrestrial tides are presented in this paper. After this necessary stage, the entire model will be accurately adjusted to the observations, providing a new determination of the parameters and, in particular (hopefully), a credible value of the lunar dissipation factor.

References

Bois, E.: 1991, 'Theory of Solid Rotation', invited paper, *Proceedings of the 21st IAU General Assembly on the JCM "Rotation of the Solar System Bodies"*, held in Buenos Aires, 24 July 1991, in *Highlights of Astronomy*.

Bois, E. and Wytrzyszczak, I.: 1990, 'The Moon's Physical Librations - Part II : Non-Rigid Moon and Direct Non-Gravitational Perturbations', *Proceedings of the NATO Advanced Study Institute on : "Predictability, Stability and Chaos in N-Body Dynamical Systems"*, held in Cortina d'Ampezzo, Italy, Aug. 90, Edited by Archie E. Roy. Series B : Physics - Vol. 272, pp. 265-271.

Bois, E., Wytrzyszczak, I. and Journet, A.: 1992, 'Planetary and Figure-figure Effects on the Moon's Rotational Motion', *Celest. Mech.* **53**, 185–201.

Borderies, N.: 1978, 'Mutual Gravitational Potential of *N* Solid Bodies', *Celest. Mech.* **18**, 295–307.

Eckhardt, D.H.: 1981, 'Theory of the Libration of the Moon', *The Moon and the Planets* **25**, 3–49.

Lambeck, K.: 1980, *The Earth's Variable Rotation*, Cambridge University Press.

Migus, A.: 1977, 'Théorie analytique de la libration physique de la Lune', *Physical Doctoral Thesis*, Université de Paris VI.

Schutz, B.E.: 1981, 'The mutual potential and gravitational torques of two bodies to fourth order', *Celest. Mech.* **24**, 173–181.

Veillet, C. and Bois, E.: 1990, 'Sélénodésie par télémétrie laser depuis la Terre', *Journée d'information sur la Lune*, CNES, Toulouse, Avril 90, in *"Initiative Lune - Rapport Préliminaire d'Orientation" - CNES*.

PASSING THROUGH RESONANCE: THE EXCITATION AND DISSIPATION OF THE LUNAR FREE LIBRATION IN LONGITUDE

DONALD H. ECKHARDT

Geophysics Directorate, Phillips Laboratory
Hanscom AFB, MA 01731, USA

Abstract. The acceleration of the mean lunar longitude has a small effect on the periods of most terms in a Fourier expansion of the longitude. There are several planetary perturbation terms that have small amplitudes, but whose periods are close to the resonant period of the lunar libration in longitude. Some of these terms are moving toward resonance, some are moving away from resonance, and the periods of those terms that do not include the Delaunay variables in their arguments are not moving. Because of its acceleration of longitude, the Moon is receding from the Earth, so the magnitude of the restoring torque that the Earth exerts on the rotating Moon is gradually attenuating; thus resonance itself is moving, but at a much slower rate than the periods of the accelerating planetary perturbations. There are five planetary perturbation terms from the ELP-2000 Ephemeris (with amplitudes of 0″.00001 or greater) that have passed through resonance in the past two million years. One of them is of special interest because it appears to be the excitation source of a supposed free libration in longitude that has been detected by the lunar laser ranging experiment. The amplitude of the term is only 0″.00021 but it could be the source of the $\sim 1''$ amplitude free libration term if the viscoelastic properties of the Moon are similar to those of the Earth.

Key words: Lunar librations, lunar dissipation, lunar Q, free librations

1. The Rigidity of the Moon

The velocity of a seismic shear wave is

$$v_s = \sqrt{\mu/\sigma} \tag{1}$$

where μ is the rigidity and σ is the density of the medium. The structure of the lunar interior is rather homogeneous. Over the depth range 60–1100 km, the shear wave velocity varies little from $v_s = 4.3$ km/s in the model (designated GDT) of Goins, Dainty and Toksoz [1981]; v_s is about five per cent higher over the same depth range in the model (designated N) of Nakamura [1983]. Let M be the mass of the Moon, R its radius, C_{11} its minimum moment of inertia, C_{33} its maximum moment of inertia, and C_{22} the moment of inertia about the third principal axis. Because $\beta = [C_{33} - C_{11}]/C_{22} = 0.000631$ and $C_{33}/MR^2 = 0.392$ [Williams, 1977], all the moments of inertia are close to that of a homogeneous sphere, so the lunar density ($\sigma = 3340$ kg m^{-3}) is fairly uniform. Thus, the lunar rigidity, calculated using (1), does not vary much from $\mu = 62 \times 10^9$ Pa (GDT model) or $\mu = 68 \times 10^9$ Pa (N model). (The lunar rigidity is somewhere between that of cast iron, $\mu = 40 \times 10^9$ Pa, and steel, $\mu = 80 \times 10^9$ Pa.)

For an incompressible non-dissipative homogeneous spherical Moon, its second degree potential-disturbance Love number (omitting the subscript in k_2) is $k = 3/[2 + 19\mu/\sigma g R]$ [Love, pp. 249–259, 1927; Jeffreys, p. 299, 1976], where g is the acceleration of gravity at the lunar surface. Using $\sigma g R = 9.5 \times 10^9$ Pa, the Love

Celestial Mechanics and Dynamical Astronomy **57**: 307–324, 1993.

number determined from seismic data is $k = 0.024$ (GDT model) or $k = 0.022$ (N model). Cheng and Toksoz [1978] used numerical integrations to calculate the Love number for two compressible Moon models in which the elastic properties are functions of depth. The differences between their estimates, $k = 0.029$ and $k = 0.034$, and those from the GDT and N models are principally due their use of earlier shear velocities which are appreciably lower than the more accurate GDT and N velocities.

The Moon deforms elastically, causing the inclination of its equator to the ecliptic to increase by $\delta I = k \times 74''$ [Eckhardt, 1981]. From analyses of lunar laser ranging data, Williams, Newhall and Dickey [1987] estimate that $k = 0.027 \pm 0.006$; that is, the deformation causes the inclination to increase by $2''$ resulting in a -17 m $\sin F$ term (F is the argument of latitude) in the latitude libration. Other libration parameters (e.g., β and C_{33}/MR^2) contribute to the $\sin F$ term in the latitude libration, but they contribute to other terms in different ways, so a thorough analysis allows the deformation term to be discriminated.

To take into account anelastic dissipation in the Moon, the Love number is replaced in the frequency domain by a complex Love number, $k(1 - i/Q)$, where $Q \gg 1$. The term $(1 - i/Q) \approx \exp(-i/Q)$ is actually a phase shifting operator that changes a periodic term of the form $\exp(i\omega t)$ to $\exp[i(\omega t - 1/Q)]$. The imaginary part of the Love number is negative because a deformation lags its inducing forces. The magnitude of the ratio of the real part to the imaginary part of $k(1 - i/Q)$ is Q, where $1/Q$ is the specific dissipation function [Munk and MacDonald, pp. 21-22, 1960] which is frequency dependent. Because $19\mu/\sigma g R \gg 2$, the following approximation is valid for the Moon

$$19\mu k \approx 3\sigma g R. \tag{2}$$

Because $Q^2 \gg 1$, (2) remains valid if k is replaced by $k(1 - i/Q)$ and μ is replaced by $\mu(1 + i/Q)$. The imaginary part of the rigidity is positive because the stress leads the strain [Ben-Menahem and Singh, pp. 856–859, 1981]. A value for k will subsequently be required in order to estimate Q; taking into account the various estimates discussed above, this paper adopts the value $k = 0.025$.

2. Equations Describing the Libration in Longitude of a Deformable Moon

In a selenocentric coordinate system, let $r_i = ru_i$ locate the Earth's center of mass, a distance r from the origin, and let $\rho_i = \rho v_i$ locate a point within the Moon, a distance ρ from the origin. The lunar gravitational potential at r_i is

$$V = (G/r)[\int_{Moon} dM + \sum_{n=2}^{\infty}(R/r)^n \int_{Moon} (\rho/R)^n P_n(u_i v_i) dM]$$

where G is the gravitational constant, P_n is the Legendre function of degree n, and the integrals are over the mass of the Moon. Through degree $n = 2$, the potential is

$$V = GM/r + (G/r^3)[\frac{3}{2}\int_{Moon}[(u_i\rho_i)^2 - \rho_i{}^2]dM + \int_{Moon}\rho_i{}^2 dM].$$

Using the moment of inertia tensor,

$$C_{ij} = \int_{Moon}(\rho_k{}^2\delta_{ij} - \rho_i\rho_j)dM,$$

where δ_{ij} is the Kronecker delta, the truncated potential is reformulated as

$$V = GM/r - (G/2r^3)(3u_iC_{ij}u_j - C_{ii}).$$

The potential and moments of inertia of an anelastic Moon are deformed by forces derived from its centrifugal potential, W_1, and from the second-degree Earth-induced tidal potential, W_2. The lunar potential at the Earth is perturbed by

$$\delta V = k(1 - i/Q)(R/r)^5(W_1 + W_2)$$

where

$$W_1 = \frac{1}{2}(\varepsilon_{ijk}\omega_j r_k)^2 = -\frac{1}{2}r_i(\omega_i\omega_j - \delta_{ij}\omega_k{}^2)r_j,$$

and

$$W_2 = (Gm/2r^3)r_i(3u_iu_j - \delta_{ij})r_j,$$

where ω_i is the lunar angular rotation rate, m is the mass of the Earth, and ε_{ijk} is the alternating tensor. Then

$$\delta V = -(G/2r^3)(3u_i\delta C_{ij}u_j - \delta C_{kk}\delta_{ij}) + k(1 - i/Q)R^5\omega_k{}^2/3r^3$$

where the perturbation of the moment of inertia tensor is

$$\begin{aligned}\delta C_{ij} &= (1 - i/Q)K[\omega_i\omega_j - 3(Gm/a^3)(a/r)^3u_iu_j] \\ &= (1 - i/Q)K[\omega_i\omega_j - 3n^2(a/r)^3u_iu_j] \\ &\approx (1 - i/Q)K(\omega_i\omega_j - 3n^2u_iu_j),\end{aligned} \tag{3}$$

a is the mean Earth-Moon distance,

$$K = kR^5/3G = \frac{k(MR^2)(R/a)^3}{(M/m)(3n^2)}, \tag{4}$$

and n is the mean rate of lunar motion.

For a small lunar rotation, $\delta\varphi_k$, the Earth's direction cosines change by $\delta u_i = \varepsilon_{ijk}u_j\delta\varphi_k$, so $\partial u_i/\partial\varphi_k = \varepsilon_{ijk}u_j$. The torque exerted on the Moon is

$$N_k = m\partial(V + \delta V)/\partial\varphi_k = (3Gm/r^3)u_i(C_{ij} + \delta C_{ij})\varepsilon_{jlk}u_l. \tag{5}$$

In an inertial frame, the time derivative of the angular momentum, $L_i + \delta L_i = (C_{ij} + \delta C_{ij})\omega_j$, is equal to the torque, so

$$d(L_i + \delta L_i)/dt = \dot{L}_i + \delta \dot{L}_i + \varepsilon_{ijk}\omega_j(L_k + \delta L_k) = N_i, \tag{6}$$

where the dot derivative is with respect to time in the lunar coordinate system which is chosen to be the same as the principal axes of C_{ij}. Setting (5) equal to (6) with $\delta C_{ij} = 0$ gives the basic Euler dynamical equations for the rotation of a rigid Moon.

If the Love number is real, the centrifugal potential deforms the Moon as an oblate spheroid whose shortest axis is the instantaneous rotation axis, so W_1 does not affect the rotation rate through the RHS of (6), only its direction. Also if the Love number is real, the tidal potential deforms the Moon as a prolate spheroid with its longest axis in the instantaneous direction of the Earth, so there can be no torque due to W_2. The real part of the Love number only has an effect on the lunar rotation rate through the change in the Moon's moment of inertia, and that is almost entirely due to the centrifugal potential. The mean lunar rotation rate n (the same as its mean rate of motion) is much larger than its perturbation τ, the libration in longitude, so the real part of the Love number affects the Moon's librations principally through a "frozen in" deformation in its moment of inertia tensor which is a permanent part of C_{ij}. The impact of the real part of the Love number on the rotation rate is essentially nil; only the imaginary part contributes significantly to perturbations from the Euler dynamical equations. This can be demonstrated mathematically by combining (3), (4), (5) and (6).

For examining the libration in longitude τ (about the 3-axis), use the approximations

$$\omega_i = \begin{bmatrix} 0 \\ 0 \\ \omega_3 \end{bmatrix}, u_i = \begin{bmatrix} 1 \\ u_2 \\ 0 \end{bmatrix},$$

$$\omega_3 = n + \dot{\tau} \quad \& \quad u_2 = s - \tau,$$

where the 1-axis is in the mean Earth direction (or nearly so [Eckhardt, 1973]) and s is comprised of the equation of the center and all inequalities in longitude. With $\gamma = (C_{22} - C_{11})/C_{33} = 0.000228$ and, approximately, ${\omega_0}^2 = 3\gamma n^2$, the bottom row of the Euler equation becomes approximately

$$\ddot{\tau} = {\omega_0}^2(s - \tau). \tag{7}$$

Excitation and dissipation require hysteresis, so only the imaginary periodic terms of the moment of inertia perturbation,

$$\delta C_{ij} = (1 - i/Q)K \begin{bmatrix} -3n^2 & -3n^2u_2 & 0 \\ -3n^2u_2 & -3n^2{u_2}^2 & 0 \\ 0 & 0 & {\omega_3}^2 \end{bmatrix},$$

are significant. Therefore, to the first order in s, τ and $\dot{\tau}$, the significant portion of δC_{ij} is

$$\delta\tilde{C}_{ij} = -i(K/Q)\begin{bmatrix} 0 & -3n^2(s-\tau) & 0 \\ -3n^2(s-\tau) & 0 & 0 \\ 0 & 0 & 2n\dot{\tau} \end{bmatrix}.$$

Dropping second order terms, the variable perturbation in the lunar angular momentum is

$$\delta\tilde{L}_i = \delta\tilde{C}_{ij}\omega_j = -2i(K/Q)n^2\dot{\tau}\begin{bmatrix} 0 \\ 0 \\ 1 \end{bmatrix},$$

and its rate of change is

$$\delta\dot{\tilde{L}}_i = \delta\dot{\tilde{C}}_{ij}\omega_j = -2i(K/Q)n^2\ddot{\tau}\begin{bmatrix} 0 \\ 0 \\ 1 \end{bmatrix}.$$

Then

$$\delta\tilde{C}_{ij}u_j = 3i(K/Q)n^2\begin{bmatrix} 0 \\ s-\tau \\ 0 \end{bmatrix}$$

and

$$\varepsilon_{kli}u_l\delta\tilde{C}_{ij}u_j = 3i(K/Q)n^2\begin{bmatrix} 0 \\ 0 \\ s-\tau \end{bmatrix}.$$

The perturbation of the torque exerted on the deformed Moon by the Earth is then

$$\delta\tilde{N}_i = m\varepsilon_{ijk}u_j[\partial(\delta\tilde{V})/\partial u_k] = 9i(K/Q)n^4\begin{bmatrix} 0 \\ 0 \\ s-\tau \end{bmatrix}.$$

The perturbation of (7) accounting for the deformable Moon model is

$$(1 - 2i\gamma\kappa/3)\ddot{\tau} = (1 + i\kappa){\omega_0}^2(s-\tau) \tag{8}$$

where $\kappa = \varsigma k/Q$ and $\varsigma = (R/a)^3(MR^2/C_{33})\gamma^{-1}(m/M) = 0.0841$. Take $k = 0.025$ and suppose that $Q > 20$ so that $\kappa < 10^{-4}$. Because $2\gamma \ll 3$, the $2i\gamma\kappa/3$ term is dropped, and (8) becomes

$$\ddot{\tau} = {\omega_0}^2(1 + i\kappa)(s-\tau). \tag{9}$$

The period of resonance, $2\pi/\omega_0$, derived by detailed semi-analytic theories has a small uncertainty. Estimates based on the LURE 2 constants [Williams, 1977] are, in sidereal months, 38.651 [Eckhardt, 1981], 38.654 [Moons, 1982] , and 38.658 [Migus, 1980]. These estimates are adjusted for the JPL constants of Williams, Newhall and Dickey [1987]. The increase in γ is $(227.951 - 227.370) \times 10^{-6} = 0.581 \times 10^{-6}$ which decreases each estimate by 0.049, but using the technique of Williams *et al.* [1973], the changes in the harmonic coefficients C_{31} and C_{33} work out to be the same as decreasing γ by 0.592×10^{-6} and the adjustments almost cancel each other. The net result is that each estimate increases by only 0.001; this reflects the fact that the lunar laser ranging experiment is more sensitive to the period of τ resonance than to the related solution parameters.

3. Analytic Solutions of the Libration Equations

Let the longitude source terms be

$$s = \sum s_k = \sum H_k \exp(i\omega_k t)$$

where each H_k is a complex coefficient. Then the solution to (9) is

$$\tau = \sum \tau_k = \sum \frac{(1 + i\kappa)s_k}{1 + i\kappa - (\omega_k/\omega_0)^2} \approx \sum \frac{s_k}{1 - (\omega_k/\omega_0)^2}. \tag{10}$$

The $i\kappa$ term has negligible effect on the steady state solution given by (10) because currently none of the s_k source terms has a frequency close enough to resonance. If, however, the argument of some source term includes an acceleration term,

$$s_k = H_k \exp[i(\omega_k t + \frac{1}{2}c_k t^2)], \tag{11}$$

the dissipation term may be important, especially near resonance. Perturbing this process is the fact that because ω_0 is proportional to n, it is a linear function of time (because of the acceleration of the mean longitude of the Moon).

Suppose that there is a term of the form (11) that carried τ through resonance in the recent past (using a geological sense of time so that a million years or so ago was recent) at $t = 0$ so that $\omega_k(0) = \omega_0(0)$. Change the time variable to the dimensionless variable η where $\dot{\eta} = \omega_0(0)$ so that η goes through 2π radians every 38.65 sidereal months, and use the prime ($'$) to denote the derivative with respect to η. The argument of s_k in (11) is then expressed in the form

$$\omega_k(0)t + \frac{1}{2}c_k t^2 = \omega_0(0)t + \frac{1}{2}c_k t^2 = \eta + \frac{1}{2}\lambda_k \eta^2,$$

in which case

$$\lambda_k = c_k/\omega_0{}^2(0).$$

In the Fourier expansion of the longitude of the Moon, many of the planetary source terms near resonance have accelerations with approximately the same magnitude as the acceleration of the mean longitude of the Moon [Williams, Newhall and Dickey, 1992],

$$\dot{n} = (-26.0 \pm 1.0)''/\text{cy}$$

or

$$\dot{n}/n = -0.150 \times 10^{-9}/\text{y}.$$

These are the only planetary terms of any significance for this study. Thus, with $\omega_0(0) = 2.17/\text{y}$,

$$|\lambda_k| \approx 26''.0\,\text{cy}^{-2}/\omega_0{}^2(0) = 2.67 \times 10^{-9}. \tag{12}$$

Let the resonance angular rate as a function of η be represented by $\nu(\eta)$; with this choice, $\nu(0) = 1$. Also define $\lambda_0 = \nu'$. Because ν is proportional to n,

$$\frac{\dot{n}}{n} = \frac{d\ln n(t)}{dt} = \omega_0(0)\frac{d\ln \nu(\eta)}{d\eta} = \omega_0(0)\lambda_0,$$

and, therefore,

$$\lambda_0 = -0.150 \times 10^{-9}\text{y}^{-1}/\omega_0(0) = -0.069 \times 10^{-9}. \tag{13}$$

Then, on dropping the negligible η^2 term,

$$\omega_0{}^2(t) = \frac{\omega_0{}^2(0)n^2(t)}{n^2(0)} = \omega_0{}^2(0)[1 + 2\frac{\dot{n}}{n(0)}t] = \omega_0{}^2(0)(1 + 2\lambda_0\eta).$$

Change the independent variable of (9) from t to η,

$$\tau_k'' + (1 + 2\lambda_0\eta)(1 + i\kappa)\tau_k = (1 + 2\lambda_0\eta)(1 + i\kappa)H_k\exp[i(\eta + \lambda_k\eta^2/2)]. \tag{14}$$

Next substitute

$$\tau_k(\eta) = \xi_k(\eta)\exp(i\eta) \tag{15}$$

into (14), where $\xi_k(\eta)$ is a function that varies slowly compared with $\exp(i\eta)$ so that $|\xi_k''(\eta)| \ll |\xi_k'(\eta)|$. Then, on neglecting ξ_k'' and all terms quadratic in κ, λ_0, H_k,

$$\xi_k' + (\kappa/2 - i\lambda_0\eta)\xi_k = -i(H_k/2)\exp(i\lambda_k\eta^2/2). \tag{16}$$

Using the integrating factor $\exp[(\kappa\eta - i\lambda_0\eta^2)/2]$, the solution to (16) at $\eta = h$ is

$$\xi_k(h) = -\frac{i}{2}H_k\exp[\frac{-\kappa h + i\lambda_0 h^2}{2}]\int_{-\infty}^{h}\exp[\frac{i(\lambda_k - \lambda_0)\eta^2 + \kappa\eta}{2}]d\eta. \tag{17}$$

From (12) and (13), it is clear that λ_k dominates the $\lambda_k - \lambda_0$ term. The integral for $\lambda_0 > \lambda_k$ (λ_k obviously being negative) is

$$\int_{-\infty}^{h}\exp[\frac{-i(\lambda_0 - \lambda_k)\eta^2 + \kappa\eta}{2}]d\eta = \exp[-i(\kappa/p_k)^2/8]\int_{-\infty}^{h}\exp(-w^2)d\eta \tag{18}$$

where

$$w = w(\eta) = (1+i)p_k\eta/2 - (1-i)\kappa/(4p_k) \quad (19)$$

and

$$p_k = |\lambda_0 - \lambda_k|^{1/2}. \quad (20)$$

(The absolute value is used in this definition so that it will apply as well when λ_k is positive.) Substituting $(1-i){p_k}^{-1}dw = d\eta$,

$$\int_{-\infty}^{h} \exp(-w^2)d\eta = \frac{1-i}{p_k}\int_{-\infty}^{w(h)} \exp(-w^2)dw = \frac{1-i}{2p_k}\sqrt{\pi}[1+\mathrm{erf}[w(h)]], \quad (21)$$

where $\mathrm{erf}[w(h)]$ is the error function with the complex argument $w(h)$.

The phase of H_k is the phase of s_k precisely at $t = 0$. There is no way of knowing the phase of H_k, so there is no way of knowing the phase of τ_k near resonance. Only the magnitude of τ_k can be calculated as a function of $|H_k|$, p_k, κ and h; combining (15), (17), (18) and (21), it is

$$|\tau_k(h)| = \sqrt{\pi/2}[|H_k|/(2p_k)]\exp(-\kappa h/2)|1+\mathrm{erf}[w(h)]|. \quad (22)$$

The integral (18) for $\lambda_k < \lambda_0$ is the complex conjugate of the integral for $\lambda_k > \lambda_0$, so (22) is valid for all values of $\lambda_0 - \lambda_k$ except, of course, for (the non-existent) $\lambda_k = \lambda_0$ which represents being parked at resonance.

If $2{p_k}^2h \gg \kappa$,

$$w(h) \approx (1+i)p_k h/2 \quad (23)$$

and

$$\frac{1-i}{2p_k}\sqrt{\pi}[1+\mathrm{erf}[w(h)]] \approx \frac{\sqrt{\pi}}{2p_k}[[1+2\mathrm{C}(p_k h/\sqrt{\pi})] - i[1+2\mathrm{S}(p_k h/\sqrt{\pi})]],$$

where $\mathrm{C}(p_k h/\sqrt{\pi})$ and $\mathrm{S}(p_k h/\sqrt{\pi})$ are the Fresnel integrals. Then

$$|\tau_k(h)| \approx \sqrt{\pi}[\frac{|H_k|}{4p_k}]\exp(-\frac{\kappa h}{2})\sqrt{[1+2\mathrm{C}(\frac{p_k h}{\sqrt{\pi}})]^2 + [1+2\mathrm{S}(\frac{p_k h}{\sqrt{\pi}})]^2}. \quad (24)$$

For $p_k h \gg \sqrt{\pi}$, $\mathrm{C}(p_k h/\sqrt{\pi}) \approx \mathrm{S}(p_k h/\sqrt{\pi}) \approx 1/2$, so (24) becomes

$$|\tau_k(h)| \approx \sqrt{\pi/2}[|H_k|/p_k]\exp(-\kappa h/2). \quad (25)$$

4. The Accelerating Arguments

Let $m = n'/n = 0.0748$ *currently* where n' is the mean rate of solar motion. Numerical integrations of the orbits of the Solar System planets demonstrate that their mean motions have inappreciable accelerations over millions of years [Quinn, Tremaine and Duncan, 1991], so $\dot{n}' = 0$ and $\dot{m}/m = -\dot{n}/n$. Using the m power series for the mean motions of the lunar node Ω and perigee $\tilde{\omega}$ [Brouwer and Clemence, pp. 322-323, 1961], their accelerations are then related to $\dot{n}$ by

$$\begin{aligned}\ddot{\Omega} = {} & (1.5m^2 - 0.84375m^3 - 8.531252m^4 - 23.91845m^5 \\ & -48.6504m^6 - 79.016m^7)\dot{n} = 0.0077\dot{n}\end{aligned}$$

and

$$\begin{aligned}\ddot{\tilde{\omega}} = {} & (-1.5m^2 - 21.03125m^3 - 127.218752m^4 - 648.17625m^5 \\ & -3130.5256m^6 - 15118.887m^7 - 80282.32m^8 \\ & -42578.3m^9)\dot{n} = -0.0236\dot{n}.\end{aligned}$$

The accelerations of all components of the planetary perturbation arguments are essentially zero except for the Delaunay variables D, l and F which have the accelerations, respectively, $\dot{n}$, $\dot{n} - \ddot{\tilde{\omega}} = 1.0236\dot{n}$, and $\dot{n} - \ddot{\Omega} = 0.9923\dot{n}$; in units of η, they are $\dot{n}/\omega_0{}^2(0)$, $1.0236\dot{n}/\omega_0{}^2(0)$, and $0.9923\dot{n}/\omega_0{}^2(0)$ where $\dot{n}/\omega_0{}^2(0) = -2.67 \times 10^{-9}$. For a Fourier term in the planetary longitude perturbation of the Moon to be of interest, its period must be near resonance and, with its argument expressed as an increasing function of time, the sum of the integer coefficients of D, l and F in the argument must be equal to (or greater than) 1 if the term's period is greater than the resonant period or the sum must be equal to (or less than) -1 if its period is less than the resonant period. The effect of the slowly changing resonance (the λ_0 term) can be overlooked in this winnowing compared with the uncertainty in the period of resonance, a value that is theory and parameter dependent. Source terms from the Chapront and Chapront-Touzé [1983] ELP-2000 lunar ephemeris that meet these criteria are given in Table I. V, T and M represent the mean longitudes of Venus, Earth and Mars.

The nominal period of resonance is chosen as 38.652 sidereal months [Eckhardt, 1981]; adjustments for the resonant periods of Migus [1980] and Moons [1982] will be considered later. Because of their amplitudes and closeness to resonance (one above and the other below), the last two terms in Table I are of principal interest. The other three are insignificant and will be neglected. For the subscripts designating these terms, use 1 for 16331 and 2 for 16018.

Term 16331 is closest to resonance and, by far, the most important of all terms that have "recently" passed through resonance. Its acceleration is $(2 - 1.0236)\dot{n}/\omega_0{}^2(0)$, so

$$\lambda_1 = 0.9764 \times (-2.67 \times 10^{-9}) = -2.61 \times 10^{-9},$$

TABLE I

Planetary perturbations in longitude for terms that have "recently" passed through resonance and whose periods differ from the resonance period by less than one per cent. These terms were selected from the ELP-2000 lunar ephemeris of Chapront and Chapront-Touzé [1983]. The sequence numers are those of ELP-2000

Sequence Number	$\lvert H_k \rvert$ (Arc-Seconds)	Argument (Excluding Phase)	Period (Sidereal Months)
17353	0.00011	$-39V + 39T + 2D$	39.080
16766	0.00003	$-24V + 30T - 4M + 2D - l$	39.039
16979	0.00001	$-26V + 29T + 4M + 2D - l$	38.765
16331	0.00021	$-21V + 23T + 2D - l$	38.666
16018	0.00066	$19V - 18T - D + l - F$	38.258

and the λ_0 term just marginally contributes to making [see (13) and (20)]

$$p_1{}^2 = 2.54 \times 10^{-9}.$$

Then, because the dimensionless angular rate at resonance is unity,

$$1 + p_1{}^2 h = 38.666/38.652 = 1 + 0.000362$$

and

$$h = 143000.$$

Resonance occurred $(2.8912/2\pi)h = 66000$ years ago. At that time, its libration amplitude was half of

$$\sqrt{\pi/2}|H_1|/p_1 = 5''.2$$

and, assuming that (25) affords a close enough approximation, its amplitude is now

$$|\tau_1(h)| \approx 5''.2 \exp(-71000\kappa). \tag{26}$$

The more precise equation (22) will be applied later. Following the same calculations for Term 16018 gives

$$|\tau_2(h)| \approx 16''.1 \exp(-1.9 \times 10^6 \kappa). \tag{27}$$

From the ratio (26)/(27),

$$|\tau_1(h)|/|\tau_2(h)| \geq 1 \text{ for } \kappa \geq 0.62 \times 10^{-6}. \tag{28}$$

From the analysis of a six year span of lunar laser ranging data, Calame [1977] estimated the free libration in longitude to have a $1''.8$ amplitude. She did not,

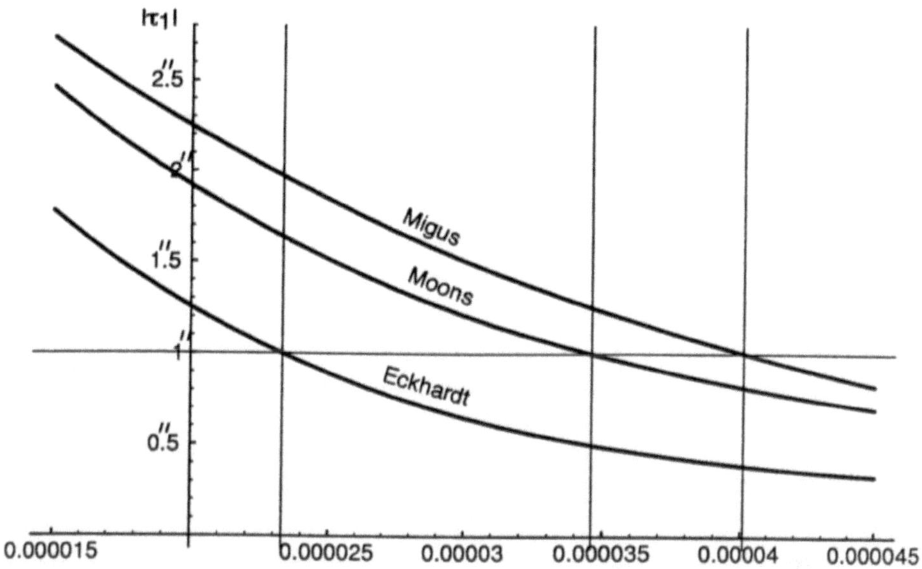

Fig. 1. Plots of $|\tau_1|$ at the current epoch as functions of κ. For the Eckhardt, Moons and Migus resonances, the solutions to $|\tau_1| = 1$ are indicated at $\kappa = 23.2 \times 10^{-6}$, $\kappa = 34.6 \times 10^{-6}$, and $\kappa = 40.1 \times 10^{-6}$.

however, allow for the planetary terms in lunar longitude. Except for $\tau_1(h)$ and $\tau_2(h)$, the neglected terms (principally ELP-2000 Term 6649 which has amplitude $0''.85$, argument $3T - 5M$ and period 39.138 sidereal months [Eckhardt, 1982]) account for only about half of her $1''.8$ estimate. Suppose that the approximately $1''.0$ residual is due to $\tau_1(h)$ or $\tau_2(h)$. If it were entirely due to $\tau_1(h)$ then, by (26),

$$\kappa = 23 \times 10^{-6}, \tag{29}$$

and if it were entirely due to $\tau_2(h)$ then, by (27), $\kappa = 1.5 \times 10^{-6}$; but, by (28), $|\tau_1(h)| > |\tau_2(h)|$ if $\kappa = 1.5 \times 10^{-6}$, so (29) provides the single estimate of the dissipation parameter for the Eckhardt [1981] resonance and $1''.0$ residual. This is only an approximate solution and modifications are also required for the Moons [1982] and Migus [1980] resonances. Moreover the $1''.0$ residual is only a nominal value. A plot of $|\tau_1(143000)|$ [Eckhardt], $|\tau_1(112000)|$ [Moons], and $|\tau_1(71000)|$ [Migus] as functions of κ by the general solution (22) is presented in Figure 1. The solutions to $|\tau_1(h)| = 1$ are indicated at

$$\kappa = 23.3 \times 10^{-6} \quad \text{[Eckhardt]}, \tag{30}$$

$$\kappa = 34.6 \times 10^{-6} \quad \text{[Moons]},$$

and

$$\kappa = 40.1 \times 10^{-6} \quad \text{[Migus]}.$$

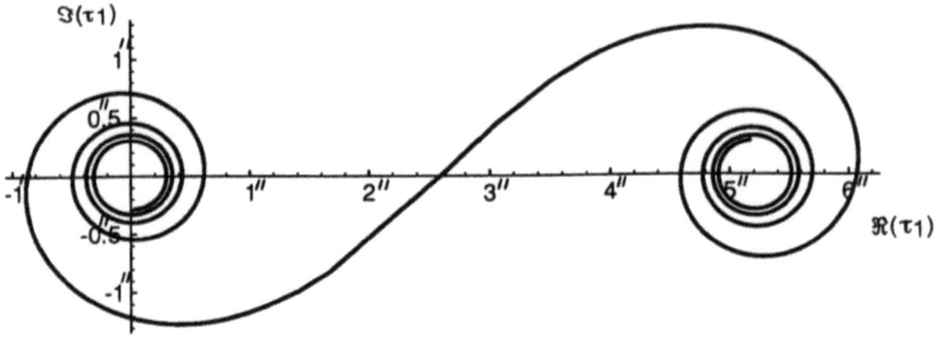

Fig. 2. The solution to (25) for ELP-2000 Term 16331 from 132000 years ago until the present. Resonance according to the adopted model occurred 66000 years ago. The coordinate units are arc-seconds. There exists no free libration with an amplitude as large as 5″ as shown in this solution, so the effect of dissipation must be included to get a plausible model.

Although Moons' resonance is closer to Eckhardt's than Migus', her solution is closer to Migus' than Eckhardt's at the current epoch. To understand how this can be, and to gain insight into the nature of the solutions, the solution paths in the complex plane are examined.

5. The Solution Paths

If $\kappa = 0$ and $\lambda_0 = 0$, Equation (16) becomes

$$\xi_k' = -i(H_k/2)\exp(i\lambda_k\eta^2/2). \tag{31}$$

Because $|\xi_k'| = |H_k/2|$, $\xi_k(\eta)$ follows a path in the complex plane at a constant rate with respect to η. The direction of the path is given by the the sum of the phase of $-iH_k$ and the argument of the exponential term. Suppose that $\lambda_k < 0$; then for $\eta < 0$, the path is clockwise, and for $\eta > 0$, the path is counter-clockwise. The angular rate is proportional to η^2 so, for $\eta < 0$, the rotation rate decreases, and for $\eta > 0$, the rotation rate increases. The net result if $-iH_k$ is real and positive is that the solution follows the Cornu spiral [Born and Wolf, pp. 430-433, 1970] depicted in Figure 2. (The rotations are in the opposite directions for $\lambda_k > 0$ and the orientation of the spiral depends on the phase of $-iH_k$.)

If $\kappa = 0$ and $H_k = 0$, Equation (16) becomes

$$\xi_k' = i\lambda_0\eta\xi_k. \tag{32}$$

In this case, $\xi_k = |\xi_k|\exp(i\lambda_0\eta^2/2)$ follows a circular path in the complex plane with an angular rate proportional to $|\lambda_0\eta|$. The path is counter-clockwise for $\lambda_0 < 0$, and it is clockwise for $\lambda_0 > 0$. A combination of (31) and (32) results in a Cornu

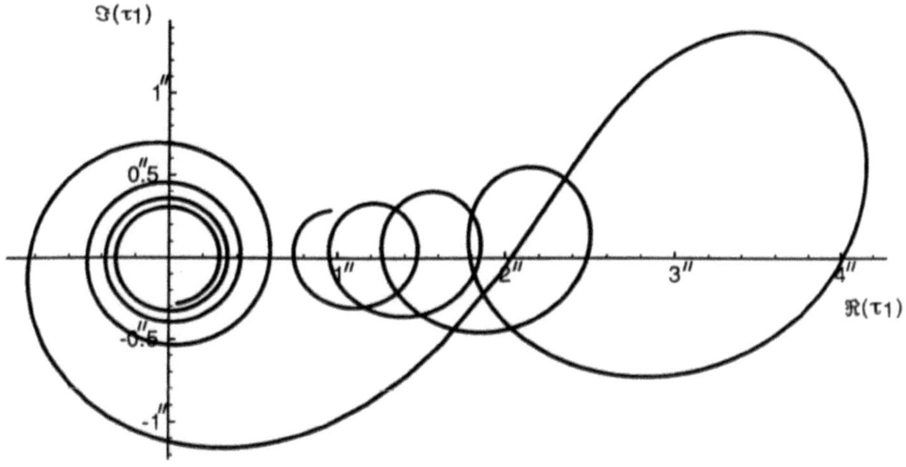

Fig. 3. The solution to (20) for ELP-2000 Term 16331 with the same units and over the same time span as Figure 2. The model is modified by including a dissipation term. The current 1″ amplitude of $\tau_1(143000)$ shown in this plot would be perceived as a free libration.

spiral with the same dimensions as as (31); the only effect of (32) is to rotate the Cornu spiral without changing its dimensions. For the current problem, (32) is only of marginal interest.

If $\lambda_0 = 0$ and $\lambda_k = 0$, Equation (16) becomes

$$\xi_k' = -(\kappa/2)\xi_k, \tag{33}$$

so $\xi_k(\eta)$ follows an exponential decay path in the complex plane that is a straight line directed toward the origin. In the general solution of (16), $\xi_k(\eta)$ both spirals as in (31) and decays as in (33). Well before passing through resonance, when $p_k\eta \ll -1$, $\xi_k(\eta)$ moves in a tight spiral around the origin and the decay term has little effect. As $p_k\eta$ becomes small and $\xi_k(\eta)$ moves through the region of the resonance inflection point, (31) dominates (33). Eventually, when $p_k\eta \gg 1$, $\xi_k(\eta)$ moves in an increasingly tight spiral around a focus that decays toward the origin, so (33) dominates (31). The general solution to (16) is given by (22) which can conveniently be evaluated and plotted using Mathematica [Wolfram, 1988]. A plot of the the $\tau(\eta)$ solution path according to (22) (retaining the relative phase from the error function) is shown in Figure 3. The dissipation parameter used is given by (30).

For the shorter time intervals and higher dissipations that are demanded by the Moons and Migus resonances to attain a $1''.0$ amplitude for $\tau_1(h)$, their solution curves, shown in Figure 4, are markedly different. The two curves in Figure 4 are currently headed in opposite directions. These directions are relatively insensitive to either κ used. As time increases, the $|\tau_1|$ in the top plot is decreasing and the $|\tau_1|$ in the bottom plot is increasing. Thus the entire curve for the Moons solution ($|\tau_1| = 1''.0$ at $\kappa = 34.6 \times 10^{-6}$) is moving downward with time whereas the entire

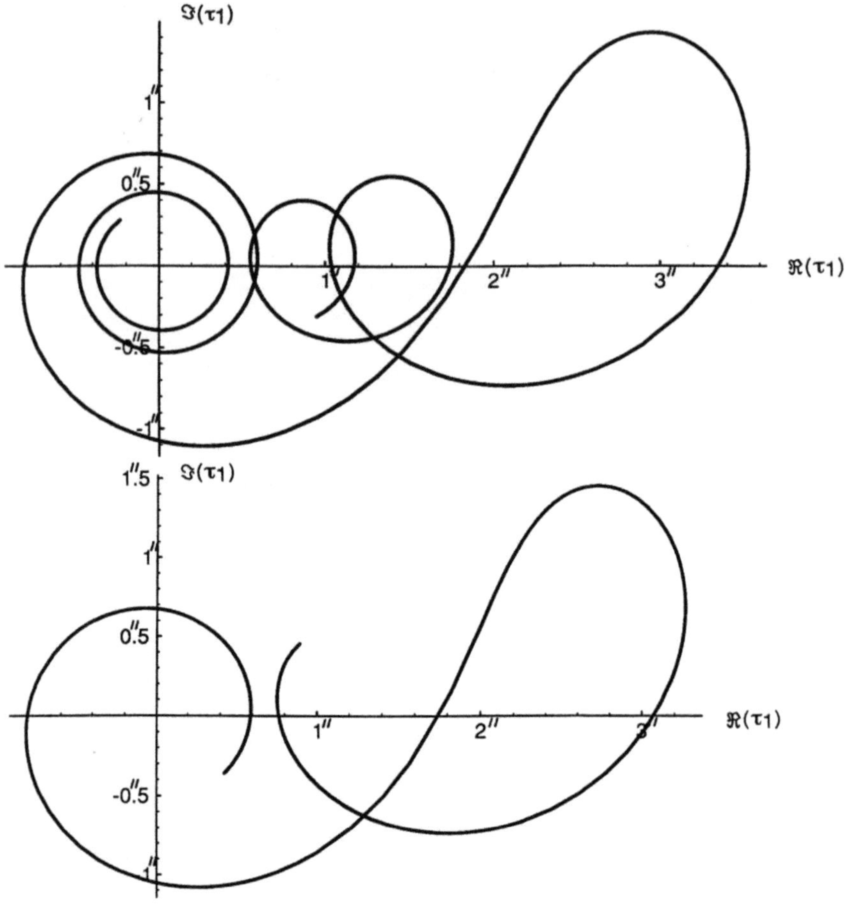

Fig. 4. The solutions to (20) for ELP-2000 Term 16331 using the Moons (top) and Migus (bottom) resonances and k given by (35.2) and (35.3). The span of the top plot is 104000 years, and that of the bottom plot is 66000 years.

curve for the Migus solution ($|\tau_1| = 1''.0$ at $\kappa = 40.1 \times 10^{-6}$) is moving upward with time. They only happen to be close to each other at the present.

Figure 5 is a plot of the libration amplitude $|\tau_1(h)|$ as a function of the dissipation parameter κ and the resonance period in sidereal months. One can see from this plot that the Figure 1 solution for the Eckhardt resonance (38.652 sidereal months) falls in a trough; the solution for the Moons resonance (38.655 sidereal months) runs along a ridge; and the solution for the Migus resonance (38.659 sidereal months) falls in a trough. A change of resonance of only 0.001 sidereal months moves a solution from a ridge to a trough. The resonance period is not known well enough to decide whether the solution runs along a ridge, in a trough, or somewhere in between.

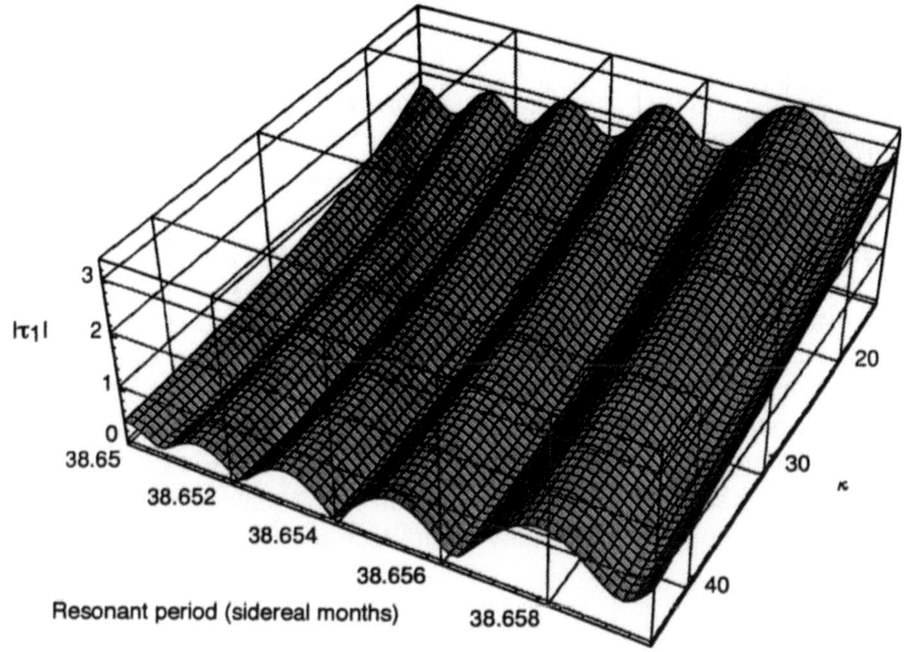

Fig. 5. The amplitude of $\tau_1(h)$ as a function of the resonance period (in sidereal months) and the dissipation parameter, κ.

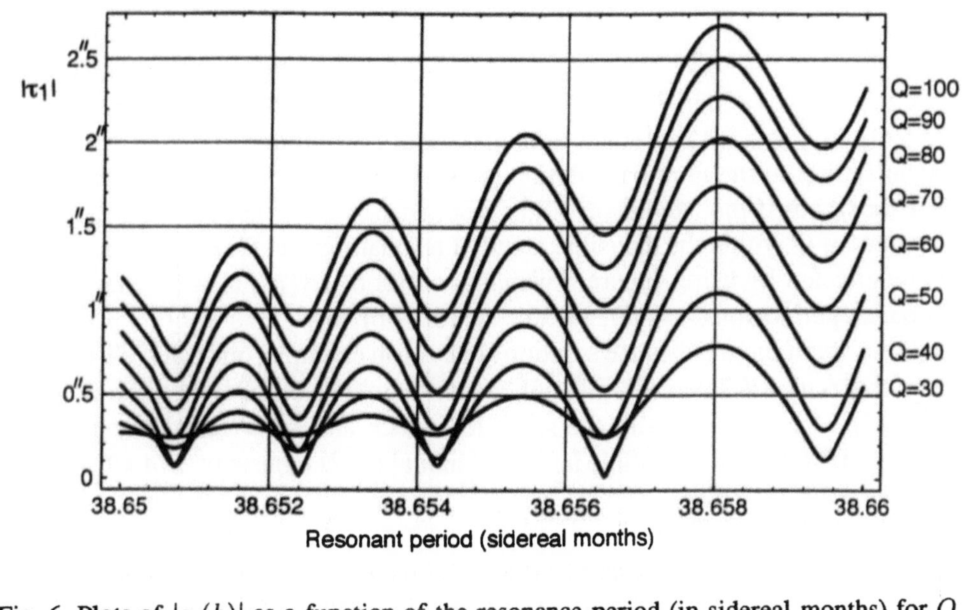

Fig. 6. Plots of $|\tau_1(h)|$ as a function of the resonance period (in sidereal months) for Q running from 30 to 100.

The steady state solution (10) for $\tau_2(h)$ has an amplitude of $0''.032$. If approximation (25) is used to estimate $|\tau_2(h)|$, the result is zero, independent of κ. If, however, (22) is used, the amplitude is $0''.032$, again independent of κ. The failure of the approximation arises from the elimination of the relatively very small term $(1-i)\kappa/(4p_k)$ in replacing (19) by (23). Because the argument of the error function is not precisely $\pi/4$ and $w(h)$ is very large, the error function has a very large real part that counters the very small $\exp(-\kappa h/2)$ attenuation term. What actually happens is that $\xi_k(\eta)$ revolves about the origin, as indicated by the solution path of (31), and the revolution rate is just the right amount to offset ω_0 to ω_2. The revolution velocity is slightly perturbed by a centripetal velocity (not acceleration) that, in effect, changes the solution phase just as the neglected $i\kappa$ term would change the solution phase in (10). The steady state solution is appropriate for ELP-2000 Term 16018, so the only term for the libration in longitude that should be modified to allow for having passed through resonance is the one due to ELP-2000 Term 16331. The steady state libration solution for this term should be replaced with a residual free libration term with unknown amplitude and phase.

6. The Q of the Moon

Seismological and astronomical techniques have been used to derive estimates of the potential-disturbance Love numbers and rigidities (or, to be more precise, their real parts) for the Earth and the Moon. The estimates are mutually consistent. This consistency unravels when it comes to estimating the imaginary parts of k and μ.

The principal effect of $\Im[k]$ in the theory of the libration of the Moon is that the node of the lunar equator on the ecliptic advances for positive $\Im[k]$. Let $\delta\Omega$ be the offset in the node; then $I\delta\Omega = \Im[k] \times 208''$ [Eckhardt, 1981]. Using (30), $\Im[k] = -\kappa/\varsigma = -0.25 \times 10^{-3}$, so, by the nominal solution, $I\delta\Omega$ decreases by $0''.052$ leading to a -45 cm $\cos F$ term in the latitude libration. Also, $Q = k\varsigma/\kappa = 100$ according to the nominal solution ($1''.0$ residual free libration and Eckhardt resonance), but if the Moons or Migus resonance is used instead, then $Q \approx 55$. Estimates of Q within the plausible range of resonances and for different residuals can be effected by using Figure 6 which is a plot of $|\tau_1(h)|$ as a function of the resonance period for $Q = 30, 40, 50, \ldots, 100$. For low Q, as at $Q = 30$, $\tau_1(h)$ oscillates about the steady state solution (10). As an example, one can see from this figure that with a resonance of 38.658 sidereal months and $Q = 40$, the residual is $|\tau_1(h)| = 1''.15$, and with the same resonance and $Q = 30$, the residual is $|\tau_1(h)| = 0''.80$. (The amplitude in the Eckhardt [1982] table is $0''.274$.) Thus a Q as low as 40, and perhaps even 30, may not be inconsistent with the observed free libration.

Williams, Newhall and Dickey [1987] estimate Q for a monthly period to be approximately 30; in effect, they detected and isolated a 1.5 m $\cos F$ term in the latitude libration. Seismic estimates of the lunar rigidity Q are depth dependent, varying from about 3000 in the region between 60 and 400 km depth to 1000 or less for depths exceeding 1100 km [Goins, Dainty and Toksoz, 1981]. The broad range of lunar Q estimates is not unlike the range of Q estimates for the Earth. Seismic rigidity (shear wave) Q's are depth dependent, varying from about 100 at the top of the mantle to about 2000 at the base of the mantle [Stacey, pp. 300-307, 1977]. Estimates of Q from polar motion data range between 30 and 600 [Munk and MacDonald, pp. 167-174, 1960; Stacey, p. 67, 1977]. The Earth has a liquid core and a more pronounced stratification than the Moon, so it is a more complicated body. It is, however, much easier to make Q-sensitive measurements of the Earth than of the Moon. That the seismic Q's of the Moon are much larger than its rotational Q's is not an inconsistency; rocks have different viscoelastic properties at $\sim$ 1 Hz than at one month or three years. The rotational Q's are roughly bracketed somewhere between 30 and 100; to some geophysicists, that's pinpoint precision.

7. Conclusion

The planetary perturbation term ELP-2000 Sequence Number 16331 [Chapront and Chapront-Touzé , 1983] modulates the lunar longitude with an amplitude of only $0''.0021$ but, because it passed through resonance 33 to 66 thousand years ago, it is the source of a $\sim 1''$ lunar libration term that has been considered a free libration. In the sense that this libration term's phase is a free parameter unrelated to other parameters of lunar motion, this is true; but in the sense that the term's amplitude is not a free parameter but, in fact, a sensitive indicator of lunar dissipation at the 38.65 sidereal month resonant period and its excitation source is known, regarding the term as a free libration is misleading by accepted standards. The definition of a free libration is hereby broadened to include this term.

Three libration theories differ in their resonance periods for the libration in longitude by less than 0.02% but the amplitude of the free libration in longitude is very sensitive to which theory is used; and, of course, it is sensitive to the imaginary part of the potential-disturbance Love number at resonance. The Moon is a fairly uniform body, and its rigidity and density are well constrained and so, therefore, is the real part of the Love number. The Q of the lunar rigidity is the dissipation parameter that, with a choice of resonance, can be estimated from the free libration amplitude. Allowing for uncertainties in the amplitude and resonance, the Moon's Q still falls within a region that is entirely reasonable by geophysical standards.

Since an astronomic excitation source for the free libration in longitude has been found, there is less justification to posit a source within the Moon, e.g., a turbulent fluid core that interacts with the mantle [Yoder, 1981]. Thus, the case for a fluid lunar core is weakened.

References

Ben-Menahem, A. and Singh, S. J.: 1981, "Seismic Waves and Sources", Springer-Verlag.
Born, E. and Wolf, E.: 1970, "Principles of Optics", Pergamon Press, Fourth Edition.

Brouwer, D. and Clemence, G. M.: 1961, "Methods of Celestial Mechanics", Academic Press.

Calame, O.: 1977, "Free Librations of the Moon from Lunar Laser Ranging", in Scientific Applications of Lunar Laser Ranging, ed. J. D. Mulholland, D. Reidel Publishing Co., 53–63.

Chapront-Touzé, M. and Chapront, J.: 1983, *Astron. Astrophys.*, **124**, 50–62.

Cheng, H. C. and Toksoz, M. N.: 1978, *J. Geophys. Res.*, **83**, 845–853.

Eckhardt, D. H.: 1973, *The Moon*, **6**, 127–134.

Eckhardt, D. H.: 1981, *Moon and the Planets*, **25**, 3–49.

Eckhardt, D. H.: 1982, "Planetary and Earth Figure Perturbations in the Librations of the Moon", in High-Precision Earth Rotation and Earth-Moon Dynamics, ed. O. Calame, D. Reidel Publishing Co., 193–198.

Goins, N. R., Dainty, A. M., and Toksoz: 1981, *J. Geophys. Res.*, **86**, 5061–5074.

Jeffreys, H.: 1976, "The Earth", Cambridge University Press.

Love, A. E. H.: 1927, "A Treatise on the Mathematical Theory of Elasticity", Fourth Edition, Cambridge University Press.

Migus, A.: 1980, *Moon and the Planets*, **23**, 391–427.

Moons, M.: *Moon and the Planets*, **27**, 257–284.

Munk, W. H. and MacDonald, G. J. F.: 1960, "The Rotation of the Earth", Sixth Edition, Cambridge University Press.

Nakamura, Y.: 1983, *J. Geophys. Res.*, **88**, 677–686.

Quinn, T. R., Tremaine, S., and Duncan, M.: 1991, *Astron. J.*, **101**, 2287–2305.

Stacey, F. D.: 1977, "Physics of the Earth", Second Edition.

Williams, J. G.: 1977, "Present Scientific Achievements from Lunar Laser Ranging", in Scientific Applications of Lunar Laser Ranging, ed. J. D. Mulholland, D. Reidel Publishing Co., 37–50.

Williams, J. G., Newhall, X X, and Dickey, J. O.: 1992, "Lunar Laser Ranging: Geophysical Results and Reference Frames", submitted to the American Geophysical Union Monograph: Space Geodesy and Geodynamics.

Williams, J. G., Newhall, X X, and Dickey, J. O.: 1987, "Lunar Gravitational Harmonics and Reflector Coordinates", in Figure and Dynamics of the Earth, Moon and Planets, ed. P. Holota, Czechoslovakia Academy of Sciences Research Institute of Geodesy, Topography and Cartography, Prague, 643–648.

Williams, J. G., Slade, M. A., Eckhardt, D. H., and Kaula, W. M.: 1973, *The Moon*, **8**, 469–483.

Wolfram, S.: 1988, "Mathematica", Addison-Wesley Publishing Co.

Yoder, C.: 1981, *Phil. Trans. R. Soc. Lond.*, **A303**, 327–338.

STABILITY OF THE SYNCHRONOUS SPIN-ORBIT RESONANCE BY CONSTRUCTION OF LIBRATIONAL TRAPPING TORI

ALESSANDRA CELLETTI
Dipartimento di Matematica Pura e Applicata, Universitá di L'Aquila
Via Vetoio - I–67010 Coppito (L'Aquila), Italy
e-mail: Celletti@vxscaq.aquila.infn.it

1. The Main Equation and the Structure of the Phase Space

We intend to study the stability of the 1:1 resonance by applying perturbation techniques. This paper is a short presentation of the method developed in (Celletti, 1993), to which we refer for a complete exposition.

Let S be a triaxial ellipsoidal satellite rotating about an internal spin–axis and at the same time revolving around a central body P. A spin–orbit resonance means an exact commensurability between the periods of rotation and revolution of the satellite. In particular a 1:1 (or synchronous resonance) implies that the period of rotation T_{rot} and the period of revolution T_{rev} are identically equal. The most familiar example of synchronous resonance is provided by our Moon which satisfies to a high degree of approximation the relation $T_{rev} \simeq T_{rot}$. This situation is not peculiar in the solar system but, on the contrary, most *evolved* satellites are trapped in a synchronous resonance.

At first we reduce the physical model making some simplifying assumptions which are generally satisfied by most satellites of the solar system. In particular we assume that $i)$ the satellite S moves on a *fixed* Keplerian ellipse around P; $ii)$ the spin–axis is perpendicular to the orbit plane and coincides with the axis of largest moment of inertia; $iii)$ dissipative torques as well as interactions with other bodies of the solar system are neglected.

Under these assumptions the equation of motion is derived from Euler's equations for a rigid body and, as presented in (Celletti, 1990) can be reduced to the form

$$\ddot{x} + \varepsilon \sum_{m\neq 0, m=N_1}^{N_2} W(\frac{m}{2}, e) \sin(2x - mt) = 0 , \tag{1}$$

where x is the angle between the longest axis of the ellipsoid and the periapsis line, ε is proportional by a factor 3/2 to the *equatorial oblateness* of the satellite and the coefficients W's have the property to decay as powers of the orbital eccentricity e as $W(\frac{m}{2}, e) \propto e^{|m-2|}$. Finally, N_1 and N_2 are some integers depending on the structure of the satellite. Equation (1) can be derived from Hamilton's equations

Celestial Mechanics and Dynamical Astronomy **57**: 325–328, 1993.

associated to the nearly integrable Hamiltonian

$$H(y,x,t) = \frac{y^2}{2} - \frac{\varepsilon}{2} \sum_{m\neq 0, m=N_1}^{N_2} W(\frac{m}{2},e) \cos(2x - mt) , \tag{2}$$

where $y \in \mathbf{R}$, $(x,t) \in (\mathbf{R}/2\pi\mathbf{Z})^2 \equiv T^2$. As it is easily seen, any surface $\{y_0\} \times T^2$ is invariant under the unperturbed flow (i.e., setting $\varepsilon = 0$ in (2)). For non–zero values of ε, most invariant surfaces persist under the perturbation. Having fixed a rotation number Ω, let $\mathcal{T}_\varepsilon(\Omega)$ be the invariant surface, for the perturbed system, with rotation number Ω. The KAM theory (Kolmogorov, 1954; Arnold, 1963; Moser, 1962) provides an explicit constructive algorithm to give a lower bound, say $\varepsilon_r = \varepsilon_r(\Omega)$, on the perturbing parameter ensuring the existence of $\mathcal{T}_\varepsilon(\Omega)$ for any $\varepsilon \leq \varepsilon_r(\Omega)$. The theorem can be applied provided the unperturbed Hamiltonian is not–degenerate and provided the rotation number Ω satisfies the diophantine condition $|\Omega - \frac{p}{q}|^{-1} \leq Cq^2, \forall p,q \in \mathbf{Z},\ q \neq 0$, for some positive constant C.

In order to obtain the stability of the synchronous periodic orbit we examine the structure of the phase–space associated to (2). Since H is one–dimensional with an explicit time dependence the corresponding phase space is three–dimensional. Therefore any invariant two–dimensional surface divides the phase space in separate regions, providing a strong stability property in the sense of confinement of the motions. A Poincarè sectio associated to our equation shows small *librational* curves surrounding the periodic orbit, whose amplitude increases as the chaotic separatrix is approached. Such separatrix divides the region of librational motion from the region in which *rotational* curves (corresponding to a complete rotation of the satellite) are found. Therefore we can confine the synchronous periodic orbit either by proving the existence of two rotational curves bounding the periodic orbit from above and below or by proving the existence of a librational curve surrounding the 1:1 resonance.

The first statement was already investigated in (Celletti, 1990). Here we intend to show a method for the construction of librational surfaces. To this end we need to compute a suitable Hamiltonian function, performing some area–preserving changes of variables as described in the following paragraph.

2. The Hamiltonian of Librational Motions

In this paragraph we introduce a Hamiltonian function which describes librational motions around the synchronous resonance. Let $a \equiv \frac{1}{2}W(1,e)$; we center the Hamiltonian (2) on the synchronous resonance and place the resonance on the (0,0) position by performing the symplectic changes of variables ($x' = 2x - 2t$, $y' = \frac{1}{2}(y-1)$), ($\overline{x} = x'$, $\overline{y} = y' - \frac{1}{2}$), so that (2) becomes

$$\begin{aligned} H(\overline{y},\overline{x},t) &= 2\,\overline{y}^2 - \varepsilon a \cos\overline{x} \\ &\quad - \frac{\mu}{2} \sum_{m\neq 0,-2} \tilde{W}(\frac{m+2}{2},e)\,[\cos\overline{x}\,\cos mt + \sin\overline{x}\sin mt] , \end{aligned} \tag{3}$$

where $\mu \equiv \varepsilon e$ and $\tilde{W}(\frac{m+2}{2}, e) = \frac{1}{e} W(\frac{m+2}{2}, e)$. One has librational motions as far as the energy is bounded by $|E| < \varepsilon a$. Therefore let us fix a level energy $\overline{\delta}$ as $2\,\overline{y}^2 - \varepsilon a \cos \overline{x} = \overline{\delta}$ with $|\overline{\delta}| < \varepsilon a$ (notice that small curves around the periodic orbit are characterized by a value of $\overline{\delta}$ very close to $-\varepsilon a$).

Now we expand (3) in Taylor series around the point (0,0) and reduce it to a standard harmonic oscillator form, performing the linear change of variables ($p = \alpha\,\overline{y}$, $q = \beta\,\overline{x}$), where $\alpha = \frac{\sqrt{2}}{(\varepsilon a)^{1/4}}$, $\beta = \frac{(\varepsilon a)^{1/4}}{\sqrt{2}}$. We denote by $\omega \equiv 2\sqrt{\varepsilon a}$ the frequency of the harmonic oscillations. The last coordinate transformation corresponds to the introduction of action–angle variables (I, ϕ) for the harmonic oscillator, i.e. $p = \sqrt{2I}\,\cos\phi$, $q = \sqrt{2I}\,\sin\phi$. After some algebra the Hamiltonian in its final form is

$$
\begin{aligned}
H(I,\phi,t) = {} & \omega I - \varepsilon a \left(\frac{I^2}{16\beta^4} - \frac{5I^3}{2\cdot 6!\beta^6} + \ldots\right) - \varepsilon a \Big[-\frac{I^2}{12\beta^4}\cos 2\phi \\
& + \frac{I^2}{48\beta^4}\cos 4\phi + \frac{I^3}{4\cdot 6!\beta^6}\cdot(15\cos 2\phi - 6\cos 4\phi + \cos 6\phi) + \cdots \Big] \\
& - \frac{\varepsilon e}{2} \sum_{m \neq 0,-2} \tilde{W}\left(\frac{m+2}{2}, e\right) \\
& \Big\{ \cos(mt) \Big[1 - \frac{I}{2!\beta^2}(1 - \cos 2\phi) \\
& \quad + \frac{I^2}{8\cdot 3!\beta^4}(3 - 4\cos 2\phi + \cos 4\phi) \\
& \quad - \frac{I^3}{4\cdot 6!\beta^6}(10 - 15\cos 2\phi + 6\cos 4\phi - \cos 6\phi) + \cdots \Big] \\
& + \sin(mt) \Big[\frac{\sqrt{2I}}{\beta}\sin\phi - \frac{\sqrt{2}\,I^{3/2}}{12\beta^3}(3\sin\phi - \sin 3\phi) \\
& \quad + \frac{\sqrt{2}I^{5/2}}{4\cdot 5!\beta^5}(10\sin\phi - 5\sin 3\phi + \sin 5\phi) + \cdots \Big] \Big\} ,
\end{aligned}
\tag{4}
$$

where we did not indicate the terms of order greater than three in the action variable. Notice that the modulus of the action variable $|I|$ represents the mean distance from the origin. More precisely the level energy $\overline{\delta}$ corresponds to a mean distance $|\overline{I}| \equiv \frac{\overline{\delta}+\varepsilon a}{2\sqrt{\varepsilon a}}$ and the expansion in (4) converges provided $|I|/\beta^2 << 1$.

3. Construction of Librational Tori

In order to construct librational tori we apply KAM theorem using the algorithm developed in (Celletti and Chierchia, 1987). However, before applying KAM theorem it is usually convenient to reduce the size of the perturbation in (4) by a Birkhoff transformation, so to conjugate (4) to a Hamiltonian of the form

$$
H_k(I', \phi', t) \equiv h_k(I'; \varepsilon) + \varepsilon^{k+1} R_k(I', \phi', t) , \tag{5}
$$

where (I', ϕ') are the transformed variables. This step can be obtained performing a close–to–identity transformation, whose generating function is

$$
I'\varphi + \sum_{j=1}^{k} \Phi_j(I', \varphi, t)\varepsilon^j .
$$

The functions $\Phi_j(I', \varphi, t)$ can be recursively determined imposing that the transformed Hamiltonian in the mixed set of variables (I', φ, t) does not depend on the angle variables up to $O(\varepsilon^{k+1})$. In order to apply the KAM theorem one needs just an estimate (and not the explicit expression, see (Celletti and Chierchia, 1987)) of the new perturbing function. This bound can be obtained through the estimates of the functions Φ_j (Celletti, 1993). Once obtained the Hamiltonian (5) one can proceed to apply the KAM theorem in order to state the existence of a librational torus with given rotation number $\Omega \equiv \frac{\partial h_k(I'_0)}{\partial I'}$ (for a suitable center I'_0). In particular, we apply the computer–assisted KAM algorithm developed in (Celletti and Chierchia, 1987) to which we refer for a complete exposition. Before concluding our discussion, we remark that increasing the order k of the Birkhoff procedure does not imply to obtain better results. Indeed, there is an optimal order, say $k = k_*$, of the Birkhoff–KAM procedure. Therefore stopping our computations at the order $k_* = 5$ for the Moon and $k_* = 4$ for Rhea, we obtain the following results.

Consider the Moon–Earth system described by Hamiltonian (2) with $N_1 = -1$, $N_2 = 5$ and let $e = 0.0549$, $\varepsilon_{obs} = 3.45 \cdot 10^{-4}$ (i.e., the real astronomical value). Then for any $\varepsilon \leq \varepsilon_{obs}/5.26$, there exists an invariant curve corresponding to a libration of about $8^o.79$. Unfortunately in this case it is not possible to obtain constructive results for parameter values consistent with the astronomical observations. On the contrary, one obtains significant results in the Rhea–Saturn system which is described again by (2) with $N_1 = 1$, $N_2 = 5$. Let $e = 0.00098$ and $\varepsilon_{obs} = 3.45 \cdot 10^{-3}$; then for any $\varepsilon \leq \varepsilon_{obs}$, there exists an invariant curve (around the synchronous orbit) corresponding to a libration of about $1^o.95$.

The details of the proof are available in (Celletti, 1993).

References

Arnold V.I.: 1963, 'Proof of a Theorem by A.N. Kolmogorov on the invariance of quasi-periodic motions under small perturbations of the Hamiltonian', *Russ. Math. Surveys*, **18**, 9.

Celletti A.: 1990, 'Analysis of resonances in the spin–orbit problem in Celestial Mechanics: The synchronous resonance (Part I)', *J. of Appl. Math. and Phys. (ZAMP)*, **41**, 174

Celletti A.: 1993,'Construction of librational invariant tori in the spin–orbit problem', J. of Appl. Math. and Phys. (ZAMP), accepted for publication.

Celletti A., Chierchia L.: 1987, 'Rigorous estimates for a computer-assisted KAM theory', *J. Math. Phys.*, **28**, 2078

Kolmogorov A.N.: 1954, 'On the conservation of conditionally periodic motions under small perturbation of the Hamiltonian', *Dokl. Akad. Nauk. SSR*, **98**, 469

Moser J.: 1962, 'On invariant curves of area-preserving mappings of an annulus', *Nach. Akad. Wiss. Göttingen, Math. Phys. Kl. II*, **1**, 1

NEBULAR DRAG AND CAPTURE INTO SPIN-ORBIT RESONANCE

OTHON C. WINTER*

Astronomy Unit, School of Mathematical Sciences, Queen Mary and Westfield College, University of London, Mile End Road, London E1 4NS, England

and

WAGNER SESSIN

Departamento de Mecânica do Vôo e Orbital, Instituto Tecnológico de Aeronáutica, 12228-900, São José dos Campos, SP, Brazil

Abstract. The majority of planetary satellites whose spin period is known are observed to be in synchronous spin-orbit resonance. The commonly accepted explanation for this observation is that it is due to the effects of tidal evolution. However, cosmogonic theories state that the formation of planetary and satellite systems occurs within a primordial solar nebula and circumplanetary nebulae, respectively. In this paper the influence of nebular drag on the capture into spin-orbit resonance is analysed. The results show that the torques generated are important for these resonances in a wide range of cases. Using the protojovian nebula model by Lunine and Stevenson (1982), conservative estimates of the despinning time scales for the Galilean satellites are computed. In comparison the despinning time scale from tidal effects are several orders of magnitude larger.

Key words: Spin-orbit resonance – nebular drag

1. Introduction

Twenty-two out of twenty-four planetary satellites whose spin rates have been determined are known to be in synchronous spin-orbit resonance (Burns, 1986). This preferred rotational state is most likely a consequence of tidal effects on the evolution of the satellite orbits.

Gravitational torques due to the permanent deviations from axial symmetry of the satellites are far larger than the tidal distortions and are responsible for the maintenance of synchronous rotation once it is attained. However, we still need to understand how the satellites were brought close to synchronous rotation.

The main theories about the rotational histories of the natural satellites have been summarized by Peale (1977,1986). According to these theories the natural satellites have evolved significantly due to the dissipation of energy in wobble motion or by the exchange of angular momentum between spin and orbital motion through tides.

Since no object in the solar system is either perfectly rigid or perfectly elastic, an arbitrary rotation of a semi-rigid satellite is expected to decay toward the minimum energy configuration corresponding to the given angular momentum (Lamy and Burns, 1972). Minimum energy thus corresponds to rotation about the principal axis of maximum moment of inertia. Adopting the tidal dissipation function Q=100 (comparable to that of the Earth's mantle), this wobble decay was calculated by Peale (1991) for most of the natural satellites in the solar system. All the resulting

* on leave of absence from Departamento de Matemática, Faculdade de Engenharia de Guaratinguetá, UNESP, CP 205, 12500-000, Guaratinguetá, SP, Brazil

Celestial Mechanics and Dynamical Astronomy **57**: 329–339. 1993.

time constants for exponential decay were small, except for Iapetus and Hyperion, with time constants of 6.7×10^8 yr and 1.5×10^9 yr, respectively.

The differential gravitational field of the planet distorts the satellite and this distortion results in the well known body tides. Dissipation of tidal energy results in a phase lag in the response of the satellite to the perturbing potential. The tidal bulge on the satellite is no longer aligned towards the planet and the same differential gravitational field which generates the tide also exerts a torque on the tidal bulge, thereby changing the spin rate. So natural satellites also lose spin angular momentum to their orbits.

If the satellite is in a circular orbit, and its spin and orbital angular momentum vectors remain parallel, the angular momentum transfer from the satellite spin to its orbit continues until the satellite keeps one face towards its planet by rotating synchronously with its orbital motion; at this stage the rotation of the satellite has reached the end-point of its tidal evolution.

For the same value of the tidal dissipation function, Q=100, Peale (1991) verified that twenty-one of the known synchronous natural satellites are tidally evolved, with times to reach synchronous rotation from an initial rotational period of 2.3 hours (the period at which a fluid, spherical body would be rotationally unstable) ranging from 5.0×10^3 yr for Io to 9.8×10^7 yr for Oberon. For the synchronous satellite Iapetus the time computed is larger than the age of the solar system. Peale states that these values must be seen as upper bounds because of the extreme value of initial spin angular velocity and the assumption of Earth-like rigidity.

When the conditions of orbit circularity and alignment of orbital and spin angular momenta are relaxed several complications in the evolution of a satellite's spin angular momentum appear. Some orbits maintain a substantial eccentricity, and the end-point of tidal evolution may not be the synchronous rotation as described above. For satellites with a permanent asymmetry about the spin axis, a resonant torque on this asymmetric mass distribution can counteract the tidal torque and stabilize the spin angular velocity at a non-synchronous value which is a half-integer multiple of the orbital angular velocity. The planet Mercury, which is in the 3:2 spin-orbit resonance, is the only known example of this end-point of tidal evolution (Pettengill and Dyce, 1965). Discussions of this type of spin-orbit coupling are given, for example, by Goldreich and Peale (1966).

In the present work we do not intend to question the effects of tidal evolution on spin-orbit resonances. Instead we consider a new factor that might have been of fundamental importance in explaining the occurrence of such rotational states. According to most cosmogonic theories, the planetary system was formed in a primordial solar nebula and the outer planets formed, at least in part, from a contracting cloud of gas and dust, with their regular satellites forming in the outer regions of this flattened nebula. In this paper we discuss the influences such primordial nebulae may have had on the capture of natural satellites into spin-orbit resonance. After the initial accretion stage planetary, and satellite, rotations were affected by the impact by planetesimals (Lissauer and Safronov, 1991; Lissauer and Kary, 1991). Such effected will not be considered in this work.

In the next section we give a brief description of theories about primordial gaseous nebulae. The gaseous drag force and the resulting torques are described in §3. In §4 we compute the despinning time scales for the Galilean satellites, the Moon and Mercury. In §5 we discuss the relaxation of some of the assumptions and in the last section we present the conclusions and discuss the sensitivity of the results to the chosen parameters from cosmogonic theories.

2. Circumplanetary Gaseous Nebulae

The coplanar and nearly circular orbits of the planets suggest planetary formation in a flattened disk revolving about the Sun (Laplace, 1796). Astrophysical evidence suggests such a disk would have been a natural consequence of the formation of the Sun from the collapse of a rotating molecular cloud (Hoyle, 1960; Cameron, 1962; Cassen and Moosman, 1981; Terebey *et al.*, 1984). The planets could have grown within this protoplanetary disk via binary accretion of solid bodies, followed, in the case of the giant planets, by accretion of hydrogen and helium envelopes once solid planetary cores had grown large enough to gravitationally capture large amounts of gas (Safronov, 1969; Mizuno, 1980; Bodenheimer and Pollack, 1986). Therefore, circumplanetary nebulae in the outer solar system are not surprising since hydrogen and helium are major components of the giant planets. However, circumplanetary nebulae for terrestrial planets are less likely. Nevertheless, some models of planetary formation (Kuyper, 1951; Cameron, 1978) postulate that the prototerrestrial planets consisted, in part, of extended gas spheres.

Lunine and Stevenson (1982) proposed and analysed a model for Galilean satellite formation, in which the satellites accrete in the presence of a dense, gaseous, disk-shaped nebula and rapidly form optically thick , gravitationally bound primordial atmospheres. The model was constructed to be consistent with the observed densities of the Galilean satellites and the current understanding of the formation of Jupiter. This protojovian nebula model appears to be physically plausible and it will be adopted in our analysis.

3. Torques

In this section we assume the orbit of the satellite to be circular and coplanar with the nebula disk. The gas drag force experienced by a body moving in a resisting medium can be expressed in different forms depending on the shape, size and velocity of the body, and the condition (density ρ, viscosity μ, velocity V and mean free path of gas molecules l) of the gas. The drag force is opposite in direction to the body's velocity and its magnitude F_{D} can generally be given in the form (Adashi *et al.*, 1976)

$$F_{\mathrm{D}} = \frac{1}{2} C_{\mathrm{D}} S \rho u^2, \tag{1}$$

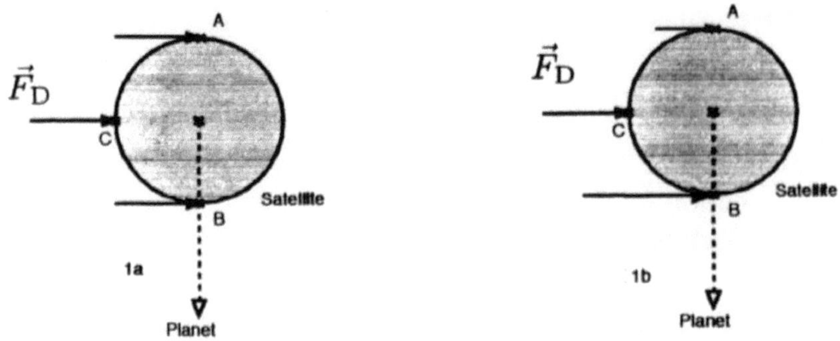

Fig. 1. The drag force vectors on the surface of a non-rotating satellite. a) The nebulae gaseous density is constant. b) The nebulae gaseous density is proportional to the radial distance from the planet.

where S is the cross-sectional area, u is the speed of the body relative to the nebula and C_D is the drag coefficient, which characterizes the kind of drag force and is a function of any two of three dimensionless quantities: the Mach number $M_a(= u/c$, where c is the sound velocity), the Knudsen number $K_n(= l/r_p$, where r_p is the mean radius of the cross-sectional area) and the Reynolds number R_e $(= 2\rho u r_p/\mu)$. For a discussion on C_D see Adashi *et al.* (1976). In the following we consider the possible torques due to the above drag force, acting on the satellite. We consider the torque due to the variation in nebula density and the torque due to the gradient of velocity resulting from spin rotation.

3.1. Torque Due to Density Variation

In theories of cosmogony which involve a disk-shaped nebula of gas and dust, it is reasonable to assume some degree of central condensation, i.e. pressure, density and temperature in the midplane decrease away from the axis. The pressure gradient in the gas causes the nebula to rotate more slowly than the keplerian orbital velocity. In some models the nebula density is assumed to decay exponentially with the radial distance (Pollack *et al.*, 1979; Hunten, 1979); others define a specific dependence between the density and the radial distance. In the case of Jupiter, Lunine and Stevenson (1982) used

$$\rho = 0.03\, x^{-2.3}\, \mathrm{g/cm^3}, \tag{2}$$

where $x = r/R_J$, r is the radial distance and R_J is the present radius of Jupiter.

The torque due to the density variation can be understood by considering a homogeneous spherical satellite which is not rotating relative to its orbital velocity vector. Due to the symmetry, no torque will result if the gaseous density is uniform in the region of the satellite's orbit (Figure 1a). However, if the density decreases as a function of the radial distance, the satellite will experience a gaseous density gradient across its cross-sectional surface (Figure 1b). The result is a torque that is perpendicular to the orbital plane and results in a spinning up of the satellite.

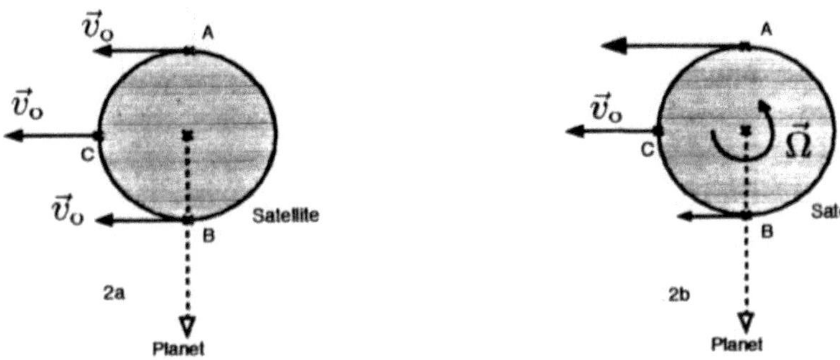

Fig. 2. The linear velocity vectors at the points A,B and C of the satellite. a) A non-rotating satellite. b) A satellite with angular velocity $\boldsymbol{\Omega}$.

However, this torque is extremely small, as will be seen in §3.2, and it will be neglected in this work.

3.2. Torque Due to Spin Rotation

If the gaseous density is constant, the drag force over a spherical shaped body produces a torque due to the gradient of velocity generated by the rotation of the body. This torque acts to stabilize the spin rotation of the satellite as long as the nebula exists since it forces the satellite to keep a fixed position relative to its orbital velocity vector. In the following analysis we simplify the problem by assuming that the interaction between the body and the gas occurs only on the cross-section of the body. In fact, the problem is more complex, the gas will flow along a larger surface of the body and turbulent eddies should appear.

Suppose that a homogeneous satellite has no initial rotation relative to its orbital velocity vector, $\mathbf{v}_o$. Its cross-section is a circle with centre at its centre of mass. All points on this surface have the same linear velocity (Figure 2a), and consequently, they will experience the same drag force intensity. Therefore no torque is produced on the satellite. If the satellite is rotating with angular velocity $\boldsymbol{\Omega}$, as in Figure 2b, its cross section will also be a circle centred on its centre of mass but the velocity will not be the same for all points over such surface. The point A in Figure 2b has velocity of magnitude $v_A = v_o + |\boldsymbol{\Omega} \times \mathbf{R}_A|$, where $\mathbf{R}_A$ is the vector from the centre of the satellite to the point A, and the symmetrically opposite point B has velocity magnitude $v_B = v_o - |\boldsymbol{\Omega} \times \mathbf{R}_A|$. Such differences in velocity occur throughout the cross-sectional surface due to the linear component of the rotational velocity and result in differences in drag forces which produce a torque in opposite direction to the angular velocity vector. Therefore, the effect of the drag on the rotational motion is to slow down the spin rate until the satellite keeps a fixed position relative to its orbital velocity vector. In other words, the torque due to spin rotation forces the satellite to the synchronous rotation state keeping the same face towards the central body.

A comparison between the magnitude of this torque and the torque due to density variation can be made by computing the differences between the drag forces per element of area at the points A and B from Figures 1b and 2b (extreme values of force per element of area over the whole satellite) in each case. So for a density variation given by (2) the difference of drag force between the two extreme points of a non-rotating satellite can be given by

$$\delta F_\rho = F_{\mathrm{D}}^*\left(\frac{2R}{r}\right), \tag{3}$$

where $F_{\mathrm{D}}^* = (1/2)C_{\mathrm{D}}\rho_o v_{\mathrm{o}}^2$, and ρ_o is the nebular density at a position equivalent to the position of the centre of mass of the satellite. The respective difference of drag force in the case of a rotating satellite, taking the nebular density to be constant, is

$$\delta F_\Omega = F_{\mathrm{D}}^*\left(\frac{4\Omega R}{v_{\mathrm{o}}}\right). \tag{4}$$

From (3) and (4), we have $\delta F_\rho/\delta F_\Omega \approx n/2\Omega$, and consequently $\delta F_\rho \ll \delta F_\Omega$, since $\Omega \gg n$, where n is the mean motion of the satellite. The corresponding torques differ only by the gradients of the forces, which vary gradually and uniformly. Therefore, from the comparison of the differences of the extreme values of the drag force per element of area for each case, we conclude that the torque due to density variation is extremely small in comparison with the torque due to the gradient of linear velocity.

For the drag force whose magnitude is given by (1) the expression for the torque due to spin rotation can be written

$$\mathbf{M} = -2R^2 S C_{\mathrm{D}}\rho u_{\mathrm{o}}\left(\frac{1}{3}\sin^3\epsilon + 1\right)\sin\epsilon\,\boldsymbol{\Omega}, \tag{5}$$

where $S = \pi R^2$ is the cross-sectional area and ϵ is the angle between $\boldsymbol{\Omega}$ and $\mathbf{v}_{\mathrm{o}}$.

Adashi *et al.* (1976) found that for planetesimals with mass $> 10^{18}$g the drag coefficient was in the range $0.5 \le C_{\mathrm{D}} \le 1.5$. Taking $C_{\mathrm{D}} = 1$, and $\mathbf{u}_{\mathrm{o}}$ perpendicular to $\boldsymbol{\Omega}$, $\epsilon = \pi/2$, as reasonable values during the formation period of the planets and satellites, (5) becomes

$$\mathbf{M} = -\frac{8}{3}\pi R^4 \rho u_{\mathrm{o}}\,\boldsymbol{\Omega}. \tag{6}$$

4. Despinning Time Scale

In order to estimate the importance of the effect described in §3, it is necessary to estimate the magnitude and the time scale for such an effect to be significant. Our analysis is based on a lifetime of the solar nebula of $\sim 10^6$ yr.

From (6), the variation on the rotational velocity is given by

$$\frac{d\Omega}{dt} = -K\Omega\,, \tag{7}$$

where $K = (8/3)(\pi R^4/I_\Omega)\rho u_o$, $I_\Omega = (2/5)mR^2$ is the moment of inertia of the satellite about its spin axis, m is the mass of the satellite and $\Omega = |\boldsymbol{\Omega}|$. Then the time scale for exponential decay is

$$\tau_e = \frac{1}{K} = 0.15\left(\frac{m}{S\rho u_o}\right). \tag{8}$$

Therefore, the despinning time scale is proportional to the cross–sectional density ($\sigma = m/S$) of the satellite , and it is inversely proportional to the nebula density ρ and to the relative velocity u_o of the satellite. In general, cosmogonic theories assume the density of the solar nebula is $10^{-13} \leq \rho \leq 10^{-8}$g/cm^3, while for the circumplanetary nebula it is $10^{-8} \leq \rho \leq 10^{-4}$g/cm^3, with relative velocity $10^{-3}v_o \leq u_o \leq 10^{-2}v_o$.

4.1. Galilean Satellites

Adopting the protojovian nebula model by Lunine and Stevenson (1982), we can compute the despinning time scale for the Galilean satellites, taking the velocity of the satellite relative to the nebula to be $10^{-2}v_o$ and the mass density of the nebula at the present region of the Galilean satellites to be in the range 10^{-5} to 10^{-4}g/cm^3 (from (2)). We have found the times for the Galilean satellites to reach synchronous rotation from an initial rotational period of 2.3 hours (an extreme value, Peale (1977)) range from 4.6 yr for Io to 21×10^2 yr for Callisto (column 3 in Table 1). These values are several orders of magnitude shorter than the times resulting from tidal theories (column 4 in Table 1).

4.2. Other Satellite Systems

Similar situations could arise for satellites of other systems. Assuming the proto-saturnian nebula to be similar to the protojovian model adopted above, keeping the due proportions (the unit of distance scale is taken to be R_S, Saturn's present radius, rather than R_J), we can compute the despinning time scales for the saturnian satellites. For example Iapetus that has been observed to be in synchronous rotation and whose time from tidal theories to reach this state is longer than the age of the solar system, would have had a despinning time scale of 2.8×10^3 yr.

We have also computed the equivalent time scales for the Moon, Phobos and Deimos. In these cases we used a nebular density of 10^{-8}g/cm^3, the lowerbound value of the circumplanetary nebulae models. This is considerably less than those we adopted for the Galilean satellites. The resulting despinning time scales (Table

TABLE I
The times to reach synchronous rotation from an initial rotation period of 2.3 hours considering nebular drag and tidal evolution are given by Δt and Δt^*, respectively. ρ is the gaseous density nebula and Δt^* are from Peale (1991).

Satellite	ρ (g/cm^3)	Δt (yr)	Δt^* (yr)
Io	5.1×10^{-4}	4.6	5.0×10^3
Europa	1.7×10^{-4}	13	1.2×10^5
Ganymede	5.9×10^{-5}	51	6.6×10^5
Callisto	1.6×10^{-5}	2.1×10^2	2.2×10^7
Iapetus	6.3×10^{-7}	2.8×10^3	1.4×10^{11}
Moon	1.0×10^{-8}	3.6×10^6	3.2×10^8
Phobos	1.0×10^{-8}	4.9×10^4	1.2×10^5
Deimos	1.0×10^{-8}	5.6×10^3	1.2×10^8
Mercury	1.0×10^{-8}	7.8×10^4	

1) are longer, but in all cases they are shorter than the respectives from tidal theories. The time for the Moon is the longest one, being of the order of the nebula lifetime.

Our analysis does not apply to the Pluto–Charon system, because of its size Pluto is not expected to have had any significant primordial nebula.

If we consider the case of Mercury orbiting in a solar nebula whose density is 10^{-8}g/cm^3 (upperbound of the solar nebula models), the despinning time scale is 7.8×10^4 yr. Then Mercury could have been left in a synchronous rotational state when the nebula dissipated. However, due to its large eccentricity (~ 0.2) Mercury would have had a large amplitude of libration in such a resonance. Then due to these circumstances and the fact that Mercury has a permanent asymmetry about its spin axis, it would have been stabilized in the 3:2 spin-orbit resonance (Goldreich and Peale, 1966).

4.3. Time Scale Dependence on the Satellite's Size

We can also consider the time scale as a function of the size of the satellite by rewriting (8) in terms of the density of the satellite, $\rho_{\rm sat}$. We have

$$\tau_{\rm e} = 0.2\rho_{\rm sat}\left(\frac{R}{\rho u_{\rm o}}\right) . \tag{9}$$

This shows the linear relationship between the radius of the satellite and the e-folding time scale. In the earlier stages of the formation of the satellites the scenario was much more complex with the satellites growing non-uniformly via accretion of smaller bodies. Nevertheless, the satellites were smaller than they are now and consequently, if we consider just the effect of the torque due to the gas drag, the time scales are shorter than those given in Table I.

Fig. 3. Satellite whose geometric centre and mass centre are not coincident. a) unstable configuration. b) stable configuration.

5. Discussion

As the actual problem is much more complex than we have considered, in this section we discuss the consequences of generalizing some of our assumptions.

Ellipsoidal Shape

The fact that satellites have an ellipsoidal shape instead of a spherical one does not change our analysis. Generally, the cross-section of a rotating ellipsoid is the area bounded by an ellipse of varying shape. For the accuracy required in our calculations, it is enough to consider an averaged cross-sectional area.

Geometric Centre and Mass Centre

If a satellite has mass distribution and shape such that its geometric centre and its mass centre are not coincident another torque component will arise due to the asymmetry of the cross-section relative to the mass centre. It forces the satellite to align the line that pass by both of the centres and the velocity direction, with the mass centre ahead of the geometric. Therefore, again, the satellite is forced to keep a fixed face toward its central body.

Orbital Decay

Kiang (1962) showed that for any reasonable nebular density dependence with orbital radius and any expected drag law the semimajor axis and eccentricity of a prograde object will all decrease , circularising the orbit, and to producing orbital decay such that the satellite spirals inward. The characteristic time scale for the orbital decay, considering the protojovian nebula model given by Lunine and Stevenson (1982), ranges from $\sim 5 \times 10^2$ yr for Io to $\sim 5 \times 10^4$ yr for Callisto. Thus the nebular gas drag implies rapid orbital evolution and the satellites must form quickly. It is possible for Jupiter to accrete $\sim$ 500 Io-sized bodies during this evolution, since this mass is comparable to the core of Jupiter (Lunine and Stevenson, 1982). In fact, at the same time tidal force would be acting in the opposite direction, expanding the orbits. Therefore it is possible that the satellites we see today could be the ones left unaccreted onto Jupiter at the time when the nebula dissipated.

Eccentric Orbit

If the satellite had some residual eccentricity after the dissipation of the nebula and it had reached its end point on spin rotation from the drag torque (to keep a fixed face relative to its orbital velocity vector), then the satellite was left librating in a synchronous rotation and the amplitude of libration was proportional to the eccentricity.

6. Conclusion

We have shown that a rotating satellite that is orbiting in a primordial gaseous nebula experiences a torque which forces it to keep a fixed face relative to its orbital velocity vector. This means that the satellite would be in synchronous spin-orbit resonance (or librating in this resonance in the case of an orbit with some residual eccentricity) in a similar way to that predicted by tidal evolution theories. The magnitude of the torque and the despinning time scale are functions of the cross-sectional density, orbital velocity and the initial spin velocity of the satellite, as well as the density and rotation velocity of the primordial gaseous nebula. Consequently, the importance of the results is tied to the validity of the primordial gaseous nebula models adopted and the initial rotational period of the object.

Nebulae gaseous drag might have been of fundamental importance for capture into spin-orbit resonances in a wide range of cases, bringing the bodies to synchronous rotation or at least reducing their spin rate drastically. However, other effects like collisions with planetesimals of comparable sizes may have spun-up the satellites, and then, after the nebula have dissipated tidal evolution would have played a major role bringing the satellites to the synchronous spin-orbit state.

Acknowledgements

The authors would like to thank Dr. Carl Murray for valuable discussions and comments on the manuscript. The first author thanks CAPES for the doctoral scholarship (Proc. 445/90-5).

References

Adashi, I., Hayashi, C., and Nakazawa, K.: 1976, *Prog. Theor. Phys.*, **56**, 1756–1771.
Burns, J.A. (Ed.):1986, *Satellites*, University of Arizona Press, Tucson.
Bodemheimer, P. and Pollack, J.B.: 1986, *Icarus*, **67**, 391–408.
Cameron, A.G.W.: 1962, *Icarus*, **1**, 13–69.
Cameron, A.G.W.: 1978, In *Origin of the Solar System*, (S. F. Dermott, Ed.), Wiley, New York, 49–74.
Cassen, P.M. and Moosman, A.: 1981, *Icarus*, **48**, 353–376.
Goldreich, P. and Peale, S.J.: 1966, *Astron. J.*, **71**, 425–438.
Hoyle, F: 1960, *Quart. J. R. Astron. Soc.*, **1**, 28–55.
Hunten, D.M.: 1979, *Icarus*, **37**, 113–123.
Kiang, T.: 1962, *Mon. Not. Roy. Astron. Soc.*, **123**, 359–382.
Kuyper, G.P.: 1951, *Proc. Nat. Acad. Sci. USA*, **37**, 717–721.
Lamy, P.L. and Burns, J.A.: 1972, *Am. J. Phys.*, **40**, 441–445.
Laplace, P.S.: 1796, *Exposition du Système du Monde*, Paris, last appendix.
Lissauer, J.J. and Safronov, V.S.: 1991,*Icarus*, **93**, 288–297.
Lissauer, J.J. and Kary, D.M.: 1991, *Icarus*, **94**, 126–159.

Lunine, J. I. and Stevenson, D.J.: 1982, *Icarus*, **52**, 14–39.
Mizuno, H.: 1980, *Prog. Theor. Phys.*, **64**, 544–557.
Peale, S.J.: 1977, In *Planetary Satellites*, (J. A. Burns, Ed.), Univ. of Arizona Press, Tucson, 87–114,
Peale, S.J.: 1986, In *Satellites* (J. A. Burns, Ed.), Univ. of Arizona Press, Tucson, 159–223,
Peale, S.J.: 1991, In *The Astronomy and Astrophysics Encyclopaedia* (S. Maran, Ed), Cambridge University Press, Cambridge, 614–618.
Pollack, J.B., Burns, J.A. and Tauber, M.E.: 1979, *Icarus*, **37**, 587–611.
Pettingill, G.H. and Dyce, R.B.: 1965, *Nature*, **206**, 1240.
Safronov, V.S.: 1969, *Evolution of the Protoplanetary Cloud & Formation of the Earth & Planets*. Moscow, Nauka; also NASA TTF–667, 1972.
Terebey, S., Shu, F.H. and Cassen, P.: 1984, *Astrophys. J.*, **286**, 529–551.

SIGNIFICANT HIGH NUMBER COMMENSURABILITIES IN THE MAIN LUNAR PROBLEM II: THE OCCURRENCE OF SAROS-LIKE NEAR PERIODICITIES

B.A. STEVES

Department of Mathematics, Glasgow Polytechnic, Cowcaddens Rd, Glasgow, U.K.

G.B. VALSECCHI

I.A.S. – C.N.R., Reparto di Planetologia, viale dell'Università 11, 00185 Roma, Italy
E-mail: GIOVANNI@IRMIAS (EARN/BITNET)

E. PEROZZI

Telespazio s.p.a., via Tiburtina 965, 00156 Roma, Italy

and

A.E. ROY

Department of Physics and Astronomy, University of Glasgow, Glasgow G12 8QQ, U.K.

Abstract. The effect of changes in the Moon's semimajor axis and the Earth's orbital eccentricity on the occurrence of Saros-like cycles is examined. The Earth-Moon-Sun dynamical system exhibits such cycles for only 25 to 30% of the time interval between -5×10^7 to $+5 \times 10^7$ years. Not only has the present Saros the smallest period during this time, but it also has one of the longest durations and the period closest to an integral number of anomalistic years, thus making it one of the most efficient Saros-like cycles for reversing solar perturbations in the main lunar problem. During the lifetime of a Saros-like cycle, variations of the Earth's orbital eccentricity cause frequent disappearances and reappearances of the cycle.

Key words: Main lunar problem – Saros like periodicities – periodic orbits

1. Introduction

In a recent paper (Perozzi et al., 1991) the authors studied the properties of a number of historically-known cycles associated with high number near commensurabilities among the synodic, the anomalistic and the nodical lunar months, and the anomalistic year. It was shown that the Earth-Moon-Sun system (within the framework of the main lunar problem) was moving in a nearly periodic orbit of period equivalent to the classical Saros cycle of 18 years and 10 or 11 days. The Saros cycle arises from a set of commensurabilities between the synodic $T_{\Delta\lambda}$, anomalistic T_M and nodical T_θ lunar months, where $223T_{\Delta\lambda} \simeq 239T_M \simeq 242T_\theta$ to within less than 0.3 d. In fact subsequent work by the authors (Roy et al., 1991 and Valsecchi et al., 1993a,b) has shown that a set of eight periodic orbits exists close to the Moon's present orbit, with values of $T_{\Delta\lambda}$, T_M and T_θ that are exactly commensurable.

The ability of a cycle such as the Saros to reverse effectively solar perturbations on the lunar orbit so that the dynamical geometry of the Earth-Moon-Sun system at the beginning of a cycle is repeated to a high degree of accuracy at the end of the cycle period (see Perozzi et al., 1991) raises questions about its precise role in the problem of the Moon's orbital stability. If the Saros cycle is relevant to the stability

Celestial Mechanics and Dynamical Astronomy **57**: 341–358, 1993.

TABLE I
Mean values of parameters relevant to the Earth-Moon system. The maximum mean eccentricity is given in (Laskar et al., 1992)

Earth's heliocentric orbital elements		
$\overline{a}'$	1.49596×10^{13} cm	mean semimajor axis
$\overline{e}'_0$	0.016726	present mean eccentricity
$\overline{e}'_{max}$	0.0588	maximum mean eccentricity
$\overline{n}'$	1.990987×10^{-7} rad/s	mean mean motion
Present Moon's geocentric mean orbital elements		
$\overline{a}$	3.844×10^{10} cm	mean semimajor axis
$\overline{e}$	0.0549	mean eccentricity
$\overline{i}$	5°.133	mean inclination
$\overline{n}$	2.6617×10^{-6} rad/s	mean motion
$\nu = \overline{n}'/\overline{n}$	0.0748133	ratio of the mean motions
Present mean duration of lunar months		
$T_{\Delta\lambda}$	29.530589 d	synodic month
T_M	27.554551 d	anomalistic month
T_θ	27.212220 d	nodical month

of the Earth-Moon system against solar perturbations, it becomes important to know how the Saros cycle is modified by long-term lunar orbital evolution and how frequently other Saros-like cycles (hereafter called SLC's) occur within the evolving lunar orbit.

The three periods, namely the synodic, the nodical and the anomalistic months which form the Saros cycle are affected by changes in the *observed mean* mean motion of the Moon $\overline{n}$, and the mean rates of change of the lunar apse and node $\dot{\overline{\omega}}$ and $\dot{\Omega}$. These in turn are affected by changes in:

- the eccentricity $\overline{e}$ and inclination $\overline{i}$ of the Moon's orbit;
- the eccentricity $\overline{e}'$ of the Earth's orbit;
- the ratio α of the semimajor axis of the Moon's orbit to that of the Earth's heliocentric orbit.

These quantities, together with other ones relevant to the Earth-Moon-Sun system, are given in Table I.

The principal causes of changes in the above orbital parameters are planetary perturbations acting on the Earth's heliocentric orbit and tidal friction acting on the Moon's orbit. The first produces periodic changes in the eccentricity of the Earth's orbit with a period approximately equal to 10^4 yr and a maximum eccentricity of about 0.0588 (Laskar et al., 1992). The second causes a transfer of angular momentum from the rotating Earth to the revolution of the Moon in its orbit, which leads to a slow spiralling outwards of the satellite. The resulting change in the Moon's orbital semimajor axis is effectively secular and acts over a very long time-scale. Changes in the Moon's mean orbital inclination and eccentricity due to tidal evolution were shown by Mignard (1981) to remain small over time intervals of the order of 10^8 yr.

If a near SLC can always be found, and the Moon's stability is in some way associated with the existence of a SLC, then it is under no threat from the destruction of the present-day Saros cycle. If the present Saros cycle is a rare occurrence, however, then it is possible that the loss of the stabilizing mechanism of the Saros cycle (see Perozzi et al., 1991) could have drastic effects on the future stability of the Earth-Moon system.

Section 2 contains criteria for discussing the properties of SLC's. In Section 3 the frequency of occurrence of SLC's is investigated on a purely numerological basis, while in Section 4 the effect of lunar evolution on the Saros cycle and other SLC's is studied. The final section draws certain conclusions from these studies.

2. Nomenclature for Discussing the Properties of Saros-Like Cycles (SLC's)

In this paper the phrase *the Saros cycle* will refer only to the present-day Saros cycle which is formed from the near commensurabilities between the current values of $T_{\Delta\lambda}$, T_M and T_θ.

A perfect SLC is one where there exists a unique Saros period which is exactly equal to integer multiples of each of the three periods

$$T_{Saros} = N_{\Delta\lambda}T_{\Delta\lambda} = N_M T_M = N_\theta T_\theta \tag{1}$$

where the N's are integers and are smaller or equal to some given maximum integer N_{max}. The Saros period T_{Saros} is taken to be the integral multiple of the synodic period given by $N_{\Delta\lambda}T_{\Delta\lambda}$.

A near SLC is one where the three combinations of the integer multiples and their corresponding periods are only approximately equal to each other to within a given accuracy q, say. This may be described as a triplet of periods ($N_{\Delta\lambda}T_{\Delta\lambda}$, $N_M T_M$, $N_\theta T_\theta$) where:

$$\begin{aligned} N_{\Delta\lambda}T_{\Delta\lambda} - N_M T_M &= \varepsilon_{12} \\ N_{\Delta\lambda}T_{\Delta\lambda} - N_\theta T_\theta &= \varepsilon_{13} \\ N_M T_M - N_\theta T_\theta &= \varepsilon_{23} \end{aligned} \tag{2}$$

with:

$$\begin{aligned} |\varepsilon_{ij}| &\leq q \\ N_{\Delta\lambda}, N_M, N_\theta &\in I \\ N_{\Delta\lambda}, N_M, N_\theta &\leq N_{max} \\ \forall i, j = 1,2,3, i &< j. \end{aligned}$$

The residuals from a near SLC (ε_{12}, ε_{13} and ε_{23}) are not all independent of each other. Note that $\varepsilon_{23} = \varepsilon_{13} - \varepsilon_{12}$. The third residual ε_{23} and the restriction placed on it $|\varepsilon_{23}| \leq q$ can always be reconstructed from the first two residuals.

Note also that in the definition of a perfect SLC and a near SLC, we have added the further restriction that the integer multiples $N_{\Delta\lambda}, N_M, N_\theta$ of the commensurability must be smaller than N_{max}. In practice it is possible, if no limit is placed on the size of the integers, to obtain a SLC as close as we please to perfection. However, with larger integers the period may be so long that solar perturbations may accumulate to a point where the system becomes unstable before the SLC has time to reverse perturbations (see Perozzi et al., 1991). For the remainder of this paper we arbitrarily choose N_{max} to be 1000 and the accuracy q to be of the order of the present Saros' accuracy at 0.3 d.

3. The Frequency of Occurrence of SLC's

Hipparchus, Newcomb, Crommelin and Stockwell, among others, discovered or described other cycles of the Saros nature (Steves, 1990), but to our knowledge no attempt has been made to study systematically the frequency of the occurrence of such cycles for any triplet of periods. This is not unexpected since until recently these cycles have been of interest only for the prediction of eclipses. But since it is possible that SLC's may be associated with some lunar stability mechanism against solar perturbations, it becomes important to know how likely the Moon's orbit is to contain a SLC, given that orbital evolution changes the period set ($T_{\Delta\lambda}$, T_M, T_θ) which forms a SLC.

Essentially the question to be asked is: How many integer triplets (N_1, N_2, N_3) exist for a given set of periods (T_1, T_2, T_3) such that Relations (2) hold true?

Without loss of generality, the periods are ordered such that $T_i > T_{i+1} \forall i = 1,2,3$ and the system parameters are divided throughout by the largest period T_1. Equations (2) then become

$$\begin{aligned} N_1 - N_2 T_2' &= \varepsilon_{12}' \\ N_1 - N_3 T_3' &= \varepsilon_{13}' \\ N_2 T_2' - N_3 T_3' &= \varepsilon_{23}' = \varepsilon_{13}' - \varepsilon_{12}' \end{aligned} \tag{3}$$

where:

$$\begin{aligned} |\varepsilon_{ij}'| &\leq q' \\ N_i &\in I \\ N_i &\leq N_{max} \\ \forall i, j = 1,2,3, i &< j \end{aligned}$$

and a prime superscript denotes a division of the original parameter by T_1. The search for a SLC then becomes the search for fractions that will approximate the rational numbers T_2/T_1 and T_3/T_1.

A straightforward procedure is to first find cycles of the required accuracy q' which involve only a commensurability between two of the periods T_1 and T_2, performing a computer search for all pairs of values of N_1 and N_2 which meet the restrictions $N_1 - N_2T_2' = \varepsilon_{12}'$ (where $|\varepsilon_{12}'| \leq q'$ and $N_1, N_2 \leq N_{max}$). The SLC's are then found by isolating those cycles having a further commensurability with the third period T_3.

To do this, the two possible integer values for N_3 that produce the nearest values of N_3T_3' above and below N_1 are found, and the two corresponding residuals between N_3T_3' and N_1 computed. The absolute minimum of the two residuals is then equal to $|\varepsilon_{13}'|$ and the integer value producing it becomes N_3. If $|\varepsilon_{13}'| \leq q'$ and if $|\varepsilon_{23}'| = |\varepsilon_{13}' - \varepsilon_{12}'| \leq q'$, then the integer set ($N_1$, N_2, N_3) found forms a SLC for the period set (T_1, T_2, T_3).

When this search is completed for the case of the present Moon with

$$\begin{aligned} T_2' &= T_M/T_{\Delta\lambda} = 0.933085 \\ T_3' &= T_\theta/T_{\Delta\lambda} = 0.921493 \\ q' &= 0.3\text{d}/T_{\Delta\lambda} \simeq 0.01 \\ N_{max} &= 1000 \end{aligned}$$

the only SLC obtained is the present Saros cycle of $N_{\Delta\lambda} = 223$, $N_M = 239$, $N_\theta = 242$.

Following the restrictions of Equations (3), we look for the probability P that $|\varepsilon_{12}'|$, $|\varepsilon_{13}'|$ and $|\varepsilon_{23}'|$ are less than or equal to q', assuming that $N_2T_2' = N_1 + \varepsilon_{12}'$ and $N_3T_3' = N_1 + \varepsilon_{13}'$ fall randomly between $N_1 - 0.5$ and $N_1 + 0.5$ (this assumption does not hold close to low-number commensurabilities).

Let R_1 be the area of the region about N_1 where ε_{12}' and ε_{13}' must be found in order for a SLC to exist and R_2 be the area of the region about an integer where ε_{12}' and ε_{13}' can be located. Then

$$P = \frac{R_1}{R_2} \tag{4}$$

Figure 1 depicts the regions described in the above fraction. The diagram is centred on the position of an integer N_1. The shaded region is the region where the restrictions for the existence of a SLC, namely

$$\begin{aligned} |\varepsilon_{12}'| &\leq q' \\ |\varepsilon_{13}'| &\leq q' \\ |\varepsilon_{23}'| = |\varepsilon_{13}' - \varepsilon_{12}'| &\leq q' \end{aligned} \tag{5}$$

hold true and the square region formed by $-0.5 \leq \varepsilon_{12}' \leq +0.5$ and $-0.5 \leq \varepsilon_{13}' \leq +0.5$ is the total region where the residuals ε_{12}' and ε_{13}' may fall about N_1.

Therefore the probability P that a Saros cycle exists near an integer N_1 is given by:

$$P = \frac{3q'^2}{1} = 3q'^2 \tag{6}$$

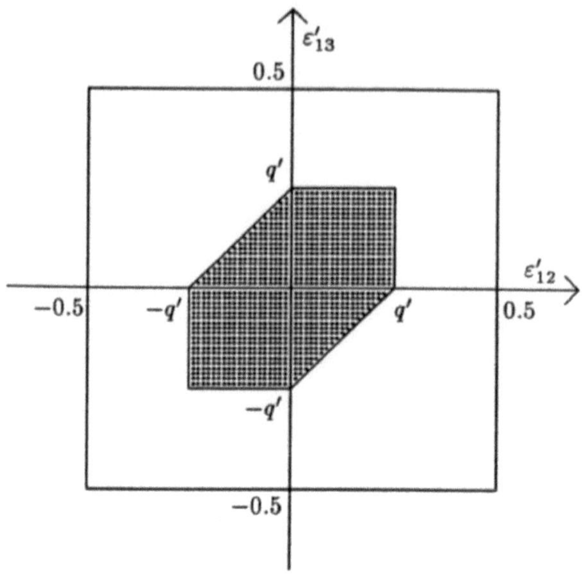

Fig. 1. The region of acceptable SLC's. The dotted part is the area in which the residuals ε'_{12} and ε'_{13} must be found in order for a Saros cycle to exist near the integer N_1, while the square encloses the area in which the residuals are always located.

The expected number J of Saros cycles is then simply:

$$J = PJ_1J_2 \tag{7}$$

where J_1 is the number of times a pair of real values $N_2T'_2$ and $N_3T'_3$ falls within 0.5 of a particular integer N_1, and J_2 is the number of integers N_1 that are available for the real values $N_2T'_2$ and $N_3T'_3$ to fall near.

The number of attempts at having $N_2T'_2$ and $N_3T'_3$ fall within 0.5 of a specific integer N_1 is $1/T'_2$ and $1/T'_3$ respectively. Thus the number J_1 of attempts at having distinct pairs of $N_2T'_2$ and $N_3T'_3$ fall within q' of each integer N_1 is given by

$$J_1 = \frac{1}{T'_2} \times \frac{1}{T'_3} \tag{8}$$

The number J_2 of integers N_1 that are accessible to the pairs $(N_2T'_2,\ N_3T'_3)$ is constrained by the restriction $N_1 < N_2 < N_3 < N_{max}$ arising from the ordering of the periods. Given that $N_1 \approx N_3T'_3$, the largest possible N_1 that $N_3T'_3$ can approach is approximately $N_{max}T'_3$. Hence,

$$J_2 \simeq T'_3N_{max} \tag{9}$$

Given q' and N_{max}, the expected number J of SLC's, on a purely random basis, becomes

$$J = \left(\frac{3q'^2}{T_2'}\right) N_{max} \tag{10}$$

Note that this expression, although dependent on the value of T_2', does not depend upon the value of T_3'. Note also that Equation (10) can give a predicted number that may vary considerably from the true number, if T_2' or T_3' lies near a low number commensurability and thus invalidates the assumption that the residuals are pseudo-random variables.

If we take $q' = 0.01$, $N_{max} = 1000$ and $T_2' = 0.93$, we find from Equation (10) that the number of expected Saros cycles is about 0.3. This implies that, during the Moon's evolutionary history, for values of T_2' not far from the current one, there is a higher probability for the Moon's orbit to be without a SLC than the opposite. It should be noted that $N_{max} = 1000$ allows cycles with periods as long as $\sim$ 80 yr to be included.

4. The Effect of Lunar Orbital Evolution on the Saros Cycle and other SLC's

At the present epoch the Earth-Moon-Sun orbital dynamics does contain a SLC; however, because of lunar orbital evolution, this situation may change.

The effects of lunar orbital evolution on the Saros cycle can be studied by considering the variations of the three lunar periods with time. Equations (11) show that the lunar synodic, nodical and anomalistic periods depend on the *observed mean* mean motion of the Moon $\overline{n}(t)$, the rate of change of the longitude of the lunar node $\dot{\Omega}(t)$, and the rate of change of the longitude of the lunar apse $\dot{\tilde{\omega}}(t)$.

$$\begin{aligned} T_{\Delta\lambda} &= \frac{2\pi}{\overline{n} - \overline{n}'} \\ T_\theta &= \frac{2\pi}{\overline{n} - \dot{\Omega}} \\ T_M &= \frac{2\pi}{\overline{n} - \dot{\tilde{\omega}}}, \end{aligned} \tag{11}$$

where $\overline{n}'$ is the mean motion of the Sun, and $\overline{n}$ differs from the *keplerian* lunar mean motion n by a small term (Brouwer and Clemence, 1961):

$$\overline{n} = n(1 - \nu^2) \tag{12}$$

where $\nu = \overline{n}'/\overline{n}$.

In turn $\dot{\Omega}$ and $\dot{\tilde{\omega}}$ may be written as functions of time, using Delaunay's (1872) analytic series for these quantities, expanded in terms of the ratio ν of the mean motions of the Moon and the Sun (for the complete expressions, see Valsecchi et al., 1993b):

$$\dot{\Omega}(t) = \dot{\Omega}(\overline{n}, \gamma, \overline{e}, \overline{e}', \nu, \alpha) \tag{13}$$

$$\dot{\omega}(t) = \dot{\omega}(\overline{n}, \gamma, \overline{e}, \overline{e}', \nu, \alpha) \tag{14}$$

where $\gamma = \sin\frac{\overline{i}}{2}$ and the other quantities have already been defined.

Hence the three lunar periods are functions of $\overline{e}$, $\overline{e}'$, $\overline{i}$ and ν which are in turn each functions of time. Note that α is related to ν through Kepler's third law:

$$\alpha = \frac{\overline{a}}{\overline{a}'} = \sqrt[3]{\frac{\mu\nu^2(1-\nu^2)^2}{1+\mu}} \tag{15}$$

μ being the ratio of the mass of the Earth-Moon system to that of the Sun.

It has already been seen in Section 1 that $\overline{e}$ and $\overline{i}$ do not change appreciably over a time span of 10^8 yr. The mean value of $\overline{e}'$, however, oscillates between 0 and 0.0588 (Laskar et al., 1992) over a time scale of $\sim 10^4$ yr.

4.1. The Effect of Planetary Perturbations on the Saros Cycle

We can study the short term effects on the present Saros cycle of variations in $\overline{e}$, $\overline{i}$ and $\overline{e}'$ by taking ν to be a constant at its present value (see Table I), choosing $\overline{e}'$ to range from 0 to its expected maximum value of 0.0588 and exploring the phase space of $\overline{e}$ and $\overline{i}$ within the vicinity of the Moon's current values of these elements.

Given values of $\overline{e}'$ and ν, Delaunay's Equations (13) and (14) combined with Equations (11) can provide values for the periods $T_{\Delta\lambda}$, T_θ and T_M for any pair of values of $\overline{e}$ and $\overline{i}$. The three periods, in turn, will have certain residuals ε'_{12}, ε'_{13} and ε'_{23} specified by Equations (3).

It is therefore possible to vary $\overline{e}$ and $\overline{i}$ to obtain boundaries in $\overline{e}$-$\overline{i}$ space within which each of the residuals meets the accuracy criteria of Equation (3). The area common to the three regions found for the three residuals is then the region of $\overline{e}$-$\overline{i}$ space within which the Saros cycle exists according to the accuracy limitations of q'.

Adopting the present value of ν, and for values of $\overline{e}' = 0.0167$ and 0.0588, Figure 2 gives examples of such regions. The cross marks the present position of the Moon in $\overline{e}$-$\overline{i}$ phase space.

Figure 2(b) with $\overline{e}'$ at its maximum value of 0.0588 shows that the Moon is not in the region which satisfies the accuracy criteria for the present Saros. In other words, when the eccentricity of the Earth increases to its maximum value, the Moon with its current values of $\overline{e}$ and $\overline{i}$ no longer has the Saros cycle within its

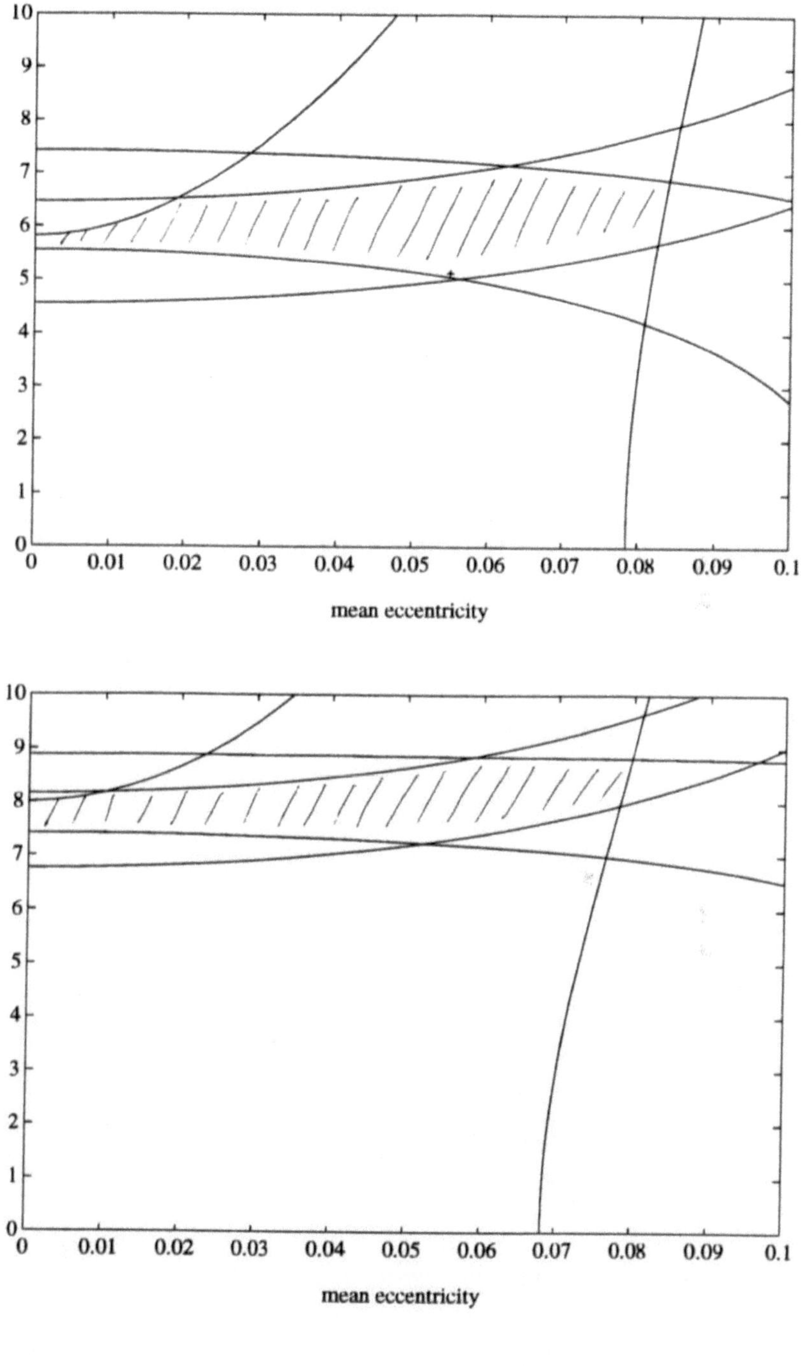

Fig. 2. The region of $\overline{e}$-$\overline{i}$ space where the present Saros cycle of $223T_{\Delta\lambda} \simeq 239T_M \simeq 242T_\theta$ with $\nu = \nu_0$ can exist to an accuracy of about 0.3 d. In (a) $\overline{e}' = 0.0167$ while in (b) $\overline{e}' = 0.0588$.

orbital dynamics. Further calculations (in which ν, $\overline{e}$ and $\overline{i}$ are taken to be constant at their present values) show that as $\overline{e}'$ proceeds through its cycle due to planetary perturbations, the Saros cycle will be lost to the Moon whenever $\overline{e}'$ becomes greater than 0.025. In fact, for values of $\overline{e}'$ between 0.025 and 0.0588, no other SLC's were found (given $q' = 0.01$ and $N_{max} = 1000$).

On the other hand, the figures show that the Moon's possession of the Saros cycle as specified by $q' \leq 0.01$ would be insensitive to wide variations in $\overline{e}$ and $\overline{i}$.

4.2. The Effect of Tidal Evolution on the Saros Cycle and other SLC's

Let us now consider only the effect of tidal friction on the Saros cycle over the much longer time interval of 10^8 yr, by keeping $\overline{e}'$ constant and allowing α and therefore ν to vary according to tidal evolution theory.

We first find the mean motion as a function of time. Goldreich and Soter (1966) give the rate of change of the Moon's semimajor axis due to tidal friction as:

$$\dot{\overline{a}} = \frac{9}{2}\left(\frac{G}{M_{earth} + M_{moon}}\right)^{\frac{1}{2}} \frac{R^5}{Q'} M_{moon} \overline{a}^{-\frac{11}{2}} \tag{16}$$

where:

$$\begin{aligned} G &= \text{constant of gravity} \\ R &= \text{radius of the Earth} \\ M_{earth}, M_{moon} &= \text{masses of the Earth and Moon respectively} \\ \overline{a} &= \text{mean semimajor axis of the Moon} \\ Q' &= Q\left(1 + \frac{19\mu}{2g\rho R}\right) \\ Q &= \frac{1}{\sin 2\epsilon} \\ \epsilon &= \text{tidal lag angle of the Earth} \\ \mu &= \text{coefficient of rigidity of the Earth} \\ g &= \text{surface gravity of the Earth} \\ \rho &= \text{mean density of the Earth} \end{aligned}$$

If we now perform a simple integration of this equation, assuming that all the other parameters are constant with respect to time and the semimajor axis, we get:

$$\overline{a} = \overline{a}_0(1 + K_1 t)^{\frac{2}{13}} \tag{17}$$

where:

$$K_1 = \frac{117}{4}\left(\frac{G}{M_{earth} + M_{moon}}\right)^{\frac{1}{2}} \frac{R^5}{Q'} M_{moon} a_0^{-\frac{13}{2}}$$

A zero subscript denotes a value of the specified parameter at $t = 0$. The equivalent relationship involving the mean motion is:

$$n = n_0(1 + K_1 t)^{-\frac{3}{13}}. \tag{18}$$

K_1 is unfortunately not accurately known because of the uncertainty in the value of Q'. However, recent work by Sonett et al. (1988) uses banded sediments tidally deposited in the PreCambrian era to derive a value for the number of synodic lunar orbits which occur in a year at that time. From their work, the Moon's mean motion during the PreCambrian era of approximately 6.8×10^8 yr ago can therefore be estimated at 0.243 rad/d. Substitution of this value of the mean motion for this particular time into Equation (18) enables an estimate of K_1 to be made of $3.1 \times 10^{-10}\text{yr}^{-1}$. In fact, examination of Equation (18) shows that for time intervals of less than 10^8 yr an error of even an order of magnitude in K_1 produces an error of less than 10% in the value of $\overline{n}$.

Substitution of Equation (18) into Equations (13) and (14) and keeping $\overline{e}$, $\overline{\imath}$ and $\overline{e}'$ fixed at their present values (see Table I) will provide the rates of change of the lunar node and perigee as functions of time. Equations (13), (14) and (18) can then be substituted into Equations (11) to give the relationship of the three periods $T_{\Delta\lambda}$, T_θ and T_M with time, assuming that only tidal effects are acting on the lunar orbit and that $\overline{e}$ and $\overline{\imath}$ do not vary from their present values over the chosen time interval.

It should be noted that for large time intervals, $T_{\Delta\lambda}$, T_θ and T_M probably do not depict the real values of these three periods at these times since values for K_1 and $\overline{e}'$ as functions of time are not well-known. On the other hand, our aim is not to find out the exact values of the period set ($T_{\Delta\lambda}$, T_M, T_θ) at any given time in order to specify exactly the SLC's that the Moon's orbit passes through during its tidal evolution. Instead, we are only interested in studying a possible sequence of SLC's in order to gain a better understanding of how the SLC's evolve and how often the Moon finds itself in a SLC.

It is worth noting however that at least for time $t = 0$ the values of $T_{\Delta\lambda}$, T_M and T_θ as functions of time are very accurate. Delaunay's expressions give:

$$\begin{aligned} T_{\Delta\lambda} &= 29.530583\,\text{d} \\ T_M &= 27.554572\,\text{d} \\ T_\theta &= 27.212229\,\text{d} \end{aligned} \tag{19}$$

while the currently accepted values are (see Table I):

$$\begin{aligned} T_{\Delta\lambda} &= 29.530589\,\text{d} \\ T_M &= 27.554551\,\text{d} \\ T_\theta &= 27.212220\,\text{d} \end{aligned} \tag{20}$$

Having found the variation of the three satellite periods with time, we can evaluate the period set ($T_{\Delta\lambda}$, T_M, T_θ) at successive intervals of time and calculate the SLC's described by the integer set $(N_{\Delta\lambda}, N_M, N_\theta)$ which exist for each period set, using a time step small enough in order to discover all the SLC's which satisfy the adopted criteria of q' and N_{max}.

For each SLC found, we also calculate the residuals $\varepsilon'_{12}, \varepsilon'_{13}, \varepsilon'_{23}$ as functions of time, the beginning t_b of the SLC (i.e. the time when $|\varepsilon'_{12}|$, $|\varepsilon'_{13}|$ and $|\varepsilon'_{23}|$ all first become less than or equal to a given accuracy q'), the end t_e of the SLC (i.e. the time when $|\varepsilon'_{12}|$ or $|\varepsilon'_{13}|$ or $|\varepsilon'_{23}|$ become greater than q'), the duration of the SLC $t_d = t_e - t_b$ and the time t_p at which the specified SLC is closest to being a perfect SLC.

To evaluate this last parameter, we minimize the quantity

$$\varepsilon_{rms} = \sqrt{{\varepsilon'_{12}}^2 + {\varepsilon'_{13}}^2 + {\varepsilon'_{23}}^2}. \tag{21}$$

In accordance with the present model, where effects due to planetary perturbations are ignored for the moment, Figure 3(a) displays the behaviour of the residuals for the current Saros cycle with time, while Figure 3(b) gives the current Saros cycle's ε_{rms} as a function of time.

According to the model used, the current Saros cycle began approximately 2.66×10^6 yr ago and is presently improving in its ability to cancel solar perturbations. It reaches its closest approach to a perfect Saros cycle in 1.83×10^6 yr hence, where the root mean squared sum of the residuals will have decreased to its minimum of $\varepsilon_{rms} = 9.31 \times 10^{-3}$. The present Saros will finally be disrupted about 7.57×10^6 yr from now, having survived for about 1.02×10^7 yr. Finally, it is the residual ε_{23} relating T'_2 and T'_3 i.e. the anomalistic and nodical months which first puts the Saros cycle outside the accuracy criterion.

Tables 2, 3 and 4 provide lists of the SLC's which the Moon passes through from $t = -5.0 \times 10^7$ yr to $t = +5.0 \times 10^7$ yr, for constant values of $\overline{e}' = 0.0, 0.016726$ (present value) and 0.0588 (maximum value), respectively. In these Tables $\overline{e}$ and $\overline{i}$ were kept constant at their present values, permitting only tidal evolution to change the Moon's orbital semimajor axis. The calculation used a step size of 10^4 yr, and we discard the SLC's that last less than 10^5 yr. The beginnings t_b of each SLC, their ends t_e, times of closest approach to a perfect SLC t_p, degree of closeness to a perfect SLC at this point in time $\varepsilon_{rms(p)}$, and the Saros period involving the synodic month $T_{Saros(p)}$ at time $t = t_p$ are also given.

Although a SLC may be good in the sense that the residuals are small, it will not closely repeat the dynamical geometry of the Earth-Moon-Sun system if the SLC period is not close to an integral number of anomalistic years. For example, it has already been seen (Perozzi et al., 1991) that, for the present Saros, the solar orbital parameters are closely repeated after one Saros period because the latter is within 10 days of an integral number of years.

The maximum possible change in the Sun's geocentric distance at the beginning and the end of the Saros period relative to the Sun's semimajor axis $(\Delta r_{max}/\overline{a}')$ is 0.0031. The last columns of Tables 3 and 4 give such maximum changes in solar geocentric distance for the other SLC's.

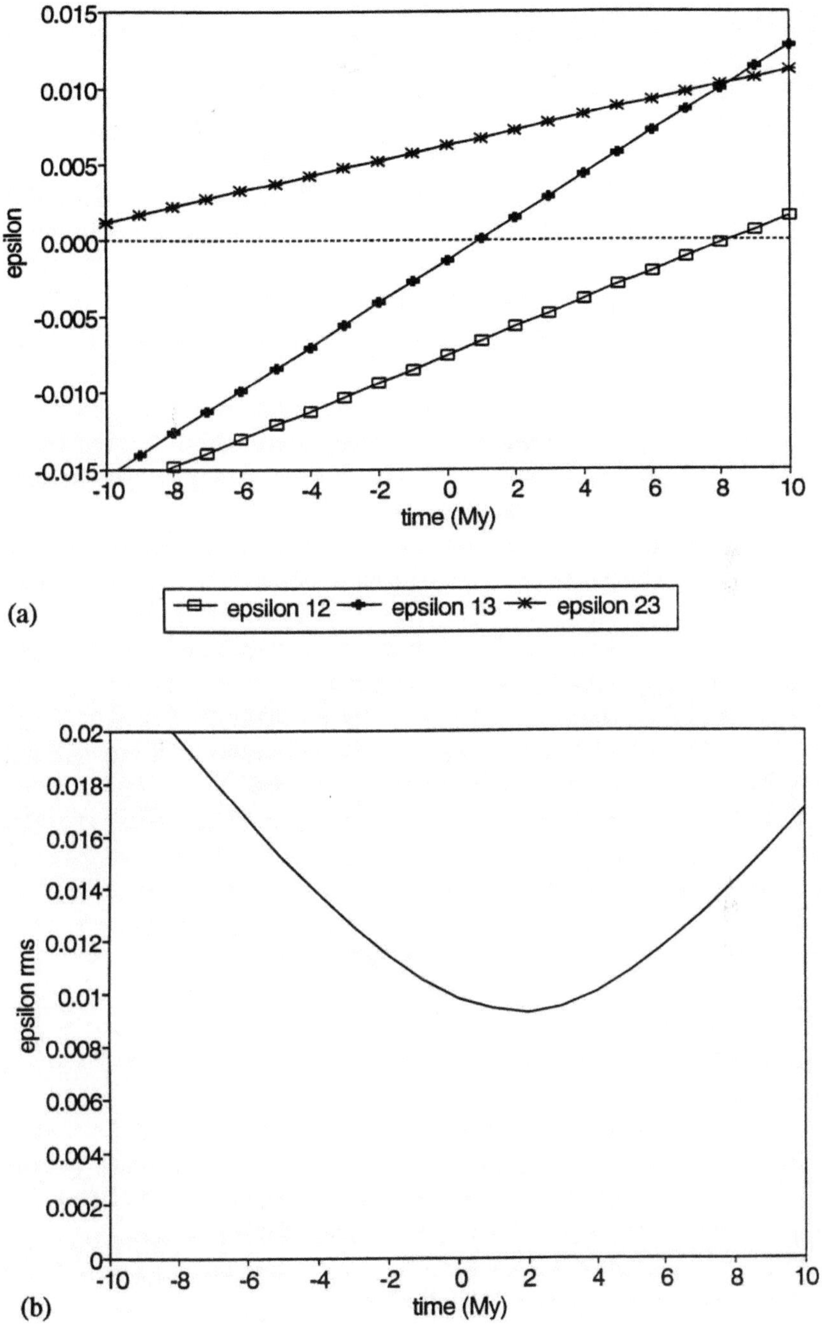

Fig. 3. The behaviour in time of (a) the residuals ε'_{12}, ε'_{13} and ε'_{23} and (b) their root mean squared sum ε_{rms} for the present Saros cycle.

TABLE II
The characteristics of SLC's which occur within the interval −50 to 50 Myr, computed for $\bar{e}' = 0$.

$N_{\Delta\lambda}$	N_M	N_θ	t_b (Myr)	t_e (Myr)	t_p (Myr)	$\varepsilon_{rms(p)}$	$T_{Saros(p)}$
447	479	485	-29.6	-24.5	-26.7	0.0093	36.06 yr
670	718	727	-19.5	-14.8	-17.3	0.0028	54.10 yr
893	957	969	-14.3	-10.8	-12.5	0.0037	72.13 yr
223	239	242	-4.6	8.2	1.8	0.0065	18.03 yr
737	790	800	40.3	44.0	42.0	0.0075	59.78 yr

TABLE III
The characteristics of SLC's which occur within the interval −50 to 50 Myr, computed for $\bar{e}' = 0.0167$.

$N_{\Delta\lambda}$	N_M	N_θ	t_b (Myr)	t_e (Myr)	t_p (Myr)	$\varepsilon_{rms(p)}$	$T_{Saros(p)}$	$(\Delta r)_{\max}$
447	479	485	-30.0	-23.0	-26.7	0.0037	36.06 yr	$0.007\bar{a}'$
670	718	727	-19.6	-15.1	-17.2	0.0056	54.10 yr	$0.010\bar{a}'$
223	239	242	-2.7	7.6	1.8	0.0093	18.03 yr	$0.003\bar{a}'$
738	791	801	22.6	26.7	24.5	0.0057	59.78 yr	$0.021\bar{a}'$

TABLE IV
The characteristics of SLC's which occur within the interval −50 to 50 Myr, computed for $\bar{e}' = 0.0588$.

$N_{\Delta\lambda}$	N_M	N_θ	t_b (Myr)	t_e (Myr)	t_p (Myr)	$\varepsilon_{rms(p)}$	$T_{Saros(p)}$	$(\Delta r)_{\max}$
742	795	805	-45.3	-42.5	-44.1	0.0097	59.78 yr	$0.074\bar{a}'$
518	555	562	-42.1	-37.5	-39.5	0.0089	41.75 yr	$0.083\bar{a}'$
517	554	561	-15.6	-15.4	-15.4	0.0137	41.75 yr	$0.083\bar{a}'$
810	868	879	-5.3	-2.3	-4.0	0.0088	65.47 yr	$0.117\bar{a}'$
293	314	318	10.6	21.3	15.6	0.0048	23.72 yr	$0.091\bar{a}'$
878	941	953	28.9	31.8	30.2	0.0083	71.15 yr	$0.054\bar{a}'$
585	627	635	36.9	37.5	37.5	0.0131	47.43 yr	$0.115\bar{a}'$

Tables 2, 3 and 4 show that, as is expected from the probability analysis of Section 3, the Moon's orbit possesses a SLC only 25 to 30% of the time (i.e. for $\overline{e}' = 0$, 30%; for $\overline{e}' = 0.016726$, 26%; and for $\overline{e}' = 0.0588$, 25%).

As was seen in Section 4.1, changing the eccentricity $\overline{e}'$ causes SLC's to disappear and new ones to appear at different times in the time interval examined. In fact, for the maximum eccentricity $\overline{e}' = 0.0588$, the present Saros cycle has not only disappeared but, for the time interval from -2.3×10^6 to $+1.06 \times 10^7$ years, no other SLC exists for the given accuracy of $q' = 0.01$ and $N_{max} = 1000$.

It is a matter of some interest that the present Saros has the smallest period (18.03 yr) of all SLC's in the Tables, one of the largest durations and the smallest $\Delta r_{max}/\overline{a}'$. Although its residuals are small, they are not the smallest. All in all, these characteristics make the Saros one of the best SLC's for compensating solar perturbations quickly and effectively.

4.3. The Effects of Both Tidal Friction and Planetary Perturbations on the Saros Cycle and other SLC's

We now consider the more realistic case where both planetary perturbations and tidal evolution operate. To this purpose we use data on the evolution of the Earth's eccentricity coming from the computations by Laskar (1990), and consider the time span of 5×10^6 years immediately preceding the current epoch (see Figure 4), during which we study the behaviour of the present Saros.

In Figure 5, the behaviour of the moduli of the three residuals $|\varepsilon'_{12}|$, $|\varepsilon'_{13}|$ and $|\varepsilon'_{23}|$ of the Saros is shown throughout the time interval; $\overline{e}$ and $\overline{i}$ are kept constant at their present values, $\overline{e}'$ is allowed to vary according to Laskar's solution, and ν is varied according to tidal evolution theory. Throughout this paper SLC's have been taken to exist only when their 3 residuals all remain less than $q' = 0.01$. Clearly in Figure 5, if this criterion is maintained, there are frequent occasions when no SLC exists at all. Even if the criterion is relaxed to a value $q' = 0.02$, there are still many occasions when no Saros exists.

Comparison of Figures 4 and 5 shows that as $\overline{e}'$ increases, so do $|\varepsilon'_{12}|$ and $|\varepsilon'_{23}|$ whereas $|\varepsilon'_{13}|$ decreases.

The fact that the time interval from -5×10^6 to the current epoch contains many oscillations of $\overline{e}'$ implies that the present Saros, for the given accuracy criterion $q' = 0.01$, disappears and reappears many times during that interval.

5. Conclusions

We have shown that, if values of 0.01 and 1000 for the accuracy criterion and upper integer limit are chosen, and if lunar tidal evolution follows the customarily adopted model, the Moon's orbit can be in a near SLC for only about 25% to 30% of the time. Taking also into account the variations of the Earth's mean eccentricity

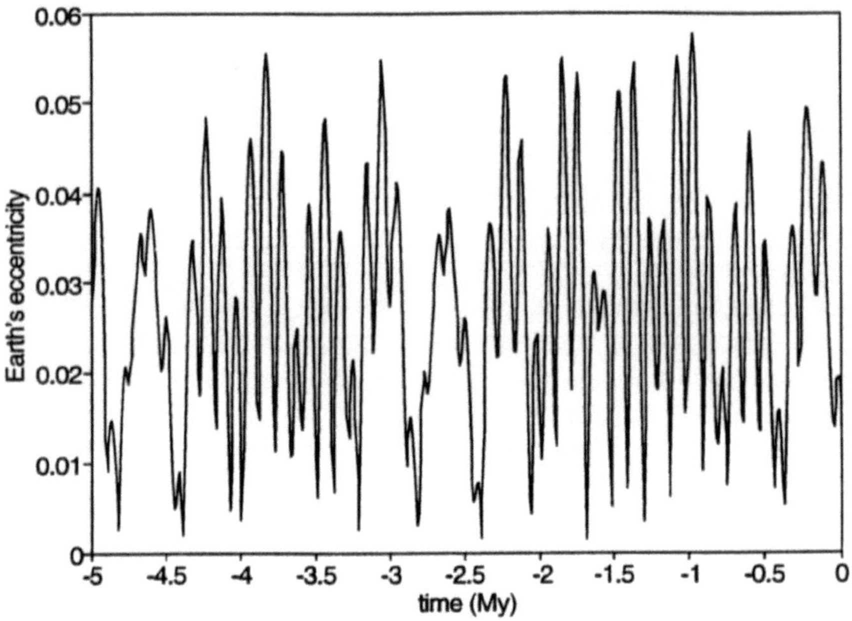

Fig. 4. The variation of the eccentricity $\overline{e}'$ of the Earth's orbit in the last 5 Myr according to Laskar (1990).

$\overline{e}'$, we find that even a SLC which appears to last for 10^7 years for constant $\overline{e}'$, exists in fact only intermittently throughout this time interval.

What happens to the Moon's orbit during epochs when tidal friction and/or the change in eccentricity of the Earth's orbit take the Moon into situations where no SLC exists in its orbital dynamics is still unknown. One possibility is that, in such a case, solar perturbations may grow to the point where they change the Moon's orbital elements irreversibly before another near SLC is found. In the past this may not have been of major importance when the Moon was nearer the Earth and solar perturbations were consequently smaller. In the future, however, as the Moon spirals outwards, these features may become more important.

There is also the question of the past evolution of our satellite. The acceptance of the occurrence of continental drift has itself cast doubts on the accuracy of previous estimates of the time span over which evolution of the Moon's orbit by tidal friction has taken place. If it turned out in addition that the Earth-Moon-Sun system has been without a quick solar perturbation-cancelling Saros mechanism more often than not, then the Moon's orbital evolutionary history may well have included hitherto unsuspected episodes when the Sun played a more direct part than it does at present.

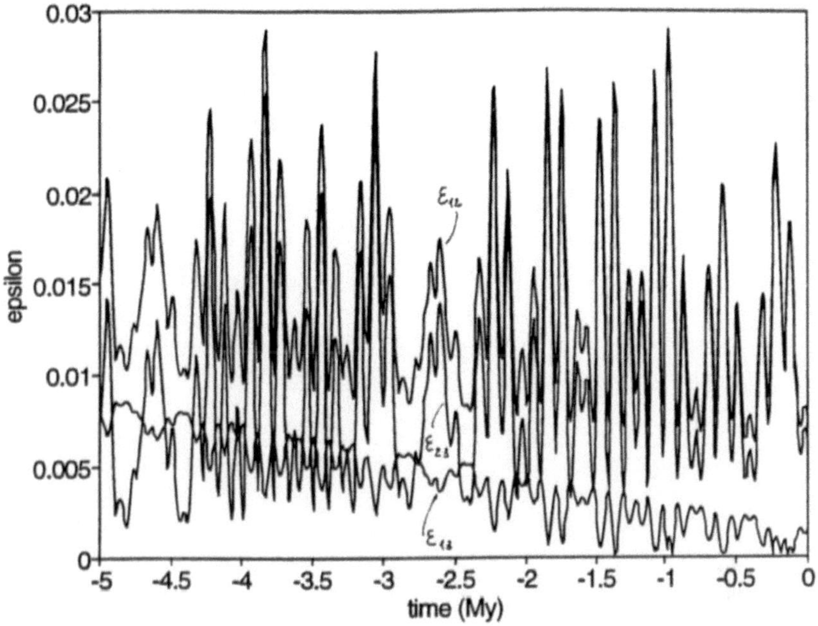

Fig. 5. The variation of the moduli of the residuals $|\varepsilon'_{12}|$, $|\varepsilon'_{13}|$ and $|\varepsilon'_{23}|$ in the last 5 Myr, due to planetary perturbations on $\overline{e}'$ and to the tidal variation of ν.

Acknowledgements

We are indebted to J. Laskar for kindly providing us his results on the long-term behaviour of the eccentricity of the Earth.

References

Brouwer, D. and Clemence, G.M.: 1961, *Methods of Celestial Mechanics*, Academic, New York

Delaunay, Ch.: 1872, 'Note sur les mouvements du périgée et du noeud de la Lune', *Compt. rend. hebdom. Acad. Sci.* **74**, 17-21

Goldreich, P. and Soter, S.: 1966, 'Q in the Solar System', *Icarus* **5**, 375-389

Laskar, J.: 1990, 'The chaotic behaviour of the solar system: A numerical estimate of the size of the chaotic zones', *Icarus* **88**, 266-291

Laskar, J., Quinn, T. and Tremaine, S.: 1992, 'Confirmation of Resonant Structure in the Solar System', *Icarus* **95**, 148-152

Mignard, F.: 1981, 'The lunar orbit revisited, III', *Moon & Planets* **24**, 189-207

Perozzi, E., Roy, A.E., Steves, B.A. and Valsecchi, G.B.: 1991, 'Significant high number commensurabilities in the main lunar problem. I: the Saros as a near periodicity of the Moon's orbit', *Celest. Mech. & Dynam. Astron.* **52**, 241-261

Roy, A.E., Steves, B.A., Valsecchi, G.B. and Perozzi, E.: 1991, in A.E. Roy, ed(s)., *Predictability, Stability and Chaos in N-Body Dynamical Systems*, Plenum, London, 273-282

Sonett, C.P., Finney, S.A. and Williams, C.R.: 1988, 'The lunar orbit in the late Precambrian and the Elatina sandstone laminae', *Nature* **335**, 806-808

Steves, B.A.: 1990, *Finite-Time Stability Criteria for Sun-Perturbed Planetary Satellites*, PhD Thesis, Univ. Glasgow

Valsecchi, G.B., Perozzi, E., Roy, A.E. and Steves, B.A.: 1993a, 'Periodic orbits close to that of the Moon', *Astron. Astrophys.*, in press

Valsecchi, G.B., Perozzi, E., Roy, A.E. and Steves, B.A.: 1993b, 'The arrangement in mean elements' space of the periodic orbits close to that of the Moon', *Celest. Mech. & Dynam. Astron.*, in press

MOTION OF THE ORBITAL PLANE OF A SATELLITE DUE TO A SECULAR CHANGE OF THE OBLIQUITY OF ITS MOTHER PLANET

HIROSHI KINOSHITA
National Astronomical Observatory
2-21-1 Osawa, Mitaka, Tokyo 181, Japan
E-Mail address:Kinoshita@c1.mtk.nao.ac.jp

Abstract. At the present state the rotational axes of Uranus and Pluto are nearly perpendicular to their orbital planes and each satellite moves in the vicinity of the equatorial plane of its mother planet. We assume that in the past a planet's equatorial plane was nearly coincident with its orbital plane and then the inclination of the equatorial plane with respect to the orbital plane began to increase secularly. Here we discuss whether a satellite that moves in its mother's equatorial plane continues to move in the equatorial plane or not. When the direct solar perturbation is neglected, the satellite continues to stay in the equatorial plane under the condition that the secular rate of change of the obliquity is slower than the precessional speed of the satellite orbital plane with respect to the equator.

Key words: Uranian Satellites–Obliquity Change

1. Introduction

The obliquities of Uranus and Pluto are almost 90 degrees (98 degrees for Uranus and 128 degrees for Pluto) and their satellites that are found so far move in the vicinities of the planetary equators. We assume here that in the early stage of the formation of planets and satellites, the rotational axes of the planets were perpendicular to their orbital planes and their satellites were moving in the vicinity of the equators. Then due to some unknown mechanism, the obliquities started to increase secularly. One possible scenario for this process was recently proposed by Tremaine (1991). What we discuss here is whether a satellite that initially moves in the vicinity of the equator continues to stay near the equator that is being twisted.

Goldreich (1965) discussed the inclination of satellite orbits about an oblate precessing planet. He investigated mainly the effect of the equatorial precession on the inclination and gave a short qualitative result on the effect due to the secular change of the obliquity: in his paper's abstract, he said "Changes in the planet's obliquity, which are slower when compared with the satellite's nodal period, will not affect a satellite's inclination to the equator plane of its planet." He did not give a quantitative result on the effect due to the secular change of the obliquity. Then this problem was discussed by Hori (1979), but he made a serious mistake in the treatment of the equatorial precession, on which a short comment is given in section 5. In section 2, we derive equations of motion of the satellite referred to the moving equator. In this problem we have three time scales: T_1 (the period of the equatorial precession), T_2 (the time of the obliquity change from 0 degree to 90 degrees), and T_3 (the period of the precession of the satellite orbital plane). In the outer planetary region T_1 is much longer than T_3. Therefore in section 3, we discuss the case where the equatorial precession is negligible. Then in section 4, we discuss the effect due to the precessional motion of the equator.

Celestial Mechanics and Dynamical Astronomy **57**: 359–368, 1993.

2. Equations of Motion Referred to the Moving Equator

The main perturbation due to the equatorial bulge of a planet is

$$V_{J_2} = -\frac{k^2 M_p R_p^2}{r^3} J_2(\frac{1}{2} - \frac{3}{2}\sin^2\phi), \tag{1}$$

where M_p and R_p are the mass and the equatorial radius of the planet, and J_2 is the second degree harmonic of the planetary potential. The secular part of V_{J_2} after elimination of short periodic terms with periods that are linear combinations orbital periods of the planet and the satellite is

$$V_{J_2}^* = -\frac{1}{4} n^2 R_p^2 (1-e^2)^{-3/2} J_2(3\cos^2 I - 1), \tag{2}$$

where the orbital inclination I is referred to the planetary equator, n is the mean motion of the satellite, and e the eccentricity. The equator is precessing due to the solar torque and here we assume that the obliquity is secularly changing. The Hamiltonian of the satellite referred to the moving equator is written (Goldreich 1965) as

$$K = V_{J_2}^* - \mathbf{R}\cdot\mathbf{G}. \tag{3}$$

The second term (3) comes from the motion of the equator. The derivation of (3) is given in Appendix 1. The vector **R** is the rotational vector of the equator, and **G** is the angular momentum vector of the orbital motion of the satellite referred to the moving equator:

$$\mathbf{R} = (\frac{di_p}{dt}, \frac{dh_p}{dt}\sin i_p, \frac{dh_p}{dt}\cos i_p), \tag{4}$$

$$\mathbf{G} = (G\sin I \sin h, -G\sin I\cos h, G\cos I). \tag{5}$$

The position of the equator is described by the two quantities i_p and h_p, where the obliquity i_p is the inclination of the equator referred to the orbital plane of the planet and h_p is the longitude of the equator's node. In the expression for **G**, G is the total angular momentum of the orbital motion of the satellite, and h is the longitude of the node that is measured from the equator's node (see figure 1).

Instead of using I and h, we describe the motion of the orbital plane using x and y:

$$x = \sin I\cos h, \qquad y = \sin I \sin h. \tag{6}$$

The equations of motion in terms of x and y are easily derived from the Hamiltonian (3):

$$\frac{dx}{dt} = qy - \frac{di_p}{dt}\cos I, \tag{7}$$

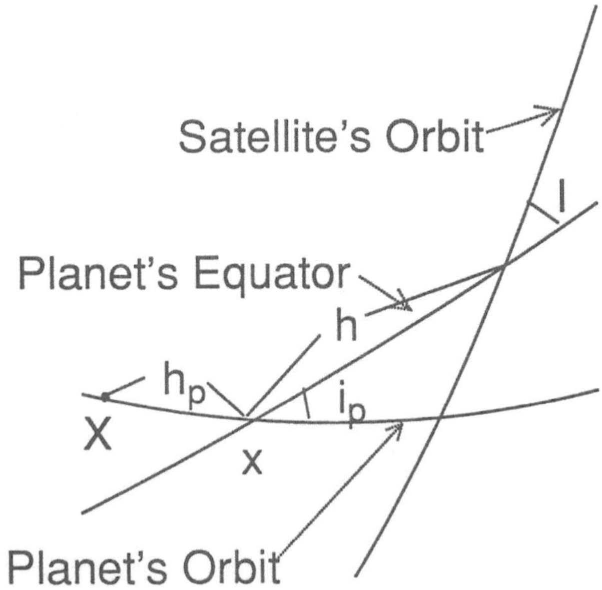

Fig. 1. Coordinate system

$$\frac{dy}{dt} = -qx - \frac{dh_p}{dt}\sin i_p \cos I, \tag{8}$$

where

$$q = c_3 \cos I + \frac{dh_p}{dt}\cos i_p, \tag{9}$$

$$c_3 = \frac{3}{2}J_2\left(\frac{R_p}{a(1-e^2)}\right)^2 n. \tag{10}$$

The precessional speed of the equator (see, for example, Smart 1953 or Kinoshita 1977) is given by

$$\frac{dh_p}{dt} = c_1 \cos i_p, \tag{11}$$

$$c_1 = -\frac{3}{2}\frac{C-A}{C}(1-e_p^2)^{-3/2}\frac{n_p^2}{\omega_p}, \tag{12}$$

where C and A are the principal moments of inertia of the planet, n_p the orbital mean motion and e_p the eccentricity of the planet around the Sun, and ω_p the spin angular velocity of the planet.

Here we assume that the obliquity i_p has a slow secular change:

$$i_p = c_2 t + i_{p0}. \tag{13}$$

Our problem has three time scales: $T_1 = 2\pi/c_1, T_2 = \pi/2c_2$, and $T_3 = 2\pi/c_3$. The time $T_1/\cos i_p$ is the period of the precession of the planetary equator, T_2 is the time of the obliquity change from 0 degree to 90 degrees, and $T_3/\cos I$ is the period of the precession of the satellite orbital plane. In the outer planetary region T_1 is much larger than T_3. For example, for Uranus T_1 is 2×10^8 years and T_3 for Oberon, which is the remotest satellite of Uranus, is 4×10^4 years. Therefore we can treat the quantity c_1 as a small quantity of first-order compared with c_3. In these estimates we use $R_p = 26000$ km, $J_2 = 3.3461 \times 10^{-3}$ (French *et al.* 1986), and adopt $M_p R_p^2/C = 4$.

3. The Motion in Case of No Precessional Motion of the Equator

As mentioned at the end of the previous section, the quantity c_1 can be treated as a small quantity of the first-order compared with c_3. In this section we discuss the case of $c_1 = 0$, where the planetary equator is not in precession and only the obliquity changes secularly. In this case the equations of motion (7) and (8) take the following form:

$$\frac{dx}{dt} = c_3 y \cos I - c_2 \cos I, \tag{14}$$

$$\frac{dy}{dt} = -c_3 x \cos I. \tag{15}$$

Here we change the independent variable from t to s that is defined by

$$ds = c_3 \cos I dt. \tag{16}$$

Now the equations of motion become linear differential equations with constant coefficients :

$$\frac{dx}{ds} = y - c_2/c_3, \qquad \frac{dy}{ds} = -x. \tag{17}$$

The general solutions of (17) are

$$x = D\cos(-s + \alpha), \qquad y = D\sin(-s + \alpha) + c_2/c_3, \tag{18}$$

where D and α are the integral constants. The relation between t and s is obtained from

$$\frac{ds}{\sqrt{1 - x^2 - y^2}} = dt. \tag{19}$$

Substituting (18) into (19) and using Jacobian elliptic functions (Byrd and Friedman 1971), we get

$$\sin(s - \alpha + \pi/2)/2 = \mathrm{sn}u/\sqrt{1 + \bar{k}^2 \mathrm{sn}^2 u}, \tag{20}$$

$$u = \frac{c_3}{2}\sqrt{1 - (D - c_2/c_3)^2} t, \tag{21}$$

$$\bar{k}^2 = \frac{4Dc_2/c_3}{1 - (D + c_2/c_3)^2}, \tag{22}$$

$$k^2 = \bar{k}^2/(1 + \bar{k}^2), \tag{23}$$

where k is the modulus of the elliptic function sn$u$. The mean motion of s is expressed as

$$(\frac{ds}{dt})_{mean} = \frac{\pi}{2K} c_3 \sqrt{1 - (D - c_2/c_3)^2}, \tag{24}$$

where K is the complete elliptic integral of the first kind with modulus k.

Now we relate the integral constants to the initial conditions. Since we can take the epoch arbitrarily, we choose the initial conditions of I and h such that

$$I = I_0, \qquad h = 0, \tag{25}$$

and we have

$$D = \sqrt{\sin^2 I_0 + (c_2/c_3)^2}. \tag{26}$$

From (18) we have the maximum orbital inclination referred to the equator as

$$\sin I_{max} = D + c_2/c_3. \tag{27}$$

If at the beginning the satellite moves in the vicinity of the equatorial plane ($I_0 \approx 0$),

$$\sin I_{max} \approx 2c_2/c_3 = T_3/2T_2. \tag{28}$$

In this case we have

$$(\frac{ds}{dt})_{mean} = \frac{\pi}{2K} c_3, \tag{29}$$

where the modulus is $k = 2c_2/c_3$. From (28), if the secular change rate of the equatorial plane c_2 is much smaller than the precessional rate of a satellite's orbital plane with respect to the equator, the satellite continues to stay in the vicinity of the equator.

Here we apply this theory to satellites of Uranus. The maximum inclination strongly depends on the semi-major axis, since c_3 is proportional to $a^{-7/2}$. Figure 2 shows the maximum inclination of a satellite against its semi-major axis. Three

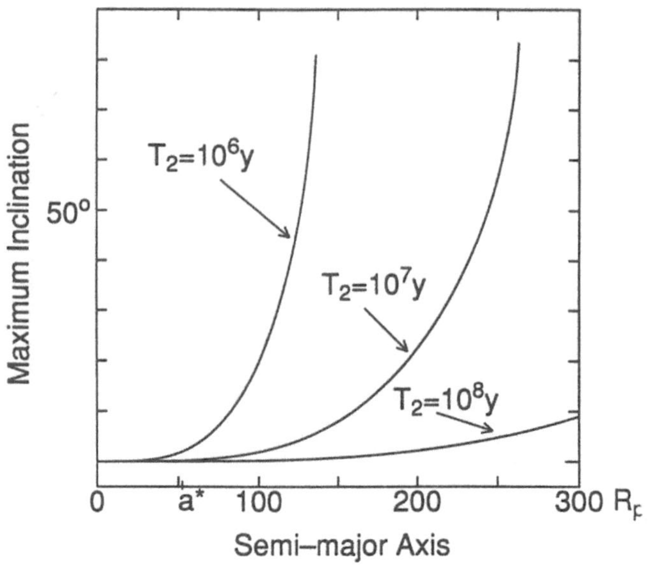

Fig. 2. Maximum inclination caused by the secular change of the obliquity

curves correspond to three values of T_2= 10^6years, 10^7years, and 10^8 years. The maximum inclination of the remotest satellite Oberon is 0.°1, 0.°01, 0.°001 for $T_2 = 10^6, 10^7, 10^8$ years, respectively. The present inclination of Oberon is less than 0.°1.

So far we have neglected the direct solar perturbation. The ratio between the direct solar perturbation $V_\odot$ and the perturbation due to the equatorial bulge V_{J_2} is given by

$$\gamma = 2(1 + \frac{M_p}{M_\odot})(\frac{n}{n_p})^2(\frac{R_p}{a})^2(1 - e_p^2)^{3/2} J_2 \approx V_{J_2}/V_\odot. \tag{30}$$

We call a satellite with $\gamma < 1$ an outer satellite (the solar perturbation is larger than the J_2 perturbation) and a satellite with $\gamma > 1$ is an inner satellite. The semi-major axis a^* with $\gamma = 1$ is given by

$$a^*/R_p = (2\frac{M_p}{M_\odot}J_2)^{1/5}(1 - e_p^2)^{3/10}(a_p/R_p)^{3/5}. \tag{31}$$

For Uranus, a^* is 52.1 R_p. As the semi-major axis of a satellite increases, c_3 becomes smaller and T_3 longer, and the maximum inclination becomes larger. At the distance $\bar{a}$ where $c_3 = 2c_2$, the maximum inclination reaches 90°, which means that the satellite does not follow the moving equator at all. The semi-major axis $\bar{a}$ depends on T_2; $\bar{a} = 135R_p, 260R_p, 500R_p$ for $T_2 = 10^6, 10^7, 10^8$ years, respectively. The

Hill radius for Uranus is $2800R_p$ and the radius of an influence of sphere is $2000R_p$. The distances $\bar{a}$ are larger than a^*, and the satellite whose semi-major axis is close to $\bar{a}$ is located in the region where the direct solar perturbation is larger than the J_2 perturbation. Therefore we have to take into account the direct solar perturbation to discuss outer satellites.

4. Effects of the Precessional Motion of the Equator

Here we examine how the precessional motion of the equator disturbs the orbital motion of the satellite that moves in the vicinity of the equator. Neglecting squares and products of x and y, the equations (7) and (8) reduce to

$$\frac{dx}{dt} = qy - \frac{di_p}{dt}, \tag{32}$$

$$\frac{dy}{dt} = -qx - \frac{dh_p}{dt}\sin i_p, \tag{33}$$

where

$$q = c_3 + \frac{dh_p}{dt}\cos i_p. \tag{34}$$

The equations (32) and (33) are non-homogeneous linear differential equations with a variable coefficient. The general solution of the homogeneous part of the differential equations is given by

$$x = D\cos(-s+\alpha), \qquad y = D\sin(-s+\alpha), \tag{35}$$

where

$$s = (c_3 + c_1/2)t + \frac{c_1}{4c_2}\sin 2i_p. \tag{36}$$

The particular solution due to the non-homogeneous part of equations (32) and (33) is easily obtained from the method of variation of constants:

$$x = \left(\frac{c_1}{4c_3} - \frac{3}{4}\frac{c_1c_3}{c_3^2 - 4c_2^2}\right)\sin 2i_p + o(c_1^2), \tag{37}$$

$$y = \frac{c_2}{c_3 + c_1/2} - \frac{3}{2}\frac{c_1c_2}{c_3^2 - 4c_2^2}\cos 2i_p + o(c_1^2). \tag{38}$$

The general solution of (32) and (33) is given by adding (37) and (38) to (35). When c_3 is not close to $2c_2$, the effect due to c_1 to the inclination I is smaller by the factor c_1/c_2 than the effect due to c_2. For inner satellites, where $c_3 >> c_2$, the precessional effect is negligible. For outer satellites, c_3 becomes close to $2c_2$ at some distance. In this region, the maximum inclination becomes 90 degrees due to the c_2 effect and also due to the second term of the particular solution (37) and (38). In this region the satellite cannot stay in the vicinity of the equator. The second effect comes from the small denominator of the particular solution, which we may call a secular resonance. However, as mentioned in section 3, in order to discuss outer satellites we have to take into account the direct solar perturbation.

5. Concluding Remarks

In this paper we take into account the indirect solar perturbation on the satellite through the equatorial precession due to the solar torque and neglect the direct solar perturbation. For inner satellites this approximation is good. However, for outer satellites we have to take into account the direct solar perturbation which causes a large change of the eccentricity and the inclination of the satellite (see Kinoshita 1991). In this case the number of degrees of freedom becomes two even after the elimination of short periodic terms. In this paper we assume that the orbital plane of the planet is fixed in space. However the orbital planes of the planets are moving periodically due to the interactions among planets, so called secular perturbations. It is probable that the long periodic motion of the orbital plane of the planet produces a secular resonance with the long time scale change of the satellite orbital plane due to the secular change of the obliquity.

Here we give a short comment on the pitfall into which Hori (1979) fell. When the obliquity does not have the secular change, the precession of the planetary equator which is the secular perturbation in the rotational motion is expressed as

$$h_p = h_{p0} + c_1 t \cos i_p. \tag{39}$$

When the time derivative of h_p is necessary to derive the equations of motion, Hori differentiated (39) with respect to time and got

$$\frac{dh_p}{dt} = c_1 \cos i_p - c_1 c_2 t \sin i_p. \tag{40}$$

As described in section 2 the time derivative of h_p derived from the original equations of motion has the following form:

$$\frac{dh_p}{dt} = c_1 \cos i_p.$$

Therefore the second term of (40) is not necessary and this term caused spurious mixed secular terms in the solutions by Hori (1979).

Appendix

The Hamiltonian Referred to a Rotating Frame of Reference

Goldreich (1965) obtained the Hamiltonian referred to a moving reference frame from the Lagrangian. Here we give another derivation with use of a generating function of a canonical transformation.

Let $\mathbf{r} = (x, y, z)$ designate the position vector referred to the inertial frame and $\mathbf{r}' = (x', y', z')$ the position vector referred to the rotating frame. The transformation between $\mathbf{r}$ and $\mathbf{r}'$ is expressed with use of an orthogonal matrix A:

$$\mathbf{r}' = A\mathbf{r}, \tag{A-1}$$

where the matrix depends explicitly on time. We introduce a canonical generating function F_2:

$$F_2 = (\mathbf{p}', A\mathbf{r}), \qquad (A-2)$$

where $\mathbf{p}'$ is the momentum vector conjugate to $\mathbf{r}'$ and the bracket means the inner product of two vectors. The generating function F_2 is also written as

$$F_2 = (\mathbf{r}, A^*\mathbf{p}'), \qquad (A-3)$$

where A^* is the transpose of A. The canonical transformations derived from F_2 are obtained from (A-2) and (A-3),

$$\mathbf{r}' = \frac{\partial F_2}{\partial \mathbf{p}'} = A\mathbf{r}, \qquad (A-1)$$

$$\mathbf{p} = \frac{\partial F_2}{\partial \mathbf{r}} = A^*\mathbf{p}'. \qquad (A-4)$$

Since the inverse of A^* is A itself, we have from (A-4)

$$\mathbf{p}' = A\mathbf{p}, \qquad (A-5)$$

which shows that the momentum vector is also obtained from the same transformation matrix A as the position vector. Since the generating function depends on time explicitly, the new Hamiltonian K referred to the rotating frame is obtained from

$$K = F + \frac{\partial F_2}{\partial t}, \qquad (A-6)$$

where F is the Hamiltonian referred to the inertial frame. Now we have to express the second term of (A-6) only in terms of $\mathbf{r}'$ and $\mathbf{p}'$:

$$\frac{\partial F_2}{\partial t} = (\mathbf{p}', \frac{dA}{dt}\mathbf{r}) = (\mathbf{p}', \dot{A}A^{-1}\mathbf{r}'). \qquad (A-7)$$

The matrix $\dot{A}A^{-1}$ is an anti-symmetric matrix and its independent three components are expressed with use of the rotational vector $\mathbf{R}$ of the rotating frame:

$$\dot{A}A^{-1} = \begin{pmatrix} 0 & R_{z'} & -R_{y'} \\ -R_{z'} & 0 & R_{x'} \\ R_{y'} & -R_{x'} & 0 \end{pmatrix}, \qquad (A-8)$$

where $R_{x'}$ is the component of the rotational vector around the x' axis. The derivation of (A-8) is found in textbooks on classical mechanics (for example Arnold 1974). Using (A-8), we get

$$\dot{A}A^{-1}\mathbf{r}' = -\mathbf{R} \times \mathbf{r}'. \qquad (A-9)$$

Using the relation of the scalar triple product, we have

$$\frac{\partial F_2}{\partial t} = -(\mathbf{p}', \mathbf{R} \times \mathbf{r}') = -(\mathbf{R}, \mathbf{r}' \times \mathbf{p}'), \qquad (A-10)$$

where $\mathbf{r}' \times \mathbf{p}'$ is the momentum vector $\mathbf{G}'$ referred to the rotating frame. Finally we have the Hamiltonian referred to the rotating frame as

$$K = F - (\mathbf{R}, \mathbf{G}'). \qquad (A-11)$$

Although we consider here a particle system, we can apply the same idea to a system of particles, and have the same form Hamiltonian for a rigid body. The textbook on classical dynamics by Landau and Lifshitz (1960) gives the Hamiltonian for a particle in a uniform rotating frame of reference in one of its exercises. Kozai and Kinoshita (1973) discussed an artificial satellite motion disturbed by nutation and precession of the Earth, in which equations of motion of the satellite referred to the equator were obtained from the equations of motion referred to the inertial reference frame.

We choose the orbital plane of the planet as the reference plane of the inertial frame and the $x'y'$ plane of the equator as the moving reference frame and take the x' axis on the orbital plane of the planet. Then we have (4) and (5) in section 2.

References

Arnold, V.: 1976, *Les Méthodes Mathématiques de la Mécanique Classique*, NAUKA, Moscou, Chap.6.

Byrd, P.F. and Friedman, M.D.: 1971, *Handbook of Elliptic Integrals for Engineers and Physicists*, Springer-Verlag,Berlin, .

French, R.G.,Elliot, J.L.,Levine, S.E.: 1986, *Icarus* **67**, 134.

Goldreich, P.: 1965, *AstronJ.* **70**, 5.

Hori, G.: 1979, in Nomura,T., ed(s)., *12 th ISAS Lunar and Planetary Symposium*, Institute of Space and Astronautical Science,Tokyo, 166-170.

Kinoshita, H.: 1977, *Celest.Mech.* **15**, 277.

Kinoshita, H.: 1991, *Celest. Mech.* **52**, 293.

Kozai, Y. and Kinoshita,H.: 1973, *Celest.Mech.* **7**, 356.

Landau, L.D. and Lifshitz, E.M.: 1960, *Mechanics*, Pergamon Press, London, p.133.

Tremaine, S.: 1991, *Icarus* **89**, 85.

Smart, W.M.: 1953, *Celestial Mechanics*, Longman,London, chap.20.

FOURIER ANALYSIS OF LOCAL LYAPUNOV CHARACTERISTIC EXPONENTS FOR SATELLITE-TYPE MOTIONS

E. LOHINGER*
Institute of Astronomy, University of Vienna, Türkenschanzstr. 17, 1180 Wien, Austria
and
C. FROESCHLÉ
Observatoire de la Côte d'Azur, B.P. 229, 06304 Nice Cedex 4, France

Abstract. For a binary star system with equal masses the transition from regular to stochastic motion has been determined by means of the Lyapunov Characteristic Numbers (LCN) in a paper of Gonczi and Froeschlé (1981), where the initial distance of the massless body to the second primary was used as parameter of integration. This time we analyse the Fourier spectra of the local LCNs ($\ln \alpha_k^i$) of this system for different intervals of time.

Key words: Lyapunov characteristic numbers, chaotic motion, binary stars

1. Introduction

It is well known that the chaoticity of a dynamical system can be measured by the exponential rate of divergence of nearby trajectories, i.e. by the Lyapunov characteristic exponents (LCEs). These numbers, which are nothing but the first moment of the distribution of the local variations of the tangent vectors to the flow (see Benettin et al.(1980), Froeschlé (1984)), have been used very often during the last years to determine whether the motion of a system is chaotic or not (see e.g. Froeschlé(1984), Gonczi and Froeschlé (1981)). For a binary star system with mass ratio $\mu = 0.5$ the distribution of these local variations, which we call local Lyapunov characteristic numbers (LLCNs), have been analysed and published in a previous paper (Lohinger et al. (1992)). Here we present a "threshold analysis" of the Fourier spectra of the LLCNs for the quasiperiodic and chaotic case. This analysis shows a similar transition from regular to chaotic motion as it was found by means of the LCEs (see Gonczi and Froeschlé (1981)).

2. The Computations

The model in consideration is the three dimensional restricted three body problem in a rotating frame of coordinates, where the masses m_1 and m_2 remain motionless in their locations on the x-axis: $(-\mu, 0, 0)$ and $(1 - \mu, 0, 0)$ and the massless body moves in their gravitational field. For detailed information about the restricted three body problem see e.g. Szebehely (1967).
The initial conditions are (see Gonczi, Froeschlé (1981)): $x = 0.5 - d_0$; $y = 0$; $z = 0$; $\dot{x} = \dot{z} = 0.235749(d_0)^{-1/2}$; $\dot{y} = -0.640312(d_0)^{-1/2} + d_0$; d_0 denotes the initial distance of the massless body to m_2. Taking different values of d_0

* Present address: Observatoire de la Côte d'Azur, Nice

Celestial Mechanics and Dynamical Astronomy **57**: 369–372, 1993.

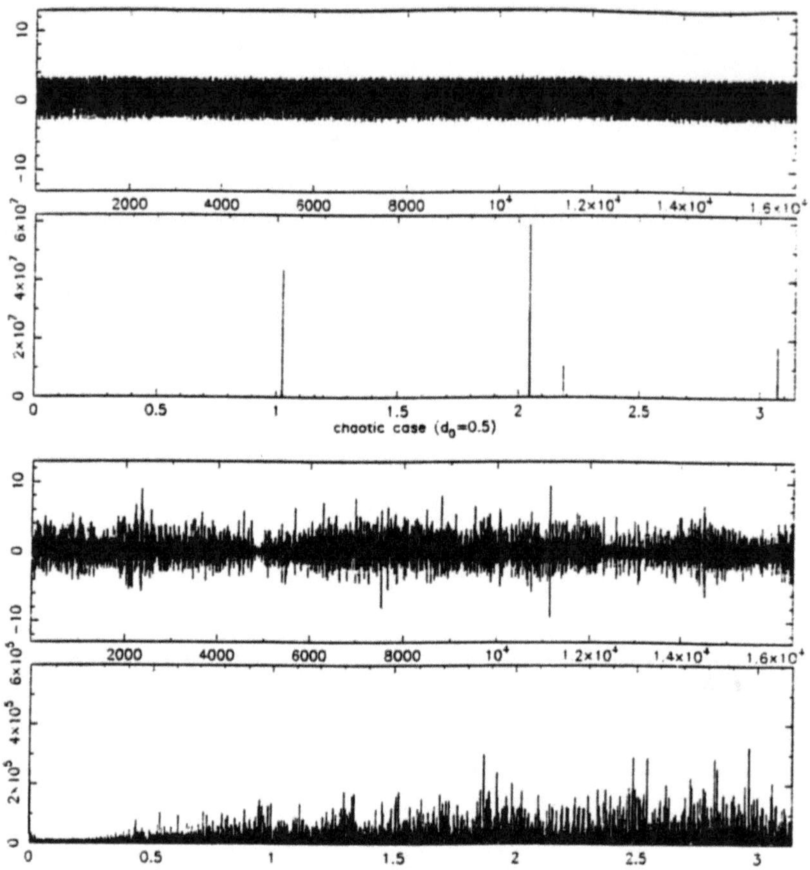

Fig. 1. Here we see the first LLCN as a function of time and the corresponding power spectrum.

(0.10; 0.13; 0.22; 0.24; 0.255; 0.27; 0.31; 0.35; 0.40; 0.50) it is possible to determine the transition from regular to stochastic motion as the perturbation due to m_1 becomes more and more important the greater d_0 is.
Using the procedure of Benettin et al. (1980), the LLCNs ($\ln \alpha_k^i$) are by-products of the computation of the LCEs (see eq. (1)) through the Gram-Schmidt procedure:

$$\chi_i(P) = \lim_{N\to\infty} \frac{1}{N\tau} \sum_{k=1}^{N} \ln \alpha_k^i \tag{1}$$

In this paper the Fourier analysis is done only for the first set of LLCNs $\ln \alpha_k^1 (k = 1, \ldots, N)$, that is the most important one as it corresponds to the greatest LCE which determines whether the motion is chaotic or not. As the model in

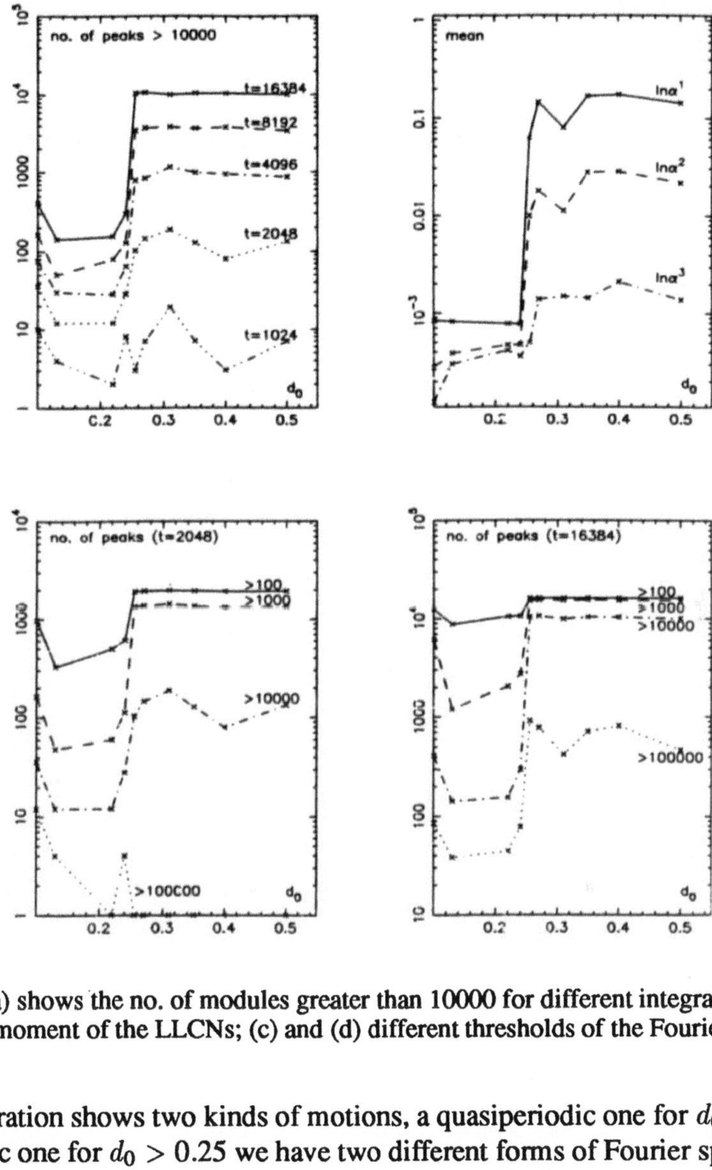

Fig. 2. (a) shows the no. of modules greater than 10000 for different integration times; (b) the first moment of the LLCNs; (c) and (d) different thresholds of the Fourier spectra

consideration shows two kinds of motions, a quasiperiodic one for $d_0 \leq 0.25$ and a chaotic one for $d_0 > 0.25$ we have two different forms of Fourier spectra for the initial conditions, an example of each is shown in Fig.1.

3. The Results

Our analysis of the Fourier spectra of the first LLCNs shows for certain thresholds the same strong increase when the motion chances to a chaotic one, as it has been found by means of the LCEs (fig.2.b).
In fig.(2.a) we plotted the numbers of peaks greater than 10000 for different integration times. Note that the integration time must be at least 2048 units to find the

desired behavior for the different initial distances d_0. Additionally, we give two examples showing different thresholds, one for an integration time of 2048 units (fig.(2.c)) and one for an integration time of 16384 units (fig.(2.d)). >From these figures we can recognize, that the transition between regular and stochastic motion is defined in the best way when we take a threshold between 1000 and 10000.

4. Conclusion

The "threshold analysis" provides a possibility to determine whether the motion is regular or chaotic. So the same study will be done for the other LLCNs and we want apply this analysis to the dynamics of Halley-like-comets.

5. Acknowledgments

The authors want to thank Dr. R. Gonczi and Ph. Bendjoya for stimulating discussions.

References

Benettin G., Galgani L., Giorgilli A., Strelcyn J. M.: 1980,"Lyapunov characteristic exponents for smooth dynamical systems; a method for computing all of them".*Meccanica* vol. **15**.

Froeschlé C.: 1984, "The Lyapunov characteristic exponents and applications". *Jounal de Mécanique théorique et appliquée*, Numéro spécial.

Gonczi R., Froeschlé C.: 1981, "The Lyapunov characteristic exponents as indicators of stochasticity in the restricted three-body problem". *Celest. Mech.* **25**, p. 271-280.

Henon M., Heiles C.: 1964, "The applicability of the third integral of motion, some numerical experiments". *Astron. Journal* **69**, p. 73-79.

Lohinger E., Froeschlé C., Dvorak R.: 1992, "Generalized Lyapunov exponent indicators in Hamiltonian dynamics - an application to a double star system". Proceedings of the 3rd Alexander v. Humboldt Colloquium, in press.

Press W.H., Flannery B.P., Teukolsky S.A., Vetterling W.T.: 1989, Numerical Recipes, Cambridge University press.

Szebehely V.: 1967, "The Theory of Orbits". Academic Press, New York and London.

CAPTURE OF GRAINS INTO RESONANCES THROUGH POYNTING-ROBERTSON DRAG

B. SICARDY
Observatoire de Paris, DAEC-EUROPA, 92195 Meudon Cédex Principal, France, Université Pierre et Marie Curie, 75005 Paris, France

C. BEAUGÉ and S. FERRAZ-MELLO
Instituto Astrónomico e Geofísico, Univ. de São Paulo, CP9638, São Paulo, SP, Brazil

D. LAZZARO
Observatório Nacional, DAF, Rio de Janeiro, RJ, Brazil, Observatoire de Paris, DAEC-EUROPA, 92195 Meudon Cédex Principal, France

and

F. ROQUES
Observatoire de Paris, DAEC-EUROPA, 92195 Meudon Cédex Principal, France

Abstract. We review here some relevant problems connected to the evolution of circumstellar dust grains, subjected to Poynting-Robertson (PR) drag, and perturbed by first-order resonances with a planet on a circular orbit. We show that only outer mean motion resonances are able to counteract the damping effect of PR drag. However, the high orbital eccentricities reached by the particle lead to orbit crossings with the planet. This is a serious difficulty for a permanent trapping to be achieved. In any case, we show that the time spent in the resonance is long enough for statistical effects (accumulation at the resonant radius) to be significant. We underline some difficulties associated with this problem, namely, the non-adiabaticity of motion in the resonance phase space and the existence of close encounters with the planet at high eccentricities.

Key words: Circumstellar dust grain – Poynting-Robertson drag – resonance

1. Introduction

Dissipative effects, when combined to resonance phenomena, may lead to complex and often counter-intuitive, dynamical behaviours. One of the first examples of this kind was given by the locking of natural satellites in mean motion resonances through the slow and irreversible tidal evolution (Goldreich, 1964, and see the review by Peale, 1986). Tidal effects can then account for most of the orbit-orbit and spin-orbit resonances now observed among planets and satellites.

Another field of interest is the influence of gas drag on planetesimal orbits in the primordial solar system. Such a drag, when combined to the perturbations of a jovian planet, can cause some trapping and eccentricity pumping of the planetesimal orbits near resonance radii (Greenberg, 1978, Weidenschilling and Davis, 1985, Patterson, 1987).

Planetary rings also exhibit resonance-dissipation coupling phenomena. It is now widely accepted that inelastic collisions, combined to collective effects and resonant perturbations from satellites, can lead to sharp edges in rings, confining of ringlets, opening of gaps, etc... (see the reviews by Goldreich and Tremaine, 1982, Borderies, Goldreich and Tremaine, 1984, Meyer-Vernet and Sicardy, 1987).

Celestial Mechanics and Dynamical Astronomy **57**: 373–390, 1993.

In this paper, we will review a new kind of dissipative process, acting together with the resonant perturbation of a planet. Namely, we investigate the effect of Poynting-Robertson (PR) drag on particles orbiting the Sun (or another star), while they enter the region of influence of a mean-motion resonance with the planet, in the course of their slow orbital decay towards the central body. Because the radiation forces responsible for the PR drag essentially act on particles with radius comparable to the wavelength (see section 2), dynamical effects will be mainly noticeable for μm-sized dust particles.

In contrast with the previous studies quoted above, PR drag vs. resonance dynamics is a relatively new subject and the full implications of the interplay between these two effects is far from being fully analyzed. Consequently, the aim of this paper is to present some review of what has been done to our knowledge on that topic, but also to underline some of the difficulties of the problem and possible future directions of investigation.

Pioneering work in this field has been achieved by Gonczi, Froeschlé and Froeschlé (1982), and by Jackson and Zook (1989). The former authors studied the passage of dust grains through the inner 2:1 mean motion resonance with Jupiter and pointed out that this passage may cause large variations in the particle osculating elements. The latter showed that particles may be trapped in outer resonances with the Earth. Jackson and Zook (1992) analyzed the orbital evolution of dust particles originating from comets or asteroids, while Dermott *et al.* (1992) reviewed the effects of planetary perturbations on the zodiacal dust cloud. Also, Scholl, Roques and Sicardy (1992) examined the collective response of a circumstellar disk perturbed by Earth-like or jovian planets showing that the global morphology of the disk may be strongly affected by resonance effects, with potential applications to the structure of the β-Pictoris circumstellar dust disk.

Without being exhaustive here, we may think of several situations of interest, where PR drag and resonances have significant dynamical consequences. First, we must note here that PR drag is a quite "unavoidable" effect, since it appears as soon as photons hit a small particle. This contrasts with gas drag, which pertains to the primordial (and thus largely unknown) solar system. Actually, PR drag is an on-going process, effective at any time around any star. Among the possible fields of investigation, we may quote:

- The evolution of dust in the solar system:
  - Zodiacal dust bands in the asteroid belt.
  - Cometary dust.
  - rings around planets.
- Cosmogonic implications of resonance trapping:
  - Confining or clearing of dusty material in the early solar system.
  - Comparison with other confining or clearing mechanisms.
- On-going processes around other stars:
  - Effect of a hypothetical planet on a circumstellar disk subjected to PR drag.

• Inverse problem: derivation of a planet mass and orbital elements from its effects on a circumstellar disk.

Although the PR drag problem bears some resemblance with the topics described at the beginning of this Section, it allows, nevertheless, the particle to reach very high orbital eccentricities, with correlated difficulties, like the crossing of the planet orbit and the non validity of the usual expansions of the disturbing potential.

We will mainly deal in this paper about orders of magnitude and also about some relevant mechanisms, even though the formalism we use is not necessarily valid for high eccentricities. We will restrict ourselves to the simplest case, i.e. a test particle orbiting a star and perturbed by a planet with circular orbit, with no mutual inclination (planar circular restricted problem), in the presence of PR drag. We will be concerned more precisely with the following points:

- Derive some order of magnitude results on the PR drag itself.
- Present a simple dynamical formulation which describes both the effect of the resonance and that of the PR drag.
- Look in this frame for the existence of periodic orbits, deriving some constraints on the planet mass.
- Discuss the difficulties inherent to PR drag (high eccentricities, orbit crossing, no adiabaticity of the motion).
- Discuss some future directions of investigation.

2. Order of Magnitude Considerations

We give here some figures which will be useful in the rest of the paper. Most of them come from the seminal paper of Burns, Lamy and Soter (1979, thereafter referred to as BLS) on radiation forces in the solar system.

The radiation force acting on a particle is:

$$\mathbf{F}_R = \beta F_g \left[(1 - \dot{r}/c)\hat{S} - \mathbf{v}/c \right], \tag{1}$$

where $\beta = SA_SQ_{RP}/cF_g$ is the (constant) ratio of the force SA_SQ_{RP}/c due to the pressure of radiation, and the gravitational force F_g due to the central star. Here, S is the stellar flux at the particle, A_S is the geometrical cross section of the latter, c is the speed of light, Q_{RP} is the radiation pressure efficiency, $\mathbf{v}$ and $\dot{r}$ are the total and radial velocities of the particle, and $\hat{S}$ is the unit vector of the incident radiation. The radiation force may be split into a main radial component, $\beta F_g \hat{S}$ (the "pressure of radiation"), and a velocity dependent component (the "Poynting-Robertson drag").

The net effect of the radiation pressure component $\beta F_g \hat{S}$ is to replace the central star of mass M_* by an "apparent" mass M'_*:

$$M'_* = (1 - \beta)M_*. \tag{2}$$

Numerical calculations (BLS) show that $\beta \sim 5.7 \times 10^{-7} Q_{RP}/\rho s$ for a spherical particle orbiting the Sun, where the radius s and the density ρ are expressed in cgs units. For typical materials of density $\rho = 3$ g cm^{-3} and in the frame of geometrical optics (Q_{RP}= 1, a reasonable assumption at this stage), one gets:

$$\beta \sim \frac{0.2}{s_{\mu m}}, \tag{3}$$

where $s_{\mu m}$ is the radius expressed in μm. For a star of luminosity L, and with a spectrum comparable to that of the Sun, one would get $\beta \sim (0.2/s_{\mu m}) \times (L/L_{\odot})$, where $L_{\odot}$ is the Sun luminosity.

An important consequence of Eq. (2) is that the mean motion, and thus the corresponding resonant semi-major axis of a particle, are changed by the radiation pressure term. This is far from being negligible, since β is a substantial fraction of unity for μm-sized particles.

The average decay rates of semi-major axis a and eccentricity e due to PR drag are (BLS):

$$\begin{cases} < \dot{a}/a >= & -\frac{\eta}{a^2}\frac{2+3e^2}{(1-e^2)^{3/2}} \\ \\ < \dot{e}/e >= & -\frac{5\eta}{2a^2}\frac{1}{(1-e^2)^{1/2}}, \end{cases} \tag{4}$$

where $\eta = \beta GM_*/c$. It is worth emphasizing that when we consider only the part of the drag proportional to **v** in Eq. (1), the results are different. The term $< \dot{a}/a >$ is affected by $\mathcal{O}(e^2)$ and the numerical coefficient of $< \dot{e}/e >$ becomes 2 (instead of $\frac{5}{2}$). In this review the drag proportional to $\dot{r}$ is also considered since the effects of the two terms of the Poynting-Robertson drag cannot be physically separated. For almost circular orbits ($e \ll 1$), one gets:

$$\begin{cases} < \dot{a}/a >= & -4C/5 \\ \\ < \dot{e}/e >= & -C, \end{cases} \tag{5}$$

where the the damping coefficient C is given by $C = 5\eta/2a^2$. The characteristic decay time around the Sun is thus:

$$t_{decay} = \frac{1}{C} \sim 3200 R_{AU}^2 s_{\mu m} \qquad years, \tag{6}$$

where R_{AU} is the orbital radius in astronomical radius. This time must be multiplied by $L_{\odot}/L$ for another star of mass comparable to that of the Sun. Note that the decay time is short compared to stellar, or planetary, ages for typical R of a few astronomical units. Only for particles of a few mm in radius, located beyond Uranus's orbit, is the decay time comparable to the age of the solar system.

Another quantity of interest is the "fractional loss" per orbit, i.e. the dimensionless ratio of C to the mean motion n of the particle:

$$\frac{C}{n} \sim \frac{5 \times 10^{-5}}{s_{\mu m}\sqrt{R_{AU}}}, \tag{7}$$

for a circumsolar grain. Thus, the relative damping of eccentricity during one revolution is small compared to unity. This contrasts with other problems like collisions in a dense ring, or gas drag in the early solar system, where the damping per orbit may be a substantial fraction of unity. This smallness allows typical dust particles to get quite high orbital eccentricities, with the associated problems discussed in Section 5.

3. Dynamical Model of PR Drag-Resonance Interaction

In order to simplify as much as possible our problem, we will consider here a massless particle orbiting a star of mass M_*, perturbed by a planet of mass M_p with circular orbit, near a first order resonance $m : m+1$, i.e. $(m+1)n_p - mn \sim 0$, where n (resp. n_p) is the particle (resp. the planet) mean motion. Furthermore, all the orbits are coplanar. Close to the resonance, the dynamics of the motion is governed by the terms in the planet perturbing potential which are slowly varying with time (averaging principle).

The analysis in this Section and the following is based on the work of Greenberg (1978). However, we extend his study by taking into account the variation of the semi-major axis a associated to the variation of the eccentricity e (while Greenberg assumed that the semi-major axis is fixed during the trapping in the resonance). The simultaneous variation of a and e is actually required by the conservation of the Jacobi constant. This results in a dependence of the particle mean motion on the eccentricity (see Eq. (15)), and allows one to obtain the phase space diagrams shown in Figs. 3 and 4. We assume here that the orbital eccentricity of the particle is small, which allows us to keep only the term to first order in eccentricity in the perturbing potential, Φ_p, due to the planet (see, e.g., Brouwer and Clemence, 1961):

$$\Phi_p = a_0^2 n_0^2 \epsilon e A.cos(\Psi_L), \tag{8}$$

where a_0 and n_0 are the reference values of the semi-major axis and mean motion of the particle at the resonance, ϵ is the (small) ratio $\epsilon = M_p/M_*$, Ψ_L is the critical argument of the (Lindblad) resonance, and A is a dimensionless coefficient which depend on the Laplace coefficients $b_{1/2}^{(m)}$:

$$A = \frac{\alpha}{2}\left[2(m+1)b_{1/2}^{(m+1)} + \alpha\frac{db_{1/2}^{(m+1)}}{d\alpha}\right] \tag{9}$$

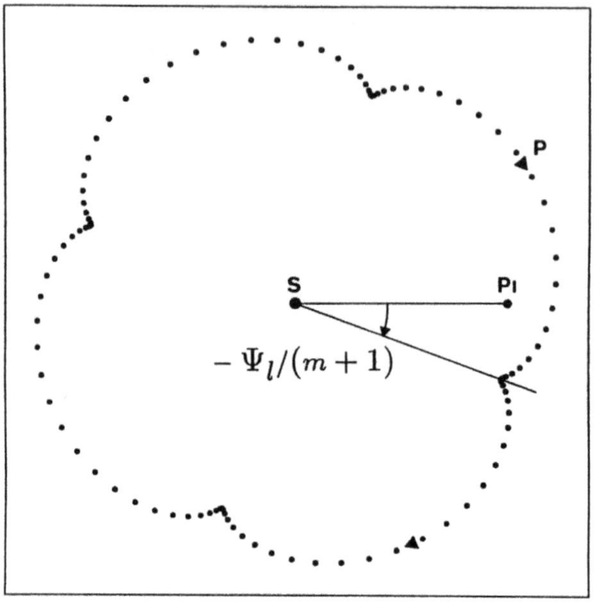

Fig. 1. The geometrical interpretation of the Lindblad resonance critical argument. The orbit of the particle **P** around the star **S** is plotted is the frame rotating with the planet **Pl**. This motion corresponds to the 5:4 outer Lindblad resonance ($m = -5$, see the text).

where $\alpha = a/a_p$ is the ratio of the particle semi-major axis to that of the planet. For numerical purposes, one can note that A is of order m. For large values of m, A/m tends to 0.802...

The critical argument Ψ_L is given by:

$$\Psi_L = (m+1)\lambda_p - m\lambda - \tilde{\omega}, \tag{10}$$

where λ is the mean longitude, $\tilde{\omega}$ is the longitude of periapsis of the particle and the subscript p refers to the planet.

The term proportional to $\epsilon e.cos(\Psi_L)$, in Eq. (8), couples the elements of the particle orbit to the gravitational influence of the planet. The angle Ψ_L can be interpreted as the orbital phase lag of the particle as observed *from the planet* (Fig. 1). This lag allows the planet to exchange energy and angular momentum with the particle and, in particular, to pump up the eccentricity of the latter (Sicardy, 1991).

If the planet orbit is elliptic, another term, proportional to the planet orbital eccentricity e_p, will appear in the perturbing potential Φ_p. This term contains a new critical argument $\Psi_c = (m+1)\lambda_p - m\lambda - \tilde{\omega}_p$, corresponding to a corotation resonance (see the discussion by Sicardy, 1991, for the geometrical and physical interpretation of Lindblad vs. corotation resonances). This more complete approach is followed by Beaugé and Ferraz-Mello (1992) who take into account the planet orbital eccentricity and, in the case of PR drag, use an asymmetric expansion of the perturbing potential valid for very high eccentricities (see Ferraz-Mello and Sato, 1989).

It is convenient to define the eccentricity (h, k) vector as:

$$\begin{cases} h = e.cos(\Psi_L) \\ k = e.sin(\Psi_L), \end{cases} \tag{11}$$

and the "distance", in mean motion, to the exact resonance by:

$$\Delta n = (m+1)n_p - mn. \tag{12}$$

The classical techniques of perturbation theory then yields:

$$\begin{cases} \dot{h} = -\Delta n.k \\ \dot{k} = \Delta n.h + \epsilon An, \end{cases} \tag{13}$$

where the dots stand for time derivatives.

It is important to note that hereafter the quantities A and n are considered as *constant*. They actually represent the values around which the equations are expanded.

From the above system, one can derive $de^2/dt = 2\epsilon Ank$. From a physical point of view, this stems from the exchange of angular momentum between the planet and the particle, at a rate which is proportional to $k = e.sin(\Psi_L)$. In other words, the torque applied to the particle orbit is proportional to the distortion e of the orbit, and to the sine of the phase lag Ψ_L (Fig. 1).

Energy is also exchanged between the two bodies, so that Δn varies at a rate:

$$\dot{\Delta n} = -3m^2n^2\epsilon Ak. \tag{14}$$

This ensures that the quantity:

$$J = 3m^2ne^2 + 2\Delta n \tag{15}$$

is conserved. Actually, J is nothing but the expansion in e^2 and Δn of the Jacobi constant, or more precisely, the Jacobi constant averaged over one sidereal period. From $J = cste$, one can derive:

$$\dot{a} = -ma\frac{de^2}{dt}. \tag{16}$$

This shows in particular that for a particle orbiting outside the planet orbit (outer Lindblad resonance, $m < 0$), $\dot{a}$ and de^2/dt have the same sign. Thus, if a particle gains energy, its eccentricity is also increased. The opposite is true for inner Lindblad resonances. This has an important consequence when dissipative processes are included: a particle at an outer Lindblad resonance will receive energy from the planet, thus increasing its eccentricity, which can compensate in principle for the damping effect of dissipation. At an inner Lindblad resonance, the planet will act to decrease the particle orbital eccentricity as it provides energy, so that no equilibrium is a priori possible. This is a particular case of a more general phenomenon, which tends to push the particle away from the planet orbit as soon as some collective or irreversible effects are introduced in the system (the so-called "shepherding mechanism", see the reviews by Goldreich and Tremaine, 1982, and Meyer-Vernet and Sicardy, 1987).

The secular effect of gas drag can be derived from Eqs. (5), assuming again small eccentricity for the particle. The complete set of equations (PR drag + resonance) then reads:

$$\begin{cases} \dot{\Delta n} = -\frac{3mn}{2}\left[2mn\epsilon Ak + \frac{4C}{5}\right] \\ \dot{h} = -\Delta n.k - Ch \\ \dot{k} = \Delta n.h + \epsilon An - Ck. \end{cases} \tag{17}$$

4. Evolution of the Particle in the Resonance

4.1. Stationary Orbits

One may ask whether the system (17) has fixed points, corresponding to stationary orbits. By setting all the time derivatives to zero, one can easily show that there are actually two fixed points:

$$\begin{cases} h_f = -\epsilon An\frac{\Delta n}{C^2+\Delta^2 n} \\ k_f = \epsilon An\frac{C}{C^2+\Delta^2 n}, \end{cases} \tag{18}$$

where Δn has the two opposite values defined by:

$$\Delta^2 n = -\left[\frac{5}{2}mn^2A^2\epsilon^2 + C^2\right]. \tag{19}$$

Since $\Delta^2 n \geq 0$, we retrieve the necessary condition that m should be negative, i.e. the particle should be in an *outer* Lindblad resonance for a fixed point to exist. Assuming from now on that $m \leq 0$, we get the final expressions for h_f and k_f:

$$\begin{cases} h_f = \pm\sqrt{\frac{2}{5|m|}}\sqrt{1 - \frac{2}{5|m|}\left(\frac{C}{\epsilon An}\right)^2} \\ k_f = \frac{2}{5|m|}\left(\frac{C}{\epsilon An}\right), \end{cases} \tag{20}$$

where the sign + (resp. −) applies to Δn positive (resp. negative). This shows that the fixed points correspond to a "universal eccentricity" $e_f = \sqrt{h_f^2 + k_f^2}$, which is *independent* of the damping coefficient C:

$$e_f = \sqrt{\frac{2}{5|m|}}. \tag{21}$$

A more exact calculation following what was done by Beaugé and Ferraz-Mello (1992) shows that the universal eccentricity is given by the root of the equation $|m|(1-e^2)^{3/2} - (|m|-1)(1+1.5e^2) = 0$. Actually, one can see that Eq. (21) is a good approximation of the root of the latter equation. For $|m|$=2 and $|m|$= 3, one gets e_f= 0.45 and 0.36 respectively, while the exact equation yields 0.48 and 0.37, respectively. For large $|m|$, the exact equation gives $e_f \sim 1/\sqrt{3|m|}$, close to the result of Eq. (21). The critical angles corresponding to each of the fixed points are:

$$\Psi_{Lf} = atan\left(-\frac{C}{\Delta n}\right), \quad \Psi_{Lf} = atan\left(-\frac{C}{\Delta n}\right) + \pi, \tag{22}$$

for Δn positive and negative, respectively, where the argument $atan$ is always taken between $-\pi/2$ and $\pi/2$.

The independence of the eccentricity e_f with respect to dissipation may appear puzzling a priori. Physically, it comes from the fact that when C increases, the critical argument Ψ_{Lf} gets closer to $-\pi/2$, i.e. farther away from the symmetry values 0 or π, so that more energy may be exchanged between the planet and the particle, everything equal besides. Thus, although more energy is dissipated through PR drag, the eccentricity corresponding to the fixed point may remain the same.

4.2. Stability of the Fixed Points

The linear stability of the fixed points is determined by the eigenfrequencies λ's of the system (17) near h_f, k_f and Δn_f. The characteristic equation of the system (17) reads:

$$(\lambda + 2C)(\lambda^2 + 3mn\Delta n_f/2) = -\left(\epsilon An/e_f\right)^2 \lambda. \tag{23}$$

For $\Delta n_f \leq 0$, the left-hand side of the equation is a third degree polynomial in λ with only one, negative, real root ($-2C$). The right hand side is a linear function of λ, with negative slope. Thus, the real roots of the complete Eq. (23) are bracketed by $-2C$ and 0. Since the sum of the three roots of Eq. (23) is $-2C$, this requires that all the real parts of the roots are negative, which ensures the linear stability of the corresponding fixed point. On the contrary, if $\Delta n_f \geq 0$, the left hand side has a positive real root and has a negative value at $\lambda = 0$. This shows that Eq. (23) has at least one positive root and, thus, that the corresponding fixed point is linearly unstable.

One can note from Eqs. (22) that the solution corresponding to $\Delta n_f \leq 0$ has Ψ_{Lf} closer to π (apocentric librator), and the solution corresponding to $\Delta n_f \geq 0$ has Ψ_{Lf} closer to zero (pericentric librator). Thus, the apocentric librating resonant orbit is linearly stable, while the pericentric one is unstable. The value of Ψ_{Lf} closer to π provides a "protection mechanism", in which the conjunction of the planet and the particle tend to occur when the two bodies are farthest away (see Fig. 1). This result is known in the conservative case and appears to be valid also in the presence of PR drag.

4.3. Minimum Planetary Mass for Stationary Orbits

A consequence of Eqs. (22) is that the critical argument corresponding to the stable stationary orbits tends to π for vanishing C, and tends to $-\pi/2$ for increasing C. This is a classical result of forced oscillators in the presence of damping: while they are in phase with the forcing term in the conservative case, they tend to be in quadrature for high dissipation. This is because more and more energy has to be provided by the perturber.

Beyond a certain value of C, however, the condition (19) cannot be achieved any more. Then, the planet is unable to provide enough energy to the particle through the Lindblad resonance, even though $\Psi_{Lf} = -\pi/2$. The condition for (19) being possible is thus:

$$\epsilon \geq \frac{C}{n|A|}\sqrt{\frac{2}{5|m|}}. \tag{24}$$

Numerical applications for a star of luminosity L yields:

$$\epsilon \gtrsim \frac{3.2 \times 10^{-5}}{|m|^{3/2}} \frac{1}{s_{\mu m}} \frac{1}{\sqrt{R_{AU}}} \frac{L}{L_\odot}. \tag{25}$$

It is interesting to note that this order of magnitude calculation provides minimum masses which lie in the range of typical planetary masses, for μm-sized particles at a few or several astronomical units from the star (recall that $\epsilon \sim 3 \times 10^{-6}$ for the Earth, $\sim 5 \times 10^{-5}$ for Uranus and Neptune and $\sim 10^{-3}$ for Jupiter).

5. Limitations and Difficulties

5.1. High Eccentricities

A first limitation of our approach is that the eccentricity corresponding to the stationary orbit, $\sqrt{\frac{2}{5|m|}}$, is *not* small compared to unity for small values of $|m|$. However, we have seen that the order of magnitude of the universal eccentricity derived with our approach was the same as that derived following Beaugé and Ferraz-Mello (1992), once second order terms in eccentricity are taken into account in the perturbing potential.

At that point, a second difficulty arises, because such a high eccentricity will cause the particle to *cross* the planet orbit. The distance in semi-major axis between the two bodies is of order $2a/3|m|$, so that orbit crossing will occur for $ae_f \geq 2a/3|m|$, i.e. $|m| \leq\sim 8/9$. In other words, *all* the outer Lindblad resonances yield stationary orbits which cross the planet orbit. Thus, although these orbits are interesting as benchmarks, which define possible attractors for the particle motion, they are nevertheless difficult to reach. More precisely, a particle may collide with the planet before it settles in the stationary orbit. This depends on the efficiency of the protection mechanism which should keep the value of Ψ_L close to π.

This situation is depicted in Fig. 2, in which we plot the osculating eccentricity e of a particle vs. its osculating semi-major axis a. The particle is assumed to orbit the star β-Pictoris ($\sim$ 1.5 solar mass, $L/L_\odot \sim 6$) and to have a radius of 3.6 μm, corresponding to a value of $\beta = 0.33$. Finally, it is perturbed by a planet of mass $\epsilon = 10^{-4}$, orbiting at a_p= 20 AU from the star. The continuous curve is the set of points verifying $a(1-e) = a_p$. Above that curve, the pericenter of the particle orbit lies inside the planet orbit, so that the two orbits intersect.

The motion of the particle is calculated exactly, through a 4th order Runge-Kutta integrator, with the full perturbing force of the planet taken into account, superimposed to the perturbing radiation force (Eq. (1)). In a first stage, the particle semi major axis decreases, due to PR drag. Then, the eccentricity increases, while the semi-major axis is locked at the value corresponding to the 2:1 outer Lindblad resonance. On its way to reach the periodic orbit defined by Eqs. (20), the particle orbit crosses that of the planet. A more detailed analysis of the motion shows that the particle escapes the libration region of the resonance and then has a close encounter with the planet, which explains its sudden removal from the resonance region. We have performed several numerical integrations, with various initial conditions. Although the particles may stay for a rather long time in the resonance region (1.6 Myears in the case of Fig. 2), they eventually all have a close encounter with the planet.

5.2. NON ADIABATICITY OF THE MOTION

Before reaching the periodic orbit, the particle must first be captured in a stable libration. Then, a slow evolution of the phase space, due to PR drag, must increase the particle orbital eccentricity to e_f, while avoiding collision with the planet by an appropriate range of values for Ψ_L. In the mean time, the particle may escape the resonance region by crossing the separatrix between libration and circulation.

This problem is much more difficult to solve than merely determining the fixed points in Eqs. (17). This requires numerical integrations rather than an analytical approach. We note here that an adiabatic invariant method could in principle be used

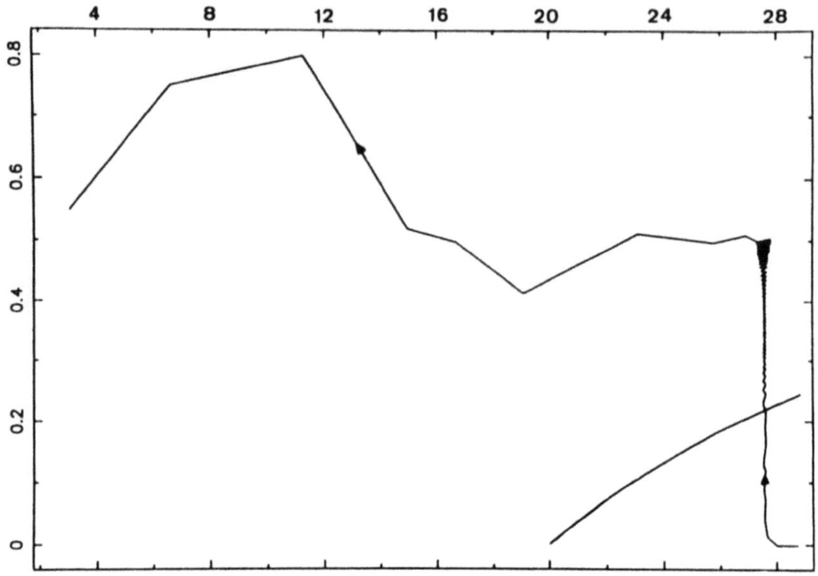

Fig. 2. The osculating eccentricity e of a particle is plotted against its semi-major axis a. The particle is perturbed by a planet of mass $\epsilon = 10^{-4}$ stellar mass, revolving on a circular orbit at 20 A.U. from the star β-Pic (1.5 $M_\odot$). The time step beween two points is 100 planetary revolutions, i.e. 7300 years. The particle spends 1.6×10^6 years in the resonance region. The curve in the lower right of the figure is the set of points where the pericenter of the particle orbit intersect the planet orbit. Above that curve, the two orbits cross. See the text for a more detailed description.

to follow the particle motion in some suitable phase space, since the dissipation term is small (Eq. (7)). Unfortunately, we will now show that this is *not* the case for a typical μm-sized particle perturbed by a typical planet.

More specifically, we want here to compare the libration period of the eccentricity vector, and the time scale over which the phase space itself evolves, due to PR drag. We turn back to the system (17), and we examine the conservative case, i.e. setting C to zero. Then, it is easy to show that besides the Jacobi constant $J = 3m^2ne^2 + 2\Delta n$, there is a second integral of motion, K:

$$e^4 - \mathcal{J}e^2 + \epsilon\mathcal{A}h = K, \tag{26}$$

where the constant coefficients are defined by:

$$\begin{cases} \mathcal{J} = 2J/3m^2n = 2(e^2 + 2\Delta n/3m^2n) \\ \mathcal{A} = -8A/3m^2. \end{cases} \tag{27}$$

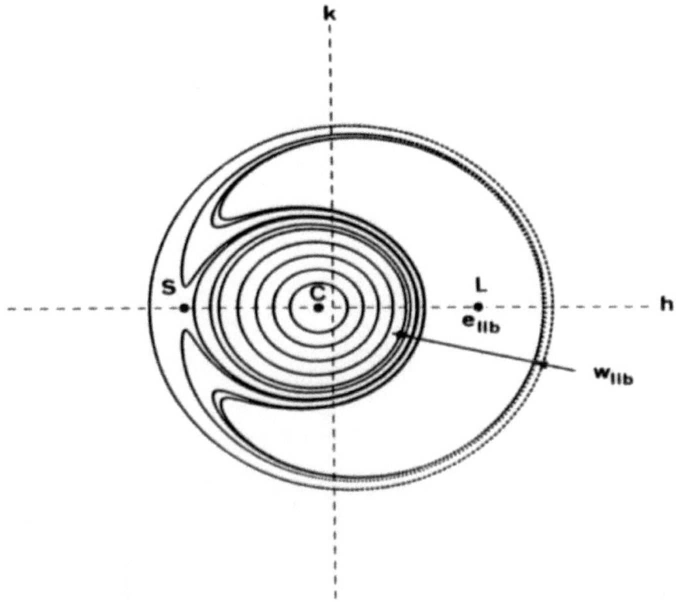

Fig. 3. Definitions of the various quantities used in the eccentricity vector (h, k) space. The quantities h and k are defined in Eqs. (11).

Note that because we are dealing with outer Lindblad resonances ($m \leq 0$), we have $\mathcal{A} \geq 0$, since A and m have the same sign. Also, because $A \sim m$, $\mathcal{A} \sim -8/(3m)$.

Eq. (26) entirely defines the trajectory followed by the eccentricity vector, once $\mathcal{J}$ and K are fixed. These curves are the projections on the (h, k) plane of the surfaces, in the (h, k, z) space, defined by $z = e^4 - \mathcal{J}e^2 = cste$ and $z = \epsilon \mathcal{A} h = cste$. The first surface is axisymmetric around the z axis, and the second surface is a plane parallel to the h axis, with a very small inclination with respect to the (h, k) plane, since ϵ is very small. A complete analysis of the morphology of the trajectories is given, e.g., by Ferraz-Mello (1985).

In particular, this analysis shows that the critical points corresponding to periodic orbits are given by $k = 0$ and $h^4 - \mathcal{J}h^2 + \epsilon \mathcal{A} h$ extremum, i.e.:

$$4h^3 - 2\mathcal{J}h + \epsilon \mathcal{A} = 0. \tag{28}$$

This third degree polynomial equation yields the classical equilibrium points that one can see in Fig. 3. One can note the stable libration point L, surrounded by bean-shaped or banana-shaped trajectories, themselves enclosed in the separatrix which connect itself through the unstable saddle point S (at least if $\mathcal{J} \geq 0$). Then, in the inner circulating region lies the third, stable, equilibrium point, C.

Due to faintness of the factor ϵ, it is easy to show that the stable libration point L is given by $h_{lib} \sim -\sqrt{\mathcal{J}/2}$, corresponding to an eccentricity of:

$$e_{lib} = |h_{lib}| = \sqrt{\mathcal{J}/2}. \tag{29}$$

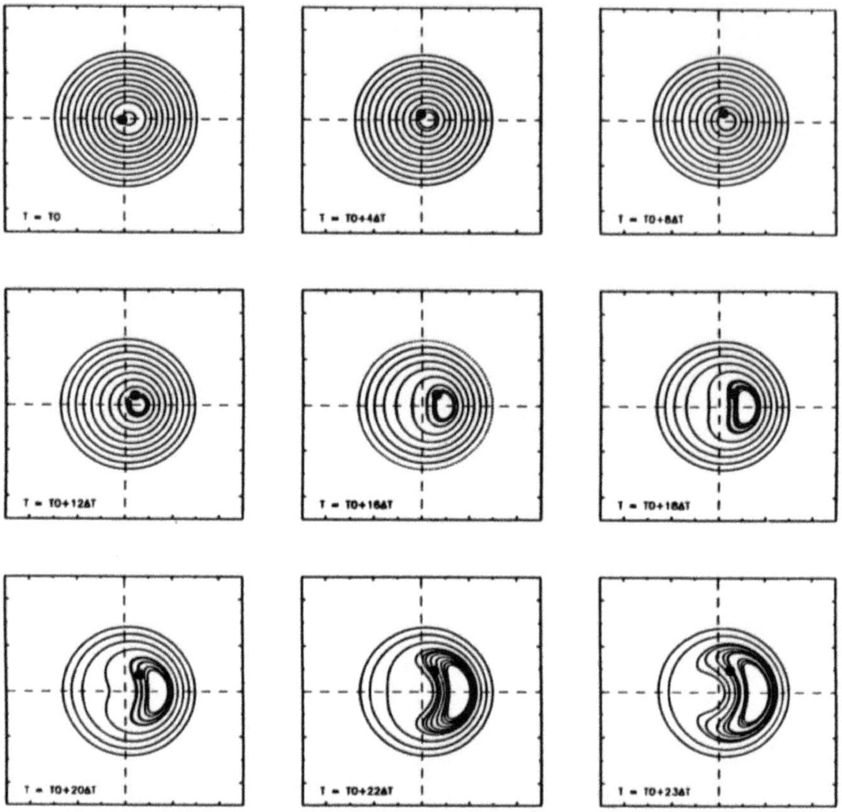

Fig. 4. The evolution of the eccentricity vector (h, k) of a particle around β-Pic, with β= 0.3, perturbed by a planet of mass $\epsilon = 10^{-5}$, orbiting at 20 A.U. from the star. The diagram corresponds to a 2:1 outer Lindblad resonance. The position of the particle is these diagrams is plotted as a dot. The time step ΔT corresponds to 10 planetary revolutions, i.e. 730 years. The units for h and k are arbitrary. See the text for details.

Moreover, the maximum width of the libration region enclosed in the separatrix is:

$$w_{lib} \sim \sqrt{2\mathcal{A}\epsilon/e_{lib}}. \tag{30}$$

Finally, a classical analysis of libration motion close to a stable equilibrium point provides the following libration period around L:

$$T_{lib} \sim \frac{4T_{orb}}{3m^2\sqrt{2\mathcal{A}\epsilon e_{lib}}}, \tag{31}$$

where T_{orb} is the orbital period of the planet.

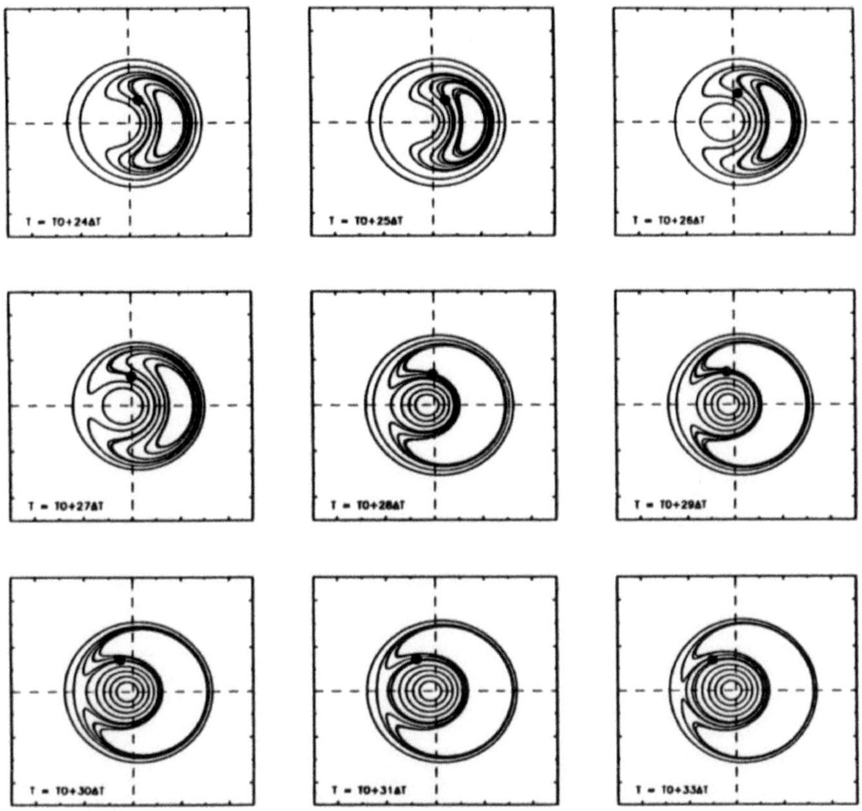

Fig. 5. Continuation of Fig. 4

Due to PR drag, the quantity $\mathcal{J}$ will slowly evolve, and so will the position of the instantaneous stable libration point L. Reintroducing the value of $\mathcal{J}$ as defined in Eq. (27), in the complete system (17) with dissipation, one can derive:

$$\dot{\mathcal{J}} \sim \frac{8C}{5|m|}, \tag{32}$$

in the approximation $e^2 \ll 1$. Since the stable libration point L corresponds to $\mathcal{J} = 2e_{lib}^2$, the forced eccentricity e_{lib} varies like:

$$e_{lib} \sim \sqrt{\frac{4Ct}{5|m|}}, \tag{33}$$

with time t. This result is interesting, because it tells us that the time scale to build up a significant eccentricity is $\sim m/C$, i.e. of the same order of magnitude as the decay time calculated in (6), $1/C$. In other words, even though the trapping in the resonance is not permanent, it is nevertheless long enough for the particles to statistically accumulate at the resonance. In this case, particles ejected through close encounters with the planet are replaced by new particles brought into the resonance from the outer regions of the system.

For the motion to be adiabatic, the libration period calculated in Eq. (31) should be very small compared to the time scale of evolution T_{evol} of the phase space itself. The latter time scale can be defined by the time it takes to the forced eccentricity e_{lib} to increase its value by an amount corresponding to the width w_{lib} of the libration region. The comparison of Eqs. (30) and (33) shows that:

$$T_{evol} \sim \frac{5|m|}{2C}\sqrt{2\mathcal{A}\epsilon e_{lib}}, \tag{34}$$

so that the motion is adiabatic if:

$$\frac{T_{lib}}{T_{evol}} \sim \frac{4CT_{orb}}{15|m|^3\mathcal{A}\epsilon e_{lib}} \ll 1. \tag{35}$$

The combination of Eq. (7) and $\mathcal{A} \sim -8/(3m)$ yields:

$$\frac{T_{lib}}{T_{evol}} \sim \frac{1}{m^2}\frac{3\times 10^{-5}}{s_{\mu m}\sqrt{R_{AU}}}\frac{1}{\epsilon e_{lib}}. \tag{36}$$

Because m is a few times unity, ϵ is in the range $10^{-6} - 10^{-3}$, and e_{lib} is small compared to unity at the entrance in the resonance, this means that for μm-sized grains, the adiabaticity condition (35) is *not* satisfied for typical situations of interest.

This problem is illustrated in Figs. 4 and 5. We consider a particle orbiting the star β-Pictoris ($L/L_\odot \sim 6$), with the coefficient β set to 0.3 (corresponding to a radius of ~ 4 μm, according to Eq. (3)). The particle is perturbed by a hypothetical planet of mass $\epsilon \sim 10^{-5}$, orbiting at 20 A.U. from the central star. The motion of the particle is integrated near the 2:1 outer Lindblad resonance, by retaining only the resonant term in the planet perturbing potential. At each time step, the "osculating conservative phase space" was plotted, according to the instantaneous value of $\mathcal{J}$ assumed by the particle (whose position in the phase space is plotted as a dot). Thus all the curves, in a given diagram in Fig. 4, or Fig. 5, have the same value of $\mathcal{J}$, but different values of K.

Following the motion of the particle, it can be seen that it lies first in the circulation region, then enters the libration zone, and finally escapes the resonance by passing in the inner circulation zone. An important feature to be noted is that the time of libration of the particle is comparable to the time of evolution of the libration zone itself. From the analysis of such a figure, it is possible to conclude that the crossing of a resonance, even if smooth, cannot be considered as adiabatic, at least with the parameters used in this case.

6. Conclusions

We have restricted ourselves to a rather specific problem, namely a massless particle orbiting a star and perturbed by (i) a planet on a circular orbit and (ii) radiation forces, including PR drag, with no mutual inclination between the particle and the planet. A simple approach, summarized by the equations of motion (17), allows us to draw several qualitative and quantitative conclusions:

- Only outer Lindblad resonances are able to counteract the dissipative effect of PR drag.
- For a given dissipation coefficient C (see Eq. (7)), there exists one linearly stable periodic orbit where the energy dissipated by the drag is balanced by the energy provided by the resonance.
- However, the eccentricity of this periodic orbit is of order $\sqrt{2/5|m|}$, where m refers to the $m : m+1$ resonance. Besides the technical difficulties associated to this high eccentricity (expansion of the perturbing potential), this means that the particle orbit will cross that of the planet at some point, a highly precarious situation for keeping the particle in the resonance region.
- A closer look at the equations of motion at small eccentricities shows that the eccentricity first increases like $\sqrt{4Ct/5|m|}$, which means that the particle may spend a large time in the resonance before colliding with the planet. This may account for statistical accumulation of particles at the resonance radii.
- For typical μm-sized particles perturbed by typical planets with masses $\sim 10^{-6} - 10^{-3}$ stellar mass, the motion of the particle eccentricity vector is not adiabatic in the phase space. This prevents the classical techniques of adiabatic invariant being used to derive the probability of capture into the resonance, the time of escape from the resonance, etc...

Several points remain to be studied at that point. The periodic orbits deserve a study of their own. After all, they could be reached directly by particles on elliptic orbits from the beginning (cometary origin). They could avoid close encounters with the planet through an efficient protection mechanism, associated to special initial conditions. Also, statistical methods, using many numerical integrations, are now necessary to derive the probability of capture in resonance of randomly distributed particles. Third, a more detailed description of the close encounters with the planet should be presented. In particular, one would like to know the percentage of particles ejected in the outer regions of the system, or sent onto the central star, during such encounters. Finally, the full problem, which takes into account the planet orbital eccentricity, as well as mutual inclinations of the planet and the particles, could exhibit a much more complicated dynamics, through the appearance of new kinds of resonances (corotation and inclined resonances).

Acknowledgements

We thank R. Greenberg, A. Jackson and an anonymous referee for several comments and suggestions during this work. D. Lazzaro was partly supported by the Brazilian National Research Council, proc. CNPQ 201053/90.0. The support of FAPESP is also acknowledged.

References

Beaugé, C. and Ferraz-Mello, S.: 1992, 'Resonance trapping in the primordial solar nebula: The case of a Stokes drag', *Icarus*, submitted

Borderies, N., Goldreich, P., and Tremaine, S.: 1984, in Planetary rings, ed(s)., *R. Greenberg and A. Brahic*, Univ. of Arizona Press, Tucson, 713-734

Brouwer, D., and Clemence, G.H.: 1961, *Methods of celestial mechanics*, Academic Press, New York, Chap. XV, 465-507

Burns, J.A., Lamy, P.L., and Soter, S.: 1979, 'Radiation forces on small particles in the solar system', *Icarus* **40**, 1-48 (paper BLS)

Dermott, S.F., Gomes, R.F., Durda, D.D., Gustafson, B.Å.S., Jayaraman, S., Xu, Y.L., and Nicholson, P.D.: 1992, in Chaos, resonance and collective dynamical phenomena in the solar system, IAU Symp. No. 152, ed(s)., *S. Ferraz-Mello*, Kluwer Academic Publishers, 333-347

Ferraz-Mello, S.: 1985, 'Resonance in regular variables, I. Morphogenetic analysis of the orbits in the case of a first order resonance', *Celest. Mech.* **35**, 209-220

Ferraz-Mello, S. and Sato, M.: 1989, 'The very-high-eccentricity asymmetric expansion of the disturbing function near resonances of any order', *Astron. Astrophys.* **225**, 541-547

Goldreich, P.: 1964, 'An explanation for the frequent occurrence of commensurable mean motions in the solar system', *Mon. Not. R. Astron. Soc.* **130**, 159-181

Goldreich, P., and Tremaine, S.: 1982, 'The dynamics of planetary rings', *Ann. Rev. Astron. Astrophys.* **20**, 249-283

Gonczi, R., Froeschlé, Ch., and Froeschlé, C.: 1982, 'Poynting-Robertson drag and orbital resonance', *Icarus* **51**, 633-654

Greenberg, R.: 1978, 'Orbital resonance in a dissipative medium', *Icarus* **33**, 62-73

Jackson, A. A., and Zook, H.A.: 1989, 'A solar system dust ring with the Earth as its shepherd', *Nature* **337**, 629-631

Jackson, A. A., and Zook, H.A.: 1992, 'Orbital evolution of dust particles from comets and asteroids', *Icarus* **97**, 70-84

Meyer-Vernet, N., and Sicardy, B.: 1987, 'On the physics of resonant disk-satellite interaction', *Icarus* **69**, 157-175

Patterson C.W.: 1987, 'Resonance capture and the evolution of the planets', *Icarus* **70**, 319-333

Peale, S.J.: 1986, in Satellites, ed(s)., *J.A. Burns and M.S. Matthews*, Univ. of Arizona Press, Tucson, 159-223

Sicardy, B.: 1991, 'Numerical exploration of planetary arc dynamics', *Icarus* **89**, 197-219

Scholl, H, Roques, F. and Sicardy, B.: 1992, 'Resonance trapping of circumstellar dust particles by an alleged planet', *Celest. Mech.*, accepted

Weidenschilling, S.J., and Davis, D.R.: 1985, 'Orbital resonances in the solar nebula: implications for planetary accretion', *Icarus* **62**, 16-29

THE GRAVITATIONAL INFLUENCE OF SATELLITE PAN ON THE RADIAL DISTRIBUTION OF RING-PARTICLES IN THE REGION OF THE ENCKE-DIVISION IN SATURN'S A RING

F. SPAHN

Working group "Nonlinear Dynamics" (MPG); University Potsdam; D-O-1571 Potsdam; Germany

J.-M. PETIT

C.N.R.S., Observatoire de Nice, B.P. 229, 06304, Nice cedex 04, France

and

PH. BENDJOYA

Observatoire de Nice, B.P. 229, 06304, Nice cedex 04, France

Abstract. Here we summarize the models and their similarities with respect to the gravitational influence of small satellites (moonlets) embedded in planetary rings, which have been developed by Petit and Hénon (1987a,b; 1988) and by Spahn (1987; Spahn et al. 1989; 1992). The results and their applicabilities are shown in the case of the Encke–gap Moon Pan located in the outer A–ring of Saturn's rings. To compare the model–profiles with the experimental ones (optical depth profile measured by Voyager–photopolarimeter), the structures embedded in the latter have been confirmed statistically using the "wavelet" analysis method. This analysis points to a size of Pan of about 13 km and an eccentricity of 10^{-4} of its orbit.

Key words: Planetary rings - Wavelets analysis

1. Motivation

Much work has been done with respect to the gravitational action of satellites - embedded and outer ones - on planetary rings, which have culminated in the prediction and detection of the satellite Pan in the Encke division of Saturn's rings (Showalter et al. 1986; Showalter 1991).

In this paper we will give a brief summary of the work done concerning the question: " What is the gravitational influence of a small satellite (moonlet) on the radial density distribution of ring–material in the close radial surrounding of its orbit ?"

In Section 2 we summarize the models developed by Petit and Hénon (1987a; b; 1988; hereafter PH 1; 2; 3), and by Spahn (1987; Spahn and Wiebicke 1989; Spahn et al. 1992; hereafter S 1; 2; 3). We enlighten the similarities of the models and the good agreement of a few of their results, which then will be taken together to look for moonlet-like structures in the Voyager occultation data. This is especially done in Section 3 where we compare the results of the models with the radial density-profiles of Saturn's rings measured by the Photopolarimeter (PPS) of Voyager 2 at the Encke division of Saturn's ring where satellite Pan has been discovered recently. This gives a possibility to prove directly the value of the results of the models. Therefore the information about the real structure in the rings has to be

Celestial Mechanics and Dynamical Astronomy **57**: 391–402, 1993.

separated from the large noise of the PPS-data. This has been realized with the wavelet analysis method which yields a noise-reduced optical depth profile.

In Section 4 the results of the comparison between model and experiment will be discussed and also a small overview of still remaining work can be found there.

2. The Models

The main common assumptions of either approaches (PH 1; 2; 3; S 1; 2; 3) are:

1. plane motion of all bodies (planet, moonlet, ring-particles);
2. a narrow radial region is considered (compared to the whole radial extension of the rings);
3. the standard-model of spherical, indestructible ring-particles has been applied.

With these assumptions, both models have been applied in order to solve the problem of the gravitational influence of the embedded moonlet on the time-evolution of the radial distribution of ring-particles in the close vicinity (region of overlapping resonances) of the orbit of the perturbing moon. That distribution is given by the function $f(m, r, t)$, $f.dm.dr$ being the number of particles in the radial interval $[r, r + dr]$ and in the mass-range $[m, m + dm]$ at a given time t. Then the evolution of $f(m, r, t)$ can be written by means of a continuity equation

$$\partial_t f(m, r, t) = \hat{\mathbf{A}}(f, r, t) \cdot f \tag{1}$$

in a very general form.

For more details concerning the meaning of Eq. (1) we refer to the derivation of equation (44) in PH2 and of equation (3) in S2.

The operator $\hat{\mathbf{A}}$ governs the radial transport of the ring–matter and is generally dependent on the gravity of the perturbing moonlet, as well as the interactions between the ring–particles - gravity and direct physical collisions - represented by the dependency of $\hat{\mathbf{A}}$ on f itself (PH2; 3). The integro-differential equation (1) is of a fairly complicated form which makes it impossible to get a general analytical solution. Therefore numerical approaches have been used (PH; S) in order to solve this equation for special cases, e.g. single and bimodal size distributions.

The advantage of the model developed by Petit and Hénon (1987a; b; 1988) is the wide physical generality taken into account in equation (1), which is presented by including:

a) arbitrary size distribution of the ring-particles;

b) particle interactions ⇒ gravity and direct inelastic collisions (fixed coefficient of restitution $\epsilon = constant < 1$; see PH1).

The price for this wide generality is a relatively low number of particles - $N = 500 \cdots 1000$ - taken for the numerical treatment of Eq. (1), which gives a fairly rough resolution of the final radial density profiles.

Such resulting profiles are shown in the fig. 1 for the special case of a bimodal size (mass) distribution for the ring- particles (one moonlet and many much smaller particles). The time evolution of the radial particle distribution is given for a

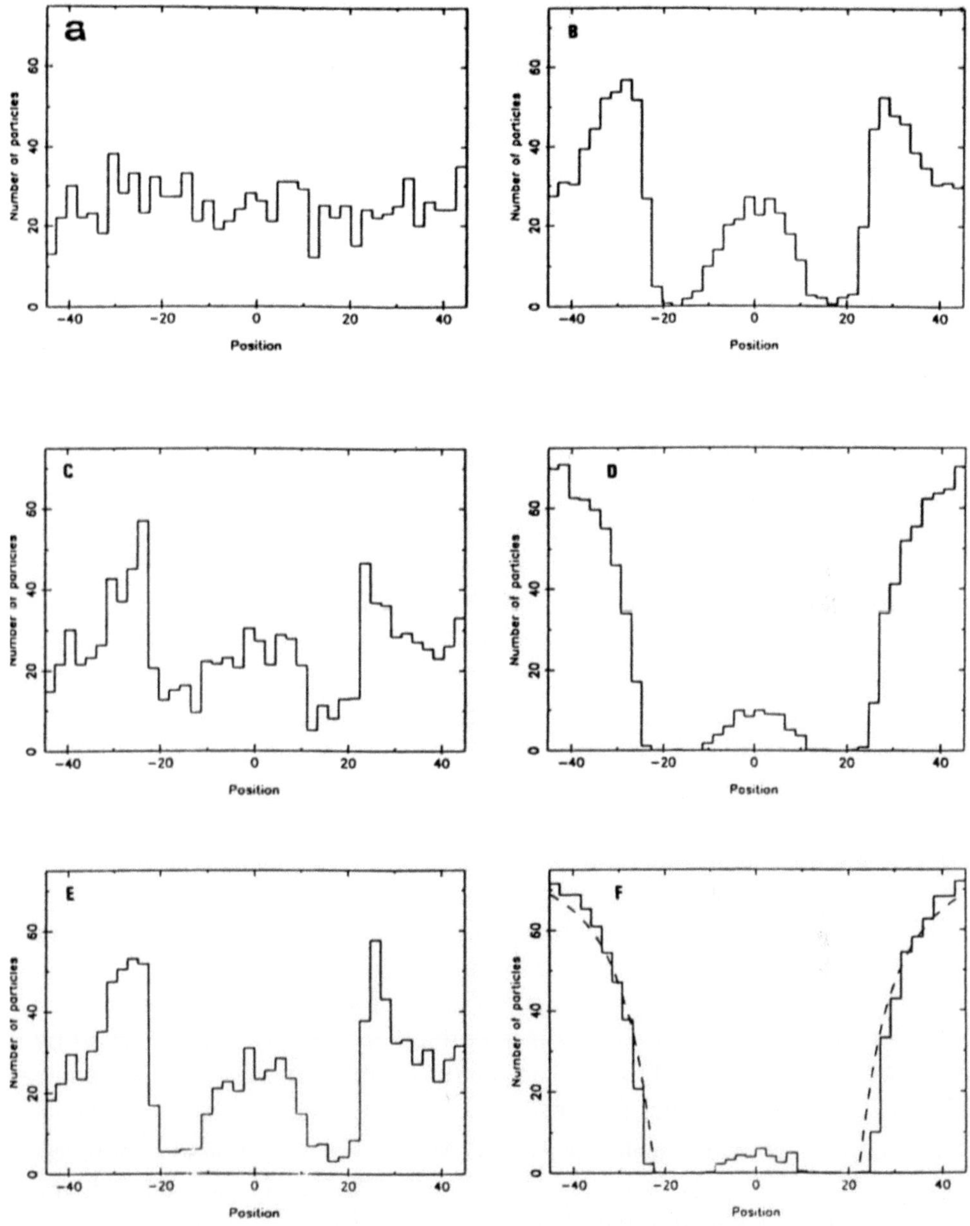

Fig. 1. Time evolution of the radial distribution of very small particles ($m \ll M_m$) around the orbit of a moonlet at the "position"$= 0.0$. The different parts belong to snapshots at different elapsed time of the computer-experiment. Clearly visible is the formation of a gap which is separated by a faint ringlet around the "position"$= 0.0$ (Figure taken from PH3).

comparatively massive moonlet. In this case the moonlet is able to create an empty gap which is divided by a faint ringlet, covering the orbit of the moonlet (fig. 1f). The ringlet consists of particles in "tadpole" and "horseshoe" orbits (Dermott and Murray 1981). Of high importance is the fact that Petit and Hénon (1988) have been able to obtain an equilibrium width of the gap created by the moonlet, that is a result of the balance between the torque exerted from the moonlet to the ring–material and the torque caused by the particle interactions itself (viscous torque; see also Lissauer et al. 1981). They obtained for that width W:

$$W \approx \begin{cases} M_m^{1/3}, \text{ for } & M_m < M_0, \\ M_m^{2/3}, \text{ for } & M_m \geq M_0, \end{cases} \tag{2}$$

where M_0 is the mass threshold depending on the different model parameter (PH3). The order of magnitude is $M_0 \approx 10^{13} kg$ which corresponds to a size of an icy moonlet of a diameter $D \approx 2.7 km$.

In the model of Spahn (1987; S2; S3) particle interactions have been neglected and only the gravitational influence of the moonlet on the ring–material near its orbit has been taken into account. This simplification has enabled to consider (S3):

1. $N = 5000 \cdots 50{,}000$ particles, giving a high radial resolution of the density profiles;
2. n (up to three) co-orbital moonlets of different masses;
3. eccentric orbits of the moonlets.

However in that model no equilibrium width of the created gap can be expected because of the neglect of the particle interactions - no viscous torque is counteracting the torque exerted by the moon. Here the advantage is to study in detail the dependency of the shape (radial profile) and the width of the inner ringlet on the orbital motion of a massive moonlet ($M_m > M_0$). Concerning the mass-dependence of the radial density profile Spahn and Wiebicke (1989) have obtained results comparable to Petit and Hénon (1988). Assuming a given random motion of the ring–particles (expressed by a mean orbital eccentricity $\overline{e}$; S2), caused by particle interactions and additional outer perturbations (Goldreich and Tremaine 1978; PH1), the inner ringlet is vanishing with decreasing mass of the moonlet M_m, while the gap-width scales as "$W \propto M_m^{1/3}$". For very massive moonlets ($h \approx [M_m/3M_p]^{1/3} \approx R_m/[km] \cdot 10^{-4} \gg \overline{e}, \quad \overline{e} \leq 10^{-6}$, M_p, R_m and h are the mass of the planet, the radius and the Hill-scale of the moonlet, respectively), the inner ringlet is surprisingly stable and exhibits a detailed structure.

Furthermore, because of the comparable large number of particles $N = 5000 \cdots 50{,}000$ it has been found (S3), that the density–profile of the inner ringlet shows a detailed structure which contains information about orbital parameters of the moonlet. This is shown in fig. 2, where the different radial density–profiles - taken after a few synodic periods when the structure changes only slightly in time - correspond to different eccentricities e_m of the orbit of the moonlet. Obviously, the larger e_m the narrower the width of the ringlet becomes, which is quite easily

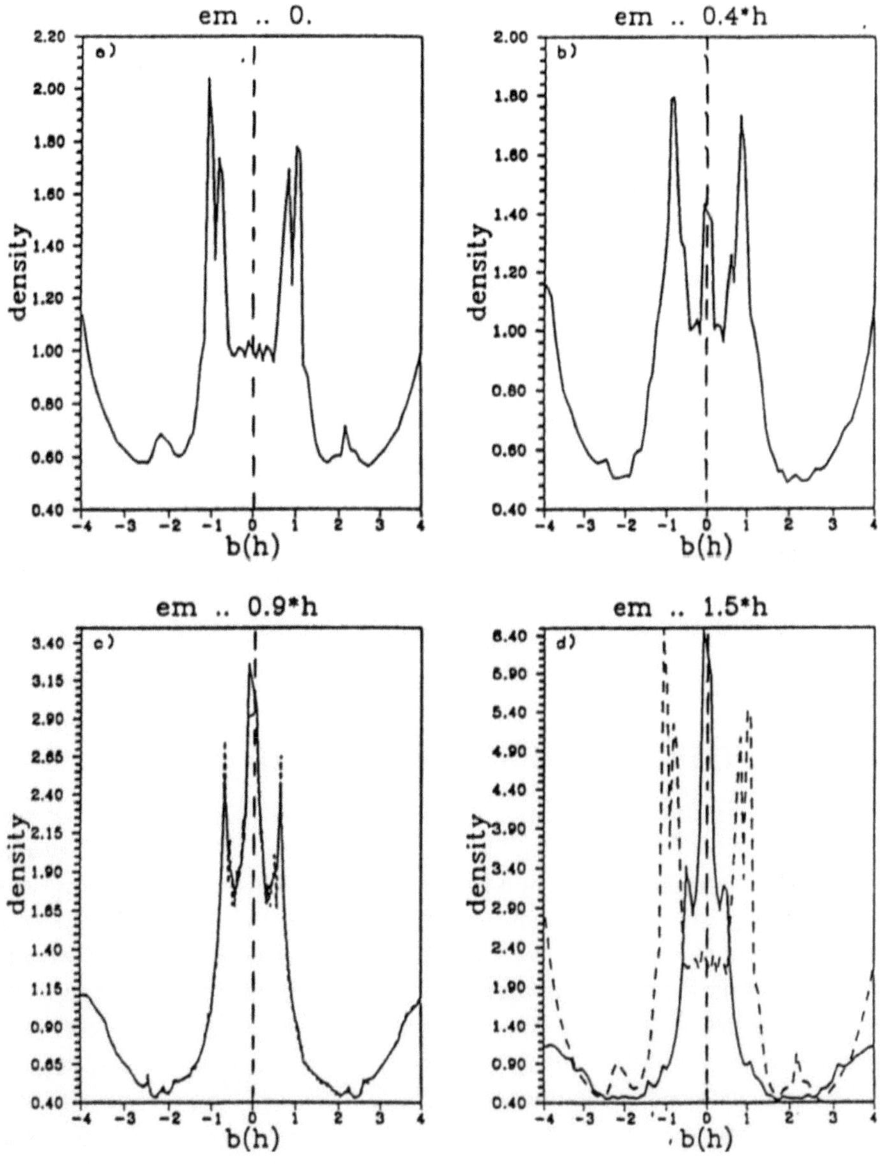

Fig. 2. Radial profiles of $\sigma(b, t = constant)$ (b - impact parameter) for different eccentricities e_m of the orbit of the moonlet (case of a "cold" ring $\Rightarrow$ mean-eccentricity of the ring-particles $\overline{e} \rightarrow 0$). A narrowing of the ringlet-width is really obvious, and furthermore the shape of the density-profile is turning from a double peaked to a triple peaked one with increasing e_m. All profiles show a quasi-symmetry $\sigma(b) \approx \sigma(-b)$ with respect to the orbit of the moonlet at $b = 0.0$.

seen in fig. 2d where the ringlets caused by moonlet in a circular orbit (solid line), and originated by a moonlet in an eccentric orbit ($e_m = 1.5h$; dashed line), are compared directly. The width W_r of the ringlet can be written

$$W_r \approx C(e_m) \cdot a_m \cdot h(M_m), \tag{3}$$

where a_m is the semi–major axis of the orbit of the moonlet. The factor $C(e_m)$ is a nearly linearly decreasing function of e_m and takes values $C(e_m = 0) \approx 3.0 \cdots C(e_m = 1.5h) \approx 1.2$. At the same time the double-peaked shape of the ringlet-profile turns to a triple peaked one with increasing eccentricity e_m .

Summarizing, the results for

1. the gap-formation and its equilibrium–width (Eq. (2)) of Petit and Hénon (1988);
2. the information about the ringlet found by Spahn et al. (1992) (the ringlet-width and the shape of its density-profile);
3. as well as the fact that all profiles obtained (PH 3; S2; 3) are quite symmetric with respect to the mean radial location of the moonlet-orbit, according to $\sigma(b) \approx \sigma(-b)$;

could now be used to search for small embedded satellites. Before doing this in a future work, the value of the above findings will be tested in the case of satellite Pan, embedded in the Encke–gap of Saturn's rings. This is presented in the next section.

3. The Action of Satellite Pan in the Encke-gap

If the results of our models are of any value for the reality, they should be visible also in the optical depth-profile of such a ring-region where a satellite is moving. This is the case for the Encke-division in the A-ring of Saturn, where Showalter (1991) has discovered recently a new satellite - named Pan. Fig. 3 shows the optical depth profile $\tau(r)$ of the Encke–division, which has been obtained with the Photopolarimeter (PPS–data) of Voyager 2. At first one recognizes a wide gap - of 315km width - around the orbit of Pan, which is one of the predictions mentioned above (Section 2). Using this value for the width and applying the relation (2) of PH3 we find that one needs a moonlet of mass $M_m \approx 1.2\ 10^{15} kg$ to create such a gap. For an icy moon the corresponding size would be about $D_m \approx 13km$. This satellite would be able to confine a ringlet of a width of $W_r \approx 25km$, if the moon is assumed to move in a circular orbit. The narrowing of the ringlet–width (we see only a 15 km wide ringlet) can be explained by an eccentric orbit of Pan, with an eccentricity of $e_{Pan} \approx h_{Pan} \approx 10^{-4}$ (according to relation (3)). Then the radial profile of the ringlet should be triple–peaked (see fig. 2).

The bottom parts of fig. 3 show enlargements of that ringlet region which correspond to the resolutions of $450m$ and $900m$ (processed by binning the PPS-data), respectively. They seem to have the required symmetry, $\tau(b) \approx \tau(-b)$ and furthermore, the τ-profile of the ringlet seems to have at all resolutions the expected

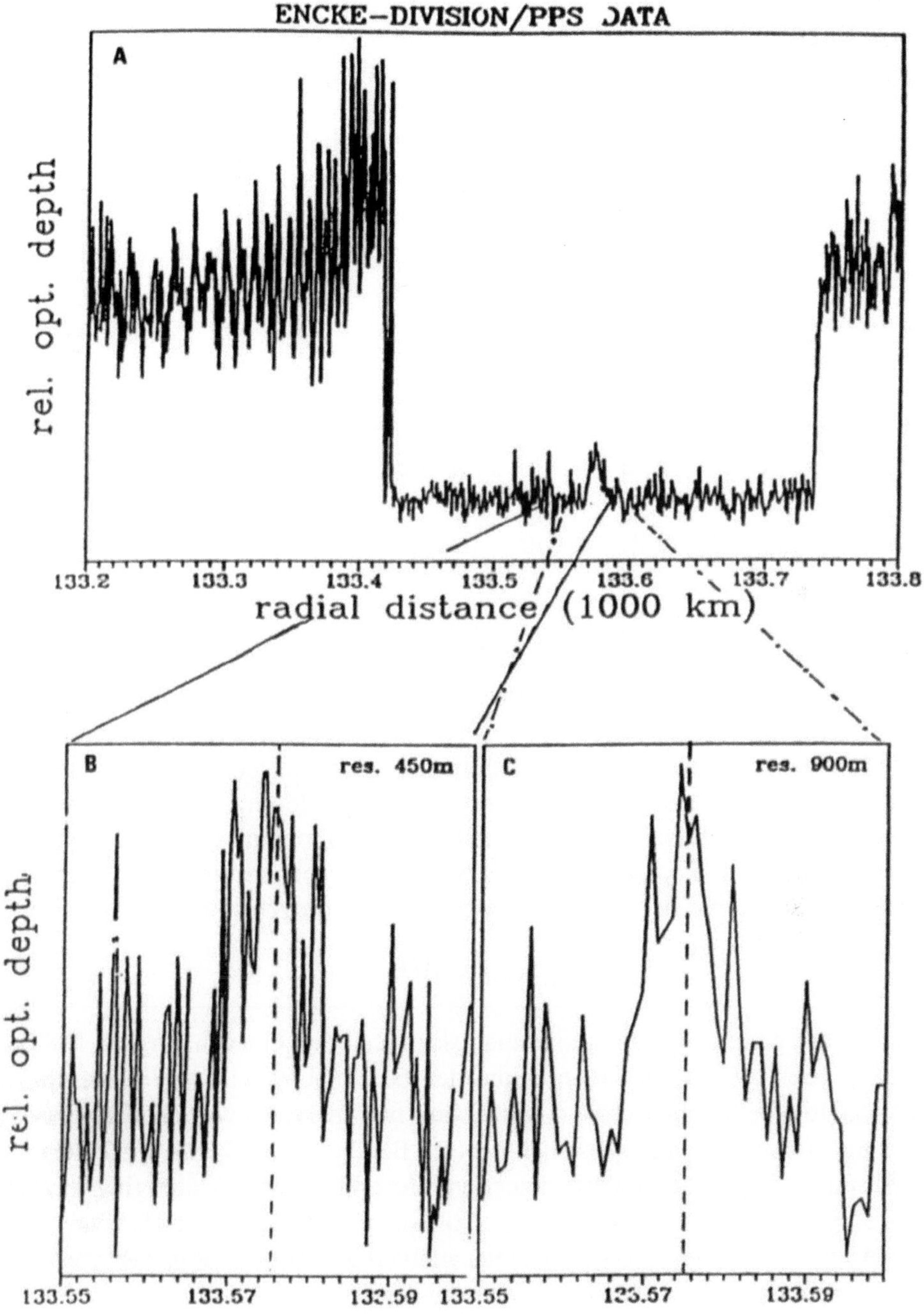

Fig. 3. a) Optical depth profile $\tau(r)$ near the Encke-division in Saturn's A-ring, measured by the Photopolarimeter (PPS) of Voyager 2. Clearly visible: a gap with a faint ringlet at its center, where the orbit of Pan is located. Both features - the gap and the ringlet - have been predicted by the models (see Section 2; PH3, and S1).

triple–peaked shape. However, the shape of the profile of the ringlet is in fact not sure because of the large fluctuations of the PPS–data in the Encke–division.

To weight the structure seen in the ringlet-profile we carried out a modified t-test, which seemed to favour the significance of the triple-peaked structure. However, this was not absolutely convincing.

Consequently, to be more sure about the nature of the structure, we have carried out an independent test by applying the wavelet-analysis method.

The wavelet transform of a signal is the decomposition of this signal onto a basis of functions obtained by dilatations and translations of a unique function called the mother wavelet. It allows to analyze a signal with respect of both a scale parameter and a location parameter, so that it is possible to focus the analysis of the signal at a given location and for a given scale.

Let $s(x)$ be a signal depending only on a space-variable, $\Psi(x)$ the mother wavelet. The wavelet transform of $s(x)$ is then a two variable function $C(a,b)$ depending on a scale variable a and on a location variable b:

$$C(a,b) = \langle s(x), \psi_{a,b}(x) \rangle \qquad \text{with} \qquad \psi_{a,b} = \frac{1}{a}\Psi(\frac{x-b}{a}) \tag{4}$$

where $\langle h(x), g(x) \rangle = \int_{-\infty}^{+\infty} h(x) g^*(x) dx$ and where $g^*(x)$ denotes the complex conjugate of $g(x)$. $C(a_0, b_0)$ can be seen as the details present in $s(x)$ at the scale a_0 in the vicinity of b_0. A wavelet is a zero mean function and rapidly vanishing, which insures from a practical point of view a good localization both in space and scale. >From the wavelet transform it is then possible to reconstruct the signal through a reconstruction formula (see Grossmann and Morlet 1985). Our aim has been to use the wavelet transform as a signal processing tool to extract the behavior of the signal in the vicinity of the ringlet, the noise having been removed. The range of considered scales has been chosen in a geometrical progression:

$$a_{i+1} = a_i * 2^{1/12}$$

which insures a practically continuous analysis in scale (see Grossmann et al. 1987), a_0 and a_{max} being defined in order to consider the signal from its step of sampling to its mean. In order to make cross comparisons, two analyzing wavelets have been chosen: the mexican hat and the french hat (see Bendjoya and Slezak 1992) which are well suited for hole or structure detections. As preliminary encouraging results we will only present here the reconstruction obtained with the mexican hat wavelet. It is to be noticed that the french hat wavelet analysis gives quite comparable results and will be published later (see Bendjoya et al. 1992).

For each scale a_i, a wavelet coefficient is computed at each point of the sampled signal and to reach our goal we have to eliminate the non significant coefficients and also the scales corresponding to the statistically estimated noise. One disposes then of a set of coefficients on a range of scale which allows to reconstruct without noise the behavior of the signal on this range of scales.

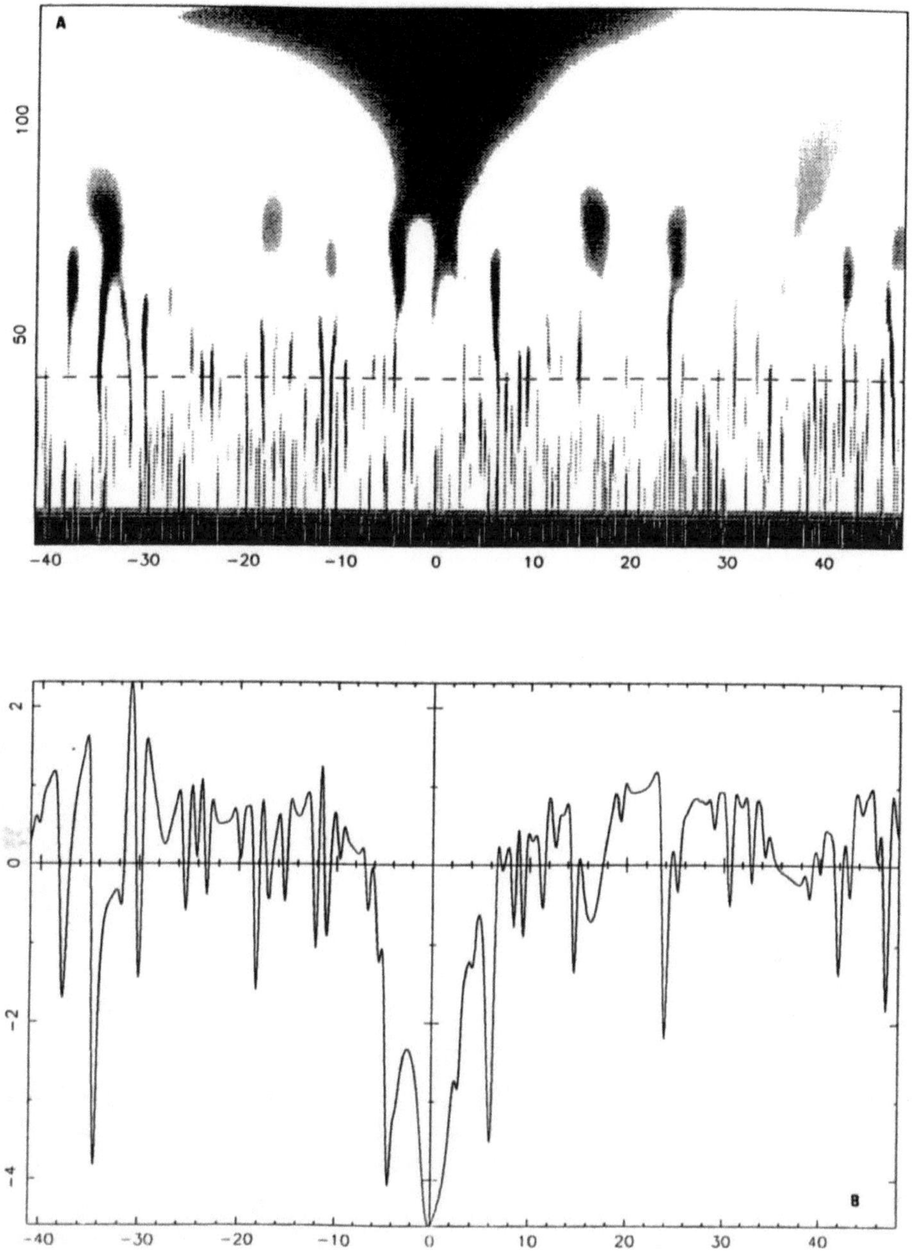

Fig. 4. a) A gray level representation of the wavelet transform of the signal. The x-axis gives the radial distance from the center of the Encke–ringlet (orbital radius of 133 576 km), and the y-axis the scale axis. The coefficient less than the threshold are in gray (black for the smallest ones). The gray levels are renormalized for each scale. The black stripe at the very small scale is due to the sampling effect. The dashed line indicates the limiting scale from which the reconstruction is done. b) The reconstructed photon–count signal. The x-axis is the same as above. The y-axis gives the deviation from the mean photon–count in the Encke–division (no extinction of the star signal). Negative value means extinction, that is presence of ring–material.

Therefore, we determine the scales corresponding to the noise and we define a threshold for the coefficients scale by scale. Then we have a set of wavelet coefficients on a range of scales not taking into account the noise, and moreover only the statistically significant coefficients are considered for each scale. The scale below which everything is considered as noise, and also the thresholds of the wavelet coefficients, are derived from the wavelet analysis of a random signal generated with the constraint to fit the distributions of the values of the PPS signal. For each scale of the wavelet transform of the generated signal we compute the distribution of the coefficients. This strategy allows to decrease significantly and quantify the statistical fluctuations in the wavelet coefficients. The minimum scale to consider for the reconstruction without noise has been evaluated to be a_{40} which corresponds to 250 m. Fig. 4a shows the wavelet transform of this signal displayed in a space-scale plane by a gray level representation. This figure is derived from the PPS–photon–count profile I, which is related to the optical depth by $\tau \propto \ln(I_0/I)$ (I_0 is the photon flux of the occulted star). Since we wanted to detect holes, for each scales the coefficients lower than the threshold are in gray (black for the smallest ones) and the higher ones in white. The dashed line visualizes the scale from which the reconstruction is made. Fig. 4b shows the reconstructed signal from the set of retained coefficients. This reconstruction is made with a threshold at 1- σ. The structure in three peaks of the ringlet appears obviously. It is to be noticed that this peculiar structure persists for more restrictive thresholds. The existence and the symmetry of the ringlet–peaks around the moonlet position as well as the width of the ringlet are in fairly good accordance with the expected results from the model recalled in the second section of this paper. Another ringlet seems also be pointed out by this analysis around the location 133 542 km from the center of Saturn.

4. Discussion

In this paper we have been able to confirm results of theoretical work concerning the gravitational influence of a small satellite, embedded in the rings, on the ring material around its orbit, by analyzing the PPS–data of Saturn's rings in the vicinity of the orbit of the satellite Pan. We have been able to estimate the size of Pan to be about $13km$ deduced from the width of the Encke–division, by assuming that the division is swept free by the satellite Pan itself (PH3).

Furthermore, we have compared the shape of PPS–photon–count profile of the Encke–gap–ringlet, as well as its width with theoretical predictions (S3). To do this it has been necessary to remove the noise from the PPS–data in the Encke division, which has been carried out using the wavelet–analysis method. The noise-reduced photon count profile shows three significant depletions inside the ringlet region, pointing to a triple–peaked density profile of the ringlet. This is in accordance with the results of the model and points to an eccentric orbit of Pan - eccentricity $e_m \approx 10^{-4}$. The width of the ringlet confirms the size estimation of Petit and Hénon (1988).

However, the data analysis presented here is only a first step in analyzing the radial density profiles of planetary rings. Future work should address the following points:

1. application of different wavelet-functions to the same ring region;
2. refinement of the estimate of the wavelet-coefficient threshold which separates the noise from "real" structures;
3. applying the wavelet-analysis to other rings (or ring regions), where also small satellites are expected to exist.

With respect to the theoretical modelling the following outstanding questions concerning moonlets in planetary rings are of interest:

1. What about the stability of the inner ringlet, if viscous transport processes, sources (the surface of the moonlets) and sinks (meteorite bombardment) of ring matter is taken into consideration?
2. How does the azimuthal distribution of the matter in the ringlet looks like (2D - simulations)?

Both sets of problems will be investigated in subsequent papers.

Acknowledgements

We would like to thank Prof. Larry Esposito, Dr. Karen Simmons, and Dr. Joshua Colwell for providing us with the Voyager-2 PPS-data of planetary rings.

References

Bendjoya, Ph., Petit, J.-M., Spahn, F.: (1992) in preparation

Bendjoya, Ph., Slezak, E.: (1992) "Wavelet analysis and applications to some dynamical system" *Celest. Mech* in press

Dermott, S.F., Murray, C.D.: (1981) "The Dynamics of Tadpole and Horseshoe Orbits I. Theory" *Icarus* **48**, pp. 1-11.

Goldreich, P., Tremaine, S.: (1978) "The velocity-dispersion in Saturn's rings" *Icarus* **34**, pp. 227-239.

Grossmann, A., Morlet, J.: 1985, in *Mathematics+Physics, Lectures on recent results*, L.Streit (Ed.), World Scientific Publishing

Grossmann, A., Kronland-Martinet, R., Morlet, J.: (1987) "Reading and understanding continuous wavelet transform" *Wavelets, Time-Frequency Methods in Phase Space* (Combes J.M., Grossmann A. and Tchamitchian Ph. Eds.) Springer-Verlag pp. 2-20

Lissauer, J.J., Shu, F.H., Cuzzi, J.N.: (1981) "Moonlets in planetary rings ?" *Nature* **292** pp. 707-711

Petit, J.M., Hénon, M.: (1987a) "A numerical simulation of planetary rings I. Binary encounters" *Astron. Astrophys* **173**, pp. 389-404. (PH1).

Petit, J.M., Hénon, M.: (1987b) "A numerical simulation of planetary rings II. Monte Carlo model" *Astron. Astrophys.* **188**, pp. 198-205. (PH2)

Petit, J.M., Hénon, M.: (1988) "A numerical simulation of planetary rings III. Mass segregation, ring confinement, and gap formation" *Astron. Astrophys.* **199**, pp. 343-356. (PH3)

Showalter, M., Cuzzi, J.N., Marouf, E.A., Esposito, L.W.: (1986) "Satellite 'wakes' and the orbit of the Encke gap moonlet" *Icarus* **66**, pp. 297-323.

Showalter, M.: (1991) "Visual detection of 1981 S13, Saturn's eighteenth satellite, and its role in the Encke gap" *Nature* **351**, pp. 709-713.

Spahn, F.: (1987) "Scattering Properties of a Moonlet (Satellite) Embedded in a Particle Ring: Application to the Rings of Saturn" *Icarus* **71**, pp. 69-77. (S1)

Spahn, F., Wiebicke, H.-J.: (1989) "Long-Term Gravitational Influence of Moonlets in Planetary Rings" *Icarus* 77, pp. 124-134. (S2)

Spahn, F., Saar, A., Schmidt, S., Schwarz, U: (1992) "The action of multiple moonlets in planetary rings" submitted to Icarus. (S3)

COLLISIONAL EVOLUTION OF THE ASTEROID SIZE DISTRIBUTION: A NUMERICAL SIMULATION

A. CAMPO BAGATIN
Dip. di Fisica, Università di Pisa, Piazza Torricelli 2, 56125 Pisa, Italy
P. FARINELLA
Dip. di Matematica, Università di Pisa, Via Buonarroti 2, 56127 Pisa, Italy
and
P. PAOLICCHI
Dip. di Fisica, Università di Pisa, Piazza Torricelli 2, 56125 Pisa, Italy

We have carried out a set of numerical simulations of the collisional evolution of the main asteroid belt, based on some models of fragmentation and cratering events consistent with the available experimental results on the outcomes of high–velocity impacts. We have taken into account such effects as gravitational self–compression, strain–rate scaling of the impact strength, ejection speed versus mass correlation for fragments, reaccumulation of fragments ejected at speeds smaller than the target's escape velocity. The asteroidal population has been divided into a sequence of discrete size bins, down from Ceres' diameter (about 1000 km) to arbitrarily small sizes; the bins have a constant logarithmic width, corresponding to a factor two in the mass of the bodies. A simple particle–in–a–box formula has been used to compute in every time step the number of collisions involving bodies belonging to any given pair of size bins. Then we have numerically integrated over a time of 4.5 billion years a set of first–order differential equations for the populations of the bins, considering the redistribution of the objects caused by collisions and including a sink term modelling the removal of small particles by non–gravitational effects (such as the Poynting–Robertson drag). The initial conditions were chosen to be consistent with a primordial asteroid population following two power–law size distributions, joined at a transition diameter of 100 km.

The results of our runs show that there are several physical effects that can affect the asteroid collisional evolution in a significant way. In particular, we have found that the evolution is very sensitive to the lower cutoff of the size distribution which is used in the simulation. This is probably due to a rapid production of very small fragments (down to 1 mm in the 90 bins case), without a competitive depletion due to the absence of even smaller projectiles. This results in a varying slope of the final size distribution, with peculiar wave–like features which propagate up to sizes of tens of km and are significantly shifted when smaller and smaller bodies are taken into account. We are further investigating this effect, in order to estimate the influence of very small bodies on the evolution of the observable asteroids and to assess the reliability of the results obtained by using a limited number of

Celestial Mechanics and Dynamical Astronomy **57**: 403–409, 1993.

size bins. We have also tested the effects of changing the fragment velocity vs. mass relationship in modelling the gravitational reaccumulation process: significant differences appear in the shape of the final size distribution for diameters exceeding ≈ 30 km.

RESONANT AND SECULAR PHENOMENA IN THE LIBRATION OF THE MOON

BARKIN Yu.V.
University of Zaragoza, Spain

Principal and subtle regularities, resonant and secular phenomena in the rotational motion of the Moon are investigated on the basis of the analytical solution of a model problem, in which the Moon and the Earth are considered as absolute rigid nonspherical bodies, and the Sun and the other planets are treated as material points. The orbital motion of all these bodies are considered as known and described by analytical theories.

An analytical solution of the problem has been built on the basis of the canonical equations of the rotational motion of the Moon in the Andoyer's variables using the method of constructing conditionally-periodic solutions of an Hamiltonian system depending upon a small parameter, and researching general solutions in their neighbourhood.

The solution so constructed is used for analyzing the principal and subtle regularities, resonant phenomena in the rotational motion of the Moon, i.e. for studying the principal departure from the Cassini's laws in the rotational motion of the Moon due to the long periodic orbital perturbations of Moon and Sun.

I discuss some questions such as: 1.) explanation of Cassini's laws; 2.) stability of the conditionally-periodic rotational motion of the Moon; 3.) constant displacement of the mean pole of the Moon; 4.) splitting of the Cassini's node; 5.) mean angular velocity of the Moon; 6.) constant angular displacement of the axes of inertia of the Moon with respect to the orbital coordinate system; 7.) periods and frequencies of resonant librations; 8.) coefficients and phases of principal resonant librations in Andoyer's variables and in classical variables; 9.) geometrical and kinematical interpretation of the resonant librations of the Moon; 10.) secular variation of the mean angle between the ecliptic plane and the lunar equatorial plane.

PRECISE ORBIT DETERMINATION WITH A SHORT-ARC TECHNIQUE

P. BONNEFOND and P. EXERTIER
Observatoire de la Côte d'Azur, Département CERGA
Avenue Copernic, 06130 Grasse, France

Satellites carrying altimeters (TOPEX/POSEIDON, ERS-1) can measure the height of the satellite above the ocean surface with an accuracy of about 2cm to 10cm. In order to use fully these data, for studies of ocean topography, it is necessary to determine the height of the satellite above a reference surface with a similar accuracy; it is also necessary to calibrate the altimetric measurements.These points require a precise orbit determination in order to obtain an absolute satellite position (with respect to the Earth center of mass) with an accuracy of a few centimeters, at least locally, and essentially for the radial component.

Considering the value of the expected accuracy of the radar altimeters (a few centimeters), the use of a short-arc technique based on a geometric approach should permit us to take into account local errors in limited geographical regions which can not be considered in a global orbit computation by a dynamical approach. Our method consists in fitting the geometrical positions of the satellite over a short period of time (10 to 20 minutes) with the tracking measurements (Satellite Laser Ranging and Doppler DORIS system) from a ground station network and considering a simplified orbital error model. We correct the reference orbit by applying along-track, across-track and radial displacements, that are represented as simple constants, or with a linear or quadratic variation in time.

The aim of this poster is to show the ability of the short-arc technique to reach the required accuracy on the radial orbit component; it depends on several conditions: the accuracy (typically 2cm for the laser data and 0.3mm/s for DORIS data) and a good repartition of the along-track measurements, but also the density and the geometrical configuration of the station network in relation to the studied arcs. Several tests are done with laser ranging and Doppler measurements, for several satellites (in particular, AJISAI and SPOT-2) with characteristics not far from those of TOPEX/POSEIDON and ERS-1 satellites. It has been demonstrated that, with the conditions previously exposed, the altitude of the satellite is determined with an accuracy better than 5cm.

SPHERICAL COORDINATES PERTURBATIONS FOR SMALL ECCENTRICITY ORBITS

ERIC BOIS
OCA/CERGA, Grasse, France
and
SŁAWOMIR BREITER
A. Mickiewicz University, Poznań, Poland
(H.Poincaré Grant-holder at OCA/CERGA, Grasse, France)

The theory of perturbed circular orbital motion developed by Bois is extended to weakly eccentric orbits. Expressions for spherical coordinates are given as functions of amplitudes of the Fourier expansion of disturbing forces whatever their nature. The solution is nonsingular with respect to zero eccentricity and inclination. Numerical modeling of the equatorial artificial satellite orbits shows, that the internal accuracy of our first-order solution remains on the level of second-order effects for the eccentricity values lower than the small parameter of disturbing forces.

DYNAMICAL FRICTION INDUCES PERTURBATIONS ON OORT CLOUD COMETS

ADRIAN BRUNINI
Instituto Astronomico e Geofisico Universidade de Sao Paulo,
Av. Miguel Stefano 4200 (01065) Sao Paulo Sp Brasil

The aphelia distribution of Oort cloud comets is recognized to be strongly modulated by tidal galactic perturbations. However, near 15% of nonrandomness in the aphelia distribution in galactic latitudes is not explained by this type of interaction between the Galaxy and the solar system. We demonstrate that this nonrandomness is a real phenomenon and is not an observational selection effect. We consider the impulse transfer of the Sun on the overall uniform distribution of mass in the solar neighborhood induced by the solar motion relative to the local standard of rest. If the perturbation is averaged and limited to the first-order terms, an analytic solution for the loss rate of comets is possible. A good agreement with the observed asymmetry is found. Furthermore, the theory predicts the observed distribution of perihelia longitudes.

POYNTING-ROBERTSON EFFECT AND ORBITAL MOTION

J. KLAČKA
Department of Astronomy and Astrophysics Comenius University,
Mlynská dolina, 842 15 Bratislava, Czech and Slovak Federal Republic

Time evolution of the meteoroid's orbit under the action of the solar electromagnetic radiation is discussed in terms of perihelion and aphelion distances. Perturbation equations for secular changes of orbital elements are written for the most simple case. Initial conditions are formulated for the obtained system of perturbation equations and simple example is presented. Results are compared with the published ones.

POYNTING-ROBERTSON EFFECT AND SMALL ECCENTRIC ORBITS

J. KLAČKA and J. KAUFMANNOVÁ
Department of Astronomy and Astrophysics Comenius University,
Mlynská dolina, 842 15 Bratislava, Czech and Slovak Federal Republic

Perturbation equations of celestial mechanics in application to solar electromagnetic radiation are investigated. Special attention is paid to nearly circular orbits. It is pointed out that real orbits of this type exhibit qualitatively different behaviour from that generally accepted, as for time evolution of orbital elements. In previous works it was assumed that as the orbit shrinks it becomes more and more circular. We have found that the eccentricity decreases only during $\approx 1/3$ of the total time of inspiralling toward the Sun and that the orbit actually becomes more and more eccentric in the subsequent orbital evolution.

DUST PARTICLE AND SOLAR RADIATION

J. KLAČKA and J. KAUFMANNOVÁ
Department of Astronomy and Astrophysics, Comenius University,
Mlynská dolina, 842 15 Bratislava, Czech and Slovak Federal Republic

Influence of the solar radiation (electromagnetic and corpuscular - solar wind) on the motion of the interplanetary dust particle (IDP) is investigated. The ratio 'time of inspiralling toward the Sun : time of inspiralling neglecting the change of mass of the particle' is presented as a function of initial eccentricity for two cases of semi-major axis
1. $A(initial) = 3\ AU \rightarrow A(final) = 0.3\ AU$,
2. $A(initial) = 1\ AU \rightarrow A(final) = 0.3\ AU$
and for two different values characterizing the decrease of the IDP's size under the action of the solar wind.

INFLUENCE OF INTERNAL LOAD ON THE NON-HYDROSTATIC FORM OF THE MOON

MARIANNE LEFFTZ and HILAIRE LEGROS
Institut de Physique du Globe de Strasbourg,
5 rue René Descartes, 67084 Strasbourg Cedex, FRANCE

The viscoelastic deformations of a graviting rotating multilayered Maxwell planet, submitted to a volumic external potential and to internal loads, are computed using a Love number formalism. First, for such a body, we calculate the relaxation modes associated with the viscoelastic Love numbers: they are due to various discontinuities in density, in rigidity or in viscosity at each interface of the multilayered planet. If an internal load is located at an interface "i", the relaxation mode due to the density jump at this interface "i" will be excited. For the Moon, such a mode creates viscoelastic deformations (in the radial displacement and in the mass redistribution potential) at the surface which relaxes with a relaxation time of about some billions years, that is, at time scale of the Moon age, quasi-constant deformations.

Finally, we show that the present non-hydrostatic form of the Moon can be explained by an internal load (appeared 2.8 billions years ago), located between the upper and the lower part of the lunar lithosphere, which creates viscoelastic surface deformations, in the selenoid, not completely relaxed.

EVOLUTION OF ASTEROIDAL ORBITS WITH HIGH INCLINATIONS

NINA A. SOLOVAYA
Celestial Mechanics Department, Sternberg State Astronomical Institute,
119 889 Moscow, Russia
and
EDUARD M. PITTICH
Astronomical Institute, Slovak Academy of Sciences,
842 28 Bratislava, Czech and Slovak Federal Republic

The 20,000 years orbital evolution of massless fictitious asteroids located at a border of the Hill's gravitational sphere has been investigated. The eleven orbits with the eccentricities from 0.0 to 0.4 in five groups of inclinations from 40° to 80° were numerically integrated with planetary perturbations of six major planets, using the numerical integration n-body program with the Everhart's integrator RA15. For each group time evolution of orbital elements of the asteroids is presented.

The orbital evolution of all low-eccentricity orbits showed a small quasi-periodic behavior of orbital elements in dependence on the commencing inclination. For the commencing inclination 40° a deep decline of some orbits' inclination took place after four millenium, for 60° after six millenium, and for 80° after ten millenium. The inclination of some of these orbits decreased close to 30° in all investigated groups. Within the investigated period eccentricities vary in all groups. For the commencing inclination 40° the eccentricity of some orbits increased up to 0.6, for 60° up to 0.8, and for 80° close to 1.0. The asteroids with the commencing inclination 40° achieved these maximum values of the eccentricity after two millenium, with 60° after six millenium, and with 80° after nine millenium.

An important feature of the investigated groups of asteroids is not only the quasi-periodical change of the form of their orbits, but also a movement of their orbits relative to the Sun. Of the asteroids with the commencing inclinations 40° only those with commencing eccentricity equal to 0.4 enter the region between Sun and Earth. But practically all asteroids with the commencing inclination 60° and higher enter this region, except those with the commencing eccentricity near to 0°. For the commencing inclination 40° the smallest perihelion distance is 0.7 AU, for 60° is 0.5 AU. The aphelion distance of the asteroids with the commencing inclination 40° is not greater than 4.2 AU, with 60° not greater than 3.9 AU, and with 80° not greater than 3.6 AU.

Some of the investigated asteroids are periodically delivered from main asteroid belt to the near-Mars space and than to the near-Earth space. These orbital changes are more frequent for the higher commencing inclinations than for the lower ones. Thus, after ten millenium, from the asteroids with strarting inclination 40°, about 20% may be delivered to the near-Earth region, while from the asteroids with starting inclination 60° about 50%, and of those with starting inclination 80° about 90%. The changes of the eccentricities and inclinations of the asteroids with investigated commencing dynamical parameters are limited within small domains. After some period the asteroids will return to the main belt.

MIX
Papier aus verantwortungsvollen Quellen
Paper from responsible sources
FSC® C105338

If you have any concerns about our products,
you can contact us on
ProductSafety@springernature.com

In case Publisher is established outside the EU,
the EU authorized representative is:
Springer Nature Customer Service Center GmbH
Europaplatz 3, 69115 Heidelberg, Germany

Printed by Libri Plureos GmbH
in Hamburg, Germany